Cracking the

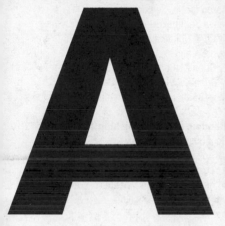

AP®

CALCULUS BC EXAM

2020 Edition

David S. Kahn

PrincetonReview.com

Penguin
Random
House

The Princeton Review
110 East 42nd St, 7th Floor
New York, NY 10017
E-mail: editorialsupport@review.com

Published in the United States by Penguin Random House LLC, New York, and in Canada by Random House of Canada, a division of Penguin Random House Ltd., Toronto.

ISBN: 978-0-525-56816-2
eBook ISBN: 978-0-525-56854-4
ISSN: 2372-2711

AP is a trademark registered and owned by the College Board, which is not affiliated with, and does not endorse this product.

The Princeton Review is not affiliated with Princeton University.

The material in this book is up-to-date at the time of publication. However, changes may have been instituted by the testing body in the test after this book was published.

If there are any important late-breaking developments, changes, or corrections to the materials in this book, we will post that information online in the Student Tools. Register your book and check your Student Tools to see if there are any updates posted there.

Editor: Aaron Riccio
Production Editors: Liz Dacey and Emma Parker
Production Artist: Deborah Weber

Printed in the United States of America.

10 9 8 7 6 5 4 3 2 1

2020 Edition

Editorial
Rob Franek, Editor-in-Chief
David Soto, Director of Content Development
Stephen Koch, Survey Manager
Deborah Weber, Director of Production
Gabriel Berlin, Production Design Manager
Selena Coppock, Managing Editor
Aaron Riccio, Senior Editor
Meave Shelton, Senior Editor
Chris Chimera, Editor
Sarah Litt, Editor
Orion McBean, Editor
Brian Saladino, Editor
Eleanor Green, Editorial Assistant

Penguin Random House Publishing Team
Tom Russell, VP, Publisher
Alison Stoltzfus, Publishing Director
Amanda Yee, Associate Managing Editor
Ellen Reed, Production Manager
Suzanne Lee, Designer

Acknowledgments

First of all, I would like to thank Arnold Feingold and Peter B. Kahn for once again doing every problem, reading every word, and otherwise lending their invaluable assistance. I also want to thank Aaron Riccio for being a terrific editor on this edition of the book, Gary King for his first-rate reading, analysis, and contributions, and the production team of Liz Dacey, Emma Parker, and Deborah Weber. Thanks to Frank, without whose advice I never would have taken this path. Thanks to Jeffrey, Miriam, and Vicki for moral support. Thanks Mom.

Finally, I would like to thank the people who really made all of this effort worthwhile—my students. I hope that I haven't omitted anyone, but if I have, the fault is entirely mine.

Aaron and Sasha, Aaron, Aaron T., Abby B., Abby F., Abby H., Abby L., Abbye, Abigail H., Addie, Aidan, Alan M., Albert G., Alec G, Alec M., Alec R., Alex and Claire, Alex A., Alex B., Alex and Daniela B., Alex F., Alex D., Alex G., Alex H., Alex I., Alex S., Alex and Gabe, Alexa, Alexandra, Natalie and Jason, Alexes, Alexis and Brittany, Ali and Jon, Ali and Amy Z., Ali H., Alice, Alice C., Alicia, Alisha, Allegra, Allie and Lauren, Allison and Andrew, Allison H., Allison and Matt S., Allison R., Ally, Ally M., Ally T., Ally W., Aly and Lauren, Alyssa C., Alyssa TDC, Alyssa and Courtney, Alvin, Amanda, Brittany, and Nick A., Amanda and Pamela B., Amanda C., Amanda E., Amanda H., Amanda M., Amanda R., Amanda S., Amanda Y., Amber and Teal, Amparo, Andrea T., Andrea V., Andrew A., Andrei, Andrew B., Andrew C., Andrew D., Andrew E., Andrew H., Andrew M., Andrew S., Andy and Sarah R., Andy and Allison, Angela, Angela F., Angela L., Anisha, Ann, Anna C-S., Anna D., Anna and Jon, Anna and Max, Anna L., Anna M., Anna S., Anna W., Annabelle M., Annerose, Annie, Annie B., Annie W., Antigone, Antonio, Anu, April, Ares, Ariadne, Ariane, Ariel, Arielle and Gabrielle, Arthur and Annie, Arya, Ashcloy and Freddy, Ashley A , Ashley K., Ashley and Sarah, Ashley and Lauren, Aubrey, Avra, Bocca A., Becky B., Becky S., Becky H., Ben S., Ben and Andrew Y., Ben D., Ben S., Benjamin D., Benjamin H., Benji, Beth and Sarah, Bethany and Lesley, Betsy and Jon, Blythe, Bianca and Isabella, Biz, Bonnie, Bonnie C., Boris, Braedan, Brendan, Brett, Brett A., Brette and Josh, Brian, Brian C., Brian N., Brian W., Brian Z., Brigid, Brin, Brittany E., Brittany S., Brooke, Devon, and Megan, Brooke and Lindsay E., Butch, Cailyn H., Caitlin, Caitlin F., Caitlin M., Caitlin and Anna S., Camilla and Eloise, Camryn, Candace, Cara, Cara G., Carinna, Caroline A., Caroline C., Caroline H., Caroline S., Caroline and Peter W., Carrie M., Catherine W., Cecilia, Celia, Chad, Channing, Charlie, Charlotte, Charlotte B., Charlotte M., Chase, Chelsea, Chloe, Chloe K., Chris B , Chris C., Chris K., Chris P., Chrissie, Christa S., Christen C., Christian, Christian C., Christina F., Christine, Christine W., CJ, Claire H., Clare, Claudia, Clio, Colette, Conor G., Corey, Corinne, Coryn, Courtney and Keith, Courtney B., Courtney F., Courtney S., Courtney W., Craig, Dan M., Dana J., Dani and Adam, Daniel, Daniel and Jen, Daniella, Daniella C., Danielle and Andreas, Danielle and Nikki D., Danielle G., Danielle H., Danny K., Dara and Stacey, Dara M., Darcy, Darya, David B., David C., David R., David S., Deborah and Matthew, Deniz and Destine, Deshaun, Deval, Devin, Devon and Jenna, Deyshaun, Diana H., Dilly, Disha, Dong Yi, Dora, Dorrien, Eairinn, Eddy, Elana, Eleanor K., Elexa and Nicky, Elisa, Eliza, Elizabeth, Elizabeth and Mary C., Elizabeth F., Ella, Ellie, Elly B., Elyssa T., Emily A., Emily B., Emily and Allison, Emily and Catie A., Emily C., Emily G., Emily H., Emily K., Emily L., Emily and Pete M., Emily R., Emily R-H., Emily S., Emily T., Emma, Emma and Sophie, Emma D., Emma S., Eric N., Erica F., Erica H., Erica R., Erica S., Eric and Lauren, Erica and Annie, Erika, Erin, Erin I., Ernesto, Esther, Ethan, Eugenie, Eva, Evan, Eve H., Eve M., Fifi, Francesca, Frank, Franki, Gabby, Geoffrey, George and Julian., George M., Gloria, Grace M., Grace P., Gracie, Graham and Will, Greg, Greg F., Griffin, Gussie, Gylianne, Haley, Halle R., Hallie, Hannah C., Hannah J., Hannah R., Hannah and Paige, Harrison, Annabel and Gillian, Harry, Hayley and Tim D., Hazel, Heather D., Heather E., Heather F., Heather and Gillian, Hernando and Vicki, Hilary and Lindsay, Hilary F., Holli, Holly G., Holly K., Honor, Ian, Ian P., India, Ingrid, Ira, Isa, Isabel, Isabella and Simone, Isabelle T., Ismini, Isobel, Ivy, Jabari, Jacob and Kara, Jack, Jack B., Jackie and

Vicky, Jackie, Jackie A., Jackie S., Jackson, Jaclyn and Adam, Jacob D., Jada, James S., Jamie, Jason P., Jason and Andrew, Jay, Jay K., Jae and Gideon, Jake T., Jarrod, Jason L., Jayne and Johnny, Jed D., Jed F., Jeffrey M., Jenna, Jen, Jen M., Jenn N., Jennifer B., Jennifer W., Jenny K, Jenny and Missy, Jeremy C., Jeremy and Caleb, Jess, Jess P., Jesse, Jessica, Jessica and Eric H., Jessica L., Jessica T., Jessie, Jessie C., Jessie and Perry N., Jill, Jillian, Daniel and Olivia, Jillian S., Jimin, Jimmy C., Jimmy P., Joanna, Joanna C., Joanna G., Joanna M., Joanna and Julia M., Joanna W., Jocelyn, Jody and Kim, Joe B., Johanna, John, John and Dan, Jonah and Zoe, Jonathan G., Jonathan P., Jonathan W., Jongwoo, Jordan C., Jordan and Blair F., Jordan G., Jordan P., Jordana, Jordyn, Jordyn K., Josh and Jesse, Josh and Noah, Judie and Rob, Julia and Caroline, Julia, Julia F., Julia G., Julia H., Julia and Charlotte P., Julia T., Julie, Julie H., Julie J., Julie and Dana, Julie P., Juliet, Juliette R., Justin L., Kara B., Kara O., Kasey, Kasia and Amy, Kat C., Kat R., Kate D., Kate F., Kate G., Kate L., Kate P., Kate S., Kate W., Katie C., Katie F., Katherine C., Kathryn, Leslie, and Travis, Katie, Katie M., Katrina, Keith M., Kelly C., Ken D., Kenzie, Keri, Kerri, Kim H., Kimberly, Kimberly H., Kirsten, Kitty and Alex, Kori, Kripali, Krista, Erika, and Karoline, Kristen, Kristen, Kristen Y., Magan, and Alexa, Laila and Olivia R., Laura B., Laura F., Laura G., Laura R., Laura S., Laura T., Laura Z., Lauren and Eric, Lauren R., Lauren S., Lauren T., Lauren and Allie, Lea K., Leah, Lee R., Leigh and Ruthie, Leigh, Lexi S., Lila, Lila M., Lilaj, Lili C., Lilla, Lily, Lily M-R., LilyHayes, Lisbeth and Charlotte, Lina, Lindsay F., Lindsay K., Lindsay L., Lindsay N., Lindsay R, Lindsay and Jessie S., Lindsey and Kari, Lisa, Liz D., Liz H., Liz M., Lizzi B., Lizzie A., Lizzie M., Lizzie W., Lizzy, Lizzy C., Lizzy R., Lizzy T., Louis, Lucas, Luke C., Lucinda, Lucy, Lucy D., Mackenzie, Maddie and JD, Maddie H., Maddie P., Maddy N., Maddy T., Maddy W., Madeline, Magnolia, Maggie L., Maggie T., Manuela, Mara and Steffie, Marcia, Margaret S., Maria B., Mariel C., Mariel L., Mariel S., Marielle K., Marielle S., Marietta, Marisa B., Marisa C., Marissa, Marlee, Marnie and Sam, Mary M., Matt, Matthew B., Matthew K., Matt F., Matt G., Matt and Caroline, Matt V., Matthew G., Mavis H., Max B., Max and Chloe K., Max M., Maxx, Ben, and Sam, Maya and Rohit, Maya N., Mckyla, Meesh, Megan G., Meliha, Melissa, Melissa and Ashley I., Meredith, Meredith and Gordie B., Meredith and Katie D., Meredith R., Meredith S-K., Michael and Dan, Michael H., Michael R., Michaela, Michal, Michele S., Mike C., Mike R., Miles, Milton, Miranda, Moira, Molly L., Molly R., Molly and Annie, Morgan, Morgan C., Morgan, Zoe, and Lila, Morgan K., Nadia and Alexis, NaEun, Nancy, Nanette, Naomi, Natalie and Andrew B., Natasha and Mikaela B., Natasha L., Nathania, Nico, Nick, Nick A., Nicky S., Nicole B., Nicole J., Nicole N., Nicole P., Nicole S., Nicolette, Nidhi, Nikki G., Nina R., Nina V., Noah, Nora, Nushien, Odean, Ogechi, Oli and Arni, Oliver, Oliver W., Olivia and Daphne, Olivia P., Omar, Oren, Pam, Paige, Pasan, Paul A., Peter, Philip and Peter, Phoebe, Pierce, Pooja, Priscilla, Quanquan, Quinn and Chris, Rachel A, Rachel B., Rachel and Adam, Rachel H., Rachel I., Rachel and Jake, Rachel K., Rachel L., Rachel and Mitchell M., Rachel S., Rachel and Eli, Rachel and Steven, Rachel T., Rachel W., Rae, Ramit, Randi and Samantha, Rasheeda, Ray B., Rayna, Rebecca A., Rebecca B., Rebecca G., Rebecca and Gaby K., Rebecca R., Rebecca W., Renee F., Resala, Richard, Tina, and Alice, Richie D., Ricky, Riley C., Rob L., Robin P., Romanah, Rose, Ryan, Ryan B., Ryan S., Saahil, Sabrina and Arianna, Sabrina O., Sally, Sam B., Sam C., Sam G., Sam L., Sam R., Sam S., Sam W., Sam Z., Samantha and Savannah, Samantha M., Samuel C., Samar, Sara G., Sara H., Sara L., Sara R., Sara S., Sara W., Sarah, Patty, and Kat, Sarah and Beth, Sarah A., Sarah C., Sarah D., Sarah F., Sarah G., Sarah and Michelle K., Sarah L., Sarah M-D., Sarah and Michael, Sarah P., Sarah S., Sascha, Samara, Saya, Sean, Selena G., Serena, Seung Woo, Shadae, Shannon, Shayne, Shayne, Shorty, Siegfried, Simon W., Simone, Sinead, Siobhan, Skye F., Skye L., Sofia G., Sofia M., Sonja, Sonja and Talya, Sophia A., Sophia B., Sophia J., Sophia S., Sophie and Tess, Sophie D., Sophie S., Sophie W., Stacey, Stacy, Steffi, Steph, Stephanie, Stephanie K., Stephanie L., Stephen C., Sumair, Sunaina, Suzie, Syd, Sydney, Sydney S., Tammy and Hayley, Tara and Max, Taryn, Tasnia, Tayler, Taylor, Tenley and Galen, Terrence, Tess, Theodora, Thomas B., Timothy H., Tina, Tita, Tom, Tori, Torri, Tracy, Tracy K., Tripp H., Tripp W., Tyler, Tyrik, Uma, Vana, Vanessa and John, Veda, Veronica M., Vicky B., Victor Z., Victoria, Victoria H., Victoria J., Victoria M., Vinny and Eric, Vivek, Vladimir, Waleed, Will A., Will, Teddy, and Ellie, William M., William Z., Wyatt and Ryan, Wyna, Xianyuan, Yakir, Yesha, Yujin, Yuou, Zach, Zach B., Zach S., Zach Y., Zachary N., Zaria, Zoë, Zoe L., and Zoey and Rachel.

Contents

Get More (Free) Content
at **PrincetonReview.com/cracking**

As easy as **1·2·3**

1 Go to PrincetonReview.com/cracking and enter the following ISBN for your book:

9780525568162

2 Answer a few simple questions to set up an exclusive Princeton Review account. *(If you already have one, you can just log in.)*

3 Enjoy access to your **FREE** content!

Once you've registered, you can...

- Get our take on any recent or pending updates to the AP Calculus BC Exam

- Take a full-length practice SAT and ACT

- Get valuable advice about the college application process, including tips for writing a great essay and where to apply for financial aid

- If you're still choosing between colleges, use our searchable rankings of *The Best 385 Colleges* to find out more information about your dream school.

- Access comprehensive study guides and a variety of printable resources, including: answer bubble sheets, a calculator appendix, and study sheets of formulas, derivatives, and integrals that you need to know

- Check to see if there have been any corrections or updates to this edition

Need to report a potential **content** issue?

Contact **EditorialSupport@review.com** and include:

- full title of the book
- ISBN
- page number

Need to report a **technical** issue?

Contact **TPRStudentTech@review.com** and provide:

- your full name
- email address used to register the book
- full book title and ISBN
- Operating system (Mac/PC) and browser (Firefox, Safari, etc.)

Look For These Icons Throughout The Book

ONLINE ARTICLES

PROVEN TECHNIQUES

APPLIED STRATEGIES

MORE GREAT BOOKS

GOING DEEPER

Part I
Using This Book to Improve Your AP Score

- Preview: Your Knowledge, Your Expectations
- Your Guide to Using This Book
- How to Begin

PREVIEW: YOUR KNOWLEDGE, YOUR EXPECTATIONS

Your route to a high score on the AP Calculus BC Exam depends a lot on how you plan to use this book. Start thinking about your plan by responding to the following questions.

1. Rate your level of confidence about your knowledge of the content tested by the AP Calculus BC Exam:

    A. Very confident—I know it all
    B. I'm pretty confident, but there are topics for which I could use help
    C. Not confident—I need quite a bit of support
    D. I'm not sure

2. If you have a goal score in mind, circle your goal score for the AP Calculus BC Exam:

    5   4   3   2   1   I'm not sure yet

3. What do you expect to learn from this book? Circle all that apply to you.

    A. A general overview of the test and what to expect
    B. Strategies for how to approach the test
    C. The content tested by this exam
    D. I'm not sure yet

YOUR GUIDE TO USING THIS BOOK

This book is organized to provide as much—or as little—support as you need, so you can use this book in whatever way will be most helpful for improving your score on the AP Calculus BC Exam.

*   The remainder of **Part I** will provide guidance on how to use this book and help you determine your strengths and weaknesses.

*   **Part II** of this book contains Practice Test 1, its answers and explanations, and a scoring guide. (Bubble sheets can be found online when you register your book—see pages x–xi for details.) We strongly recommend that you take this test before going any further, in order to realistically determine:
    o   your starting point right now
    o   which question types you're ready for and which you might need to practice
    o   which content topics you are familiar with and which you will want to carefully review

    Once you have nailed down your strengths and weaknesses with regard to this exam, you can focus your test preparation, build a study plan, and be efficient with your time.

- **Part III** of this book will:
  - o provide information about the structure, scoring, and content of the AP Calculus BC Exam
  - o help you to make a study plan
  - o point you towards additional resources

- **Part IV** of this book will explore the following strategies:
  - o how to attack multiple-choice questions
  - o how to write a high scoring free-response answer
  - o how to manage your time to maximize the number of points available to you

- **Part V** of this book covers the content you need for your exam.

- **Part VI** of this book contains Practice Tests 2 and 3, their answers and explanations, and a scoring guide. If you skipped Practice Test 1, we recommend that you use it, in conjunction with Tests 2 and 3 (with at least a day or two between them) to chart your progress and identify lingering issues: if you get a certain type of question wrong every time, you probably need to review it. If you only got it wrong once, you may have run out of time or been distracted by something. In either case, this will allow you to focus on the factors that caused the discrepancy in scores and to be as prepared as possible on the day of the test.

When you register your book online, you can download and print out bubble sheets for each Practice Test in this book. You can also find a handy Appendix that contains all the formulas you should know. See pages x–xi for details!

You may choose to use some parts of this book over others, or you may work through the entire book. This will depend on your needs and how much time you have. Let's now look at how to make this determination.

HOW TO BEGIN

1. **Take a Test**

   Before you can decide how to use this book, you need to take a practice test. Doing so will give you insight into your strengths and weaknesses, and the test will also help you make an effective study plan. If you're feeling test-phobic, remind yourself that a practice test is a tool for diagnosing yourself—it's not how well you do that matters but how you use information gleaned from your performance to guide your preparation.

   So, before you read further, take the AP Calculus BC Practice Test 1 starting on page 9. Be sure to do so in one sitting, following the instructions that appear before the test.

2. **Check Your Answers**

   Using the answer key on page 39, count how many multiple-choice questions you got right and how many you missed. Don't worry about the explanations for now, and don't worry about why you missed questions. We'll get to that soon.

3. **Reflect on the Test**

    After you take your first test, respond to the following questions:

    *   How much time did you spend on the multiple-choice questions?

    *   How much time did you spend on each free-response question?

    *   How many multiple-choice questions did you miss?

    *   Do you feel you had the knowledge to address the subject matter of the free-response questions?

    *   Do you feel your free responses were well organized and thoughtful?

    *   Circle the content areas that were most challenging for you and draw a line through the ones in which you felt confident/did well.

        o   Functions, Graphs, and Limits

        o   Differential Calculus

        o   Integral Calculus

        o   Polynomial Approximations and Series

        o   Applications of Derivatives

        o   Applications of Integrals

4. **Read Part III, and Complete the Self-Evaluation**

    Part III will provide information on how the test is structured and scored. It will also set out areas of content that are tested.

    As you read Part III, reevaluate your answers to the questions above. At the end of Part III, you will revisit and refine the questions you answer above. You will then be able to make a study plan, based on your needs and time available, that will allow you to use this book most effectively.

5. **Engage with Parts IV and V as Needed**

    Notice the word *engage*. You'll get more out of this book if you use it intentionally than if you read it passively, hoping for an improved score through osmosis.

    Strategy chapters will help you think about your approach to the question types on this exam. Part IV will open with a reminder to think about how you approach questions now and then close with a reflection section asking you to think about how/whether you will change your approach in the future.

The content chapters in Part V are designed to provide a review of the content tested on the AP Calculus BC Exam, including the level of detail you need to know and how the content is tested. You will have the opportunity to assess your mastery of the content of each chapter through test-appropriate questions and a reflection section.

6. **Take Test 2 and Assess Your Performance**

Once you feel you have developed the strategies you need and gained the knowledge you lacked, you should take Test 2. You should do so in one sitting, following the instructions at the beginning of the test.

When you are done, check your answers to the multiple-choice sections. See if a teacher will read your long-form calculus responses and provide feedback.

Once you have taken the test, reflect on what areas you still need to work on, and revisit the chapters in this book that address those topics. Through this type of reflection and engagement, you will continue to improve.

7. **Keep Working**

After you have revisited certain chapters in this book, continue the process of testing, reflecting, and engaging with Practice Test 3. You want to be increasing your readiness by carefully considering the types of questions you are getting wrong and how you can change your strategic approach to different parts of the test.

As discussed in Part III, there are other resources available to you, including a wealth of information on AP Central. You can continue to explore areas that can stand to improve and engage in those areas right up to the day of the test.

Part II
Practice Test 1

- Practice Test 1
- Practice Test 1 Answers and Explanations

Practice Test 1

AP® Calculus BC Exam

DO NOT OPEN THIS BOOKLET UNTIL YOU ARE TOLD TO DO SO.

At a Glance

Total Time
1 hour and 45 minutes
Number of Questions
45
Percent of Total Grade
50%
Writing Instrument
Pencil required

Instructions

Section I of this examination contains 45 multiple-choice questions. Fill in only the ovals for numbers 1 through 45 on your answer sheet.

CALCULATORS MAY NOT BE USED IN THIS PART OF THE EXAMINATION.

Indicate all of your answers to the multiple-choice questions on the answer sheet. No credit will be given for anything written in this exam booklet, but you may use the booklet for notes or scratch work. After you have decided which of the suggested answers is best, completely fill in the corresponding oval on the answer sheet. Give only one answer to each question. If you change an answer, be sure that the previous mark is erased completely. Here is a sample question and answer.

Sample Question   Sample Answer

Chicago is a
(A) state
(B) city
(C) country
(D) continent

Use your time effectively, working as quickly as you can without losing accuracy. Do not spend too much time on any one question. Go on to other questions and come back to the ones you have not answered if you have time. It is not expected that everyone will know the answers to all the multiple-choice questions.

About Guessing

Many candidates wonder whether or not to guess the answers to questions about which they are not certain. Multiple choice scores are based on the number of questions answered correctly. Points are not deducted for incorrect answers, and no points are awarded for unanswered questions. Because points are not deducted for incorrect answers, you are encouraged to answer all multiple-choice questions. On any questions you do not know the answer to, you should eliminate as many choices as you can, and then select the best answer among the remaining choices.

CALCULUS BC

SECTION I, Part A

Time—60 Minutes

Number of questions—30

A CALCULATOR MAY NOT BE USED ON THIS PART OF THE EXAMINATION

<u>Directions</u>: Solve each of the following problems, using the available space for scratchwork. After examining the form of the choices, decide which is the best of the choices given and fill in the corresponding oval on the answer sheet. No credit will be given for anything written in the test book. Do not spend too much time on any one problem.

<u>In this test</u>: Unless otherwise specified, the domain of a function f is assumed to be the set of all real numbers x for which $f(x)$ is a real number.

1. Evaluate $\int_1^\infty 4xe^{-x^2}dx$.

(A) $\dfrac{2}{e^2}$

(B) $\dfrac{2}{e}$

(C) 2

(D) The integral diverges.

GO ON TO THE NEXT PAGE.

2. The table below gives some values of a function f and its first three derivatives. What is the third degree Taylor Polynomial for f about $x = 1$?

X	$f(x)$	$f'(x)$	$f''(x)$	$f'''(x)$
0	5	3	1	2
1	1	2	4	18
2	3	7	16	32

(A) $3x^3 + 7x^2 + 16x + 32$

(B) $3x^3 + 7x^2 + 7x + 2$

(C) $3x^3 - 7x^2 + 7x - 2$

(D) $3x^3 - 7x^2 + 16x - 32$

---

3. Which of the following is true about the series $\displaystyle\sum_{n-1}^{\infty} \frac{(-1)^n}{n!}$?

(A) The series converges conditionally.

(B) The series diverges.

(C) The series converges absolutely.

(D) None of the above.

GO ON TO THE NEXT PAGE.

4. The position of a particle is given by the parametric equations $x(t) = 8\sqrt{3t+1}$ and $y(t) = 9\sqrt[3]{t^2+2}$. Find the slope of the tangent line to the path of the particle at the time $t = 5$ seconds.

(A) 30

(B) 10

(C) $\dfrac{10}{3}$

(D) $\dfrac{10}{9}$

5. Evaluate $\int \dfrac{9x-13}{(x-1)(x-3)} dx$.

(A) $9\ln|x-1| + 16\ln|x-3| + C$

(B) $2\ln|x-1| + 7\ln|x-3| + C$

(C) $7\ln|x-1| + 2\ln|x-3| + C$

(D) $9\ln|x-1| - 16\ln|x-3| + C$

GO ON TO THE NEXT PAGE.

6. Which of the following equations gives the length of the curve $f(x) = 6\sin(2x)$ from $x = 0$ to $x = 4$?

(A) $\displaystyle\int_0^4 \sqrt{1+36\sin^2(2x)}\,dx$

(B) $\displaystyle\int_0^4 \sqrt{1+144\sin^2(2x)}\,dx$

(C) $\displaystyle\int_0^4 \sqrt{1+36\cos^2(2x)}\,dx$

(D) $\displaystyle\int_0^4 \sqrt{1+144\cos^2(2x)}\,dx$

7. Which of the following is the particular solution to $\dfrac{dy}{dx} = 3x + e^{2x}$ with the initial condition $y(0) = 5$?

(A) $\dfrac{3x^2}{2} + \dfrac{e^{2x}}{2} + \dfrac{9}{2}$

(B) $\dfrac{3x^2}{2} + \dfrac{e^{2x}}{2} + \dfrac{11}{2}$

(C) $\dfrac{3x^2}{2} + \dfrac{e^{2x}}{2} + 3$

(D) $\dfrac{3x^2}{2} + \dfrac{e^{2x}}{2} - \dfrac{9}{2}$

GO ON TO THE NEXT PAGE.

8. Let $y = f(x)$ be the solution of the differential equation $\dfrac{dy}{dx} = 6y - 2$ with the initial condition $f(0) = 2$. Use Euler's Method to approximate $f(1)$ starting at $x = 0$ with the step size of 0.5.

(A) 7

(B) 10

(C) 22

(D) 40

9. Evaluate $\displaystyle \int 3x\sin\left(\frac{x}{3}\right)dx$.

(A) $-9x\cos\left(\dfrac{x}{3}\right) + C$

(B) $-9x\cos\left(\dfrac{x}{3}\right) + 27x\sin\left(\dfrac{x}{3}\right) + C$

(C) $-9x\cos\left(\dfrac{x}{3}\right) - 27x\sin\left(\dfrac{x}{3}\right) + C$

(D) $-x\cos\left(\dfrac{x}{3}\right) + C$

GO ON TO THE NEXT PAGE.

10. Find the area between the curve $y = 6x - x^2$ and the x-axis.

    (A) 324

    (B) 180

    (C) 72

    (D) 36

11. Find $\dfrac{dy}{dx}$ if $y = \cos^4 x$.

    (A) $-4\sin^3 x$

    (B) $4\sin^3 x$

    (C) $-4\cos^3 x \sin x$

    (D) $4\cos^3 x \sin x$

12. If R is the region between the curves $y = x^2 - 4x$ and $y = x + 6$, find the area of R.

    (A) $48\dfrac{1}{6}$

    (B) $57\dfrac{1}{6}$

    (C) $57\dfrac{5}{6}$

    (D) $60\dfrac{1}{6}$

GO ON TO THE NEXT PAGE.

13. The MacLaurin series for the function f is given by $f(x) = \sum\limits_{n=0}^{\infty} \left(\dfrac{x}{5}\right)^n$. What is the value of $f(2)$?

(A) 0

(B) $\dfrac{5}{7}$

(C) 1

(D) $\dfrac{5}{3}$

14. Which of the following is the differential equation of the slope field below?

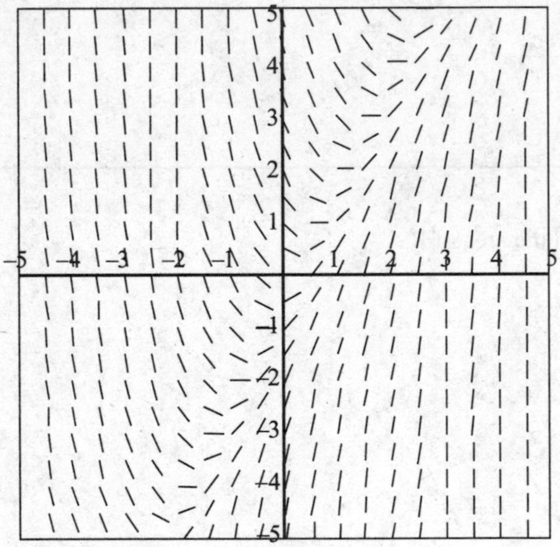

(A) $\dfrac{dy}{dx} = 2x + y$

(B) $\dfrac{dy}{dx} = 2x - y$

(C) $\dfrac{dy}{dx} = x^2 + y$

(D) $\dfrac{dy}{dx} = x^2 - y$

GO ON TO THE NEXT PAGE.

15. A curve is defined by the parametric equations $x = t^4 - t^2 + 1$ and $y = t^3$. Which of the following is the equation of the line tangent to the graph at $(13, 8)$?

    (A) $7x - 3y = -17$

    (B) $3x - 7y = 17$

    (C) $7x - 3y = 17$

    (D) $3x - 7y = -17$

16. The graph of the function f is shown below.

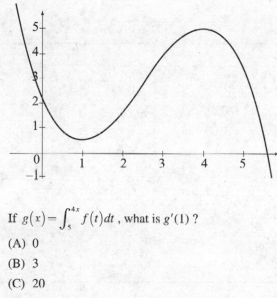

    If $g(x) = \int_5^{4x} f(t)\,dt$, what is $g'(1)$?

    (A) 0

    (B) 3

    (C) 20

    (D) 27

GO ON TO THE NEXT PAGE.

17. The sum of the infinite geometric series $5 + \dfrac{10}{\pi} + \dfrac{20}{\pi^2} + \dfrac{40}{\pi^3} + \ldots$ is

(A) $\dfrac{5\pi}{\pi - 2}$

(B) $\dfrac{5}{\pi - 2}$

(C) $\dfrac{5\pi}{\pi + 2}$

(D) $\dfrac{5}{\pi + 2}$

18. The average value of $\dfrac{4}{1 + x^2}$ on the interval $[0, 1]$ is

(A) π

(B) 2π

(C) 4π

(D) 16π

GO ON TO THE NEXT PAGE.

19. The function f is given by $f(x) = 9x^4 - 8x^3 - 96x^2 + 10$. On which of the following intervals is f increasing?

(A) $(-\infty, -2)$

(B) $(0, 2)$

(C) $\left(2, \dfrac{8}{3}\right)$

(D) $\left(\dfrac{8}{3}, \infty\right)$

20. What are all values of x for which $\dfrac{(x-1)^k}{k\left(4^k\right)}$ converges?

(A) $-3 < x < 5$

(B) $-3 \le x \le 5$

(C) $-3 \le x < 5$

(D) The series diverges.

GO ON TO THE NEXT PAGE.

21. Find the area between the polar curves $r = \cos\theta$ and $r = 2\cos\theta$ from $\theta = 0$ to $\theta = 2\pi$.

   (A) $\dfrac{\pi}{4}$

   (B) $\dfrac{\pi}{2}$

   (C) $\dfrac{3\pi}{4}$

   (D) $\dfrac{3\pi}{2}$

22. Evaluate $\displaystyle\int_3^4 \frac{dx}{(x-3)^2}$.

   (A) -1

   (B) 0

   (C) $\ln 4$

   (D) Divergent

23. A car is accelerating at $20\dfrac{\text{m}}{\text{s}^2}$ from an initial velocity of $10\dfrac{\text{m}}{\text{s}}$. How far does the car travel in the first six seconds?

   (A) 120 m

   (B) 130 m

   (C) 420 m

   (D) 780 m

GO ON TO THE NEXT PAGE.

24. The value of c that satisfies the Mean Value Theorem for Derivatives on the interval $[3, 4]$ for the function $f(x) = x + \dfrac{1}{x}$ is

(A) $-\sqrt{12}$

(B) -1

(C) 1

(D) $\sqrt{12}$

25. For $x > 0$, the power series $x + x^2 + \dfrac{x^3}{2!} + \dfrac{x^4}{3!} + \ldots + \dfrac{x^n}{(n-1)!} + \ldots$ converges to which of the following?

(A) xe^x

(B) $\dfrac{e^x}{x}$

(C) e^{x^2}

(D) The series does not converge.

GO ON TO THE NEXT PAGE.

26. If $f(x) = x^2 \ln x$, for what value of x is the tangent line horizontal?

    (A) $-\dfrac{1}{2}$

    (B) $e^{-\frac{1}{2}}$

    (C) 1

    (D) $e^{\frac{1}{2}}$

27. Given $x^3 + 2xy = y^3 + x$, find $\dfrac{d^2y}{dx^2}$ at $(2, 1)$.

    (A) -1054
    (B) -13
    (C) 13
    (D) 1054

28. If $\displaystyle\int_0^4 f(x)\,dx = 20$ and $\displaystyle\int_4^2 f(x)\,dx = 11$, then $\displaystyle\int_0^2 f(x)\,dx = ?$

    (A) -31
    (B) -9
    (C) 9
    (D) 31

GO ON TO THE NEXT PAGE.

29. Evaluate $\displaystyle\lim_{h\to 0}\frac{\ln(4+h)-\ln 4}{h}$.

(A) 0

(B) $\dfrac{1}{4}$

(C) 4

(D) The limit does not exist.

---

30. Evaluate $\displaystyle\int_1^\infty \frac{dx}{\sqrt[3]{x^5}}$.

(A) $-\dfrac{3}{2}$

(B) 0

(C) 1

(D) $\dfrac{3}{2}$

END OF PART A, SECTION I
IF YOU FINISH BEFORE TIME IS CALLED, YOU MAY CHECK YOUR WORK ON PART A ONLY.
DO NOT GO ON TO PART B UNTIL YOU ARE TOLD TO DO SO.

CALCULUS BC

SECTION I, Part B

Time—45 Minutes

Number of questions—15

A GRAPHING CALCULATOR IS REQUIRED FOR SOME QUESTIONS ON THIS PART OF THE EXAMINATION

Directions: Solve each of the following problems, using the available space for scratchwork. After examining the form of the choices, decide which is the best of the choices given and fill in the corresponding oval on the answer sheet. No credit will be given for anything written in the test book. Do not spend too much time on any one problem.

In this test:

1. The **exact** numerical value of the correct answer does not always appear among the choices given. When this happens, select from among the choices the number that best approximates the exact numerical value.

2. Unless otherwise specified, the domain of a function f is assumed to be the set of all real numbers x for which $f(x)$ is a real number.

---

31. Approximate the area under $y = \sin x$ from $x = 0$ to $x = 2$, using $n = 4$ right endpoint rectangles.

    (A) 0.044
    (B) 1.614
    (C) 2.318
    (D) 4.636

GO ON TO THE NEXT PAGE.

32. The radius of a circle is increasing at $1.5 \frac{cm}{s}$. How fast is the area increasing when $r = 4$ cm?

    (A)   7.069
    (B)   25.133
    (C)   37.699
    (D)   75.398

---

33. For what value of x is the slope of the tangent line to $y_1 = 2\sin x$ equal to the slope of the tangent line to $y_2 = \tan x$ on the interval $[0, 2\pi]$?

    (A)   0
    (B)   0.654
    (C)   1.047
    (D)   There is no value of x.

---

34. Using the Taylor series about $x = 0$ for e^x, approximate $e^{0.2}$ to three decimal places.

    (A)   1.220
    (B)   1.249
    (C)   1.250
    (D)   7.389

GO ON TO THE NEXT PAGE.

35. What is the maximum value of $y = \dfrac{\ln x}{x}$ on the interval $[1, 5]$?

    (A) 0
    (B) 0.322
    (C) 0.368
    (D) 2.718

36. If $f(x) = \dfrac{d}{dx} \displaystyle\int_{4}^{6x} \sin t \, dt$, find $f(0.1)$.

    (A) 0.099
    (B) 0.565
    (C) 3.388
    (D) 6.600

37. Evaluate $\displaystyle\int \sin^3 x \cos^2 x \, dx$.

    (A) $\dfrac{\cos^3 x}{3} - \dfrac{\cos^5 x}{5} + C$

    (B) $-\dfrac{\cos^3 x}{3} + \dfrac{\cos^5 x}{5} + C$

    (C) $-\dfrac{\sin^3 x}{3} + \dfrac{\sin^5 x}{5} + C$

    (D) $\dfrac{\sin^3 x}{3} - \dfrac{\sin^5 x}{5} + C$

GO ON TO THE NEXT PAGE.

38. What is the average value of $y = \cos 2x$ on the interval $[1, 5]$?

    (A) −1.453
    (B) −0.182
    (C)   0.182
    (D)   1.453

---

39. Let R be the region between $y = e^{-x^2}$ and the x-axis from $x = -1$ to $x = 1$. Find the volume of the solid that results when R is revolved about the x-axis.

    (A)   1.196
    (B)   3.758
    (C)   7.516
    (D) 14.124

---

40. Evaluate $\int e^{2x} \cos 2x \, dx$.

    (A) $e^{2x} \sin 2x - e^{2x} \cos 2x + C$

    (B) $e^{2x} \sin 2x + e^{2x} \cos 2x + C$

    (C) $\frac{1}{4} e^{2x} \sin 2x - \frac{1}{4} e^{2x} \cos 2x + C$

    (D) $\frac{1}{4} e^{2x} \sin 2x + \frac{1}{4} e^{2x} \cos 2x + C$

GO ON TO THE NEXT PAGE.

41. Evaluate $\lim\limits_{x \to 0} \dfrac{x - \sin x}{e^x - 1}$.

   (A) 0

   (B) 1

   (C) $\dfrac{1}{e}$

   (D) The limit does not exist.

42. The equation of the line normal to the graph of $y = 6^{x^2}$ at $x = 1$ is

   (A) $y - 6 = \dfrac{1}{12 \ln 6}(x - 1)$

   (B) $y - 6 = -\dfrac{1}{12 \ln 6}(x - 1)$

   (C) $y - 6 = 12 \ln 6 (x - 1)$

   (D) $y - 6 = -12 \ln 6 (x - 1)$

43. What is the length of the curve $y = e^{-4x}$ from $x = -1$ to $x = 1$?

   (A) 1.056

   (B) 1.111

   (C) 4.505

   (D) 55.174

GO ON TO THE NEXT PAGE.

44. Evaluate $\displaystyle\int \frac{dx}{\sqrt{36-x^2}}$.

   (A) $\sin^{-1} 6x + C$

   (B) $\sin^{-1} \dfrac{x}{6} + C$

   (C) $\dfrac{1}{2x} \sqrt{36-x^2} + C$

   (D) $-\dfrac{1}{2x} \sqrt{36-x^2} + C$

---

45. Evaluate $\displaystyle\int x^3 \sqrt{x^2-1}\,dx$.

   (A) $\dfrac{\left(x^2-1\right)^{\frac{3}{2}}}{3} + \dfrac{\left(x^2-1\right)^{\frac{5}{2}}}{5} + C$

   (B) $\dfrac{x^4}{4}\left(\dfrac{2}{3}\left(x^2-1\right)^{\frac{3}{2}}\right) + C$

   (C) $3x^2\left(-\dfrac{1}{2}\left(x^2-1\right)^{-\frac{1}{2}}\right) + C$

   (D) $\dfrac{x^4}{4} + \dfrac{2}{3}\left(x^2-1\right)^{\frac{3}{2}} + C$

STOP

END OF PART B, SECTION I

IF YOU FINISH BEFORE TIME IS CALLED, YOU MAY CHECK YOUR WORK ON PART B ONLY.

DO NOT GO ON TO SECTION II UNTIL YOU ARE TOLD TO DO SO.

SECTION II
GENERAL INSTRUCTIONS

You may wish to look over the problems before starting to work on them, since it is not expected that everyone will be able to complete all parts of all problems. All problems are given equal weight, but the parts of a particular problem are not necessarily given equal weight.

A GRAPHING CALCULATOR IS REQUIRED FOR SOME PROBLEMS OR PARTS OF PROBLEMS ON THIS SECTION OF THE EXAMINATION.

- You should write all work for each part of each problem in the space provided for that part in the booklet. Be sure to write clearly and legibly. If you make an error, you may save time by crossing it out rather than trying to erase it. Erased or crossed-out work will not be graded.

- Show all your work. You will be graded on the correctness and completeness of your methods as well as your answers. Correct answers without supporting work may not receive credit.

- Justifications require that you give mathematical (noncalculator) reasons and that you clearly identify functions, graphs, tables, or other objects you use.

- You are permitted to use your calculator to solve an equation, find the derivative of a function at a point, or calculate the value of a definite integral. However, you must clearly indicate the setup of your problem, namely the equation, function, or integral you are using. If you use other built-in features or programs, you must show the mathematical steps necessary to produce your results.

- Your work must be expressed in standard mathematical notation rather than calculator syntax. For example, $\int_{1}^{5} x^2 \, dx$ may not be written as fnInt (X^2, X, 1, 5).

- Unless otherwise specified, answers (numeric or algebraic) need not be simplified. If your answer is given as a decimal approximation, it should be correct to three places after the decimal point.

- Unless otherwise specified, the domain of a function f is assumed to be the set of all real numbers x for which $f(x)$ is a real number.

GO ON TO THE NEXT PAGE.

SECTION II, PART A
Time—30 minutes
Number of problems—2

A graphing calculator is required for some problems or parts of problems.

During the timed portion for Part A, you may work only on the problems in Part A.

On Part A, you are permitted to use your calculator to solve an equation, find the derivative of a function at a point, or calculate the value of a definite integral. However, you must clearly indicate the setup of your problem, namely the equation, function, or integral you are using. If you use other built-in features or programs, you must show the mathematical steps necessary to produce your results.

1. Grain is being loaded into a silo at the rate of $G(t) = 400e^{\frac{-t^2}{4}}$ ft^3/hr, where t is the number of hours that it is being loaded,

    $0 \le t \le 8$. At time $t = 0$, there is 100 ft^3 of grain in the silo. Grain is also being removed through the base of the silo at the

    following rates, where $R(t)$ is the amount of grain being removed in ft^3/hr, $0 \le t \le 8$:

t	0	2	5	7	8
$R(t)$	60	90	110	120	125

(a) Estimate the total amount of grain removed from the silo at $t = 8$ hrs, using a left-hand Riemann sum and 4 subintervals.

(b) Estimate the amount of grain in the silo at the end of 8 hours, using your answer from part (a).

(c) Estimate $R'(5)$, showing your work. Indicate the units of measure.

2. Consider the function given by $f(x) = x^2 e^{-4x}$.

(a) Find $\lim_{x \to \infty} f(x)$.

(b) Find the maximum value of f on the interval $[0, \infty]$. Justify your answer.

(c) Evaluate $\int_0^{\infty} f(x)dx$, or show that the integral diverges.

GO ON TO THE NEXT PAGE.

SECTION II, PART B
Time—1 hour
Number of problems—4

No calculator is allowed for these problems.

During the timed portion for Part B, you may continue to work on the problems in Part A without the use of any calculator.

3. Let R be the region in the first quadrant bounded from above by $g(x) = 19 - x^2$ and from below by $f(x) = x^2 + 1$.

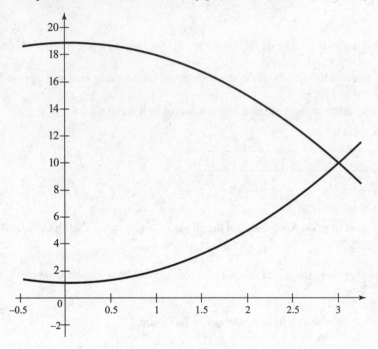

(a) Find the area of R.

(b) A solid is formed by revolving R around the x-axis. Find the volume of the solid.

(c) A solid has its base as the region R, whose cross-sections perpendicular to the x-axis are squares. Find the volume of the solid.

GO ON TO THE NEXT PAGE.

4. A particle begins on the y-axis at the point $(0, 4)$ at time $t = 0$, and travels along a straight line. For $0 \leq t \leq 20$, the particle's velocity, in ft/sec can be modeled by the piecewise-linear function in the graph below.

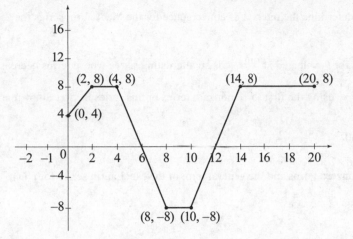

(a) At what times in the interval $0 \leq t \leq 20$ does the particle change direction? Explain your answer.

(b) Find the total displacement of the particle in the interval $0 \leq t \leq 20$.

(c) (i) Write an expression for the particle's velocity, $v(t)$, in the time interval $10 < t < 14$.

(ii) Write an expression for the particle's acceleration, $a(t)$, in the time interval $10 < t < 14$.

(d) Write an expression for the particle's distance, $s(t)$, in the time interval $10 < t < 14$.

5. Consider the curve $y^2 - 3xy = -2$.

(a) Find an equation of the tangent line to the curve at the point $(1, 1)$.

(b) Find all x-coordinates where the slope of the tangent line to the curve is undefined.

(c) Evaluate $\dfrac{d^2y}{dx^2}$ at the point $(1, 1)$.

GO ON TO THE NEXT PAGE.

6. The function f has derivatives of all orders and the MacLaurin series for f is $\displaystyle\sum_{n=0}^{\infty}(-1)^n\frac{x^{2n}}{2n+2}=\frac{1}{2}-\frac{x^2}{4}+\frac{x^4}{6}-\cdots$

(a) Using the Ratio Test, determine the interval of convergence for the MacLaurin series for f.

(b) The MacLaurin series for f evaluated at $x=\dfrac{1}{2}$ is an alternating series whose terms decrease in absolute value to 0. The approximation for $f\left(\dfrac{1}{2}\right)$ using the first three nonzero terms of this series is $\dfrac{19}{96}$. Show that this approximation differs from $f\left(\dfrac{1}{2}\right)$ by less than $\dfrac{1}{100}$.

(c) Write the first three nonzero terms and the general term of the MacLaurin series for $f'(x)$.

STOP

END OF EXAM

_____

Practice Test 1
Answers and
Explanations

ANSWER KEY TO SECTION I

Question Number	Question Answer	Correct?	See Chapter #
1	B		16
2	C		22
3	C		22
4	D		12
5	B		20
6	D		20
7	A		21
8	C		21
9	B		16
10	D		18
11	C		17
12	B		18
13	D		22
14	B		21
15	D		12
16	C		14
17	A		22
18	A		14
19	D		9
20	C		22
21	D		20
22	D		13
23	C		13

Question Number	Question Answer	Correct?	See Chapter #
24	D		8
25	A		22
26	B		8
27	D		7
28	D		14
29	B		5
30	D		20
31	B		14
32	C		10
33	B		8
34	A		22
35	C		9
36	C		14
37	B		17
38	B		14
39	B		19
40	D		16
41	A		12
42	B		8
43	D		20
44	B		17
45	A		17

ANSWERS AND EXPLANATIONS TO SECTION I

1. **B** First, rewrite the integral as a limit: $\lim\limits_{a \to \infty} \int_1^a 4xe^{-x^2}\,dx$. Next evaluate the indefinite

    integral. You get $\int 4xe^{-x^2}\,dx = -2e^{-x^2}$. Now you can evaluate the definite integral:

    $\int_1^a 4xe^{-x^2}\,dx = -2e^{-x^2}\Big|_1^a = -2e^{-a^2} + 2e^{-1} = \dfrac{2}{e^{a^2}} + \dfrac{2}{e}$. Finally, take the limit: $\lim\limits_{a \to \infty} \dfrac{2}{e^{a^2}} + \dfrac{2}{e} = \dfrac{2}{e}$.

2. **C** The formula for the Taylor Polynomial about $x = a$ is

    $$f(a) + f'(a)(x - a) + \frac{f''(a)(x - a)^2}{2!} + \frac{f'''(a)(x - a)^3}{3!}.$$

    Here, you get $f(1) + f'(1)(x - 1) + \dfrac{f''(1)(x - 1)^2}{2!} + \dfrac{f'''(1)(x - 1)^3}{3!}$. Plugging in, you get

    $$1 + 2(x - 1) + \frac{4(x - 1)^2}{2!} + \frac{18(x - 1)^3}{3!}.$$

    This simplifies to $1 + 2x - 2 + 2(x^2 - 2x + 1) + 3(x^3 - 3x^2 + 3x - 1) = 3x^3 - 7x^2 + 7x - 2$.

3. **C** First, let's test the series for absolute convergence: the absolute value of the terms is $\sum\limits_{n=1}^{\infty} \dfrac{1}{n!}$.

    You should be able to tell by inspection that this converges but if not, use the Ratio Test. The

    $\lim\limits_{k \to \infty} \dfrac{a_{k+1}}{a_k} = \lim\limits_{k \to \infty} \dfrac{\dfrac{1}{(k+1)!}}{\dfrac{1}{k!}} = \lim\limits_{k \to \infty} \dfrac{k!}{(k+1)!} = \lim\limits_{k \to \infty} \dfrac{1}{k+1} = 0$. Therefore, the series converges absolutely.

4. **D** First, find $x'(t)$ and $y'(t)$. You get

    $$x'(t) = 8\left(\frac{1}{2}\right)(3t + 1)^{-\frac{1}{2}}(3) = \frac{12}{\sqrt{3t + 1}}$$

    $$y'(t) = 9\left(\frac{1}{3}\right)(t^2 + 2)^{-\frac{2}{3}}(2t) = \frac{6t}{\sqrt[3]{(t^2 + 2)^2}}$$

    Next, plug in $t = 5$: $x'(5) = \dfrac{12}{\sqrt{3(5) + 1}} = 3$ and $y'(t) = \dfrac{6t}{\sqrt[3]{(5)^2 + 2}} = \dfrac{10}{3}$.

    The slope of the tangent line $\dfrac{dy}{dx}$ is found by $\dfrac{dy}{dx} = \dfrac{y'(t)}{x'(t)} = \dfrac{\dfrac{10}{3}}{3} = \dfrac{10}{9}$.

5. **B** You can evaluate this integral using the Method of Partial Fractions. First, let

$\dfrac{A}{x-1}+\dfrac{B}{x-3}=\dfrac{9x-13}{(x-1)(x-3)}$. Next, multiply each term by $(x-1)(x-3)$. You get

$\dfrac{A(x-1)(x-3)}{x-1}+\dfrac{B(x-1)(x-3)}{x-3}=\dfrac{(9x-13)(x-1)(x-3)}{(x-1)(x-3)}$, which simplifies to $A(x-3)+B(x-1)$

$=9x-13$. Now, distribute the terms on the left side: $Ax-3A+Bx-B=9x-13$. Because the

left side equals the right side, you know that $Ax+Bx=9x$ and $-3A-B=-13$, so $A+B=9$ and

$-3A-B=-13$. Add these equations to get $-2A=-4$, so $A=2$ and $B=7$.

Now, rewrite the integral as $\displaystyle\int\dfrac{9x-13}{(x-1)(x-3)}dx=\int\dfrac{2}{x-1}dx+\int\dfrac{7}{x-3}dx$. Integrating, you get

$\displaystyle\int\dfrac{2}{x-1}dx+\int\dfrac{7}{x-3}dx=2\ln|x-1|+7\ln|x-3|+C$.

6. **D** The formula for the length of the arc L, from $x=a$ to $x=b$, is $\displaystyle\int_a^b\sqrt{1+\left(\dfrac{dy}{dx}\right)^2}\,dx$. So here, you first

need to find $\dfrac{dy}{dx}$: $\dfrac{dy}{dx}=6\cos(2x)(2)=12\cos(2x)$. If you plug this into the formula, you get

$\displaystyle\int_0^4\sqrt{1+144\cos^2(2x)}\,dx$.

7. **A** You can solve this differential equation using separation of variables. You get $dy=\left(3x+e^{2x}\right)dx$.
    Next, integrate both sides:

$$\int dy=\int\left(3x+e^{2x}\right)dx$$

$y=\dfrac{3x^2}{2}+\dfrac{e^{2x}}{2}+C$. Now use the initial condition to solve for C:

$5=\dfrac{3(0)^2}{2}+\dfrac{e^{2(0)}}{2}+C=\dfrac{1}{2}+C$. Therefore, $C=\dfrac{9}{2}$, so $y=\dfrac{3x^2}{2}+\dfrac{e^{2x}}{2}+\dfrac{9}{2}$.

8. **C** You are given that $f(0)=2$, so let $x_0=0$ and $y_0=2$. The slope is found by plugging $y_0=2$ into

$\dfrac{dy}{dx}=6y-2$: $\dfrac{dy}{dx}=6(2)-2=10$. So your initial slope is $y_0'=10$.

Now find the next set of points.

Step 1: Increase x_0 by h (the step size) to get x_1: $x_1=0.5$.

Step 2: Multiply h by $y_0'=10$ and add to y_0 to get y_1: $y_1=(0.5)(10)+2=7$.

Step 3: Find y_1' by plugging y_1 into $\dfrac{dy}{dx}=6y-2$: $\dfrac{dy}{dx}=6(7)-2=40$.

Repeat

Step 1: Increase x_1 by h to get x_2: $x_1 = 0.5 + 0.5 = 1$.

Step 2: Multiply h by $y_1' = 40$ and add to y_1 to get y_2: $y_2 = (0.5)(40) + 2 = 22$.

So the approximate value of $f(1) = 22$.

9.   **B**   Evaluate this integral using Integration By Parts. Let $u = 3x$ and $dv = \sin\left(\dfrac{x}{3}\right) dx$. Then $du = 3\,dx$

and $v = -3\cos\left(\dfrac{x}{3}\right)$. This gives you $\displaystyle\int 3x \sin\left(\dfrac{x}{3}\right) dx = (3x)\left(-3\cos\left(\dfrac{x}{3}\right)\right) - \int -3\cos\left(\dfrac{x}{3}\right)(3\,dx)$,

which simplifies to $\displaystyle\int 3x \sin\left(\dfrac{x}{3}\right) dx = -9x\cos\left(\dfrac{x}{3}\right) + \int 9\cos\left(\dfrac{x}{3}\right) dx$. Now integrate the second term

to get $\displaystyle\int 3x \sin\left(\dfrac{x}{3}\right) dx = -9x\cos\left(\dfrac{x}{3}\right) + 27\sin\left(\dfrac{x}{3}\right) + C$.

10.   **D**   First, graph the curve:

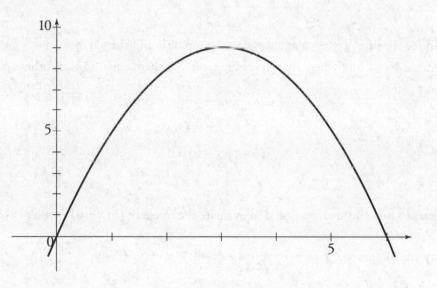

In order to find the area, you need to evaluate the integral $\displaystyle\int_0^6 6x - x^2\,dx$. You get

$$\int_0^6 6x - x^2\,dx = \left(3x^2 - \frac{x^3}{3}\right)\Bigg|_0^6 = \left(3(6)^2 - \frac{6^3}{3}\right) - 3(0)^2 - \frac{0^3}{3} = 36$$

11. **C** You need to use the Chain Rule. The derivative is $\frac{dy}{dx} = 4\cos^3 x(-\sin x) = -4\cos^3 x \sin x$.

12. **B** Sketch the region:

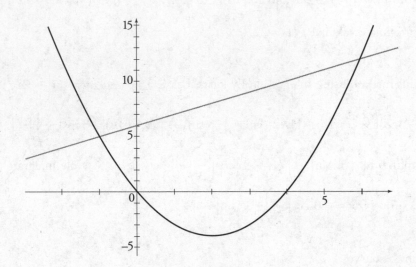

You can see the line $y = x + 6$ is always on the top of R and the parabola $y = x^2 - 4x$ is always on the bottom. Next, find where the curves intersect. Set them equal to each other and solve:

$$x^2 - 4x = x + 6$$

$$x^2 - 5x - 6 = 0$$

$$(x+1)(x-6) = 0$$

$$x = -1 \text{ or } x = 6$$

Therefore, to find the area, you need to evaluate the integral $\int_{-1}^{6} (x+6) - (x^2 - 4x)\,dx$.

Simplify the integrand: $\int_{-1}^{6} x + 6 - x^2 + 4x\,dx = \int_{-1}^{6} -x^2 + 5x + 6\,dx$.

Evaluate the integral:

$$\int_{-1}^{6} -x^2 + 5x + 6\,dx = -\frac{x^3}{3} + \frac{5x^2}{2} + 6x \Big|_{-1}^{6} = (-72 + 90 + 36) - \left(\frac{1}{3} + \frac{5}{2} - 6\right) = 57\frac{1}{6} \text{ or } \frac{343}{6}$$

13. **D** The value of the MacLaurin series is $f(2) = \sum_{n=0}^{\infty} \left(\frac{2}{5}\right)^n$. This is a geometric series, with a common

ratio of $r = \frac{2}{5}$. The sum of an infinite geometric series, with $-1 < r < 1$, is $S = \frac{t_1}{1-r}$, where t_1 is

the first term of the series. Here, $t_1 = \left(\frac{2}{5}\right)^0 = 1$, so the sum is $S = \frac{1}{1 - \frac{2}{5}} = \frac{1}{\frac{3}{5}} = \frac{5}{3}$.

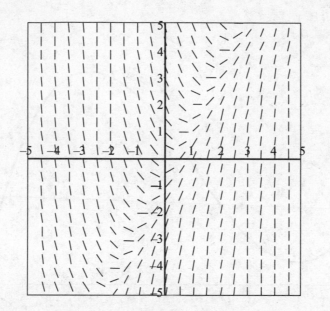

14. **B**   You can find the correct differential equation by testing some values for x and y and seeing what the slope of the graph is at those values. At the origin, the slope is 0 which, unfortunately, does not eliminate any choices. At (1, 1) the slope is positive, which eliminates (D). Note that at (1, 2), the slope is 0, which only leaves (B).

15. **D**   For a parametric equation, $\dfrac{dy}{dx} = \dfrac{\dfrac{dy}{dt}}{\dfrac{dx}{dt}}$. Here, $\dfrac{dy}{dt} = 3t^2$ and $\dfrac{dx}{dt} = 4t^3 - 2t$. This means that the slope

of the tangent line is $\dfrac{dy}{dx} = \dfrac{3t^2}{4t^3 - 2t}$. Now you just have to figure out which value of t to plug in.

You know that our equation for y is $y = t^3$ and you wish to find the equation of the line at (13, 8).

You get $y = t^3 = 8$, so $t = 2$. Check the x-value. When $t = 2$, $x = 2^4 - 2^2 + 1 = 13$, so you have the

correct value of t. Now find the slope of the tangent line by plugging $t = 2$ into $\dfrac{dy}{dx} = \dfrac{3t^2}{4t^3 - 2t}$. You

get $\dfrac{dy}{dx} = \dfrac{3(2)^2}{4(2)^3 - 2(2)} = \dfrac{12}{28} = \dfrac{3}{7}$. Therefore, the equation of the tangent line is $y - 8 = \dfrac{3}{7}(x - 13)$,

which can be rewritten as $3x - 7y = -17$.

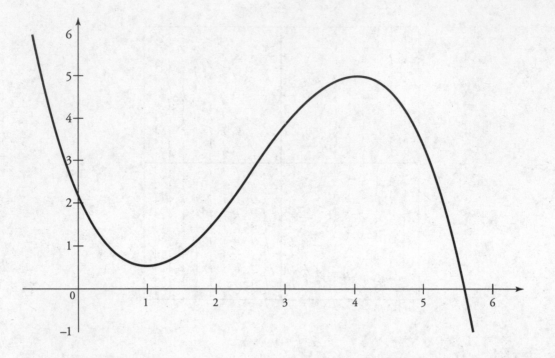

16. **C** The Second Fundamental Theorem of Calculus says that $\dfrac{d}{dx}\displaystyle\int_c^x f(t)\,dt = f(x)$. Here you get

$g'(x) = \dfrac{d}{dx}\displaystyle\int_5^{4x} f(t)\,dt = f(4x)\cdot 4$ (the 4 comes from applying the Chain Rule to the upper

limit of the integral). Therefore, $g'(1) = f(4)\cdot 4$. From the graph, you see that $f(4) = 5$, so

$g'(1) = 5\cdot 4 = 20$.

17. **A** The sum of the infinite series $a + ar + ar^2 + ar^3 + \dots$, with the common ratio r, and where

$-1 < r < 1$, is $S_\infty = \dfrac{a}{1-r}$. Here, $a = 5$ and $r = \dfrac{2}{\pi}$, so the sum is $S_\infty = \dfrac{5}{1-\dfrac{2}{\pi}} = \dfrac{5}{\dfrac{\pi-2}{\pi}} = \dfrac{5\pi}{\pi-2}$.

18. **A** You can find the average value of the function $f(x)$ on the interval $[a,\ b]$ by eval-

uating $\dfrac{1}{b-a}\displaystyle\int_a^b f(x)\,dx$. So you need to evaluate $\dfrac{1}{1-0}\displaystyle\int_0^1 \dfrac{4}{1+x^2}\,dx$. You get

$\dfrac{1}{1-0}\displaystyle\int_0^1 \dfrac{4}{1+x^2}\,dx = 4\tan^{-1} x\Big|_0^1 = 4\left(\tan^{-1}(1) - \tan^{-1}(0)\right) = 4\left(\dfrac{\pi}{4} - 0\right) = \pi$.

19. **D** A function is increasing where the first derivative is positive. So, first take the derivative:

$f'(x) = 36x^3 - 24x^2 - 192x$. Next, to find where the derivative is positive and negative, solve for where the derivative is zero. You get $36x^3 - 24x^2 - 192x = 0$

$$3x^3 - 2x^2 - 16x = 0$$

$$x(3x^2 - 2x - 16) = 0$$

$$x(3x - 8)(x + 2) = 0$$

The derivative is zero at $x = 0$, $x = -2$, and $x = \frac{8}{3}$. Now you need to test each of the intervals to see where the derivative is positive/negative. Pick a number in the interval $(-\infty, -2)$, say -3. If you plug that into the derivative, you get $f'(-3) < 0$, so f is decreasing on that interval. Next, pick a number in the interval $(-2, 0)$, say -1. If you plug that into the derivative, you get $f'(-1) > 0$, so f is increasing on that interval. Next, pick a number in the interval $\left(0, \frac{8}{3}\right)$, say 1. If you plug that into the derivative, you get $f'(1) < 0$, so f is decreasing on that interval. Finally, pick a number in the interval $\left(\frac{8}{3}, \infty\right)$, say 3. If you plug that into the derivative, you get $f'(3) > 0$, so f is increasing on that interval. Therefore, f is increasing on the intervals $\left(\frac{8}{3}, \infty\right)$ and $(-2, 0)$.

20. **C** Find the values using the Ratio Test. First, find R, which is the ratio of the terms $\frac{t_{k+1}}{t_k}$. You get

$R = \dfrac{\frac{(x-1)^{k+1}}{(k+1)(4^{k+1})}}{\frac{(x-1)^k}{k(4^k)}}$. This can be simplified to $R = \dfrac{\frac{(x-1)^{k+1}}{(k+1)(4^{k+1})}}{\frac{(x-1)^k}{k(4^k)}} = \frac{(x-1)^{k+1}}{(k+1)(4^{k+1})}\frac{k(4^k)}{(x-1)^k} = \frac{k}{k+1}\frac{x-1}{4}$.

According to the Ratio Test, the series converges if $-1 < \lim_{k \to \infty} R < 1$. Take the limit:

$\lim_{k \to \infty} \frac{k}{k+1}\frac{x-1}{4} = \frac{x-1}{4}$. Thus, find the values of x where $-1 < \frac{x-1}{4} < 1$. You get $-3 < x < 5$. You also need to check the endpoints. At $x = 5$, the series becomes $\frac{1}{k}$, which diverges. At $x = -3$, the series becomes $\frac{(-1)^k}{k}$, which converges. Therefore, the series converges for $-3 \le x < 5$.

21. **D** The formula for the area between two polar curves r_1 and r_2 from θ_1 to θ_2 is $\dfrac{1}{2}\displaystyle\int_{\theta_1}^{\theta_2} r_1^2 - r_2^2 \, d\theta$, where r_1 is outside of r_2. Graph the two curves

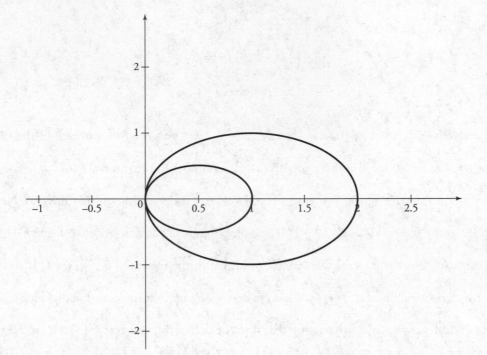

Here $r_1 = 2\cos\theta$ and $r_2 = \cos\theta$. You need to evaluate the integral $\dfrac{1}{2}\displaystyle\int_0^{2\pi} \left(2\cos\theta\right)^2 - \left(\cos\theta\right)^2 \, d\theta$,

which can be simplified to $\dfrac{1}{2}\displaystyle\int_0^{2\pi} 4\cos^2\theta - \cos^2\theta \, d\theta = \dfrac{1}{2}\displaystyle\int_0^{2\pi} 3\cos^2\theta \, d\theta$. Next, use the trig identity

$\cos^2\theta = \dfrac{1}{2}\left(1 + \cos 2\theta\right)$ to get $\dfrac{1}{2}\displaystyle\int_0^{2\pi} \dfrac{3}{2}\left(1 + \cos 2\theta\right) d\theta = \dfrac{3}{4}\displaystyle\int_0^{2\pi} \left(1 + \cos 2\theta\right) d\theta$. Evaluate the integral:

$\dfrac{3}{4}\displaystyle\int_0^{2\pi} \left(1 + \cos 2\theta\right) d\theta = \dfrac{3}{4}\left(\theta + \dfrac{\sin 2\theta}{2}\right)\Bigg|_0^{2\pi} = \dfrac{3}{4}\left[\left(2\pi + \dfrac{\sin 4\pi}{2}\right) - \left(0 + \dfrac{\sin 0}{2}\right)\right] = \dfrac{3\pi}{2}$.

22. **D** Note that the integrand is undefined at $x = 3$ so this is an improper integral. First,

rewrite the integral as a limit: $\displaystyle\lim_{a \to 3^+} \int_a^4 \dfrac{dx}{\left(x-3\right)^2}$. Next evaluate the indefinite inte-

gral. You get $\displaystyle\int \dfrac{dx}{\left(x-3\right)^2} = \int \left(x-3\right)^{-2} dx = \dfrac{-1}{x-3}$. Now evaluate the definite integral:

$\displaystyle\int_a^4 \dfrac{dx}{\left(x-3\right)^2} = \dfrac{-1}{x-3}\Bigg|_a^4 = \dfrac{-1}{4-3} + \dfrac{1}{a-3} = -1 + \dfrac{1}{a-3}$. Finally, take the limit: $\displaystyle\lim_{a \to 3^+} -1 + \dfrac{1}{a-3} = \infty$.

Therefore, the integral is divergent.

23. **C** If you know that the acceleration of the car is $a(t) = 20$, then the velocity is $v(t) = \int 20dt = 20t + C$.

Next, use the information that the initial velocity is $10\frac{m}{s}$ to get $10 = 20(0) + C$, so $C = 10$ and

thus $v(t) = 20t + 10$. Finally, to find the distance traveled, evaluate the integral $\int_0^6 20t + 10dt$. You

get $\int_0^6 20t + 10dt = 10t^2 + 10t\Big|_0^6 = \left[\left(10(6)^2 + 60\right) - \left(10(0)^2 + 0\right)\right] = 420$.

24. **D** The Mean Value Theorem for Derivatives states that if you have a function, f, that is

continuous on the interval $[a, b]$ and differentiable on the interval (a, b), then there exists

a value, c, such that $f'(c) = \dfrac{f(b) - f(a)}{b - a}$. Here, $f'(x) = 1 - \dfrac{1}{x^2}$, so $f'(c) = 1 - \dfrac{1}{c^2}$. Next,

$\dfrac{f(4) - f(3)}{4 - 3} = \dfrac{\left(4 + \frac{1}{4}\right) - \left(3 + \frac{1}{3}\right)}{1} = \dfrac{11}{12}$.

Thus, $1 - \dfrac{1}{c^2} = \dfrac{11}{12}$.

Solve for c: $1 - \dfrac{1}{c^2} = \dfrac{11}{12}$

$\dfrac{1}{12} = \dfrac{1}{c^2}$

$c = \pm\sqrt{12}$

The negative value of c is not in the interval so, $c = \sqrt{12}$.

25. **A** If you factor out x from the power series, you get $x\left(1 + x + \dfrac{x^2}{2!} + \dfrac{x^3}{3!} + \ldots + \dfrac{x^n}{n!} + \ldots\right)$, which is x

times the power series for e^x. Therefore, the power series is xe^x.

26. **B** A tangent line is horizontal when its slope is 0. Take the derivative and set it equal to zero.

$$f'(x) = x^2\left(\frac{1}{x}\right) + 2x \ln x = 0$$

$$x + 2x \ln x = 0$$

Factor out x: $x(1 + 2\ln x) = 0$.

So $x = 0$ or $1 + 2\ln x = 0$, which gives $x = e^{-\frac{1}{2}}$. You can throw out the solution $x = 0$ because it is not part of the domain of the original function. Therefore, the tangent line is horizontal for $x = e^{-\frac{1}{2}}$.

27. **D** First, find $\dfrac{dy}{dx}$ using implicit differentiation.

$$3x^2 + 2x\frac{dy}{dx} + 2y = 3y^2\frac{dy}{dx} + 1$$

Group the terms containing $\dfrac{dy}{dx}$ on the left and those without $\dfrac{dy}{dx}$ on the right:

$$2x\frac{dy}{dx} - 3y^2\frac{dy}{dx} = 1 - 3x^2 - 2y$$

Factor out $\dfrac{dy}{dx}$: $\left(2x - 3y^2\right)\dfrac{dy}{dx} = 1 - 3x^2 - 2y$.

And divide: $\dfrac{dy}{dx} = \dfrac{1 - 3x^2 - 2y}{2x - 3y^2}$. At $(2,1)$, $\dfrac{dy}{dx} = \dfrac{1 - 12 - 2}{4 - 3} = -13$.

Next, find $\dfrac{d^2y}{dx^2}$, again using implicit differentiation. Using the Quotient Rule, you get

$$\frac{d^2y}{dx^2} = \frac{\left(2x - 3y^2\right)\left(-6x - 2\dfrac{dy}{dx}\right) - \left(1 - 3x^2 - 2y\right)\left(2 - 6y\dfrac{dy}{dx}\right)}{\left(2x - 3y^2\right)^2}$$

Now, plug in $(2, 1)$ and $\dfrac{dy}{dx} = -13$:

$$\frac{d^2y}{dx^2} = \frac{\left(2(2) - 3(1)^2\right)\left(-6(2) - 2(-13)\right) - \left(1 - 3(2)^2 - 2(1)\right)\left(2 - 6(1)(-13)\right)}{\left(2(2) - 3(1)^2\right)^2}$$

$$\frac{d^2y}{dx^2} = \frac{(1)(14) - (-13)(80)}{(1)^2} = 1054$$

28. **D** The Fundamental Theorem of Calculus says that $\int_a^b f(x)\,dx + \int_b^c f(x)\,dx = \int_a^c f(x)\,dx$ and that

$\int_a^b f(x)\,dx = -\int_b^a f(x)\,dx$. This gives $\int_0^2 f(x)\,dx + \int_2^4 f(x)\,dx = \int_0^4 f(x)\,dx$. Plugging in, you get

$\int_0^2 f(x)\,dx - 11 = 20$, so $\int_0^2 f(x)\,dx = 31$.

29. **B** This limit is actually the formula for the definition of the derivative, where $f(x) = \ln x$ evaluated at $x = 4$. You know that the derivative of $\ln x = \dfrac{1}{x}$, so the limit is $\dfrac{1}{4}$.

30. **D** This is an improper integral. First, rewrite the integral as a limit: $\displaystyle\lim_{a \to \infty} \int_1^a \dfrac{dx}{\sqrt[3]{x^5}}$. Next, evaluate the

   indefinite integral. You get $\displaystyle\int \dfrac{dx}{\sqrt[3]{x^5}} = \int x^{-\frac{5}{3}}\, dx = \dfrac{x^{-\frac{2}{3}}}{-\frac{2}{3}} = -\dfrac{3x^{-\frac{2}{3}}}{2}$. Now evaluate the definite integral:

   $\displaystyle\int_1^a \dfrac{dx}{\sqrt[3]{x^5}} = -\dfrac{3x^{-\frac{2}{3}}}{2}\Bigg|_1^a = -\dfrac{3a^{-\frac{2}{3}}}{2} + \dfrac{3}{2} = -\dfrac{3}{2a^{\frac{2}{3}}} + \dfrac{3}{2}$. Finally, take the limit: $\displaystyle\lim_{a \to \infty} -\dfrac{3}{2a^{\frac{2}{3}}} + \dfrac{3}{2} = \dfrac{3}{2}$.

Part B

31. **B** The width of each of the rectangles is $\dfrac{2 - 0}{4} = \dfrac{1}{2}$. You can approximate the area with the formula

   $A = \dfrac{1}{2}\left[y\left(\dfrac{1}{2}\right) + y(1) + y\left(\dfrac{3}{2}\right) + y(2) \right]$. You get $A = \dfrac{1}{2}[0.4794 + 0.8415 + 0.9975 + 0.9093] = 1.614$.

32. **C** The area of a circle is $A = \pi r^2$ and you are given that $\dfrac{dr}{dt} = 1.5$. You need to find $\dfrac{dA}{dt}$. If you

   differentiate both sides of the equation for the area, you get $\dfrac{dA}{dt} = 2\pi r \dfrac{dr}{dt}$. Plugging in, you get

   $\dfrac{dA}{dt} = 2\pi(4)(1.5) \approx 37.699$.

33. **B** The slope of the tangent line at a point is the derivative of the function evaluated at that point. If you take the derivatives, you get $\dfrac{dy_1}{dx} = 2\cos x$ and $\dfrac{dy_2}{dx} = \sec^2 x$. Set them equal to each other and solve:

$$2\cos x = \sec^2 x$$

$$2\cos x = \frac{1}{\cos^2 x}$$

$$\cos^3 x = \frac{1}{2}$$

$$\cos x = \sqrt[3]{\frac{1}{2}}$$

$$x = \cos^{-1}\sqrt[3]{\frac{1}{2}} \approx 0.654$$

34. **A** The Taylor series for e^x about $x = 0$ is $e^x = 1 + x + \dfrac{x^2}{2!} + \dfrac{x^3}{3!} + \dfrac{x^4}{4!} + \dots$ All you have to do is plug 0.2 in for x and use as many terms as necessary until the third decimal place doesn't change. You get $e^x = 1 + (0.2) + \dfrac{(0.2)^2}{2!} + \dfrac{(0.2)^3}{3!} + \dfrac{(0.2)^4}{4!} \approx 1.220$.

35. **C** In order to find the maximum value, first find the critical points of y. Take the derivative and set it equal to zero:

$$\frac{dy}{dx} = \frac{x\left(\dfrac{1}{x}\right) - \ln x}{x^2} = \frac{1 - \ln x}{x^2}$$

$$\frac{1 - \ln x}{x^2} = 0 \text{ at } x = e$$

Next, in order to find the maximum value of y, plug in the endpoints of the interval and e and the largest value will be the answer.

$$y(1) = \frac{\ln 1}{1} = 0$$

$$y(5) = \frac{\ln 5}{5} \approx 0.322$$

$$y(e) = \frac{\ln e}{e} = \frac{1}{e} \approx 0.368$$

36. **C** You can find the derivative of the integral using the Second Fundamental Theorem of Calculus.

You get $\dfrac{d}{dx}\displaystyle\int_4^{6x} \sin t\,dt = 6\sin(6x)$. Now plug in $x = 0.1$: $f(0.1) = 6\sin(0.6) = 3.388$.

37. **B** You can integrate this using some trigonometric substitutions. First, break up the integrand

into $\displaystyle\int \sin x \sin^2 x \cos^2 x\,dx$. Next, replace $\sin^2 x$ with $1 - \cos^2 x$: $\displaystyle\int \sin x\left(1 - \cos^2 x\right)\cos^2 x\,dx$.

Now use u-substitution. Let $u = \cos x$ and $du = -\sin x$. Substituting, you get

$-\displaystyle\int\left(1 - u^2\right)u^2\,du = \int -u^2 + u^4\,du$. Integrate: $\displaystyle\int -u^2 + u^4\,du = -\dfrac{u^3}{3} + \dfrac{u^5}{5} + C$. And substitute back:

$-\dfrac{\cos^3 x}{3} + \dfrac{\cos^5 x}{5} + C$.

38. **B** The average value of $f(x)$ on the interval $[a, b]$ is $\dfrac{1}{b-a}\displaystyle\int_a^b f(x)\,dx$. Here, you get

$\dfrac{1}{5-1}\displaystyle\int_1^5 \cos(2x)\,dx = \dfrac{1}{4}\int_1^5 \cos(2x)\,dx$.

Integrate: $\dfrac{1}{4}\displaystyle\int_1^5 \cos(2x)\,dx = \dfrac{1}{4}\left(\dfrac{1}{2}\sin(2x)\right)\Big|_1^5 = \left(\dfrac{1}{8}\sin(2x)\right)\Big|_1^5 = \dfrac{1}{8}(\sin 10 - \sin 2) = -0.182$.

39. **B** First, sketch the region:

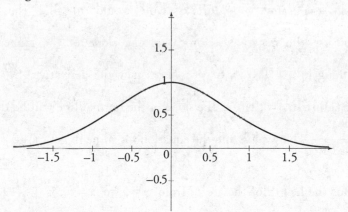

You can find the volume with the Disc Method. You get $V = \pi\displaystyle\int_{-1}^1 \left(e^{-x^2}\right)^2 dx$. You need a calculator

to evaluate the integral: $V = \pi\displaystyle\int_{-1}^1 \left(e^{-x^2}\right)^2 dx \approx \pi(.196) \approx 3.758$.

40. **D** You can integrate this integral using Integration By Parts. Let $u = e^{2x}$ and $dv = \cos 2x\, dx$. Then $du = 2e^{2x}\, dx$ and $v = \dfrac{1}{2}\sin 2x\, dx$. This gives $\int e^{2x}\cos 2x\, dx = \dfrac{1}{2}e^{2x}\sin 2x - \int e^{2x}\sin 2x\, dx$. You are going to need Integration By Parts again to do the second integral. Let $u = e^{2x}$ and $dv = \sin 2x\, dx$. Then $du = 2e^{2x}\, dx$ and $v = -\dfrac{1}{2}\cos 2x\, dx$. This gives $\int e^{2x}\cos 2x\, dx = \dfrac{1}{2}e^{2x}\sin 2x - \left(-\dfrac{1}{2}e^{2x}\cos 2x + \int e^{2x}\cos 2x\, dx\right)$, which simplifies to $\int e^{2x}\cos 2x\, dx = \dfrac{1}{2}e^{2x}\sin 2x + \dfrac{1}{2}e^{2x}\cos 2x - \int e^{2x}\cos 2x\, dx$. Now, it may look as if you are back where you started, but use the trick of adding $\int e^{2x}\cos 2x\, dx$ to both sides to get $2\int e^{2x}\cos 2x\, dx = \dfrac{1}{2}e^{2x}\sin 2x + \dfrac{1}{2}e^{2x}\cos 2x$. Now, simply divide both sides by 2: $\int e^{2x}\cos 2x\, dx = \dfrac{1}{4}e^{2x}\sin 2x + \dfrac{1}{4}e^{2x}\cos 2x + C$.

41. **A** If you plug in $x = 0$, you get a limit of the form $\dfrac{0}{0}$, which is an indeterminate form. Therefore, use L'Hôpital's Rule. Take the derivative of the numerator and the denominator: $\lim\limits_{x \to 0} \dfrac{x - \sin x}{e^x - 1} = \lim\limits_{x \to 0} \dfrac{1 - \cos x}{e^x}$. Now take the limit: $\lim\limits_{x \to 0} \dfrac{1 - \cos x}{e^x} = \dfrac{1 - \cos 0}{e^0} = \dfrac{0}{1} = 0$.

42. **B** Here, you do the same procedure as you would to find the equation of the tangent line, but use the negative reciprocal of the slope of the tangent line. First, find the y-coordinate. Plug in $x = 1$: $y = 6^1 = 6$. Next, take the derivative: $\dfrac{dy}{dx} = 2x \ln 6 \left(6^{x^2}\right)$. Plug in $x = 1$: $\dfrac{dy}{dx} = 2(1)\ln 6\left(6^1\right) = 12 \ln 6$. The slope of the normal line will be the negative reciprocal of this slope: $-\dfrac{1}{12 \ln 6}$. Now plug into the equation for a line to get $y - 6 = -\dfrac{1}{12 \ln 6}(x - 1)$.

43. **D** The formula for the length of the arc L, from $x = a$ to $x = b$, is $\int_a^b \sqrt{1 + \left(\dfrac{dy}{dx}\right)^2}\, dx$. So here, you first need to find $\dfrac{dy}{dx}$: $\dfrac{dy}{dx} = -4e^{-4x}$. If you plug into the formula, you get $\int_{-1}^1 \sqrt{1 + \left(-4e^{-4x}\right)^2}\, dx$.

You need to use a calculator to evaluate this integral: $\int_{-1}^1 \sqrt{1 + \left(-4e^{-4x}\right)^2}\, dx = 55.174$.

44. **B** If you factor out 36 from the radical, you can rewrite the integral as

$$\int \frac{dx}{\sqrt{36\left(1-\frac{x^2}{36}\right)}} = \int \frac{dx}{6\sqrt{\left(1-\frac{x^2}{36}\right)}} = \frac{1}{6}\int \frac{dx}{\sqrt{\left(1-\frac{x^2}{36}\right)}}.$$

Now use u-substitution. Let $u = \frac{x}{6}$ and $du = \frac{1}{6}dx$, or $6du = dx$. Substituting into the integrand, you get $\frac{1}{6}\int \frac{dx}{\sqrt{\left(1-\frac{x^2}{36}\right)}} = \frac{1}{6}\int \frac{6du}{\sqrt{1-u^2}} = \int \frac{du}{\sqrt{1-u^2}} = \sin^{-1}u + C$. Now, just substitute back:

$\sin^{-1}\frac{x}{6} + C$.

45. **A** You can evaluate this integral using a trigonometric substitution. Let $x = \sec\theta$ and $dx = \sec\theta\tan\theta$.

Substituting into the integrand, you get $\int \sec^3\theta\sqrt{\sec^2\theta-1}\ \sec\theta\tan\theta d\theta$. Next use the trigonometric identity $\tan^2\theta = \sec^2\theta - 1$: $\int \sec^3\theta\sqrt{\tan^2\theta}\ \sec\theta\tan\theta d\theta = \int \sec^4\theta\tan^2\theta d\theta$. Now break up

the integrand into $\int \sec^2\theta\sec^2\theta\tan^2\theta d\theta$. Now use the trigonometric identity $1 + \tan^2\theta = \sec^2\theta$:

$\int \sec^2\theta\left(1+\tan^2\theta\right)\tan^2\theta d\theta$. Next, use u-substitution. Let $u = \tan\theta$ and $du = \sec^2\theta d\theta$. Substi-

tuting into the integrand, you get $\int\left(1+u^2\right)u^2 du = \int u^2 + u^4 du = \frac{u^3}{3} + \frac{u^5}{5} + C$. Now substitute back:

$\frac{u^3}{3} + \frac{u^5}{5} + C = \frac{\tan^3\theta}{3} + \frac{\tan^5\theta}{5} + C$. Finally, because $x = \sec\theta$, you know that $\tan\theta = \sqrt{x^2-1}$.

Therefore, you get $\frac{\tan^3\theta}{3} + \frac{\tan^5\theta}{5} + C = \frac{\left(x^2-1\right)^{\frac{3}{2}}}{3} + \frac{\left(x^2-1\right)^{\frac{5}{2}}}{5} + C$.

ANSWERS AND EXPLANTIONS TO SECTION II

1. Grain is being loaded into a silo at the rate of $G(t) = 400e^{\frac{-t^2}{4}}$ ft^3/hr, where t is the number of hours that it is being loaded, $0 \le t \le 8$. At time $t = 0$, there is 100 ft^3 of grain in the silo. Grain is also being removed through the base of the silo at the following rates, where $R(t)$ is the amount of grain being removed in ft^3/hr, $0 \le t \le 8$:

t	0	2	5	7	8
$R(t)$	60	90	110	120	125

(a) Estimate the total amount of grain removed from the silo at $t = 8$ hrs, using a left-hand Riemann sum and 4 subintervals.

The Riemann sum is found by multiplying the width of each interval by R at each left endpoint of an interval. You get

$$(2-0)(60) + (5-2)(90) + (7-5)(110) + (8-7)(120) =$$

$$(2)(60) + (3)(90) + (2)(110) + (1)(120) = 120 + 270 + 220 + 120 = 730$$

(b) Estimate the amount of grain in the silo at the end of 8 hours, using your answer from part (a).

You can estimate the amount of grain in the silo by taking the starting amount, 100, adding the amount that comes in, and subtracting the amount that is being removed (which was found in part (a). You can find the amount that comes in by $\int_0^8 400e^{\frac{-t^2}{4}} \, dt$. If you plug this in your calculator, you get 708.982. (If you are unsure how to evaluate the integral on your calculator, check the online Appendix for some tips on using a TI-84 calculator). Therefore, the amount in the silo at the end of 8 hours is $100 + 708.982 - 730 = 78.982$.

(c) Estimate $R'(5)$, showing your work. Indicate the units of measure.

You can estimate $R'(5)$ by finding the slope of the secant line from $t = 2$ to $t = 5$, and from $t = 5$ to $t = 7$ and averaging the two. The slope of the secant line from $t = 2$ to $t = 5$ is $\dfrac{110-90}{5-2} = \dfrac{20}{3}$,

and the slope of the secant line from $t = 5$ to $t = 7$ is $\dfrac{120-110}{7-5} = 5$, so a good estimate for $R'(5)$ is $\dfrac{5+\dfrac{20}{3}}{2} = \dfrac{35}{6}$.

2. Consider the function given by $f(x) = x^2 e^{-4x}$.

    (a) Find $\lim_{x \to \infty} f(x)$.

    $\lim_{x \to \infty} f(x) = \lim_{x \to \infty} x^2 e^{-4x} = \lim_{x \to \infty} \dfrac{x^2}{e^{4x}}$. At infinity, the numerator is infinite and the denominator is

    also infinite, so you have an Indeterminate Form. You can evaluate this limit using L'Hôpital's

    Rule. Differentiate the numerator and denominator: $\lim_{x \to \infty} \dfrac{x^2}{e^{4x}} = \lim_{x \to \infty} \dfrac{2x}{4e^{4x}}$. You still have an

    Indeterminate Form, so use L'Hôpital's Rule again. Differentiate the numerator and denominator:

    $\lim_{x \to \infty} \dfrac{2x}{4e^{4x}} = \lim_{x \to \infty} \dfrac{2}{16e^{4x}} = 0$.

    (b) Find the maximum value of f on the interval $[0, \infty]$. Justify your answer.

    First, differentiate $f(x)$: $f'(x) = x^2 \left(-4e^{-4x} \right) + 2xe^{-4x} = e^{-4x} \left(-4x^2 + 2x \right)$.

    Next, find the critical values by setting the derivative equal to zero. The exponential function is always positive, so just solve

    $$-4x^2 + 2x = 0$$

    $$-2x(2x+1) = 0$$

    $$x = 0 \text{ or } x = -\dfrac{1}{2}$$

    Use the Second Derivative test to justify which critical value is the x-coordinate of the maximum.

    First, find the second derivative:

    $$f''(x) = e^{-4x} \left(-8x + 2 \right) - 4e^{-4x} \left(-4x^2 + 2x \right) = e^{-4x} \left(-8x + 2 + 16x^2 - 8x \right) = e^{-4x} \left(16x^2 - 16x + 2 \right)$$

    At $x = 0$, $f''(0) > 0$ (minimum).

    At $x = -\dfrac{1}{2}$, $f''(x) < 0$ (maximum).

    Finally, to find the maximum value, plug in $x = \dfrac{1}{2}$: $f\left(\dfrac{1}{2} \right) = \left(\dfrac{1}{2} \right)^2 e^{-4\left(\frac{1}{2} \right)} = \dfrac{e^{-2}}{4} = \dfrac{1}{4e^2}$.

(c) Evaluate $\int_0^\infty f(x)\,dx$, or show that the integral diverges.

Note that the integrand is undefined at infinity, so this is an improper integral. First,

rewrite the integral as a limit: $\lim_{a\to\infty}\int_0^a x^2 e^{-4x}\,dx$. Next evaluate the indefinite integral $\int x^2 e^{-4x}\,dx$

using Integration by Parts. Let $u = x^2$ and $dv = e^{-4x}\,dx$. Then $du = 2x\,dx$ and $v = -\dfrac{e^{-4x}}{4}$. This

gives $\int x^2 e^{-4x}\,dx = -\dfrac{x^2 e^{-4x}}{4} + \int 2x\dfrac{e^{-4x}}{4}\,dx = -\dfrac{x^2 e^{-4x}}{4} + \dfrac{1}{2}\int xe^{-4x}\,dx$. You will need to use Inte-

gration by Parts again. Let $u = x$ and $dv = e^{-4x}\,dx$. Then $du = dx$ and $v = -\dfrac{e^{-4x}}{4}$. This gives

$\int x^2 e^{-4x}\,dx = -\dfrac{x^2 e^{-4x}}{4} + \dfrac{1}{2}\left(-\dfrac{xe^{-4x}}{4} + \dfrac{1}{4}\int e^{-4x}\,dx\right) = -\dfrac{x^2 e^{-4x}}{4} - \dfrac{xe^{-4x}}{8} + \dfrac{1}{8}\int e^{-4x}\,dx$.

Now evaluate the final integral: $\int x^2 e^{-4x}\,dx = -\dfrac{x^2 e^{-4x}}{4} - \dfrac{xe^{-4x}}{8} + \dfrac{1}{8}\int e^{-4x}\,dx = -\dfrac{x^2 e^{-4x}}{4} - \dfrac{xe^{-4x}}{8} - \dfrac{e^{-4x}}{32}$,

which you can rewrite as $\int x^2 e^{-4x}\,dx = -\dfrac{x^2}{4e^{4x}} - \dfrac{x}{8e^{4x}} - \dfrac{1}{32e^{4x}}$. Now evaluate the definite inte-

gral: $\int_0^a x^2 e^{-4x}\,dx = -\dfrac{x^2}{4e^{4x}} - \dfrac{x}{8e^{4x}} - \dfrac{1}{32e^{4x}}\bigg|_0^a = -\dfrac{a^2}{4e^{4a}} - \dfrac{a}{8e^{4a}} - \dfrac{1}{32e^{4a}} + \dfrac{1}{32}$. Finally, take the limit:

$\lim_{a\to\infty} -\dfrac{a^2}{4e^{4a}} - \dfrac{a}{8e^{4a}} - \dfrac{1}{32e^{4a}} + \dfrac{1}{32}$ is infinite. Therefore, the integral is divergent.

3. Let R be the region in the first quadrant bounded from above by $g(x) = 19 - x^2$ and from below by $f(x) = x^2 + 1$.

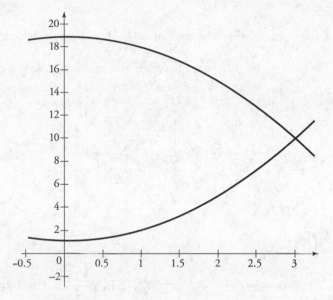

(a) Find the area of R.

First, find where the curves intersect. Set them equal to each other:

$$19 - x^2 = x^2 + 1$$

$$2x^2 = 18$$

$$x^2 = 9$$

$$x = \pm 3$$

You are only interested in the first quadrant, so in order to find the area, you need to evaluate the integral $\int_0^3 (19 - x^2) - (x^2 + 1)\, dx$. You get

$$\int_0^3 (19 - x^2) - (x^2 + 1)\, dx = \int_0^3 (18 - 2x^2)\, dx = \left(18x - \frac{2x^3}{3} \right)\Bigg|_0^3 = (54 - 18) - (0 - 0) = 36$$

(b) A solid is formed by revolving R around the x-axis. Find the volume of the solid.

In order to find the volume, evaluate the integral $\pi \int_0^3 (19 - x^2)^2 - (x^2 + 1)^2\, dx$. You get

$$\pi \int_0^3 (361 - 38x^2 + x^4) - (x^4 + 2x^2 + 1)\, dx =$$

$$\pi \int_0^3 (360 - 40x^2)\, dx =$$

$$\pi \int_0^3 (360 - 40x^2)\, dx = \pi \left(360x - \frac{40x^3}{3} \right)\Bigg|_0^3 = \pi (1080 - 360) - \pi (0 - 0) = 720\pi$$

(c)   A solid has its base as the region R, whose cross-sections perpendicular to the x-axis are squares. Find the volume of the solid.

In order to find the volume, find the area of each cross-section, which is $side^2$ (because it is a square), where the side is the difference between the upper and lower curve. Thus, you need to evaluate the integral $\int_0^3 \left[\left(19 - x^2\right) - \left(x^2 + 1\right) \right]^2 dx$. You get

$$\int_0^3 \left[\left(19 - x^2\right) - \left(x^2 + 1\right) \right]^2 dx =$$

$$\int_0^3 \left(18 - 2x^2\right)^2 dx =$$

$$\int_0^3 \left(324 - 72x^2 + 4x^4\right) dx =$$

$$\left(324x - 24x^3 + \frac{4x^5}{5}\right)\Bigg|_0^3 = \left(972 - 648 + \frac{324}{5}\right) - (0 - 0) = 324 + \frac{324}{5} = \frac{1944}{5}$$

4.   A particle begins on the y-axis at the point $(0, 4)$ at time $t = 0$, and it travels along a straight line. For $0 \le t \le 20$, the particle's velocity, in ft/sec can be modeled by the piecewise-linear function in the graph below.

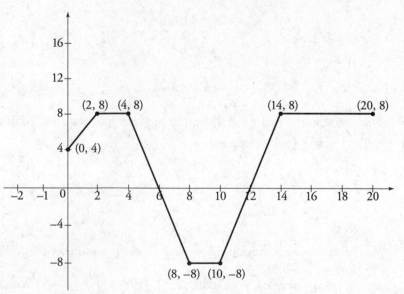

(a)   At what times in the interval $0 \le t \le 20$ does the particle change direction? Explain your answer.

The particle changes direction at points where the velocity changes sign. The particle's velocity switches from positive to negative at $t = 6$, and from negative to positive at $t = 12$.

(b) Find the total displacement of the particle in the interval $0 \le t \le 20$.

You can find the distance traveled by finding the area between the curve and the x-axis. Thus, from $t = 0$ to $t = 6$, the particle travels $\frac{1}{2}(4+8)(2)+(8)(2)+\frac{1}{2}(8)(2)=36$. From $t = 6$ to $t = 12$, the particle travels $-\frac{1}{2}(6+2)(8)=-32$. Note that this area is negative because the area is below the x-axis. From $t = 12$ to $t = 20$, the particle travels $\frac{1}{2}(2)(8)+(6)(8)=56$. Thus, the total distance that the particle travels is $36-32+56=60$ ft.

(c) (i) Write an expression for the particle's velocity, $v(t)$, in the time interval $10 < t < 14$.

Note that at time $t = 10$, the particle is at $y = -8$. Next, its slope in the interval $10 < t < 14$ is $\frac{8+8}{14-10}=4$. Therefore, the particle's velocity is $v(t)=-8+4t$.

(ii) Write an expression for the particle's acceleration, $a(t)$, in the time interval $10 < t < 14$.

The acceleration of the particle in the interval $10 < t < 14$ is simply the slope of the line. Thus, the particle's acceleration is $a(t) = 4$.

(d) Write an expression for the particle's distance, $s(t)$, in the time interval $10 < t < 14$.

First, note that the particle's position is the distance that it has traveled in the first 10 seconds is $36-24=12$. Thus, the particle's distance can be found by evaluating $12+\int_{10}^{t}-8+4x\,dx$. (Here, x is a dummy variable that we use for the purposes of integration. Any letter will suffice.) You get

$$12+\int_{10}^{t}-8+4x\,dx=12+\left(-8x+2x^{2}\right)\Big|_{10}^{t}=12+\left(-8t+2t^{2}\right)-(120)=2t^{2}-8t+120.$$

5. Consider the curve $y^2 - 3xy = -2$.

(a) Find an equation of the tangent line to the curve at the point (1, 1).

First, you need to use Implicit Differentiation to find $\frac{dy}{dx}$. You get $2y\frac{dy}{dx}-3\left(x\frac{dy}{dx}+y\right)=0$. Isolate $\frac{dy}{dx}$ because you will need it in part (c). Otherwise, you would plug in (1, 1) here to solve for the slope.

$$2y\frac{dy}{dx} - 3x\frac{dy}{dx} - 3y = 0$$

$$\frac{dy}{dx}(2y - 3x) = 3y$$

$$\frac{dy}{dx} = \frac{3y}{2y - 3x}$$

Now plug in (1, 1) to solve for the slope: $\dfrac{dy}{dx} = \dfrac{3}{2-3} = -3.$

Therefore, the equation of the tangent line is $y - 1 = -3(x - 1)$.

(b) Find all x-coordinates where the slope of the tangent line to the curve is undefined.

The tangent line will be undefined where the denominator is zero. That is,

$$2y - 3x = 0$$

$$2y = 3x$$

$$y = \frac{3x}{2}$$

Plug $y = \dfrac{3x}{2}$ into the original equation: $\left(-\dfrac{3x}{2}\right)^2 - 3x\left(\dfrac{3x}{2}\right) = -2.$ And solve:

$$\frac{9x^2}{4} - \frac{9x^2}{2} = -2$$

$$\frac{9x^2}{4} = 2$$

$$x^2 = \frac{8}{9}$$

$$x = \pm\frac{\sqrt{8}}{3}$$

(c) Evaluate $\dfrac{d^2y}{dx^2}$ at the point (1, 1).

You need to take the derivative of $\dfrac{dy}{dx}$ that was found in part (a). Using the Quotient Rule and

Implicit Differentiation, you get

$$\frac{d^2y}{dx^2} = \frac{(2y - 3x)\left(3\dfrac{dy}{dx}\right) - 3y\left(2\dfrac{dy}{dx} - 3\right)}{(2y + 3x)^2}$$

Note that at (1, 1), $\dfrac{dy}{dx} = -3$. Therefore, at (1, 1),

$$\dfrac{d^2 y}{dx^2} = \dfrac{(2-3)(3(-3)) - 3(2(-3)-3)}{(2-3)^2} = \dfrac{(-1)(-9)-(3)(-9)}{1} = 36$$

6. The function f has derivatives of all orders and the MacLaurin series for f is

$$\sum_{n=0}^{\infty} (-1)^n \dfrac{x^{2n}}{2n+2} = \dfrac{1}{2} - \dfrac{x^2}{4} + \dfrac{x^4}{6} - \dots$$

(a) Using the Ratio Test, determine the interval of convergence for the MacLaurin series for f.

For the Ratio Test, you will need to find $R = \left| \dfrac{a_{n+1}}{a_n} \right|$, where a_n is the nth term of the series. You get

$$R = \left| \dfrac{\dfrac{x^{2(n+1)}}{2(n+1)+2}}{\dfrac{x^{2n}}{2n+2}} \right| = \left| \dfrac{\dfrac{x^{2n+2}}{2n+2+2}}{\dfrac{x^{2n}}{2n+2}} \right|, \text{ which can be simplified to}$$

$$R = \left| \dfrac{\dfrac{x^{2n+2}}{2n+4}}{\dfrac{x^{2n}}{2n+2}} \right| = \left| \dfrac{x^{2n+2}}{x^{2n}} \dfrac{2n+2}{2n+4} \right| = \left| x^2 \dfrac{2n+2}{2n+4} \right|. \text{ Next, take } \lim_{n \to \infty} R = \lim_{n \to \infty} \left| \dfrac{2n+2}{2n+4} \right| x^2 = x^2. \text{ Thus, the}$$

series will converge when $x^2 < 1$, or $-1 < x < 1$.

When $x = -1$, the series is $\dfrac{1}{2} - \dfrac{1}{4} + \dfrac{1}{6} - \dots$, which converges (Alternating Series Test).

When $x = 1$, the series is again $\dfrac{1}{2} - \dfrac{1}{4} + \dfrac{1}{6} - \dots$, which converges (Alternating Series Test).

Therefore, the interval of convergence is $-1 \leq x \leq 1$.

(b)   The MacLaurin series for f evaluated at $x = \dfrac{1}{2}$ is an alternating series whose terms decrease in absolute value to 0. The approximation for $f\left(\dfrac{1}{2}\right)$ using the first three nonzero terms of this series is $\dfrac{19}{96}$. Show that this approximation differs from $f\left(\dfrac{1}{2}\right)$ by less than $\dfrac{1}{100}$.

The error will be less than the next term in the MacLaurin series, which is the fourth term $\dfrac{x^8}{8}$. At $x = \dfrac{1}{2}$, this is $\dfrac{\left(\dfrac{1}{2}\right)^8}{8} = \dfrac{1}{2048}$, which is less than $\dfrac{1}{1000}$.

(c)   Write the first three nonzero terms and the general term of the MacLaurin series for $f'(x)$.

Simply take the derivatives of the first three terms to get: $f'(x) = -\dfrac{2x}{4} + \dfrac{4x^3}{6} - \dfrac{6x^5}{8}$. The general term is $\displaystyle\sum_{n=0}^{\infty} (-1)^{n+1} \dfrac{(2n+2)x^{2n+1}}{2n+4}$.

Part III
About the
AP Calculus BC
Exam

- AB Calculus vs. BC Calculus
- Structure of the Exam
- Overview of Content Topics
- General Overview of This Book
- How AP Exams Are Used
- Other Resources
- Designing Your Study Plan

AB CALCULUS VS. BC CALCULUS

AP Calculus is divided into two types: AB and BC. The former is supposed to be the equivalent of a semester of college calculus; the latter, a year. In truth, AB calculus covers closer to three-quarters of a year of college calculus. In fact, the main difference between the two is that BC calculus tests some more theoretical aspects of calculus and it covers a few additional topics. In addition, BC calculus is harder than AB calculus. The AB exam usually tests straightforward problems in each topic. They're not too tricky, and they don't vary very much. The BC Exam asks harder questions. But neither exam is tricky, in the sense that the SAT is. Nor do they test esoteric aspects of calculus. Rather, both tests tend to focus on testing whether you've learned the basics of differential and integral calculus. The tests are difficult because of the breadth of topics that they cover, not the depth. You will probably find that many of the problems in this book seem easier than the problems you've had in school. This is because your teacher is giving you problems that are harder than those on the AP.

> Please note that this book will focus only on the topics that will appear on the AP Calculus BC test. If you are looking to prepare for the AP Calculus AB Exam, please purchase *Cracking the AP Calculus AB Exam* in stores now!

STRUCTURE OF THE EXAM

Now, some words about the test itself. The AP Exam comes in two parts. First, there is a section of multiple-choice questions covering a variety of calculus topics. The multiple-choice section has two parts. Part A consists of 30 questions; you are not permitted to use a calculator on this section. Part B consists of 15 questions; you are permitted to use a calculator on this part. These two parts comprise a total of 45 questions.

After that, there is a free-response section consisting of six questions, each of which requires you to write out the solutions and the steps by which you solved it. You are permitted to use a calculator for the first two problems but not for the four other problems. Partial credit is given for various steps in the solution of each problem. You'll usually be required to sketch a graph in one of the questions. The College Board does you a big favor here: you may use a graphing calculator. In fact, The College Board recommends it! And they allow you to use programs as well. But here's the truth about calculus: most of the time, you don't need the calculator anyway. Remember, these are the people who bring you the SAT. Any gift from them should be regarded skeptically!

HOW THE AP CALCULUS BC EXAM IS SCORED

A numeral score of 1 to 5 is going to be assigned to your test, based on the number of questions you've answered correctly.

5 = Extremely Well Qualified
4 = Well Qualified
3 = Qualified
2 = Possibly Qualified
1 = No Recommendation

Colleges decide for themselves the minimum score they will accept for college credit and/or advanced placement. The American Council on Education recommends the acceptance of grades 3 or above, and many colleges adhere to these standards. About 80 percent of students who take the AP Calculus BC Exam receive a score of 3 or higher. Check the website for each college you plan to apply to so that you know their policy on granting credit or advanced placement.

AP Calculus—2018 Score Distributions		
Score	Number of Test Takers	Percentage
5	56,324	40.4%
4	25,982	18.6%
3	28,891	20.7%
2	20,349	14.6%
1	7,830	5.6%
Total	139,376	100%

Data from The College Board, May 2018 AP Exam administrations

OVERVIEW OF CONTENT TOPICS

This list is drawn from the topical outline for AP Calculus furnished by the College Board. You might find that your teacher covers some additional topics, or omits some, in your course. Some of the topics are very broad, so we cannot guarantee that this book covers these topics exhaustively.

I. Functions, Graphs, and Limits
   A. Analysis of Graphs

- You should be able to analyze a graph based on "the interplay between geometric and analytic information." The preceding phrase comes directly from the College Board. Don't let it scare you. What the College Board really means is that you should have covered graphing in precalculus, and you should know (a) how to graph and (b) how to read a graph. **This is a precalculus topic, and we won't cover it in this book.**

High Scores, Explained
You may have noticed that the percentage of those who get a 5 on the BC Exam is rather high. This doesn't mean the test is easy. Instead, it suggests that the BC test is selective: those who feel less confident may take the AB instead. Make sure you're taking the right test for you!

Same Song, Slightly Different Verse
The 2019–2020 administration of the course will be modifying this outline. This book will still cover all the material, but in a potentially different order, one that's designed to help maximize your test-prep. To stay up-to-date with changes that might impact your test, register your book and check your online Student Tools. You can also go to the College Board's AP Central for this course (https://apcentral.collegeboard.org/courses/ap-calculus/course/updates-2019–20).

B. Limits

- You should be able to calculate limits algebraically or to estimate them from a graph or from a table of data.

- You do **not** need to find limits using the Delta-Epsilon definition of a limit.

C. Asymptotes

- You should understand asymptotes graphically and be able to compare the growth rates of different types of functions (namely, polynomial functions, logarithmic functions, and exponential functions). **This is a topic that should have been covered in precalculus, and we won't cover it in this book.**

- You should understand asymptotes in terms of limits involving infinity.

D. Continuity

- You should be able to test the continuity of a function in terms of limits, and you should understand continuous functions graphically.

- You should understand the Intermediate Value Theorem (IVT) and the Extreme Value Theorem (EVT).

E. Parametric, Polar, and Vector Functions

- You should be able to analyze plane curves given in any of these three forms. Usually, you will be asked to convert from one of these three forms back to Rectangular Form (also known as Cartesian Form).

II. Differential Calculus

A. The Definition of the Derivative

- You should be able to find a derivative by finding the limit of the difference quotient.

- You should also know the relationship between differentiability and continuity. That is, if a function is differentiable at a point, it's continuous there. But if a function is continuous at a point, it's not necessarily differentiable there.

B. Derivative at a Point

- You should know the Power Rule, the Product Rule, the Quotient Rule, and the Chain Rule.

- You should be able to find the slope of a curve at a point and the tangent and normal lines to a curve at a point.

- You should also be able to use local linear approximation and differentials to estimate the tangent line to a curve at a point.

- You should be able to find the instantaneous rate of change of a function using the derivative or the limit of the average rate of change of a function.

- You should be able to approximate the rate of change of a function from a graph or from a table of values.

- You should be able to find Higher-Order Derivatives and to use Implicit Differentiation.

C. Derivative of a Function

- You should be able to relate the graph of a function to the graph of its derivative, and vice-versa.

- You should know the relationship between the sign of a derivative and whether the function is increasing or decreasing (positive derivative means increasing; negative means decreasing).

- You should know how to find relative and absolute maxima and minima.

- You should know the Mean Value Theorem for derivatives and Rolle's Theorem.

D. Second Derivative

- You should be able to relate the graph of a function to the graph of its derivative and its second derivative, and vice-versa. This is tricky.

- You should know the relationship between concavity and the sign of the second derivative (positive means concave up; negative means concave down).

- You should know how to find points of inflection.

E. Applications of Derivatives

- You should be able to sketch a curve using first and second derivatives and be able to analyze the critical points.

- You should be able to solve Optimization problems (Max/Min problems) and Related Rates problems.

- You should be able to find the derivative of the inverse of a function.

- You should be able to solve Rectilinear Motion problems.

F. More Applications of Derivatives

- You should be able to analyze planar curves in parametric, polar, and vector form, including velocity and acceleration vectors.

- You should be able to use Euler's Method to find numerical solutions of differential equations.

- You should know L'Hôpital's Rule.

G. Computation of Derivatives

- You should be able to find the derivatives of trig functions, logarithmic functions, exponential functions, and inverse trig functions.

- You should be able to find the derivatives of parametric, polar, and vector functions.

III. Integral Calculus
A. Riemann Sums

- You should be able to find the area under a curve using left, right, and midpoint evaluations and the Trapezoid Rule.

- You should know the fundamental theorem of calculus:

$$\int_a^b f(x)\,dx = F(b) - F(a)$$

B. Applications of Integrals

- You should be able to find the area of a region, the volume of a solid of known cross-section, the volume of a solid of revolution, and the average value of a function.

- You should be able to solve acceleration, velocity, and position problems.

- You should be able to find the length of a curve (including a curve in parametric form) and the area of a region bounded by polar curves.

C. Fundamental Theorem of Calculus

- You should know the first and second fundamental theorems of calculus and be able to use them both to find the derivative of an integral and for the analytical and graphical analysis of functions.

D. Techniques of Antidifferentiation

- You should be able to integrate using the power rule and u-substitution.

- You should be able to do integration by parts and simple partial fractions.

- You should also be able to evaluate Improper Integrals as limits of definite integrals.

E. Applications of Antidifferentiation

- You should be able to find specific antiderivatives using initial conditions.

- You should be able to solve separable differential equations and logistic differential equations.

- You should be able to interpret differential equations via slope fields. Don't be intimidated. These look harder than they are.

IV. Polynomial Approximations and Series
A. The Concept of a Series

- You should know that a series is a sequence of partial sums and that convergence is defined as the limit of the sequence of partial sums.

B. Series Concepts

- You should understand and be able to solve problems involving Geometric Series, the Harmonic Series, Alternating Series, Conditional Convergence, and p-Series.

- You should know the Integral Test, the Ratio Test, Comparison Test, and the Limit Comparison Test, and how to use them to determine whether a series converges (absolutely or conditionally) or diverges.

C. Taylor Series

- You should know Taylor Polynomial Approximation; the general Taylor Series centered at $x = a$; and the MacLaurin Series for e^x, $\sin x$, $\cos x$, and $\dfrac{1}{1-x}$.

- You should know how to differentiate and antidifferentiate Taylor Series and how to form new series from known series.

- You should know functions defined by power series and radius of convergence.

- You should know the Lagrange error bound for Taylor Series.

GENERAL OVERVIEW OF THIS BOOK

The key to doing well on the exam is to memorize a variety of techniques for solving calculus problems and to recognize when to use them. There's so much to learn in AP Calculus that it's difficult to remember everything. Instead, you should be able to derive or figure out how to do certain things based on your mastery of a few essential techniques. In addition, you'll be expected to remember a lot of the math that you did before calculus—particularly trigonometry. You should be able to graph functions, find zeros, derivatives, and integrals with the calculator.

Furthermore, if you can't derive certain formulas, you should memorize them! A lot of students don't bother to memorize the trigonometry special angles and formulas because they can do them on their calculators. This is a big mistake. You'll be expected to be very good with these angles and formulas in calculus, and if you can't recall them easily, you'll be slowed down and the problems will seem much harder. Make sure that you're also comfortable with analytic geometry. If you rely on your calculator to graph for you, you'll get a lot of questions wrong because you won't recognize the curves when you see them.

This advice is going to seem backward compared with what your teachers are telling you. In school you're often yelled at for memorizing things. Teachers tell you to understand the concepts, not just memorize the answers. Well, things are different here. The understanding will come later, after you're comfortable with the mechanics. In the meantime, you should learn techniques and practice them, and, through repetition, you will ingrain them in your memory.

Each chapter is divided into three types of problems: examples, solved problems, and practice problems. The first type is contained in the explanatory portion of the unit. The examples are designed to further your understanding of the subject and to show you how to get the problems right. Each step of the solution to the example is worked out, except for some simple algebraic and arithmetic steps that should come easily to you at this point.

The second type is solved problems. The solutions are worked out in approximately the same detail as the examples. Before you start work on each of these, cover the solution with an index card or something, then check the solution afterward. And you should read through the solution, not just assume that you knew what you were doing because your answer was correct.

The third type is practice problems. Only the answer explanations to these are given. We hope you'll find that each chapter offers enough practice problems for you to be comfortable with the material. The topics that are emphasized on the exam have more problems; those that are de-emphasized have fewer. In other words, if a chapter has only a few practice problems, it's not an important topic on the AP Exam and you shouldn't worry too much about it. These practice problems come in a variety of forms throughout the book (and in your online Student Tools). Practice Sets cover a set amount of material from a chapter in an open-ended format, while End-of-Chapter Drills present multiple-choice questions from topics across that whole chapter.

Check out our free, handy guide to using the TI-84 series calculator. You can download this Calculator Appendix from your Student Tools, once you register your book online. See pages x–xi for details.

There's More Online
You can find a full list of all the formulas in this book in your online Student Tools, which also contains a wealth of other resources, like calculator shortcuts.

No, Really, There's Even More Online
Once you've registered your copy of the book, you can also access drills on fundamental math skills, and also work on challenging End-of-Unit Drills that put everything to the test.

HOW AP EXAMS ARE USED

Different colleges use AP Exams in different ways, so it is important that you go to a particular college's website to determine how it uses AP Exams. The three items below are the main ways in which AP Exam scores can be used.

- **College Credit**. Some colleges will give you college credit if you score well on an AP Exam. These credits count towards your graduation requirements, meaning that you can take fewer courses while in college. Given the cost of college, this could be quite a benefit, indeed.
- **Satisfy Requirements**. Some colleges will allow you to "place out" of certain requirements if you do well on an AP Exam, even if they do not give you actual college credits. For example, you might not need to take an introductory-level course, or perhaps you might not need to take a class in a certain discipline at all.
- **Admissions Plus**. Even if your AP Exam will not result in college credit or even allow you to place out of certain courses, most colleges will respect your decision to push yourself by taking an AP Course or even an AP Exam outside of a course. A high score on an AP Exam shows mastery of more difficult content than is taught in many high school courses, and colleges may take that into account during the admissions process.

OTHER RESOURCES

There are many resources available to help you improve your score on the AP Calculus BC Exam, not the least of which are your **teachers**. If you are taking an AP class, you may be able to get extra attention from your teacher, such as obtaining feedback on your free-response questions. If you are not in an AP course, reach out to a teacher who teaches calculus, and ask if the teacher will review your free-response questions or otherwise help you with content.

Another wonderful resource is **AP Central**, the official site of the AP Exams. The scope of the information at this site is quite broad and includes:

- a course description, which includes details on what content is covered and sample questions
- sample test questions
- free-response prompts from previous years

The AP Central home page address is: **http://apcentral.collegeboard.com/**.

The AP Calculus BC Exam Course home page address is: https://apcentral.collegeboard.org/courses/ap-calculus-bc?course=ap-calculus-bc.

Finally, **The Princeton Review** offers tutoring and small group instruction. Our expert instructors can help you refine your strategic approach and add to your content knowledge. For more information, call 1-800-2REVIEW.

DESIGNING YOUR STUDY PLAN

As part of the Introduction, you identified some areas of potential improvement. Let's now delve further into your performance on Test 1, with the goal of developing a study plan appropriate to your needs and time commitment.

Read the answers and explanations associated with the multiple-choice questions (starting at page 39). After you have done so, respond to the following:

Break up your review into manageable portions. Download our helpful study guide for this book, once you register online.

- Review the Overview of Content Topics on pages 65–68. Next to each topic, indicate your rank of the topic as follows: "1" means "I need a lot of work on this," "2" means "I need to beef up my knowledge," and "3" means "I know this topic well."

- How many days/weeks/months away is your exam?

- What time of day is your best, most focused study time?

- How much time per day/week/month will you devote to preparing for your exam?

- When will you do this preparation? (Be as specific as possible: Mondays & Wednesdays from 3 to 4 PM, for example.)

- Based on the answers above, will you focus on strategy (Part IV) or content (Part V) or both?

- What are your overall goals in using this book?

Part IV
Test-Taking Strategies for the AP Calculus BC Exam

PREVIEW ACTIVITY

Review your responses to the first three questions on page 2 of the Introduction, and then respond to the following questions:

- How many multiple-choice questions did you miss even though you knew the answer?

- On how many multiple-choice questions did you guess blindly?

- How many multiple-choice questions did you miss after eliminating some answers and guessing based on the remaining answers?

- Did you find any of the free-response questions easier or harder than the others— and, if so, why?

HOW TO USE THE CHAPTERS IN THIS PART

For the following Strategy chapters, think about what you are doing now before you read the chapters. As you read and engage in the directed practice, be sure to appreciate the ways you can change your approach. At the end of each chapter in Part Three, you will have the opportunity to reflect on how you will change your approach.

Chapter 1
How to Approach
Multiple-Choice
Questions

Some of the AP tests are changing for the 2019–2020 school year. As of the printing of this book, no such changes were scheduled for the AP Calculus BC Exam. However, please check your online student tools, as if there is any late-breaking news, we will be sharing updates there.

CRACKING THE MULTIPLE-CHOICE QUESTIONS

Section I of the AP Calculus Exam consists of 45 multiple-choice questions, which you're given 105 minutes to complete. This section is worth 50 percent of your grade.

All the multiple-choice questions will have a similar format: each will be followed by four answer choices. At times, it may seem that there could be more than one possible correct answer. There is only one! Remember that the committee members who write these questions are calculus teachers. So, when it comes to calculus, they know how students think and what kind of mistakes they make. Answers resulting from common mistakes are often included in the four answer choices to trap you.

Use the Answer Sheet

For information on how to access your free online Student Tools, see pages x-xi.

For the multiple-choice section, you write the answers not in the test booklet but on a separate answer sheet (very similar to the ones we've supplied online in your Student Tools as printable documents). Four oval-shaped bubbles follow the question number, one for each possible answer. *Don't* forget to fill in all your answers on the answer sheet. Don't just mark them in the test booklet. Marks in the test booklet will not be graded. Also, make sure that your filled-in answers correspond to the correct question numbers! Check your answer sheet after every four answers to make sure you haven't skipped any bubbles by mistake.

Should You Guess?

Use Process of Elimination (POE) to rule out answer choices you know are wrong and increase your chances of guessing the right answer. Read all the answer choices carefully. Eliminate the ones that you know are wrong. If you only have one answer choice left, *choose it,* even if you're not completely sure why it's correct. Remember, questions in the multiple-choice section are graded by a computer, so it doesn't care *how* you arrived at the correct answer.

Even if you can't eliminate answer choices, go ahead and guess. AP Exams no longer include a guessing penalty of a quarter of a point for each incorrect answer. You will be assessed only on the total number of correct answers, so be sure to fill in all the bubbles even if you have no idea what the correct answers are. When you get to questions that are too time-consuming, or that you don't know the answer to (and can't eliminate any options), don't just fill in any answer. Use what we call "your letter of the day" (LOTD). Selecting the same answer choice each time you guess will increase your odds of getting a few of those skipped questions right.

Use the Two-Pass System

Remember that you have about two and a quarter minutes per question on this section of the exam. Do not waste time by lingering too long over any single question. If you're having trouble, move on to the next question. After you finish all the questions, you can come back to the ones you skipped.

The best strategy is to go through the multiple-choice section twice. The first time, do all the questions that you can answer fairly quickly—the ones where you feel confident about the correct answer. On this first pass, skip the questions that seem to require more thinking or the ones you need to read two or three times before you understand them. Circle the questions that you've skipped in the question booklet so that you can find them easily in the second pass. You must *be very careful* with the answer sheet by making sure the filled-in answers correspond correctly to the questions.

Once you have gone through all the questions, go back to the ones that you skipped in the first pass. But don't linger too long on any one question even in the second pass. Spending too much time wrestling over a hard question can cause two things to happen. One, you may run out of time and miss out on answering easier questions in the later part of the exam. Two, your anxiety might start building up, and this could prevent you from thinking clearly, which would make answering other questions even more difficult. If you simply don't know the answer, or can't eliminate any of them, just use your LOTD and move on.

REFLECT

Respond to the following questions:

- How long will you spend on multiple-choice questions?

- How will you change your approach to multiple-choice questions?

- What is your multiple-choice guessing strategy?

Chapter 2
How to Approach
Free-Response
Questions

CRACKING FREE RESPONSE QUESTIONS

Section II is worth 50 percent of your grade on the AP Calculus Exam. This section is composed of two parts. Part A contains two free-response questions (you may use a calculator on this part); Part B contains four free-response questions where there are no calculators allowed. You're given a total of 90 minutes for this section.

Clearly Explain and Justify Your Answers

Remember that your answers to the free-response questions are graded by *readers* and not by computers. Communication is a very important part of AP Calculus. Compose your answers in precise sentences. Just getting the correct numerical answer is not enough. You should be able to *explain* your reasoning behind the technique that you selected and *communicate* your answer in the context of the problem. Even if the question does not explicitly say so, always explain and *justify* every step of your answer, including the final answer. Do not expect the graders to read between the lines. Explain everything as though somebody with no knowledge of calculus is going to read it. Be sure to present your solution in a systematic manner using solid logic and appropriate language. And remember that although you won't earn points for neatness, the graders can't give you a grade if they can't read and understand your solution!

Use Only the Space You Need

Do not try to fill up the space provided for each question. The space given is usually more than enough. The people who design the tests realize that some students write in big letters or make mistakes and need extra space for corrections. So if you have a complete solution, don't worry about the extra space. Writing more will not earn you extra credit. In fact, many students tend to go overboard and shoot themselves in the foot by making a mistake after they've already written the right answer.

Read the Whole Question!

Some questions might have several subparts. Try to answer them all, and don't give up on the question if one part is giving you trouble. For example, if the answer to part (b) depends on the answer to part (a), but you think you got the answer to part (a) wrong, you should still go ahead and do part (b) using your answer to part (a) as required. Chances are that the grader will not mark you wrong twice, unless it is obvious from your answer that you should have discovered your mistake.

Use Common Sense

Always use your common sense in answering questions. For example, on one free-response question that asked students to compute the mean weight of newborn babies from given data, some students answered 70 pounds. It should have been immediately obvious that the answer was probably off by a decimal point. A 70-pound baby would be a giant! This is an important mistake that should be easy to fix. Some mistakes may not be so obvious from the answer. However, the grader will consider simple, *easily recognizable errors* to be *very important*.

REFLECT

Respond to the following questions:

- How much time will you spend on each free-response question?

- How will you change your approach to the free-response questions?

- Will you seek further help, outside of this book (such as a teacher, tutor, or AP Central), on how to approach the AP Calculus Exam?

Part V
Content Review for the AP Calculus BC Exam

HOW TO USE THE CHAPTERS IN THIS PART

You may need to review the following content chapters more than once. Your goal is to obtain mastery of the content you are missing, and a single read of a chapter may not be sufficient. At the end of each chapter, you will have an opportunity to reflect on whether you truly have mastered the content of that chapter.

Unit 1
Differential
Calculus
Essentials

Chapter 3
Limits

WHAT IS A LIMIT?

In order to understand calculus, you need to know what a "limit" is. A limit is the value a function (which usually is written "$f(x)$" on the AP Exam) approaches as the variable within that function (usually "x") gets nearer and nearer to a particular value. In other words, when x is very close to a certain number, what is $f(x)$ very close to? As far as the AB test is concerned, that's all you have to know about evaluating limits. There's a more technical method that you BC students have to learn, but we won't discuss it until we have to—at the end of this chapter.

Let's look at an example of a limit: What is the limit of the function $f(x) = x^2$ as x approaches 2? In limit notation, the expression "the limit of $f(x)$ as x approaches 2" is written like this: $\lim\limits_{x \to 2} f(x)$. In order to evaluate the limit, let's check out some values of $\lim\limits_{x \to 2} f(x)$ as x increases and gets closer to 2 (without ever exactly getting there).

When $x = 1.9$, $f(x) = 3.61$.
When $x = 1.99$, $f(x) = 3.9601$.
When $x = 1.999$, $f(x) = 3.996001$.
When $x = 1.9999$, $f(x) = 3.99960001$.

As x increases and approaches 2, $f(x)$ gets closer and closer to 4. This is called the **left-hand limit** and is written: $\lim\limits_{x \to 2^-} f(x)$. Notice the little minus sign!

What about when x is bigger than 2?

When $x = 2.1$, $f(x) = 4.41$.
When $x = 2.01$, $f(x) = 4.0401$.
When $x = 2.001$, $f(x) = 4.004001$.
When $x = 2.0001$, $f(x) = 4.00040001$.

As x decreases and approaches 2, $f(x)$ still approaches 4. This is called the **right-hand limit** and is written like this: $\lim\limits_{x \to 2^+} f(x)$. Notice the little plus sign!

We got the same answer when evaluating both the left- and right-hand limits, because when x is 2, $f(x)$ is 4. You should always check both sides of the independent variable because, as you'll see shortly, sometimes you don't get the same answer. Therefore, we write that $\lim\limits_{x \to 2} x^2 = 4$.

We didn't really need to look at all of these decimal values to know what was going to happen when x got really close to 2. But it's important to go through the exercise because, typically, the answers get a lot more complicated. Let's do a few examples.

Example 1: Find $\lim\limits_{x \to 5} x^2$.

The approach is simple: Plug in 5 for x, and you get 25.

Example 2: Find $\lim\limits_{x \to 3} x^3$.

Here the answer is 27.

There are some simple algebraic rules of limits that you should know. These are:

$$\lim_{x \to a} kf(x) = k \lim_{x \to a} f(x)$$

Example: $\lim_{x \to 5} 3x^2 = 3 \lim_{x \to 5} x^2 = 75$

$$\text{If } \lim_{x \to a} f(x) = L_1 \text{ and } \lim_{x \to a} g(x) = L_2 \text{, then } \lim_{x \to a} \left[f(x) + g(x) \right] = L_1 + L_2$$

Example: $\lim_{x \to 5} \left[x^2 + x^3 \right] = \lim_{x \to 5} x^2 + \lim_{x \to 5} x^3 = 150$

$$\text{If } \lim_{x \to a} f(x) = L_1 \text{ and } \lim_{x \to a} g(x) = I_2 \text{, then } \lim_{x \to a} \left[f(x) \cdot g(x) \right] = L_1 \cdot L_2$$

Example: $\lim_{x \to 5} \left[\left(x^2 + 1 \right) \sqrt{x - 1} \right] = \lim_{x \to 5} \left(x^2 + 1 \right) \lim_{x \to 5} \sqrt{x - 1} = 52$

Example 3: Find $\lim_{x \to 0} \left(x^2 + 5x \right)$.

Plug in 0, and you get 0.

So far, so good. All you do to find the limit of a simple polynomial is plug in the number that the variable is approaching and see what the answer is. Naturally, the process can get messier—especially if x approaches zero.

Example 4: Find $\lim_{x \to 0} \dfrac{1}{x^2}$.

If you plug in some very small values for x, you'll see that this function approaches ∞. And it

doesn't matter whether x is positive or negative, you still get ∞. Look at the graph of $y = \dfrac{1}{x^2}$:

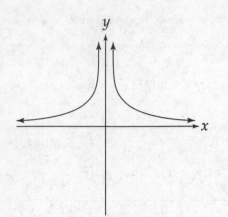

On either side of $x = 0$ (the y-axis), the curve approaches ∞.

Example 5: Find $\lim\limits_{x \to 0} \dfrac{1}{x}$.

Here you have a problem. If you plug in some very small positive values for x (0.1, 0.01, 0.001, and so on), you approach ∞. In other words, $\lim\limits_{x \to 0^+} \dfrac{1}{x} = \infty$. But, if you plug in some very small negative values for x (–0.1, –0.01, –0.001, and so on) you approach $-\infty$. That is, $\lim\limits_{x \to 0^-} \dfrac{1}{x} = -\infty$. Because the right-hand limit is not equal to the left-hand limit, the limit does not exist.

Look at the graph of $\dfrac{1}{x}$.

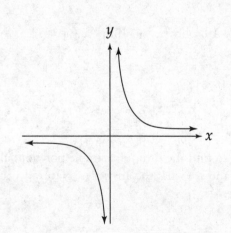

You can see that on the left side of $x = 0$, the curve approaches $-\infty$, and on the right side of $x = 0$, the curve approaches ∞. There are some very important points that we need to emphasize from the last two examples.

Why do we state the limit in Example 4 but not for Example 5? Because when we have $\dfrac{k}{x^2}$, the function is always positive no matter what the sign of x is and thus the function has the same limit from the left and the right. But when we have $\dfrac{k}{x}$, the function's sign depends on the sign of x, and you get a different limit from each side.

Let's look at a few examples in which the independent variable approaches infinity.

Example 6: Find $\lim\limits_{x \to \infty} \dfrac{1}{x}$.

As x gets bigger and bigger, the value of the function gets smaller and smaller. Therefore, $\lim\limits_{x \to \infty} \dfrac{1}{x} = 0$.

Example 7: Find $\lim\limits_{x \to -\infty} \dfrac{1}{x}$.

It's the same situation as the one in Example 6; as x decreases (approaches negative infinity), the value of the function increases (approaches zero). We write the following:

$$\lim\limits_{x \to -\infty} \dfrac{1}{x} = 0$$

We don't have the same problem here that we did when x approached zero because "positive zero" is the same thing as "negative zero," whereas positive infinity is different from negative infinity.

Here's another rule.

> If k and n are constants with $n > 0$, then $\lim\limits_{x \to \infty} \dfrac{k}{x^n} = 0$.

---

(1) If the left-hand limit of a function is not equal to the right-hand limit of the function, then the limit does not exist.

(2) A limit equal to infinity is not the same as a limit that does not exist, but sometimes you will see the expression "no limit," which serves both purposes. If $\lim\limits_{x \to a} f(x) = \infty$, the limit, technically, does not exist.

(3) If k is a positive constant, then $\lim\limits_{x \to 0^+} \dfrac{k}{x} = \infty$, $\lim\limits_{x \to 0^-} \dfrac{k}{x} = -\infty$, and $\lim\limits_{x \to 0} \dfrac{k}{x}$ does not exist.

(4) If k is a positive constant, then

$$\lim\limits_{x \to 0^+} \dfrac{k}{x^2} = \infty, \ \lim\limits_{x \to 0^-} \dfrac{k}{x^2} = \infty, \text{ and}$$

$$\lim\limits_{x \to 0} \dfrac{k}{x^2} = \infty.$$

Example 8: Find $\lim\limits_{x \to \infty} \dfrac{3x+5}{7x-2}$.

When you have a rational expression (that is, a fraction with a polynomial in both the numerator and the denominator), you can't just plug ∞ into the expression. You'll get $\dfrac{\infty}{\infty}$. We solve this by using the following technique:

> When an expression consists of a polynomial divided by another polynomial, divide each term of the numerator and the denominator by the highest power of x that appears in the expression.

The highest power of x in this case is x^1, so we divide every term in the expression (both top and bottom) by x, like so:

$$\lim_{x \to \infty} \frac{3x+5}{7x-2} = \lim_{x \to \infty} \frac{\dfrac{3x}{x}+\dfrac{5}{x}}{\dfrac{7x}{x}-\dfrac{2}{x}} = \lim_{x \to \infty} \frac{3+\dfrac{5}{x}}{7-\dfrac{2}{x}}$$

Now when we take the limit, the two terms containing x approach zero. We're left with $\dfrac{3}{7}$.

Example 9: Find $\lim\limits_{x \to \infty} \dfrac{8x^2 - 4x + 1}{16x^2 + 7x - 2}$.

Divide each term by x^2. You get

$$\lim_{x \to \infty} \frac{8 - \dfrac{4}{x} + \dfrac{1}{x^2}}{16 + \dfrac{7}{x} - \dfrac{2}{x^2}} = \frac{8}{16} = \frac{1}{2}$$

Example 10: Find $\lim\limits_{x \to \infty} \dfrac{-3x^{10} - 70x^5 + x^3}{33x^{10} + 200x^8 - 1000x^4}$.

> Remember to focus your attention on the highest power of x.

Divide each term by x^{10}.

$$\lim_{x \to \infty} \frac{-3x^{10} - 70x^5 + x^3}{33x^{10} + 200x^8 - 1000x^4} = \lim_{x \to \infty} \frac{-3 - \dfrac{70}{x^5} + \dfrac{1}{x^7}}{33 + \dfrac{200}{x^2} - \dfrac{1000}{x^6}} = -\frac{3}{33} = -\frac{1}{11}$$

The other powers don't matter, because they're all going to disappear. Now we have three new rules for evaluating the limit of a rational expression as x approaches infinity.

---

(1) If the highest power of x in a rational expression is in the numerator, then the limit as x approaches infinity is infinity.

Example: $\lim\limits_{x\to\infty}\dfrac{5x^7 - 3x}{16x^6 - 3x^2} = \infty$

---

(2) If the highest power of x in a rational expression is in the denominator, then the limit as x approaches infinity is zero.

Example: $\lim\limits_{x\to\infty}\dfrac{5x^6 - 3x}{16x^7 - 3x^2} = 0$

---

(3) If the highest power of x in a rational expression is the same in both the numerator and denominator, then the limit as x approaches infinity is the coefficient of the highest term in the numerator divided by the coefficient of the highest term in the denominator.

Example: $\lim\limits_{x\to\infty}\dfrac{5x^7 - 3x}{16x^7 - 3x^2} = \dfrac{5}{16}$

LIMITS OF TRIGONOMETRIC FUNCTIONS

At some point during the exam, you'll have to find the limit of certain trig expressions, usually as x approaches either zero or infinity. There are four standard limits that you should memorize—with those, you can evaluate all of the trigonometric limits that appear on the test. As you'll see throughout this book, calculus requires that you remember all of your trig from previous years.

Rule No. 1: $\lim\limits_{x\to 0}\dfrac{\sin x}{x}=1$ (x is in radians, *not* degrees)

This may seem strange, but if you look at the graphs of $f(x) = \sin x$ and $f(x) = x$, they

Remember that the $\lim\limits_{x\to 0}\sin x = 0$.

have approximately the same slope near the origin (as x gets closer to zero). Because x and the sine of x are about the same as x approaches zero, their quotient will be very close to one. Furthermore, because $\lim\limits_{x\to 0}\cos x = 1$ (review cosine values if you don't get this!), we know that $\lim\limits_{x\to 0}\tan x = \lim\limits_{x\to 0}\dfrac{\sin x}{\cos x}=0$.

Now we will find a second rule. Let's evaluate the limit $\lim\limits_{x\to 0}\dfrac{\cos x-1}{x}$. First, multiply the top and bottom by $\cos x+1$. We get $\lim\limits_{x\to 0}\left(\dfrac{\cos x-1}{x}\right)\left(\dfrac{\cos x+1}{\cos x+1}\right)$. Now simplify the limit to $\lim\limits_{x\to 0}\dfrac{\cos^2 x-1}{x(\cos x+1)}$. Next, we can use the trigonometric identity $\sin^2 x = 1-\cos^2 x$ and rewrite the limit as $\lim\limits_{x\to 0}\dfrac{-\sin^2 x}{x(\cos x+1)}$. Now, break this into two limits: $\lim\limits_{x\to 0}\dfrac{-\sin x}{x}\dfrac{\sin x}{(\cos x+1)}$. The first limit is -1 (see Rule No. 1) and the second is 0, so the limit is 0.

Rule No. 2: $\lim\limits_{x\to 0}\dfrac{\cos x-1}{x}=0$

Example 11: Find $\lim\limits_{x\to 0}\dfrac{\sin 3x}{x}$.

Use a simple trick: Multiply the top and bottom of the expression by 3. This gives us $\lim\limits_{x\to 0}\dfrac{3\sin 3x}{3x}$. Next, substitute a letter for $3x$; for example, a. Now, we get the following:

$$\lim\limits_{a\to 0}\dfrac{3\sin a}{a}=3\lim\limits_{a\to 0}\dfrac{\sin a}{a}=3(1)=3$$

Example 12: Find $\displaystyle\lim_{x\to 0}\frac{\sin 5x}{\sin 4x}$.

Now we get a bit more sophisticated. First, divide both the numerator and the denominator by x, like so:

$$\lim_{x\to 0}\frac{\dfrac{\sin 5x}{x}}{\dfrac{\sin 4x}{x}}$$

Next, multiply the top and bottom of the numerator by 5, and the top and bottom of the denominator by 4, which gives us

$$\lim_{x\to 0}\frac{\dfrac{5\sin 5x}{5x}}{\dfrac{4\sin 4x}{4x}}$$

From the work we did in Example 11, we can see that this limit is $\dfrac{5}{4}$.

Guess what! You have two more rules!

Rule No. 3: $\displaystyle\lim_{x\to 0}\frac{\sin ax}{x}=a$

Rule No. 4: $\displaystyle\lim_{x\to 0}\frac{\sin ax}{\sin bx}=\frac{a}{b}$

Example 13: Find $\displaystyle\lim_{x\to 0}\frac{x^2}{1-\cos^2 x}$.

Using trigonometric identities, you can replace $(1-\cos^2 x)$ with $\sin^2 x$.

$$\lim_{x\to 0}\frac{x^2}{1-\cos^2 x}=\lim_{x\to 0}\frac{x^2}{\sin^2 x}=\lim_{x\to 0}\left(\frac{x}{\sin x}\cdot\frac{x}{\sin x}\right)=1\cdot 1=1$$

Notice that in the above, we used the idea that $\displaystyle\lim_{x\to 0}\frac{x}{\sin x}=1$, even though previously we had only established that $\displaystyle\lim_{x\to 0}\frac{\sin x}{x}=1$. Because the number 1 is its own reciprocal, if $\dfrac{\sin x}{x}$ approaches 1, then $\dfrac{x}{\sin x}$ will as well.

> If plugging in the value of x results in the denominator equaling zero and you cannot factor the quotient anymore, then check the left and right hand limits to find the limit of the expression.

Here are other examples for you to try, with answers right beneath them. Give 'em a try, and check your work.

PROBLEM 1. Find $\lim\limits_{x \to 3} \dfrac{x-3}{x+2}$.

Answer: If you plug in 3 for x, you get $\lim\limits_{x \to 3} \dfrac{3-3}{3+2} = \dfrac{0}{5} = 0$.

PROBLEM 2. Find $\lim\limits_{x \to 3} \dfrac{x+2}{x-3}$.

Answer: The left-hand limit is: $\lim\limits_{x \to 3^-} \dfrac{x+2}{x-3} = -\infty$.

The right-hand limit is: $\lim\limits_{x \to 3^+} \dfrac{x+2}{x-3} = \infty$.

These two limits are not the same. Therefore, the limit does not exist.

PROBLEM 3. Find $\lim\limits_{x \to 3} \dfrac{x+2}{(x-3)^2}$.

Answer: The left-hand limit is: $\lim\limits_{x \to 3^-} \dfrac{x+2}{(x-3)^2} = \infty$.

The right-hand limit is: $\lim\limits_{x \to 3^+} \dfrac{x+2}{(x-3)^2} = \infty$.

These two limits are the same, so the limit is ∞.

PROBLEM 4. Find $\lim\limits_{x \to -4} \dfrac{x^2+6x+8}{x+4}$.

> If $\lim\limits_{x \to a} f(x) = \dfrac{0}{0}$, then $x - a$ is a factor of the top and the bottom.

Answer: If you plug -4 into the top and bottom, you get $\dfrac{0}{0}$. You have to factor the top into $(x+2)(x+4)$ to get this: $\lim\limits_{x \to -4} \dfrac{(x+2)(x+4)}{(x+4)}$.

Now it's time to cancel like terms: $\lim\limits_{x \to -4} \dfrac{(x+2)(x+4)}{(x+4)} = \lim\limits_{x \to -4} (x+2) = -2$.

PROBLEM 5. Find $\lim\limits_{x \to \infty} \dfrac{15x^2-11x}{22x^2+4x}$.

Answer: Divide each term by x^2.

$$\lim\limits_{x \to \infty} \dfrac{15x^2-11x}{22x^2+4x} = \lim\limits_{x \to \infty} \dfrac{15-\dfrac{11}{x}}{22+\dfrac{4}{x}} = \dfrac{15}{22}$$

PROBLEM 6. Find $\lim\limits_{x\to 0}\dfrac{4x}{\tan x}$.

Answer: Replace $\tan x$ with $\dfrac{\sin x}{\cos x}$, which changes the expression into

$$\lim_{x\to 0}\frac{4x}{\tan x}=\lim_{x\to 0}\frac{4x}{\dfrac{\sin x}{\cos x}}=\lim_{x\to 0}\frac{4x\cos x}{\sin x}$$

Because $\lim\limits_{x\to 0}\dfrac{\sin x}{x}=1$, the $\lim\limits_{x\to 0}\dfrac{x}{\sin x}=1$ as well. Thus, because $\lim\limits_{x\to 0}\dfrac{x}{\sin x}=1$ and $\lim\limits_{x\to 0}\cos x=1$,

the answer is 4.

PROBLEM 7. Find $\lim\limits_{h\to 0}\dfrac{(5+h)^2-25}{h}$. Answer: First, expand and simplify the numerator.

$$\lim_{h\to 0}\frac{(5+h)^2-25}{h}=\lim_{h\to 0}\frac{25+10h+h^2-25}{h}=\lim_{h\to 0}\frac{10h+h^2}{h}$$

> Note: Pay careful attention to this next solved problem. It will be very important when you work on problems in Chapter 5.

Next, factor h out of the numerator and the denominator.

$$\lim_{h\to 0}\frac{10h+h^2}{h}=\lim_{h\to 0}\frac{h(10+h)}{h}=\lim_{h\to 0}(10+h)$$

Taking the limit you get $\lim\limits_{h\to 0}(10+h)=10$.

PRACTICE PROBLEM SET 1

Try these 22 problems to test your skill with limits. The answers are in Chapter 23, starting on page 442.

1. $\lim\limits_{x\to 8}\left(x^2-5x-11\right)=$

2. $\lim\limits_{x\to 5}\left(\dfrac{x+3}{x^2-15}\right)=$

3. $\lim\limits_{x\to 3}\left(\dfrac{x^2-2x-3}{x-3}\right)=$

4. $\lim\limits_{x\to\infty}\left(\dfrac{x^4-8}{10x^2+25x+1}\right)=$

5. $\lim\limits_{x \to \infty}\left(\dfrac{x^4 - 8}{10x^4 + 25x + 1}\right) =$

6. $\lim\limits_{x \to 6^+}\left(\dfrac{x + 2}{x^2 - 4x - 12}\right) =$

7. $\lim\limits_{x \to 6^-}\left(\dfrac{x + 2}{x^2 - 4x - 12}\right) =$

8. $\lim\limits_{x \to 6}\left(\dfrac{x + 2}{x^2 - 4x - 12}\right) =$

9. $\lim\limits_{x \to 0^+}\left(\dfrac{x}{|x|}\right) =$

10. $\lim\limits_{x \to 7^+}\left(\dfrac{x}{x^2 - 49}\right) =$

11. $\lim\limits_{x \to 7}\left(\dfrac{x}{x^2 - 49}\right) =$

12. Let $f(x) = \begin{cases} x^2 - 5, & x \le 3 \\ x + 2, & x > 3 \end{cases}$

Find: (a) $\lim\limits_{x \to 3^-} f(x)$; (b) $\lim\limits_{x \to 3^+} f(x)$; and (c) $\lim\limits_{x \to 3} f(x)$

13. Let $f(x) = \begin{cases} x^2 - 5, & x \le 3 \\ x + 1, & x > 3 \end{cases}$

Find: (a) $\lim\limits_{x \to 3^-} f(x)$; (b) $\lim\limits_{x \to 3^+} f(x)$; and (c) $\lim\limits_{x \to 3} f(x)$

14. Find $\lim\limits_{x \to \frac{\pi}{4}} 3\cos x$.

15. Find $\lim\limits_{x \to 0} 3\dfrac{x}{\cos x}$.

16. Find $\lim\limits_{x \to 0} 3\dfrac{x}{\sin x}$.

17. Find $\lim\limits_{x \to 0} \dfrac{\tan 7x}{\sin 5x}$.

18. Find $\lim\limits_{x \to \infty} \sin x$.

19. Find $\lim\limits_{x \to \infty} \sin\dfrac{1}{x}$.

20. Find $\lim\limits_{x \to 0} \dfrac{\sin^2 7x}{\sin^2 11x}$.

21. Find $\lim\limits_{h \to 0} \dfrac{(3+h)^2 - 9}{h}$.

22. Find $\lim\limits_{h \to 0} \dfrac{\dfrac{1}{x+h} - \dfrac{1}{x}}{h}$.

End of Chapter 3 Drill

The answers are in Chapter 24.

1. Evaluate $\lim_{x \to 2}\left(x^4 - 3x^2 + 5\right)$.

    (A) 0
    (B) 2
    (C) 9
    (D) The limit does not exist.

2. Evaluate $\lim_{x \to 2}\dfrac{x^2 - x - 2}{x^2 - 8x + 15}$.

    (A) 0
    (B) $\dfrac{2}{3}$
    (C) 3
    (D) The limit does not exist.

3. Evaluate $\lim_{x \to 7}\dfrac{x^2 - 3x - 28}{x^2 - 5x - 14}$.

    (A) 0
    (B) $\dfrac{11}{9}$
    (C) 2
    (D) The limit does not exist.

4. Evaluate $\lim_{x \to \infty}\dfrac{4x^3 - 8x^2 + 11}{5x^3 + 3x - 9}$.

    (A) $-\dfrac{11}{9}$
    (B) 0
    (C) $\dfrac{4}{5}$
    (D) The limit does not exist.

5. Evaluate $\lim_{x \to 0}\dfrac{4\sin 3x}{2\tan 5x}$.

    (A) 0
    (B) $\dfrac{3}{5}$
    (C) 2
    (D) The limit does not exist.

6. Evaluate $\lim_{x \to 5}\dfrac{11}{x^2 - 25}$.

    (A) 0
    (B) $\dfrac{11}{25}$
    (C) ∞
    (D) The limit does not exist.

REFLECT

Respond to the following questions:

- For which topics discussed in this chapter do you feel you have achieved sufficient mastery to answer multiple-choice questions correctly?

- For which topics discussed in this chapter do you feel you have achieved sufficient mastery to answer open-ended questions correctly?

- For which topics discussed in this chapter do you feel you need more work before you can answer multiple-choice questions correctly?

- For which topics discussed in this chapter do you feel you need more work before you can answer open-ended questions correctly?

- What parts of this chapter are you going to re-review?

Chapter 4
Continuity

Every AP Exam has a few questions on continuity, so it's important to understand the basic idea of what it means for a function to be continuous. The concept is very simple: if the graph of the function doesn't have any breaks or holes in it within a certain interval, the function is continuous over that interval.

Simple polynomials are continuous everywhere; it's the other ones—trigonometric, rational, piecewise—that might have continuity problems. Most of the test questions concern these last types of functions. In order to learn how to test whether a function is continuous, you'll need some more mathematical terminology.

THE DEFINITION OF CONTINUITY

In order for a function $f(x)$ to be continuous at a point $x = c$, it must fulfill *all three* of the following conditions:

Condition 1: $f(c)$ exists.

Condition 2: $\lim_{x \to c} f(x)$ exists.

Condition 3: $\lim_{x \to c} f(x) = f(c)$

Let's look at a simple example of a continuous function.

Example 1: Is the function $f(x) = \begin{cases} x+1, & x < 2 \\ 2x-1, & x \geq 2 \end{cases}$ continuous at the point $x = 2$?

Condition 1: Does $f(2)$ exist?

Yes. It's equal to $2(2) - 1 = 3$.

Condition 2: Does $\lim_{x \to 2} f(x)$ exist?

You need to look at the limit from both sides of 2. The left-hand limit is $\lim_{x \to 2^-} f(x) = 2 + 1 = 3$. The right-hand limit is $\lim_{x \to 2^+} f(x) = 2(2) - 1 = 3$.

Because the two limits are the same, the limit exists.

Condition 3: Does $\lim_{x \to 2} f(x) = f(2)$?

The two equal each other, so yes; the function is continuous at $x = 2$.

A simple and important way to check whether a function is continuous is to sketch the function. If you can't sketch the function without lifting your pencil from the paper at some point, then the function is not continuous.

Now let's look at some examples of functions that are not continuous.

Polynomials are continuous for all reals in their domains.

Example 2: Is the function $f(x) = \begin{cases} x+1, & x < 2 \\ 2x-1, & x > 2 \end{cases}$ continuous at $x = 2$?

Condition 1: Does $f(2)$ exist?

Nope. The function of x is defined if x is greater than or less than 2, but not if x is equal to 2. Therefore, the function is not continuous at $x = 2$. Notice that we don't have to bother with the other two conditions. Once you find a problem, the function is automatically not continuous, and you can stop.

Example 3: Is the function $f(x) = \begin{cases} x+1, & x < 2 \\ 2x+1, & x \geq 2 \end{cases}$ continuous at $x = 2$?

Condition 1: Does $f(x)$ exist?

Yes. It is equal to $2(2) + 1 = 5$.

Condition 2: Does $\lim_{x \to 2} f(x)$ exist?

The left-hand limit is $\lim_{x \to 2^-} f(x) = 2 + 1 = 3$.

The right-hand limit is $\lim_{x \to 2^+} f(x) = 2(2) + 1 = 5$.

The two limits don't match, so the limit doesn't exist and the function is not continuous at $x = 2$.

Example 4: Is the function $f(x) = \begin{cases} x+1, & x < 2 \\ x^2, & x = 2 \\ 2x-1, & x > 2 \end{cases}$ continuous at $x = 2$?

Condition 1: Does $f(2)$ exist?

Yes. It's equal to $2^2 = 4$.

Condition 2: Does $\lim_{x \to 2} f(x)$ exist?

The left-hand limit is $\lim_{x \to 2^-} f(x) = 2 + 1 = 3$.

The right-hand limit is $\lim_{x \to 2^+} f(x) = 2(2) - 1 = 3$.

Because the two limits are the same, the limit exists.

Condition 3: Does $\lim_{x \to 2} f(x) = f(2)$?

The $\lim_{x \to 2} f(x) = 3$, but $f(2) = 4$. Because these aren't equal, the answer is "no" and the function is not continuous at $x = 2$.

TYPES OF DISCONTINUITIES

There are three types of discontinuities you have to know: jump, essential, and removable.

Intermediate Learning

The intermediate value theorem says that if a function is continuous on the interval [a, b], then for any value c in the interval (a, b), there exists a value $f(c)$ between $f(a)$ and $f(b)$ inclusive. In other words, if the function is continuous in some interval, then all of the possible values of f between the two end values of f exist in that interval.

A **jump** discontinuity occurs when the curve "breaks" at a particular place and starts somewhere else. The limits from the left and the right will both exist, but they will not match.

An example of jump discontinuity looks like this.

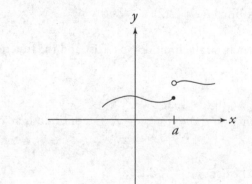

An **essential** discontinuity occurs when the curve has a vertical asymptote.

This is an example of an essential discontinuity.

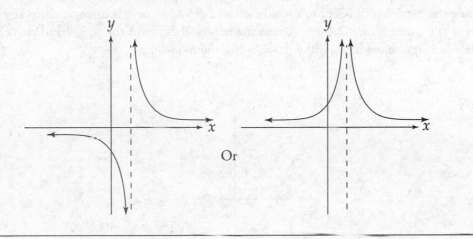

Or

A **removable** discontinuity occurs when an otherwise continuous curve has a "hole" in it. This type of discontinuity is called "removable" because one can remove the discontinuity simply by filling the hole.

There are two types of removable discontinuity. As the following graphs illustrate, both curves are continuous everywhere except at the "hole."

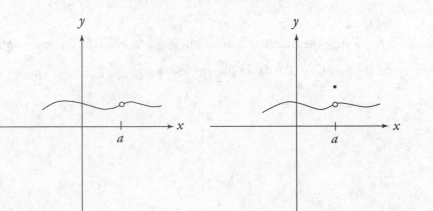

In the left graph, the function is not defined at all at the x-value of the hole, while in the right graph, the function is defined at the x-value of the hole, but the function value does not match the y-coordinate of the hole. In both cases, $\lim_{x \to a^-} f(x) = \lim_{x \to a^+} f(x)$ which means that $\lim_{x \to a} f(x)$ exists, and it is equal to the y-coordinate of the hole.

Now that you know what these three types of discontinuities look like, let's see what types of functions are not everywhere continuous.

Example 5: Consider the following function:

$$f(x) = \begin{cases} x + 3, & x \le 2 \\ x^2, & x > 2 \end{cases}$$

The left-hand limit is 5 as x approaches 2, and the right-hand limit is 4 as x approaches 2. Because the curve has different values on each side of 2, the curve is discontinuous at $x = 2$. We say that the curve "jumps" at $x = 2$ from the left-hand curve to the right-hand curve because the left and right-hand limits differ. It looks like the following:

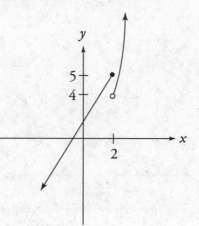

This is an example of a jump discontinuity.

Example 6: Consider the following function:

$$f(x) = \begin{cases} x^2, & x \ne 2 \\ 5, & x = 2 \end{cases}$$

Because $\lim\limits_{x \to 2} f(x) \ne f(2)$; the function is discontinuous at $x = 2$. The curve is continuous everywhere except at the point $x = 2$. It looks like the following:

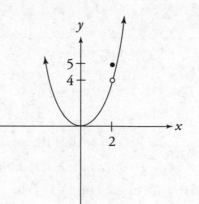

This is an example of a point discontinuity.

Example 7: Consider the following function: $f(x) = \dfrac{5}{x-2}$.

The function is discontinuous because it's possible for the denominator to equal zero (at $x = 2$). This means that $f(2)$ doesn't exist, and the function has an asymptote at $x = 2$. In addition, $\lim\limits_{x \to 2^-} f(x) = -\infty$ and $\lim\limits_{x \to 2^+} f(x) = \infty$.

The graph looks like the following:

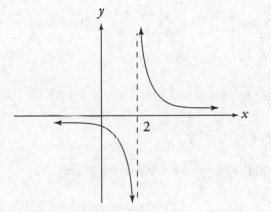

This is an example of an essential discontinuity.

Example 8: Consider the following function:

$$f(x) = \frac{x^2 - 8x + 15}{x^2 - 6x + 5}$$

If you factor the top and bottom, you can see where the discontinuities are.

$$f(x) = \frac{x^2 - 8x + 15}{x^2 - 6x + 5} = \frac{(x-3)(x-5)}{(x-1)(x-5)}$$

The function has a zero in the denominator when $x = 1$ or $x = 5$, so the function is discontinuous at those two points. But you can cancel the term $(x - 5)$ from both the numerator and the denominator, leaving you with

$$f(x) = \frac{x-3}{x-1}$$

Now the reduced function *is* continuous at $x = 5$. Thus, the original function has a removable discontinuity at $x = 5$. Furthermore, if you now plug $x = 5$ into the reduced function, you get

$$f(5) = \frac{2}{4} = \frac{1}{2}$$

The discontinuity is at $x = 5$, and there's a hole at $\left(5, \frac{1}{2}\right)$. In other words, if the original function were continuous at $x = 5$, it would have the value $\frac{1}{2}$. Notice that this is the same as $\lim_{x \to 5} f(x)$.

These are the types of discontinuities that you can expect to encounter on the AP Exam. Here are some sample problems and their solutions. Cover the answers as you work, then check your results.

PROBLEM 1. Is the function $f(x) = \begin{cases} 2x^3 - 1, & x < 2 \\ 6x - 3, & x \geq 2 \end{cases}$ continuous at $x = 2$?

Answer: Test the conditions necessary for continuity.

Condition 1: $f(2) = 9$, so we're okay so far.

Condition 2: The $\lim_{x \to 2^-} f(x) = 15$ and the $\lim_{x \to 2^+} f(x) = 9$. These two limits don't agree, so the $\lim_{x \to 2} f(x)$ doesn't exist and the function is not continuous at $x = 2$.

PROBLEM 2. Is the function $f(x) = \begin{cases} x^2 + 3x + 5, & x < 1 \\ 6x + 3, & x \geq 1 \end{cases}$ continuous at $x = 1$?

Answer: Condition 1: $f(1) = 9$.

Condition 2: The $\lim_{x \to 1^-} f(x) = 9$ and the $\lim_{x \to 1^+} f(x) = 9$.

Therefore, the $\lim_{x \to 1} f(x)$ exists and is equal to 9.

Condition 3: $\lim_{x \to 1} f(x) = f(1) = 9$.

The function satisfies all three conditions, so it is continuous at $x = 1$.

PROBLEM 3. For what value of a is the function $f(x) = \begin{cases} ax + 5, & x < 4 \\ x^2 - x, & x \geq 4 \end{cases}$ continuous at $x = 4$?

Answer: Because $f(4) = 12$, the function passes the first condition.

For Condition 2 to be satisfied, the $\lim_{x \to 4^-} f(x) = 4a + 5$ must equal the $\lim_{x \to 4^+} f(x) = 12$. So, set $4a + 5 = 12$. If $a = \frac{7}{4}$, the limit will exist at $x = 4$ and the other two conditions will also be fulfilled. Therefore, the value $a = \frac{7}{4}$ makes the function continuous at $x = 4$.

PROBLEM 4. Where does the function $f(x) = \dfrac{2x^2 - 7x - 15}{x^2 - x - 20}$ have: (a) an essential discontinuity; and (b) a removable discontinuity?

Answer: If you factor the top and bottom of this fraction, you get

$$f(x) = \frac{2x^2 - 7x - 15}{x^2 - x - 20} = \frac{(2x+3)(x-5)}{(x+4)(x-5)}$$

Thus, the function has an essential discontinuity at $x = -4$. If we then cancel the term $(x-5)$, and substitute $x = 5$ into the reduced expression, we get $f(5) = \dfrac{13}{9}$. Therefore, the function has a removable discontinuity at $\left(5, \dfrac{13}{9}\right)$.

> **Note:** Don't confuse coordinate parentheses with interval notation. In interval notation, square brackets include endpoints and parentheses do not. For example, the interval $2 \le x \le 4$ is written $[2, 4]$ and the interval $2 < x < 4$ is written $(2, 4)$.

PRACTICE PROBLEM SET 2

Now try these problems. The answers are in Chapter 23, starting on page 448.

1. Is the function $f(x) = \begin{cases} x+7, & x < 2 \\ 9, & x = 2 \\ 3x + 3, & x > 2 \end{cases}$ continuous at $x = 2$?

2. Is the function $f(x) = \begin{cases} 4x^2 - 2x, & x < 3 \\ 10x - 1, & x = 3 \\ 30, & x > 3 \end{cases}$ continuous at $x = 3$?

3. Is the function $f(x) = \sec x$ continuous everywhere?

4. Is the function $f(x) = \sec x$ continuous on the interval $\left[-\dfrac{\pi}{2}, \dfrac{\pi}{2}\right]$?

5. Is the function $f(x) = \sec x$ continuous on the interval $\left(-\dfrac{\pi}{2}, \dfrac{\pi}{2}\right)$?

6. For what value(s) of k is the function $f(x) = \begin{cases} 3x^2 - 11x - 4, & x \le 4 \\ kx^2 - 2x - 1, & x > 4 \end{cases}$ continuous at $x = 4$?

7. At what point is the removable discontinuity for the function $f(x) = \dfrac{x^2 + 5x - 24}{x^2 - x - 6}$?

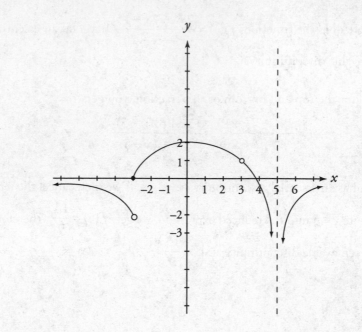

8.   Given the graph of $f(x)$ above, find

(a) $\lim\limits_{x \to -\infty} f(x)$

(b) $\lim\limits_{x \to \infty} f(x)$

(c) $\lim\limits_{x \to 3^-} f(x)$

(d) $\lim\limits_{x \to 3^+} f(x)$

(e) $f(3)$

(f)   Any discontinuities

End of Chapter 4 Drill

The answers are in Chapter 24.

1. Does the function $f(x) = \dfrac{x^2 - 8x + 15}{x^2 - 9}$ have any discontinuities? If so, where and what type?

   (A) An essential discontinuity at $x = 3$ and at $x = -3$
   (B) A removable discontinuity at $x = 3$ and at $x = 3$
   (C) A removable discontinuity at $x = 3$ and an essential discontinuity at $x = -3$.
   (D) The function is continuous for all real numbers.

2. Does the function $f(x) = \begin{cases} 5x - 11; & x \geq 4 \\ 2x + 1; & x < 4 \end{cases}$ have any discontinuities? If so, where and what type?

   (A) An essential discontinuity at $x = 4$
   (B) A removable discontinuity at $x = 4$
   (C) A jump discontinuity at $x = 4$
   (D) The function is continuous for all real numbers.

3. Does the function $f(x) = \begin{cases} 6x + 1; & x \geq 2 \\ 5x + 2; & x < 2 \end{cases}$ have any discontinuities? If so, where and what type?

   (A) An essential discontinuity at $x = 2$
   (B) A removable discontinuity at $x = 2$
   (C) A jump discontinuity at $x = 2$
   (D) The function is continuous for all real numbers.

4. For what value of k is $f(x) = \begin{cases} x^2 + kx - 4; & x \geq 1 \\ 6kx + 7; & x < 1 \end{cases}$ continuous for all real numbers?

   (A) $k = -2$
   (B) $k = -\dfrac{4}{5}$
   (C) $k = \dfrac{4}{5}$
   (D) $k = 2$

5. For what values of a and b is
   $$f(x) = \begin{cases} ax^2 + 2bx - 3; & x > 2 \\ 4bx^2 - 3ax + 3; & -2 \leq x \leq 2 \\ 4ax^2 + 5bx + 25; & x < -2 \end{cases}$$
   continuous for all real numbers?

   (A) $a = 2$ and $b = 3$
   (B) $a = 3$ and $b = 2$
   (C) $a = 1$ and $b = -1$
   (D) $a = -1$ and $b = 1$

REFLECT

Respond to the following questions:

- For which topics discussed in this chapter do you feel you have achieved sufficient mastery to answer multiple-choice questions correctly?

- For which topics discussed in this chapter do you feel you have achieved sufficient mastery to answer open-ended questions correctly?

- For which topics discussed in this chapter do you feel you need more work before you can answer multiple-choice questions correctly?

- For which topics discussed in this chapter do you feel you need more work before you can answer open-ended questions correctly?

- What parts of this chapter are you going to re-review?

Chapter 5
The Definition of the Derivative

The main tool that you'll use in differential calculus is called the **derivative**. All of the problems that you'll encounter in differential calculus make use of the derivative, so your goal should be to become an expert at finding, or "taking," derivatives by the end of Chapter 6. However, before you learn a simple way to take a derivative, your teacher will probably make you learn how derivatives are calculated by teaching you something called the "Definition of the Derivative."

DERIVING THE FORMULA

The best way to understand the definition of the derivative is to start by looking at the simplest continuous function: a line. As you should recall, you can determine the slope of a line by taking two points on that line and plugging them into the slope formula.

$$m = \frac{y_2 - y_1}{x_2 - x_1}$$ *m* stands for slope.

Notice that you can use the coordinates in reverse order and still get the same result. It doesn't matter in which order you do the subtraction as long as you're consistent.

For example, suppose a line goes through the points (3, 7) and (8, 22). First, you subtract the *y*-coordinates $(22 - 7) = 15$. Next, subtract the corresponding *x*-coordinates $(8 - 3) = 5$. Finally, divide the first number by the second: $\frac{15}{5} = 3$. The result is the slope of the line: $m = 3$.

Let's look at the graph of an arbitrary line. The slope measures the steepness of the line, which looks like the following:

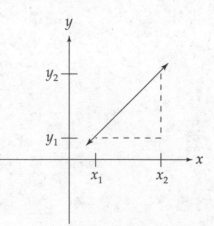

You probably remember your teachers referring to the slope as the "rise" over the "run." The rise is the difference between the *y*-coordinates, and the run is the difference between the *x*-coordinates. The slope is the ratio of the two.

Now for a few changes in notation. Instead of calling the x-coordinates x_1 and x_2, we're going to call them x_1 and $x_1 + h$, where h is the difference between the two x-coordinates. Second, instead of using y_1 and y_2, we use $f(x_1)$ and $f(x_1 + h)$. So now the graph looks like the following:

Sometimes, instead of h, some books use Δx.

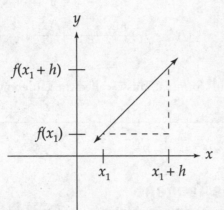

The picture is exactly the same—only the notation has changed.

The Slope of a Curve

Suppose that instead of finding the slope of a line, we wanted to find the slope of a curve. Here the slope formula no longer works because the distance from one point to the other is along a curve, not a straight line. But we could find the approximate slope if we took the slope of the line between the two points. This is called the **secant line**.

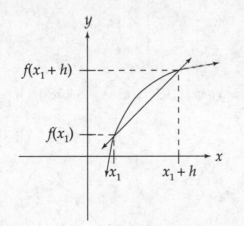

While at first the Difference Quotient may seem like another formula to memorize, keep in mind that it's just another way of writing the slope formula.

The formula for the slope of the secant line is

$$\frac{f(x_1 + h) - f(x_1)}{h}$$

Remember this formula! This is called the **Difference Quotient**.

The Secant and the Tangent

As you can see, the farther apart the two points are, the less the slope of the line corresponds to the slope of the curve.

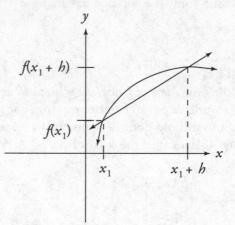

Conversely, the closer the two points are, the more accurate the approximation is.

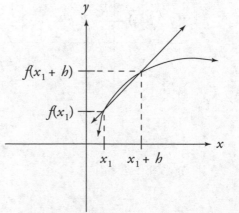

Keep in mind that there are infinitely many tangents for any curve because there are infinitely many points on the curve.

In fact, there is one line, called the **tangent line**, that touches the curve at exactly one point. The slope of the tangent line is equal to the slope of the curve at exactly this point. The object of using the above formula, therefore, is to shrink h down to an infinitesimally small amount. If we could do that, then the difference between $(x_1 + h)$ and x_1 would be a point.

Graphically, it looks like the following:

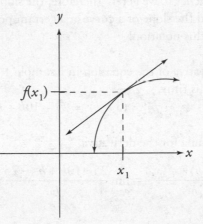

How do we perform this shrinking act? By using the limits we discussed in Chapter 3. We set up a limit during which h approaches zero, like the following:

$$\lim_{h \to 0} \frac{f(x_1 + h) - f(x_1)}{h}$$

This is the **definition of the derivative**, and we call it $f'(x)$.

Notice that the expression above is simply the difference quotient, evaluated at the limit as h (the horizontal distance between the points) approaches zero. We are finding the slope of the line between two points, and asking what that slope approaches as the point on the right is allowed to move infinitesimally close to the fixed point on the left.

Example 1: Find the slope of the curve $f(x) = x^2$ at the point $(2, 4)$.

This means that $x_1 = 2$ and $f(2) = 2^2 = 4$. If we can figure out $f(x_1 + h)$, then we can find the slope. Well, how did we find the value of $f(x)$? We plugged x_1 into the equation $f(x) = x^2$. To find $f(x_1 + h)$ we plug $x_1 + h$ into the equation, which now looks like this

$$f(x_1 + h) = (2 + h)^2 = 4 + 4h + h^2$$

Now plug this into the slope formula.

$$\lim_{h \to 0} \frac{f(x_1 + h) - f(x_1)}{h} = \lim_{h \to 0} \frac{4 + 4h + h^2 - 4}{h} = \lim_{h \to 0} \frac{4h + h^2}{h}$$

Next simplify by factoring h out of the top.

$$\lim_{h \to 0} \frac{4h + h^2}{h} = \lim_{h \to 0} \frac{h(4 + h)}{h} = \lim_{h \to 0}(4 + h)$$

Taking the limit as h approaches 0, we get 4. Therefore, the slope of the curve $y = x^2$ at the point (2, 4) is 4. Now we've found the slope of a curve at a certain point, and the notation looks like this: $f'(2) = 4$. Remember this notation!

Example 2: Find the derivative of the equation in Example 1 at the point (5, 25). This means that $x_1 = 5$ and $f(x) = 25$. This time,

$$(x_1 + h)^2 = (5 + h)^2 = 25 + 10h + h^2$$

Now plug this into the formula for the derivative.

$$\lim_{h \to 0} \frac{f(x_1 + h) - f(x_1)}{h} = \lim_{h \to 0} \frac{25 + 10h + h^2 - 25}{h} = \lim_{h \to 0} \frac{10h + h^2}{h}$$

Once again, simplify by factoring h out of the top.

$$\lim_{h \to 0} \frac{10h + h^2}{h} = \lim_{h \to 0} \frac{h(10 + h)}{h} = \lim_{h \to 0}(10 + h)$$

Taking the limit as h goes to 0, you get 10. Therefore, the slope of the curve $y = x^2$ at the point (5, 25) is 10, or: $f'(5) = 10$.

Using this pattern, let's forget about the arithmetic for a second and derive a formula.

Example 3: Find the slope of the equation $f(x) = x^2$ at the point $\left(x_1, \ x_1^2\right)$.

Follow the steps in the last two problems, but instead of using a number, use x_1. This means that $f(x_1) = x_1^2$ and $(x_1 + h)^2 = x_1^2 + 2x_1 h + h^2$. Then the derivative is

$$\lim_{h \to 0} \frac{x_1^2 + 2x_1 h + h^2 - x_1^2}{h} = \lim_{h \to 0} \frac{2x_1 h + h^2}{h}$$

Factor h out of the top.

$$\lim_{h \to 0} \frac{h(2x_1 + h)}{h} = \lim_{h \to 0}(2x_1 + h)$$

Now take the limit as h goes to 0: you get $2x_1$. Therefore, $f'(x_1) = 2x_1$.

This example gives us a general formula for the derivative of this curve. Now we can pick any point, plug it into the formula, and determine the slope at that point. For example, the derivative at the point $x = 7$ is 14. At the point $x = \dfrac{7}{3}$, the derivative is $\dfrac{14}{3}$.

DIFFERENTIABILITY

One of the important requirements for the differentiability of a function is that the function be continuous. But, even if a function is continuous at a point, the function is not necessarily differentiable there. Check out the graph below.

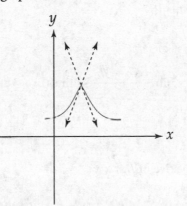

If a function has a "sharp corner," you can draw more than one tangent line at that point, and because the slopes of these tangent lines are not equal, the function is not differentiable there.

Another possible problem occurs when the tangent line is vertical (which can also occur at a cusp) because a vertical line has an infinite slope. For example, if the derivative of a function is $\dfrac{1}{x+1}$, it doesn't have a derivative at $x = -1$.

Try these problems on your own, then check your work against the answers immediately beneath each problem.

PROBLEM 1. Find the derivative of $f(x) = 3x^2$ at (4, 48).

Answer: $f(4 + h) = 3(4 + h)^2 = 48 + 24h + 3h^2$. Use the definition of the derivative.

$$f'(4) = \lim_{h \to 0} \frac{48 + 24h + 3h^2 - 48}{h}$$

Simplify.

$$\lim_{h \to 0} \frac{24h + 3h^2}{h} = \lim_{h \to 0}(24 + 3h) = 24$$

The slope of the curve at the point (4, 48) is 24.

PROBLEM 2. Find the derivative of $f(x) = 3x^2$.

Answer: $f(x + h) = 3(x + h)^2 = 3x^2 + 6xh + 3h^2$. Use the definition of the derivative.

$$f'(x) = \lim_{h \to 0} \frac{3x^2 + 6xh + 3h^2 - 3x^2}{h}$$

Simplify.

$$\lim_{h \to 0} \frac{6xh + 3h^2}{h} = \lim_{h \to 0}(6x + 3h) = 6x$$

The derivative is $6x$.

PROBLEM 3. Find the derivative of $f(x) = x^3$.

Answer: $f(x + h) = (x + h)^3 = x^3 + 3x^2h + 3xh^2 + h^3$. First, use the definition of the derivative.

$$f'(x) = \lim_{h \to 0} \frac{x^3 + 3x^2h + 3xh^2 + h^3 - x^3}{h}$$

And simplify.

$$\lim_{h \to 0} \frac{3x^2h + 3xh^2 + h^3}{h} = \lim_{h \to 0}\left(3x^2 + 3xh + h^2\right) = 3x^2$$

The derivative is $3x^2$.

This next one will test your algebraic skills. Don't say we didn't warn you!

PROBLEM 4. Find the derivative of $f(x) = \sqrt{x}$.

Answer: $f(x + h) = \sqrt{x + h}$.

Use the definition of the derivative.

$$f'(x) = \lim_{h \to 0} \frac{\sqrt{x + h} - \sqrt{x}}{h}$$

Notice that this one doesn't cancel as conveniently as the other problems did. In order to simplify this expression, we have to multiply both the top and the bottom of the expression by $\sqrt{x+h}+\sqrt{x}$ (the conjugate of the numerator).

$$f'(x)=\lim_{h\to 0}\frac{\sqrt{x+h}-\sqrt{x}}{h}\left(\frac{\sqrt{x+h}+\sqrt{x}}{\sqrt{x+h}+\sqrt{x}}\right)=\lim_{h\to 0}\frac{x+h-x}{h\left(\sqrt{x+h}+\sqrt{x}\right)}=\lim_{h\to 0}\frac{h}{h\left(\sqrt{x+h}+\sqrt{x}\right)}$$

Simplify.

$$\lim_{h\to 0}\frac{1}{\left(\sqrt{x+h}+\sqrt{x}\right)}=\frac{1}{2\sqrt{x}}$$

The derivative is $\dfrac{1}{2\sqrt{x}}$.

PRACTICE PROBLEM SET 3

Now find the derivative of the following expressions. The answers are in Chapter 23, starting on page 451.

1. $f(x)=5x$ at $x=3$

2. $f(x)=4x$ at $x=8$

3. $f(x)=5x^2$ at $x=-1$

4. $f(x)=8x^2$

5. $f(x)=-10x^2$

6. $f(x)=20x^2$ at $x=a$

7. $f(x)=2x^3$ at $x=-3$

8. $f(x)=-3x^3$

9. $f(x)=x^5$

10. $f(x) = 2\sqrt{x}$ at $x = 9$

11. $f(x) = 5\sqrt{2x}$ at $x = 8$

12. $f(x) = \sin x$ at $x = \dfrac{\pi}{3}$

13. $f(x) = x^2 + x$

14. $f(x) = x^3 + 3x + 2$

15. $f(x) = \dfrac{1}{x}$

16. $f(x) = ax^2 + bx + c$

End of Chapter 5 Drill

The answers are in Chapter 24.

1. Use the Definition of the Derivative to find $f'(x)$ for $f(x) = 8x - 11$

   (A) -11
   (B) 0
   (C) 8
   (D) $8x$

2. Use the Definition of the Derivative to find $f'(x)$ for $f(x) = 3x^2 + 7x - 2$

   (A) $6x$
   (B) $6x - 2$
   (C) $6x + 7$
   (D) $3x + 7$

3. Use the Definition of the Derivative to find $f'(x)$ for $f(x) = x^3 - 4x^2 + 11$

   (A) $3x^2 - 4$
   (B) $3x^2 - 4x + 11$
   (C) $3x^2 - 8x + 11$
   (D) $3x^2 - 8x$

4. Use the Definition of the Derivative to find $f'(25)$, if $f(x) = 6\sqrt{x}$.

   (A) $\dfrac{3}{5}$

   (B) $\dfrac{3}{25}$

   (C) $\dfrac{6}{25}$

   (D) $\dfrac{6}{5}$

5. Use the Definition of the Derivative to find $f'(2)$, if $f(x) = \dfrac{1}{4x^2}$.

   (A) $-\dfrac{1}{16}$

   (B) $-\dfrac{1}{8}$

   (C) $\dfrac{1}{8}$

   (D) $\dfrac{1}{16}$

REFLECT

Respond to the following questions:

- For which topics discussed in this chapter do you feel you have achieved sufficient mastery to answer multiple-choice questions correctly?

- For which topics discussed in this chapter do you feel you have achieved sufficient mastery to answer open-ended questions correctly?

- For which topics discussed in this chapter do you feel you need more work before you can answer multiple-choice questions correctly?

- For which topics discussed in this chapter do you feel you need more work before you can answer open-ended questions correctly?

- What parts of this chapter are you going to re-review?

Chapter 6

Basic
Differentiation

In calculus, you'll be asked to do two things: differentiate and integrate. In this section, you're going to learn differentiation. Integration will come later, in the second half of this book. Before we go about the business of learning how to take derivatives, however, here's a brief note about notation. Read this!

NOTATION

There are several different notations for derivatives in calculus. We'll use two different types interchangeably throughout this book, so get used to them now.

We'll refer to functions three different ways: $f(x)$, u or v, and y. For example, we might write: $f(x) = x^3$, $g(x) = x^4$, $h(x) = x^5$. We'll also use notation like: $u = \sin x$ and $v = \cos x$. Or we might use $y = \sqrt{x}$. Usually, we pick the notation that causes the least confusion.

The derivatives of the functions will use notation that depends on the function, as shown in the following table:

Function	First Derivative	Second Derivative
$f(x)$	$f'(x)$	$f''(x)$
$g(x)$	$g'(x)$	$g''(x)$
y	y' or $\dfrac{dy}{dx}$	y'' or $\dfrac{d^2y}{dx^2}$

In addition, if we refer to a derivative of a function in general (for example, $ax^2 + bx + c$), we might enclose the expression in parentheses and use either of the following notations:

$$\left(ax^2 + bx + c\right)', \text{ or } \frac{d}{dx}\left(ax^2 + bx + c\right)$$

Sometimes math books refer to a derivative using either D_x or f_x. We're not going to use either of them.

THE POWER RULE

In the last chapter, you learned how to find a derivative using the definition of the derivative, a process that is very time-consuming and sometimes involves a lot of complex algebra. Fortunately, there's a shortcut to taking derivatives, so you'll never have to use the definition again—except when it's a question on an exam!

The basic technique for taking a derivative is called the **Power Rule**.

> Rule No. 1: If $y = x^n$, then $\dfrac{dy}{dx} = nx^{n-1}$

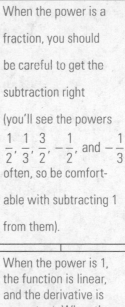

That's it. Wasn't that simple? Of course, this and all of the following rules can be derived easily from the definition of the derivative. Look at these next few examples of the Power Rule in action.

Example 1: If $y = x^5$, then $\dfrac{dy}{dx} = 5x^4$.

Example 2: If $y = x^{20}$, then $\dfrac{dy}{dx} = 20x^{19}$.

Example 3: If $f(x) = x^{-5}$, then $f'(x) = -5x^{-6}$.

> Notice that when the power of the function is negative, the power of the derivative is more negative.

Example 4: If $u = x^{\frac{1}{2}}$, then $\dfrac{du}{dx} = \dfrac{1}{2}x^{-\frac{1}{2}}$.

> When the power is a fraction, you should be careful to get the subtraction right (you'll see the powers $\dfrac{1}{2}, \dfrac{1}{3}, \dfrac{3}{2}, -\dfrac{1}{2}$, and $-\dfrac{1}{3}$ often, so be comfortable with subtracting 1 from them).

Example 5: If $y = x^1$, then $\dfrac{dy}{dx} = 1x^0 = 1$. (Because x^0 is 1!)

Example 6: If $y = x^0$, then $\dfrac{dy}{dx} = 0 \cdot x^{-1} = 0$.

> When the power is 1, the function is linear, and the derivative is a constant. When the power is 0, the function is a constant, and the derivative is 0.

Here are two important rules that you should bear in mind when taking derivatives.

(1) **Constant Multiple Rule.** If you have a constant multiplied by a function, and you would like the derivative of the entire expression, you can "pull the constant out." The derivative will be the constant times the derivative of the function: $[k \cdot f(x)]' = k \cdot f'(x)$. In the other notation, this is $\frac{d}{dx}[k \cdot f(x)] = k \cdot \frac{d}{dx}[f(x)]$.

(2) **Sum Rule.** If you have a sum of two or more functions, and would like the derivative of the entire expression, you can take derivatives "one at a time." The derivative will be the sum of the individual derivatives: $[f(x) + g(x)]' = f'(x) + g'(x)$. In the other notation, this is $\frac{d}{dx}[f(x) + g(x)] = \frac{d}{dx}[f(x)] + \frac{d}{dx}[g(x)]$.

Note: For future reference, a, b, c, n, and k always stand for constants.

Example 7: If $y = 8x^4$, then $\frac{dy}{dx} = 8 \cdot \frac{d}{dx}x^4 = 8 \cdot 4x^3 = 32x^3$.

Example 8: If $y = 5x^{100}$, then $y' = 500x^{99}$.

Example 9: If $y = -3x^{-5}$, then $\frac{dy}{dx} = 15x^{-6}$.

Example 10: If $f(x) = 7x^{\frac{1}{2}}$, then $f'(x) = \frac{7}{2}x^{-\frac{1}{2}}$.

Example 11: If $y = x\sqrt{15}$, then $\frac{dy}{dx} = \sqrt{15}$.

Example 12: If $y = 12$, then $\frac{dy}{dx} = 0$.

If you have any questions about any of these 12 examples (especially the last two), review the rules. Now for one last rule.

THE ADDITION RULE

This handy rule works for subtraction too.

If $y = ax^n + bx^m$, where a and b are constants, then
$$\frac{dy}{dx} = a\left(nx^{n-1}\right) + b\left(mx^{m-1}\right)$$

Example 13: If $y = 3x^4 + 8x^{10}$, then $\dfrac{dy}{dx} = 12x^3 + 80x^9$.

Example 14: If $y = 7x^{-4} + 5x^{-\frac{1}{2}}$, then $\dfrac{dy}{dx} = -28x^{-5} - \dfrac{5}{2}x^{-\frac{3}{2}}$.

Example 15: If $y = 5x^4(2 - x^3) = 10x^4 - 5x^7$, then $\dfrac{dy}{dx} = 40x^3 - 35x^6$.

Example 16: If $y = (3x^2 + 5)(x - 1)$, then

$$y = 3x^3 - 3x^2 + 5x - 5 \text{ and } \frac{dy}{dx} = 9x^2 - 6x + 5$$

Example 17: If $y = ax^3 + bx^2 + cx + d$, then $\dfrac{dy}{dx} = 3ax^2 + 2bx + c$.

After you've worked through all 17 of these examples, you should be able to take the derivative of any polynomial with ease.

As you may have noticed from the examples above, in calculus, you are often asked to convert from fractions and radicals to negative powers and fractional powers. In addition, don't freak out if your answer doesn't match any of the answer choices. Because answers to problems are often presented in simplified form, your answer may not be simplified enough.

There are two basic expressions that you'll often be asked to differentiate. You can make your life easier by memorizing the following derivatives:

$$\text{If } y = \frac{k}{x} \text{, then } \frac{dy}{dx} = -\frac{k}{x^2}.$$

$$\text{If } y = k\sqrt{x} \text{, then } \frac{dy}{dx} = \frac{k}{2\sqrt{x}}.$$

HIGHER ORDER DERIVATIVES

This may sound like a big deal, but it isn't. This term refers only to taking the derivative of a function more than once. You don't have to stop at the first derivative of a function; you can keep taking derivatives. The derivative of a first derivative is called the second derivative. The derivative of the second derivative is called the third derivative, and so on.

Generally, you'll have to take only first and second derivatives.

Notice how we simplified the derivatives in the latter example? You should be able to do this mentally.

Function	First Derivative	Second Derivative
x^6	$6x^5$	$30x^4$
$8\sqrt{x}$	$\dfrac{4}{\sqrt{x}}$	$-2x^{-\frac{3}{2}}$

Here are some sample problems involving the rules we discussed above. As you work, cover the answers with an index card, and then check your work after you're done. By the time you finish them, you should know the rules by heart.

PROBLEM 1. If $y = 50x^5 + \dfrac{3}{x} - 7x^{-\frac{5}{3}}$, then $\dfrac{dy}{dx} =$

Answer: $\dfrac{dy}{dx} = 50\left(5x^4\right) + \left(-\dfrac{3}{x^2}\right) - 7\left(-\dfrac{5}{3}\right)x^{-\frac{8}{3}} = 250x^4 - \dfrac{3}{x^2} + \dfrac{35}{3}x^{-\frac{8}{3}}$

PROBLEM 2. If $y = 9x^4 + 6x^2 - 7x + 11$, then $\dfrac{dy}{dx} =$

Answer: $\dfrac{dy}{dx} = 9\left(4x^3\right) + 6\left(2x\right) - 7\left(1\right) + 0 = 36x^3 + 12x - 7$

PROBLEM 3. If $f(x) = 6x^{\frac{3}{2}} - 12\sqrt{x} - \dfrac{8}{\sqrt{x}} + 24x^{-\frac{3}{2}}$, then $f'(x) =$

Answer: $f'(x) = 6\left(\dfrac{3}{2}x^{\frac{1}{2}}\right) - \left(\dfrac{12}{2\sqrt{x}}\right) - 8\left(-\dfrac{1}{2}x^{-\frac{3}{2}}\right) + 24\left(-\dfrac{3}{2}x^{-\frac{5}{2}}\right) = 9\sqrt{x} - \dfrac{6}{\sqrt{x}} + 4x^{-\frac{3}{2}} - 36x^{-\frac{5}{2}}$

How'd you do? Did you notice the changes in notation? How about the fractional powers, radical signs, and x's in denominators? You should be able to switch back and forth between notations, between fractional powers and radical signs, and between negative powers in a numerator and positive powers in a denominator.

PRACTICE PROBLEM SET 4

Find the derivative of each expression and simplify. The answers are in Chapter 23, starting on page 460.

1. $(4x^2 + 1)^2$

2. $(x^5 + 3x)^2$

3. $11x^7$

4. $18x^3 + 12x + 11$

5. $\dfrac{1}{2}(x^{12} + 17)$

6. $-\dfrac{1}{3}(x^9 + 2x^3 - 9)$

7. π^5

8. $-8x^{-8} + 12\sqrt{x}$

9. $6x^{-7} - 4\sqrt{x}$

10. $x^{-5} + \dfrac{1}{x^8}$

11. $(6x^2 + 3)(12x - 4)$

12. $(3 - x - 2x^3)(6 + x^4)$

13. $e^{10} + \pi^3 - 7$

14. $\left(\dfrac{1}{x} + \dfrac{1}{x^2}\right)\left(\dfrac{4}{x^3} - \dfrac{6}{x^4}\right)$

15. $\sqrt{x} + \dfrac{1}{\sqrt{3}}$

16. 0

17. $(x + 1)^3$

18. $\sqrt{x} + \sqrt[3]{x} + \sqrt[3]{x^2}$

19. $x(2x + 7)(x - 2)$

20. $\sqrt{x}\left(\sqrt[3]{x} + \sqrt[5]{x}\right)$

THE PRODUCT RULE

Now that you know how to find derivatives of simple polynomials, it's time to get more complicated. What if you had to find the derivative of this?

$$f(x) = (x^3 + 5x^2 - 4x + 1)(x^5 - 7x^4 + x)$$

You could multiply out the expression and take the derivative of each term, like

$$f(x) = x^8 - 2x^7 - 39x^6 + 29x^5 - 6x^4 + 5x^3 - 4x^2 + x$$

And the derivative is

$$f'(x) = 8x^7 - 14x^6 - 234x^5 + 145x^4 - 24x^3 + 15x^2 - 8x + 1$$

Needless to say, this process is messy. Naturally, there's an easier way. When a function involves two terms multiplied by each other, we use the **Product Rule**.

> The Product Rule: If $f(x) = uv$, then $f'(x) = u\dfrac{dv}{dx} + v\dfrac{du}{dx}$

With the product rule the order of these two operations doesn't matter. It does matter with other rules, though, so it helps to use the same order each time.

To find the derivative of two things multiplied by each other, you multiply the first function by the derivative of the second, and add that to the second function multiplied by the derivative of the first.

Let's use the Product Rule to find the derivative of our example.

$$f'(x) = (x^3 + 5x^2 - 4x + 1)(5x^4 - 28x^3 + 1) + (x^5 - 7x^4 + x)(3x^2 + 10x - 4)$$

If we were to simplify this, we'd get the same answer as before. But here's the best part: we're not going to simplify it. One of the great things about the AP Exam is that when it's difficult to simplify an expression, you almost never have to.

Nonetheless, you'll often need to simplify expressions when you're taking second derivatives, or when you use the derivative in some other equation. Practice simplifying whenever possible.

Example 1: $f(x) = (9x^2 + 4x)(x^3 - 5x^2)$

$$f'(x) = (9x^2 + 4x)(3x^2 - 10x) + (x^3 - 5x^2)(18x + 4)$$

Example 2: $y = \left(\sqrt{x} + 4\sqrt[3]{x}\right)\left(x^5 - 11x^8\right)$

$$y' = \left(\sqrt{x} + 4\sqrt[3]{x}\right)\left(5x^4 - 88x^7\right) + \left(x^5 - 11x^8\right)\left(\frac{1}{2\sqrt{x}} + \frac{4}{3\sqrt[3]{x^2}}\right)$$

Example 3: $y = \left(\dfrac{1}{x} + \dfrac{1}{x^2} - \dfrac{1}{x^3} \right)\left(\dfrac{1}{x} - \dfrac{1}{x^3} + \dfrac{1}{x^5} \right)$

$$y' = \left(\dfrac{1}{x} + \dfrac{1}{x^2} - \dfrac{1}{x^3} \right)\left(-\dfrac{1}{x^2} + \dfrac{3}{x^4} - \dfrac{5}{x^6} \right) + \left(\dfrac{1}{x} - \dfrac{1}{x^3} + \dfrac{1}{x^5} \right)\left(-\dfrac{1}{x^2} - \dfrac{2}{x^3} + \dfrac{3}{x^4} \right)$$

THE QUOTIENT RULE

What happens when you have to take the derivative of a function that is the quotient of two other functions? You guessed it: Use the **Quotient Rule**.

The Quotient Rule: If $f(x) = \dfrac{u}{v}$, then $f'(x) = \dfrac{v\dfrac{du}{dx} - u\dfrac{dv}{dx}}{v^2}$

In this rule, as opposed to the Product Rule, the order in which you take the derivatives is very important because you're subtracting instead of adding. It's always the bottom function times the derivative of the top minus the top function times the derivative of the bottom. Then divide the whole thing by the bottom function squared. A good way to remember this is to say the following:

$$\dfrac{"LoDeHi - HiDeLo"}{(Lo)^2}$$

> For the Quotient Rule, remember that order matters!

Here are some more examples.

Example 4: $f(x) = \dfrac{\left(x^5 - 3x^4 \right)}{\left(x^2 + 7x \right)}$

$$f'(x) = \dfrac{\left(x^2 + 7x \right)\left(5x^4 - 12x^3 \right) - \left(x^5 - 3x^4 \right)(2x + 7)}{\left(x^2 + 7x \right)^2}$$

Example 5: $y = \dfrac{\left(x^{-3} - x^{-8}\right)}{\left(x^{-2} + x^{-6}\right)}$

$$\frac{dy}{dx} = \frac{\left(x^{-2} + x^{-6}\right)\left(-3x^{-4} + 8x^{-9}\right) - \left(x^{-3} - x^{-8}\right)\left(-2x^{-3} - 6x^{-7}\right)}{\left(x^{-2} + x^{-6}\right)^2}$$

We're not going to simplify these, although the Quotient Rule often produces expressions that simplify more readily than those involving the Product Rule. Sometimes it's helpful to simplify, but avoid it otherwise. When you have to find a second derivative, however, you do have to simplify the quotient. If this is the case, the AP Exam usually will give you a simple expression to deal with, such as in the example below.

Example 6: $y = \dfrac{3x + 5}{5x - 3}$

$$\frac{dy}{dx} = \frac{(5x - 3)(3) - (3x + 5)(5)}{(5x - 3)^2} = \frac{(15x - 9) - (15x + 25)}{(5x - 3)^2} = \frac{-34}{(5x - 3)^2}$$

In order to take the derivative of $\dfrac{dy}{dx}$, you have to use the Chain Rule.

THE CHAIN RULE

The most important rule in this chapter (and sometimes the most difficult one) is called the **Chain Rule**. It's used when you're given composite functions—that is, a function inside of another function. You'll always see one of these on the AP Exam, so it's important to know the Chain Rule cold.

A composite function is usually written as: $f(g(x))$.

For example: If $f(x) = \dfrac{1}{x}$ and $g(x) = \sqrt{3x}$, then $f\left(g(x)\right) = \dfrac{1}{\sqrt{3x}}$.

We could also find $g\left(f(x)\right) = \sqrt{\dfrac{3}{x}}$.

When finding the derivative of a composite function, we take the derivative of the "outside" function, with the inside function g considered as the variable, leaving the "inside" function alone. Then, we multiply this by the derivative of the "inside" function, with respect to its variable x.

Here is another way to write the Chain Rule.

> The Chain Rule: If $y = f\big(g(x)\big)$, then $y' = f'\big(g(x)\big) \cdot g'(x)$

In other words, y' is found by taking the derivative of f, putting $g(x)$ as the input of the derivative function $f'(x)$, and multiplying the result by $g'(x)$.

Chain Rule—always work from the outside in, kind of like Russian dolls or peeling an onion.

This rule is tricky, so here are several examples. The last couple incorporate the Product Rule and the Quotient Rule.

Example 7: If $y = (5x^3 + 3x)^5$, then $\dfrac{dy}{dx} = 5(5x^3 + 3x)^4(15x^2 + 3)$.

We just dealt with the derivative of something to the fifth power, like this

$$y = (g)^5, \text{ so } \frac{dy}{dg} = 5(g)^4, \text{ where } g = 5x^3 + 3x$$

Then we multiplied by the derivative of g: $(15x^2 + 3)$.

Always do it this way. The process has several successive steps, like peeling away the layers of an onion until you reach the center.

Example 8: If $y = \sqrt{x^3 - 4x}$, then $\dfrac{dy}{dx} = \dfrac{1}{2}\big(x^3 - 4x\big)^{\frac{1}{2}}\big(3x^2 - 4\big)$.

Again, we took the derivative of the outside function and evaluated it using the inside function as input. Then we multiplied by the derivative of the inside function.

Example 9: If $y = \sqrt{\big(x^5 - 8x^3\big)\big(x^2 + 6x\big)}$, then

$$\frac{dy}{dx} = \frac{1}{2}\Big[\big(x^5 - 8x^3\big)\big(x^2 + 6x\big)\Big]^{-\frac{1}{2}}\Big[\big(x^5 - 8x^3\big)(2x + 6) + \big(x^2 + 6x\big)\big(5x^4 - 24x^2\big)\Big]$$

Messy, isn't it? That's because we used the Chain Rule and the Product Rule. Now for one with the Chain Rule and the Quotient Rule.

Example 10: If $y = \left(\dfrac{2x + 8}{x^2 - 10x}\right)^5$, then

$$\frac{dy}{dx} = 5\left[\frac{2x + 8}{x^2 - 10x}\right]^4\left[\frac{\big(x^2 - 10x\big)(2) - (2x + 8)(2x - 10)}{\big(x^2 - 10x\big)^2}\right]$$

Example 11: If $y = \sqrt{5x^3 + x}$, then $\dfrac{dy}{dx} = \dfrac{1}{2}\left(5x^3 + x\right)^{-\frac{1}{2}}\left(15x^2 + 1\right)$

Now we use the Product Rule and the Chain Rule to find the second derivative.

$$\dfrac{d^2 y}{dx^2} = \dfrac{1}{2}\left(5x^3 + x\right)^{-\frac{1}{2}}(30x) + \left(15x^2 + 1\right)\left[-\dfrac{1}{4}\left(5x^3 + x\right)^{-\frac{3}{2}}\left(15x^2 + 1\right)\right]$$

You can also simplify this further, if necessary.

There's another representation of the Chain Rule that you need to learn.

> If $y = y(v)$ and $v = v(x)$, then $\dfrac{dy}{dx} = \dfrac{dy}{dv}\dfrac{dv}{dx}$.

As you can see, these grow quite complex, so we simplify these only as a last resort. If you must simplify, the AP Exam will have only a very simple Chain Rule problem.

Example 12: $y = 8v^2 - 6v$ and $v = 5x^3 - 11x$, then

$$\dfrac{dy}{dx} = (16v - 6)(15x^2 - 11)$$

Then substitute for v.

$$\dfrac{dy}{dx} = \left(16\left(5x^3 - 11x\right) - 6\right)\left(15x^2 - 11\right) = \left(80x^3 - 176x - 6\right)\left(15x^2 - 11\right)$$

Here are some solved problems. Cover the answers first, then check your work.

PROBLEM 1. Find $\dfrac{dy}{dx}$ if $y = \left(5x^4 + 3x^7\right)\left(x^{10} - 8x\right)$.

Answer: $\dfrac{dy}{dx} = \left(5x^4 + 3x^7\right)\left(10x^9 - 8\right) + \left(x^{10} - 8x\right)\left(20x^3 + 21x^6\right)$

PROBLEM 2. Find $\dfrac{dy}{dx}$ if $y = \left(x^3 + 3x^2 + 3x + 1\right)\left(x^2 + 2x + 1\right)$.

Answer: $\dfrac{dy}{dx} = \left(x^3 + 3x^2 + 3x + 1\right)(2x + 2) + \left(x^2 + 2x + 1\right)\left(3x^2 + 6x + 3\right)$

PROBLEM 3. Find $\dfrac{dy}{dx}$ if $y = \left(\sqrt{x} + \dfrac{1}{x} \right)\left(\sqrt[3]{x^2} - \dfrac{1}{x^3} \right)$.

Answer: $\dfrac{dy}{dx} = \left(\sqrt{x} + \dfrac{1}{x} \right)\left(\dfrac{2}{3}x^{-\frac{1}{3}} + \dfrac{3}{x^4} \right) + \left(\sqrt[3]{x^2} - \dfrac{1}{x^3} \right)\left(\dfrac{1}{2\sqrt{x}} - \dfrac{1}{x^2} \right)$

PROBLEM 4. Find $\dfrac{dy}{dx}$ if $y = \left(x^3 + 1 \right)\left(x^2 + 5x - \dfrac{1}{x^5} \right)$.

Answer: $\dfrac{dy}{dx} = \left(x^3 + 1 \right)\left(2x + 5 + \dfrac{5}{x^6} \right) + \left(x^2 + 5x - \dfrac{1}{x^5} \right)\left(3x^2 \right)$

PROBLEM 5. Find $\dfrac{dy}{dx}$ if $y = \dfrac{2x - 4}{x^2 - 6}$.

Answer: $\dfrac{dy}{dx} = \dfrac{\left(x^2 - 6 \right)(2) - (2x - 4)(2x)}{\left(x^2 - 6 \right)^2} = \dfrac{-2x^2 + 8x - 12}{\left(x^2 - 6 \right)^2}$

PROBLEM 6. Find $\dfrac{dy}{dx}$ if $y = \dfrac{x^2 + 1}{x^2 + x + 4}$.

Answer: $\dfrac{dy}{dx} = \dfrac{\left(x^2 + x + 4 \right)(2x) - \left(x^2 + 1 \right)(2x + 1)}{\left(x^2 + x + 4 \right)^2} = \dfrac{x^2 + 6x - 1}{\left(x^2 + x + 4 \right)^2}$

PROBLEM 7. Find $\dfrac{dy}{dx}$ if $y = \dfrac{x + 5}{x - 5}$.

Answer: $\dfrac{dy}{dx} = \dfrac{(x - 5)(1) - (x + 5)(1)}{(x - 5)^2} = \dfrac{-10}{(x - 5)^2}$

PROBLEM 8. Find $\dfrac{dy}{dx}$ if $y = (x^4 + x)^2$.

Answer: $\dfrac{dy}{dx} = 2\left(x^4 + x \right)\left(4x^3 + 1 \right)$

PROBLEM 9. Find $\dfrac{dy}{dx}$ if $y = \left(\dfrac{x+3}{x-3}\right)^3$.

Answer: $\dfrac{dy}{dx} = 3\left(\dfrac{x+3}{x-3}\right)^2\left(\dfrac{(x-3)(1)-(x+3)(1)}{(x-3)^2}\right) = -18\left(\dfrac{(x+3)^2}{(x-3)^4}\right)$

PROBLEM 10. Find $\dfrac{dy}{dx}$ at $x = 1$ if $y = \left[\left(x^3 + x\right)\left(x^4 - x^2\right)\right]^2$.

Answer: $\dfrac{dy}{dx} = 2\left[\left(x^3 + x\right)\left(x^4 - x^2\right)\right]\left[\left(x^3 + x\right)\left(4x^3 - 2x\right) + \left(x^4 - x^2\right)\left(3x^2 + 1\right)\right]$

Once again, plug in right away. Never simplify until after you've substituted.

At $x = 1$, $\dfrac{dy}{dx} = 0$.

PRACTICE PROBLEM SET 5

Simplify when possible. The answers are in Chapter 23, starting on page 464.

1. Find $f'(x)$ if $f(x) = \left(\dfrac{4x^3 - 3x^2}{5x^7 + 1}\right)$.

2. Find $f'(x)$ if $f(x) = \left(x^2 - 4x + 3\right)(x+1)$.

3. Find $f'(x)$ if $f(x) = 8\sqrt{\left(x^4 - 4x^2\right)}$.

4. Find $f'(x)$ if $f(x) = \left(\dfrac{x}{x^2 + 1}\right)^3$.

5. Find $f'(x)$ if $f(x) = \sqrt[4]{\left(\dfrac{2x - 5}{5x + 2}\right)}$.

6. Find $f'(x)$ if $f(x) = \left(x + \dfrac{1}{x}\right)\left(x^2 - \dfrac{1}{x^2}\right)$.

7. Find $f'(x)$ if $f(x) = \left(\dfrac{x}{x+1}\right)^4$.

8. Find $f'(x)$ if $f(x) = (x^2 + x)^{100}$.

9. Find $f'(x)$ if $f(x) = \sqrt{\dfrac{x^2 + 1}{x^2 - 1}}$.

10. Find $f'(x)$ at $x = 2$ if $f(x) = \dfrac{(x+4)(x-8)}{(x+6)(x-6)}$.

11. Find $f'(x)$ at $x = 1$ if $f(x) = \dfrac{x^6 + 4x^3 + 6}{(x^4 - 2)^2}$.

12. Find $f'(x)$ if $f(x) = \dfrac{x^2 - 3}{(x - 3)}$.

13. Find $f'(x)$ at $x = 1$ if $f(x) = (x^4 - x^2)(2x^3 + x)$.

14. Find $f'(x)$ at $x = 2$ if $f(x) = \dfrac{x^2 + 2x}{x^4 - x^3}$.

15. Find $f'(x)$ if $f(x) = \sqrt{x^4 + x^2}$.

16. Find $\dfrac{dy}{dx}$ if $y = u^2 - 1$ and $u = \dfrac{1}{x - 1}$.

17. Find $\dfrac{dy}{dx}$ at $x = 1$ if $y = \dfrac{t^2 + 2}{t^2 - 2}$ and $t = x^3$.

18. Find $\dfrac{dy}{dt}$ if $y = (x^6 - 6x^5)(5x^2 + x)$ and $x = \sqrt{t}$.

19. Find $\dfrac{dy}{dx}$ at $x = 1$ if $y = \dfrac{1 + u}{1 + u^2}$ and $u = x^2 - 1$.

20. Find $\dfrac{du}{dv}$ if $u = y^3$, $y = \dfrac{x}{x + 8}$, and $x = v^2$.

DERIVATIVES OF TRIG FUNCTIONS

There are a lot of trigonometry problems in calculus and on the AP Calculus Exam. You'll need to remember your trig formulas, the values of the special angles, and the trig ratios, among other stuff.

In addition, angles are *always* referred to in radians. You can forget all about using degrees.

You should know the derivatives of all six trig functions. The good news is that the derivatives are pretty easy, and all you have to do is memorize them. Because the AP Exam might ask you about this, though, let's use the definition of the derivative to figure out the derivative of sin x.

Trigon(line)ometry
If you're not sure of your trig, log on to your free online student tools and download the Appendix, which contains a focused review of Prerequisite Math.

If $f(x) = \sin x$, then $f(x + h) = \sin(x + h)$.

Substitute this into the definition of the derivative.

$$\lim_{h \to 0} \frac{f(x+h) - f(x)}{h} = \lim_{h \to 0} \frac{\sin(x+h) - \sin x}{h}$$

Remember that $\sin(x + h) = \sin x \cos h + \cos x \sin h$. Now simplify it.

$$\lim_{h \to 0} \frac{\sin x \cos h + \cos x \sin h - \sin x}{h}$$

Next, rewrite this as

$$\lim_{h \to 0} \frac{\sin x (\cos h - 1) + \cos x \sin h}{h} = \lim_{h \to 0} \frac{\sin x (\cos h - 1)}{h} + \lim_{h \to 0} \frac{\cos x \sin h}{h}$$

Next, use some of the trigonometric limits that you memorized back in Chapter 3. Specifically,

$$\lim_{h \to 0} \frac{(\cos h - 1)}{h} = 0 \text{ and } \lim_{h \to 0} \frac{\sin h}{h} = 1$$

This gives you

$$\lim_{h \to 0} \frac{\sin x (\cos h - 1)}{h} + \lim_{h \to 0} \frac{\cos x \sin h}{h} = (\sin x)(0) + (\cos x)(1) = \cos x$$

$$\frac{d}{dx} \sin x = \cos x$$

Example 1: Find the derivative of $\sin\left(\dfrac{\pi}{2}-x\right)$.

$$\frac{d}{dx}\sin\left(\frac{\pi}{2}-x\right)=\cos\left(\frac{\pi}{2}-x\right)(-1)=-\cos\left(\frac{\pi}{2}-x\right)$$

Use some of the rules of trigonometry you remember from last year. Because

$$\sin\left(\frac{\pi}{2}-x\right)=\cos x \text{ and } \cos\left(\frac{\pi}{2}-x\right)=\sin x,$$

you can substitute into the above expression and get

$$\frac{d}{dx}\cos x=-\sin x$$

Now, let's derive the derivatives of the other four trigonometric functions.

Example 2: Find the derivative of $\dfrac{\sin x}{\cos x}$.

Use the Quotient Rule.

$$\frac{d}{dx}\frac{\sin x}{\cos x}=\frac{(\cos x)(\cos x)-(\sin x)(-\sin x)}{(\cos x)^2}=\frac{\cos^2 x+\sin^2 x}{\cos^2 x}=\frac{1}{\cos^2 x}=\sec^2 x$$

Because $\dfrac{\sin x}{\cos x}=\tan x$, you should get

$$\frac{d}{dx}\tan x=\sec^2 x$$

Example 3: Find the derivative of $\dfrac{\cos x}{\sin x}$.

Use the Quotient Rule.

$$\frac{d}{dx}\frac{\cos x}{\sin x} = \frac{(\sin x)(-\sin x) - (\cos x)(\cos x)}{(\sin x)^2} = \frac{-\left(\cos^2 x + \sin^2 x\right)}{\sin^2 x} = -\frac{1}{\sin^2 x} = -\csc^2 x$$

Because $\dfrac{\cos x}{\sin x} = \cot x$, you get: $\dfrac{d}{dx}\cot x = -\csc^2 x$.

Example 4: Find the derivative of $\dfrac{1}{\cos x}$.

Use the rule mentioned in the box on page 135.

$$\frac{d}{dx}\frac{1}{\cos x} = \frac{-1}{(\cos x)^2}(-\sin x) = \frac{\sin x}{\cos^2 x} = \frac{1}{\cos x}\frac{\sin x}{\cos x} = \sec x \tan x$$

Because $\dfrac{1}{\cos x} = \sec x$, you get

$$\frac{d}{dx}\sec x = \sec x \tan x$$

Example 5: Find the derivative of $\dfrac{1}{\sin x}$.

You get the idea by now.

$$\frac{d}{dx}\frac{1}{\sin x} = \frac{-1}{(\sin x)^2}(\cos x) = \frac{-\cos x}{\sin^2 x} = \frac{-1}{\sin x}\frac{\cos x}{\sin x} = -\csc x \cot x$$

Because $\dfrac{1}{\sin x} = \csc x$, you get

$$\frac{d}{dx}\csc x = -\csc x \cot x$$

There you go. We have now found the derivatives of all six of the trigonometric functions. (A chart of them appears at the end of the book.) Now memorize them. You'll thank us later.

Let's do some more examples.

Example 6: Find the derivative of sin(5x).

$$\frac{d}{dx}\sin(5x)=\cos(5x)(5)=5\cos(5x)$$

Example 7: Find the derivative of sec(x^2).

$$\frac{d}{dx}\sec\left(x^2\right)=\sec\left(x^2\right)\tan\left(x^2\right)(2x)$$

Example 8: Find the derivative of csc($x^3 - 5x$).

$$\frac{d}{dx}\csc\left(x^3-5x\right)=-\csc\left(x^3-5x\right)\cot\left(x^3-5x\right)\left(3x^2-5\right)$$

These derivatives are almost like formulas. You just follow the pattern and use the Chain Rule when appropriate.

Here are some solved problems. Do each problem, covering the answer first, then checking your answer.

PROBLEM 1. Find $f'(x)$ if $f(x) = \sin(2x^3)$.

Answer: Follow the rule: $f'(x) = \cos(2x^3)6x^2$.

PROBLEM 2. Find $f'(x)$ if $f(x) = \cos\left(\sqrt{3x}\right)$.

Answer: $f'(x)=-\sin\left(\sqrt{3x}\right)\left[\frac{1}{2}(3x)^{-\frac{1}{2}}(3)\right]=\dfrac{-3\sin\left(\sqrt{3x}\right)}{2\sqrt{3x}}$

PROBLEM 3. Find $f'(x)$ if $f(x)=\tan\left(\dfrac{x}{x+1}\right)$.

Answer: $f'(x)=\sec^2\left(\dfrac{x}{x+1}\right)\left[\dfrac{(x+1)-x}{(x+1)^2}\right]=\left(\dfrac{1}{(x+1)^2}\right)\sec^2\left(\dfrac{x}{x+1}\right)$

PROBLEM 4. Find $f'(x)$ if $f(x) = \csc(x^3 + x + 1)$.

Answer: Follow the rule: $f'(x) = -\csc(x^3 + x + 1)\cot(x^3 + x + 1)(3x^2 + 1)$

GRAPHICAL DERIVATIVES

Sometimes, we are given the graph of a function and we are asked to graph the derivative. We do this by analyzing the sign of derivative at various places on the graph and then sketching a graph of the derivative from that information. Let's start with something simple. Suppose we have the graph of $y = f(x)$ below and we are asked to sketch the graph of its derivative.

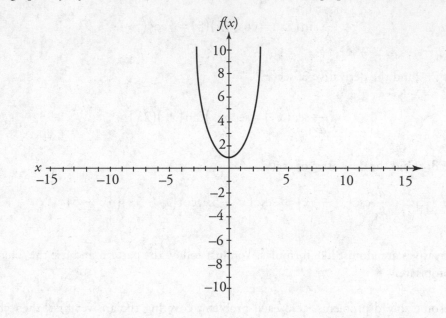

First, note that the derivative is zero at the point (0, 1) because the tangent line is horizontal there. Next, the derivative is negative for all $x < 0$ because the tangent lines to the curve have negative slopes everywhere on the interval $(-\infty, 0)$. Finally, the derivative is positive for all $x > 0$ because the tangent lines to the curve have positive slopes everywhere on the interval $(0, \infty)$. Now we can make a graph of the derivative. It will go through the origin (because the derivative is 0 at $x = 0$), it will be negative on the interval $(-\infty, 0)$, and it will be positive on the interval $(0, \infty)$. The graph looks something like the following:

> Note that it's not important for your graph to be exact. All we are doing here is sketching the derivative. The important parts of the graph are where the derivative is positive, negative, and zero.

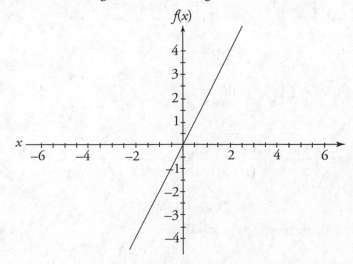

Now let's try something a little harder. Suppose we have the following graph and we are asked to sketch the graph of the derivative:

Maxima and minima on the graph of *f* become zeroes on the graph of *f'*.

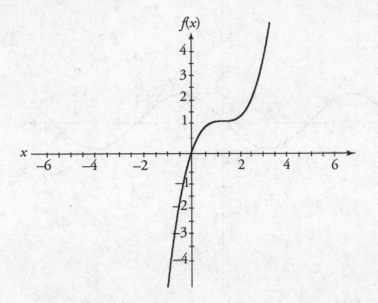

Notice that the tangent line looks as if it's horizontal at $x = 1$. This means that the graph of the derivative is zero there. Next, the curve is increasing for all other values of x, so the graph of the derivative will be positive. As we go from left to right on the graph, notice that the slope starts out very steep, so the derivatives are large positive numbers. As we approach $x = 1$, the curve starts to flatten out, so the derivatives will approach zero but will still be positive. Then, the slope is zero at $x = 1$. Then the curve gets steep again. If we sketch the derivative, we get something like the following:

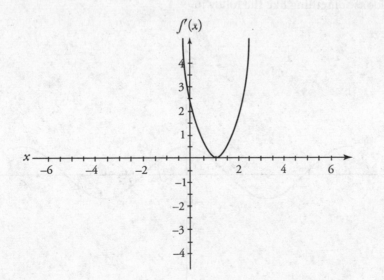

Here is an important one to understand. Suppose we have the graph of $y = \sin x$.

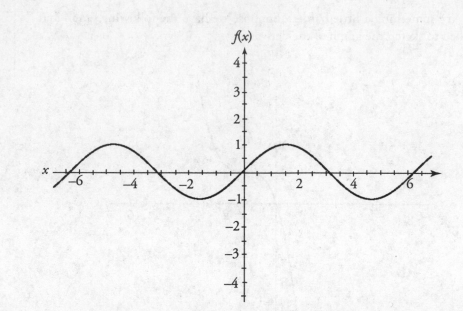

Notice that the slope of the tangent line will be horizontal at all of the maxima and minima of the graph. Because the slope of a horizontal line is zero, this means that the derivative will be zero at those values $(\pm\dfrac{\pi}{2}, \dfrac{3\pi}{2}, \ldots)$. Next, notice that the slope of the curve is about 1 as the curve goes through the origin. This should make sense if you recall that $\lim\limits_{x\to 0}\dfrac{\sin x}{x} = 1$. The slope of the curve is about −1 as the curve goes through $x = \pi$, and so on. If we now sketch the derivative, it looks something like the following:

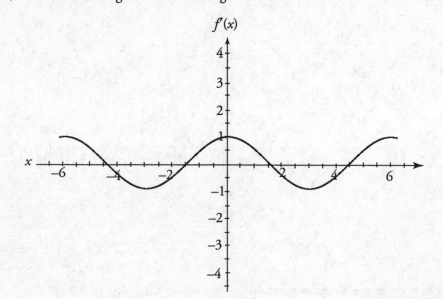

Notice that this is the graph of $y = \cos x$. This should be obvious because the derivative of sin x is cos x.

Now let's do a hard one. Suppose we have the following graph:

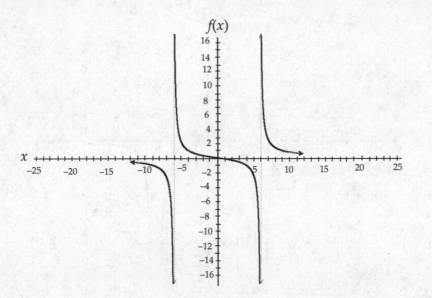

First, notice that we have two vertical asymptotes at $x = 6$ and $x = -6$. This means that the graph of the derivative will also have vertical asymptotes at $x = 6$ and $x = -6$. Next, notice that the curve is always decreasing. This means that the graph of the derivative will always be negative. Moving from left to right, the graph starts out close to flat, so the derivative will be close to zero. Then, the graph gets very steep and points downward, so the graph of the derivative will be negative and getting more negative. Then, we have the asymptote $x = -6$. Next, the graph begins very steep and negative and starts to flatten out as we approach the origin. At the origin, the slope of the graph is approximately $-\dfrac{1}{2}$. This means that the graph of the derivative will increase until it reaches $\left(0, -\dfrac{1}{2}\right)$. Then, the graph starts to get steep again as we approach the other asymptote $x = 6$. Thus, the graph will get more negative again. Finally, to the right of the asymptote $x = 6$ the graph starts out steep and negative and flattens out, approaching zero. This means that the graph of the derivative will start out very negative and will approach zero.

If we now sketch the derivative, it looks something like the following:

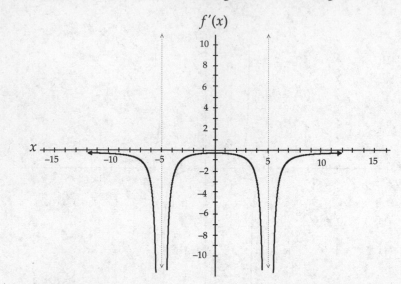

PRACTICE PROBLEM SET 6

Now try these problems. The answers are in Chapter 23, starting on page 472.

1. Find $\dfrac{dy}{dx}$ if $y = \sin^2 x$.

2. Find $\dfrac{dy}{dx}$ if $y = \cos x^2$.

3. Find $\dfrac{dy}{dx}$ if $y = (\tan x)(\sec x)$.

4. Find $\dfrac{dy}{dx}$ if $y = \sqrt{\sin 3x}$.

5. Find $\dfrac{dy}{dx}$ if $y = \dfrac{1 + \sin x}{1 - \sin x}$.

6. Find $\dfrac{dy}{dx}$ if $y = \csc^2 x^2$.

7. Find $\dfrac{d^4 y}{dx^4}$ if $y = \sin 2x$.

8. Find $\dfrac{dy}{dx}$ if $y = \sin t - \cos t$ and $t = 1 + \cos^2 x$.

9. Find $\dfrac{dy}{dx}$ if $y = \left(\dfrac{\tan x}{1 - \tan x}\right)^2$.

10. Find $\dfrac{dr}{d\theta}$ if $r = \cos(1 + \sin\theta)$.

11. Find $\dfrac{dr}{d\theta}$ if $r = \dfrac{\sec\theta}{1 + \tan\theta}$.

12. Find $\dfrac{dy}{dx}$ if $y = \left(1 + \cot\left(\dfrac{2}{x}\right)\right)^{-2}$.

End of Chapter 6 Drill

The answers are in Chapter 24.

1. Find $\dfrac{dy}{dx}$ if $y = 5x^3 - 7x^2 + 11x + 2$.

(A) $\dfrac{dy}{dx} = 15x^2 - 14x$

(B) $\dfrac{dy}{dx} = 35x^2 - 27x + 11$

(C) $\dfrac{dy}{dx} = 5x^2 - 7x + 11$

(D) $\dfrac{dy}{dx} = 15x^2 - 14x + 11$

2. Find $\dfrac{dy}{dx}$ if $y = \tan(4x)$.

(A) $\dfrac{dy}{dx} = \sec^2(4x)$

(B) $\dfrac{dy}{dx} = 4\sec^2(4x)$

(C) $\dfrac{dy}{dx} = 4\tan(4x)$

(D) $\dfrac{dy}{dx} = 4\cot(4x)$

3. Find $\dfrac{dy}{dx}$ if $y = \cos^2(x^2)$.

(A) $\dfrac{dy}{dx} = 4x\cos(x^2)$

(B) $\dfrac{dy}{dx} = \sin^2(x^2)$

(C) $\dfrac{dy}{dx} = -4x\cos(x^2)\sin(x^2)$

(D) $\dfrac{dy}{dx} = -4x\sin^2(x^2)$

4. Find $\dfrac{dy}{dx}$ if $y = \dfrac{4}{\sqrt{x-1}}$.

(A) $\dfrac{dy}{dx} = -\dfrac{2}{(x-1)^{\frac{3}{2}}}$

(B) $\dfrac{dy}{dx} = \dfrac{2}{(x-1)^{\frac{3}{2}}}$

(C) $\dfrac{dy}{dx} = -\dfrac{8}{(x-1)^{\frac{3}{2}}}$

(D) $\dfrac{dy}{dx} = \dfrac{8}{(x-1)^{\frac{3}{2}}}$

5. Find $\dfrac{dy}{dx}$ if $y = \sqrt{\dfrac{x^2}{\sec x}}$.

(A) $\dfrac{dy}{dx} = \dfrac{1}{2}\left(\dfrac{2x}{\sec x \tan x}\right)^{-\frac{1}{2}}$

(B) $\dfrac{dy}{dx} = \left(\dfrac{2x}{\sec x \tan x}\right)^{\frac{1}{2}}$

(C) $\dfrac{dy}{dx} = \dfrac{1}{2}\left(\dfrac{\sec x(2x) - (x^2)\sec x \tan x}{\sec^2 x}\right)^{-\frac{1}{2}}$

(D) $\dfrac{dy}{dx} = \dfrac{1}{2}\left(\dfrac{x^2}{\sec x}\right)^{-\frac{1}{2}}\left(\dfrac{\sec x(2x) - (x^2)\sec x \tan x}{\sec^2 x}\right)$

REFLECT

Respond to the following questions:

- For which topics discussed in this chapter do you feel you have achieved sufficient mastery to answer multiple-choice questions correctly?

- For which topics discussed in this chapter do you feel you have achieved sufficient mastery to answer open-ended questions correctly?

- For which topics discussed in this chapter do you feel you need more work before you can answer multiple-choice questions correctly?

- For which topics discussed in this chapter do you feel you need more work before you can answer open-ended questions correctly?

- What parts of this chapter are you going to re-review?

Chapter 7
Implicit
Differentiation

HOW TO DO IT

By now, it should be easy for you to take the derivative of an equation such as $y = 3x^5 - 7x$. If you're given an equation such as $y^2 = 3x^5 - 7x$, you can still figure out the derivative by taking the square root of both sides, which gives you y in terms of x. This is known as finding the derivative **explicitly**. It's messy, but possible.

If you have to find the derivative of $y^2 + y = 3x^5 - 7x$, you don't have an easy way to get y in terms of x, so you can't differentiate this equation using any of the techniques you've learned so far. That's because each of those previous techniques needs to be used on an equation in which y is in terms of x. When you can't isolate y in terms of x (or if isolating y makes taking the derivative a nightmare), it's time to take the derivative **implicitly**.

Implicit differentiation is one of the simpler techniques you need to learn to do in calculus, but for some reason it gives many students trouble. Suppose you have the equation $y^2 = 3x^5 - 7x$. This means that the value of y is a function of the value of x. When we take the derivative, $\dfrac{dy}{dx}$, we're looking at the rate at which y changes as x changes. Thus, given $y = x^2 + x$, when we write

$$\frac{dy}{dx} = 2x + 1$$

we're saying that "the rate" at which y changes, with respect to how x changes, is $2x + 1$.

Now, suppose you want to find $\dfrac{dx}{dy}$. As you might imagine,

$$\frac{dx}{dy} = \frac{1}{\frac{dy}{dx}}$$

So here, $\dfrac{dx}{dy} = \dfrac{1}{2x+1}$. But notice that this derivative is in terms of x, not y, and you need to find the derivative with respect to y. This derivative is an **implicit** one. When you can't isolate the variables of an equation, you often end up with a derivative that is in terms of both variables.

Another way to think of this is that there is a hidden term in the derivative, $\dfrac{dx}{dx}$, and when we take the derivative, what we really get is

$$\frac{dy}{dx} = 2x\left(\frac{dx}{dx}\right) + 1\left(\frac{dx}{dx}\right)$$

A fraction that has the same term in its numerator and denominator is equal to 1, so we write

$$\frac{dy}{dx} = 2x(1) + 1(1) = 2x + 1$$

Every time we take a derivative of a term with x in it, we multiply by the term $\frac{dx}{dx}$, but because this is 1, we ignore it. Suppose however, that we wanted to find out how y changes with respect to t (for time). Then we would have

$$\frac{dy}{dt} = 2x\left(\frac{dx}{dt}\right) + 1\left(\frac{dx}{dt}\right)$$

If we wanted to find out how y changes with respect to r, we would have

$$\frac{dy}{dr} = 2x\left(\frac{dx}{dr}\right) + 1\left(\frac{dx}{dr}\right)$$

and if we wanted to find out how y changes with respect to y, we would have

$$\frac{dy}{dy} = 2x\left(\frac{dx}{dy}\right) + 1\left(\frac{dx}{dy}\right) \text{ or } 1 = 2x\frac{dx}{dy} + \frac{dx}{dy}$$

This is how we really do differentiation. Remember the following:

$$\frac{dx}{dy} = \frac{1}{\frac{dy}{dx}}$$

When you have an equation of x in terms of y, and you want to find the derivative with respect to y, simply differentiate. But if the equation is of y in terms of x, find $\frac{dy}{dx}$ and take its reciprocal to find $\frac{dx}{dy}$. Go back to our original example.

$$y^2 + y = 3x^5 - 7x$$

To take the derivative according to the information in the last paragraph, you get

$$2y\left(\frac{dy}{dx}\right) + 1\left(\frac{dy}{dx}\right) = 15x^4\left(\frac{dx}{dx}\right) - 7\left(\frac{dx}{dx}\right)$$

Notice how each variable is multiplied by its appropriate $\frac{d}{dx}$. Now, remembering that $\frac{dx}{dx} = 1$, rewrite the expression this way: $2y\left(\frac{dy}{dx}\right) + 1\left(\frac{dy}{dx}\right) = 15x^4 - 7$.

You should use implicit differentiation any time you can't write a function explicitly in terms of the variable that we want to take the derivative with respect to.

Next, factor $\dfrac{dy}{dx}$ out of the left-hand side: $\dfrac{dy}{dx}(2y+1)=15x^4-7$.

Isolating $\dfrac{dy}{dx}$ gives you $\dfrac{dy}{dx}=\dfrac{15x^4-7}{(2y+1)}$.

This is the derivative you're looking for. Notice how the derivative is defined in terms of y **and** x. Up until now, $\dfrac{dy}{dx}$ has been strictly in terms of x. This is why the differentiation is "implicit."

Confused? Let's do a few examples and you will get the hang of it.

Example 1: Find $\dfrac{dy}{dx}$ if $y^3 - 4y^2 = x^5 + 3x^4$.

Using implicit differentiation, you get

$$3y^2\left(\frac{dy}{dx}\right)-8y\left(\frac{dy}{dx}\right)=5x^4\left(\frac{dx}{dx}\right)+12x^3\left(\frac{dx}{dx}\right)$$

Remember that $\dfrac{dx}{dx}=1$: $\dfrac{dy}{dx}(3y^2-8y)=5x^4+12x^3$.

After you factor out $\dfrac{dy}{dx}$, divide both sides by $3y^2-8y$.

$$\frac{dy}{dx}=\frac{5x^4+12x^3}{(3y^2-8y)}$$

Note: Now that you understand that the derivative of an x term with respect to x will always be multiplied by $\dfrac{dx}{dx}$, and that $\dfrac{dx}{dx}=1$, we won't write $\dfrac{dx}{dx}$ anymore. You should understand that the term is implied.

Example 2: Find $\dfrac{dy}{dx}$ if $\sin y^2 - \cos x^2 = \cos y^2 + \sin x^2$.

Use implicit differentiation.

$$\cos y^2\left(2y\frac{dy}{dx}\right)+\sin x^2\left(2x\right)=-\sin y^2\left(2y\frac{dy}{dx}\right)+\cos x^2\left(2x\right)$$

Then simplify.

$$2y\cos y^2\left(\frac{dy}{dx}\right)+2x\sin x^2=-2y\sin y^2\left(\frac{dy}{dx}\right)+2x\cos x^2$$

Next, put all of the terms containing $\dfrac{dy}{dx}$ on the left and all of the other terms on the right.

$$2y\cos y^2\left(\dfrac{dy}{dx}\right)+2y\sin y^2\left(\dfrac{dy}{dx}\right)=-2x\sin x^2+2x\cos x^2$$

Next, factor out $\dfrac{dy}{dx}$.

$$\dfrac{dy}{dx}\left(2y\cos y^2+2y\sin y^2\right)=-2x\sin x^2+2x\cos x^2$$

And isolate $\dfrac{dy}{dx}$.

$$\dfrac{dy}{dx}=\dfrac{-2x\sin x^2+2x\cos x^2}{\left(2y\cos y^2+2y\sin y^2\right)}$$

This can be simplified further to the following:

$$\dfrac{dy}{dx}=\dfrac{-x\left(\sin x^2-\cos x^2\right)}{y\left(\cos y^2+\sin y^2\right)}$$

Example 3: Find $\dfrac{dy}{dx}$ if $3x^2+5xy^2-4y^3=8$.

Implicit differentiation should result in the following:

$$6x+\left[5x\left(2y\dfrac{dy}{dx}\right)+(5)y^2\right]-12y^2\left(\dfrac{dy}{dx}\right)=0$$

> Did you notice the use of the Product Rule to find the derivative of $5xy^2$? The AP Exam loves to make you do this. All of the same differentiation rules that you've learned up until now still apply. We're just adding another technique.

You can simplify this to

$$6x+10xy\dfrac{dy}{dx}+5y^2-12y^2\dfrac{dy}{dx}=0$$

Next, put all of the terms containing $\dfrac{dy}{dx}$ on the left and all of the other terms on the right.

$$10xy\dfrac{dy}{dx}-12y^2\dfrac{dy}{dx}=-6x-5y^2$$

Next, factor out $\dfrac{dy}{dx}$.

$$\left(10xy-12y^2\right)\dfrac{dy}{dx}=-6x-5y^2$$

Then, isolate $\dfrac{dy}{dx}$.

$$\frac{dy}{dx} = \frac{-6x - 5y^2}{\left(10xy - 12y^2\right)}$$

Example 4: Find the derivative of $3x^2 - 4y^2 + y = 9$ at $(2, 1)$.

You need to use implicit differentiation to find $\dfrac{dy}{dx}$.

$$6x - 8y\left(\frac{dy}{dx}\right) + \left(\frac{dy}{dx}\right) = 0$$

Now, instead of rearranging to isolate $\dfrac{dy}{dx}$, plug in $(2, 1)$ immediately and solve for the derivative.

$$6(2) - 8(1)\left(\frac{dy}{dx}\right) + \left(\frac{dy}{dx}\right) = 0$$

Be smart about your problem solving. Just because you can simplify something doesn't mean that you should. In a case like this, plugging into this form of the derivative is more effective.

Simplify: $12 - 7\left(\dfrac{dy}{dx}\right) = 0$, so $\dfrac{dy}{dx} = \dfrac{12}{7}$.

Getting the hang of implicit differentiation yet? We hope so, because these next examples are slightly harder.

Example 5: Find the derivative of $\dfrac{2x - 5y^2}{4y^3 - x^2} = -x$ at $(1, 1)$.

First, cross-multiply.

$$2x - 5y^2 = -x\left(4y^3 - x^2\right)$$

Distribute.

$$2x - 5y^2 = -4xy^3 + x^3$$

Take the derivative.

$$2 - 10y\frac{dy}{dx} = -4x\left(3y^2\frac{dy}{dx}\right) - 4y^3 + 3x^2$$

Do not simplify now. Rather, plug in $(1, 1)$ right away. This will save you from the algebra.

$$2 - 10(1)\frac{dy}{dx} = -4(1)\left(3(1)^2\frac{dy}{dx}\right) - 4(1)^3 + 3(1)^2$$

Now solve for $\dfrac{dy}{dx}$.

$$2 - 10\frac{dy}{dx} = -12\frac{dy}{dx} - 1$$

$$2\frac{dy}{dx} = -3$$

$$\frac{dy}{dx} = \frac{-3}{2}$$

SECOND DERIVATIVES

Sometimes, you'll be asked to find a second derivative implicitly.

Example 6: Find $\dfrac{d^2 y}{dx^2}$ if $y^2 + 2y = 4x^2 + 2x$.

Differentiating implicitly, you get

$$2y\frac{dy}{dx} + 2\frac{dy}{dx} = 8x + 2$$

> Remember: When it is required to take a second derivative, the first derivative should be simplified first.

Next, simplify and solve for $\dfrac{dy}{dx}$.

$$\frac{dy}{dx} = \frac{4x+1}{y+1}$$

Now, it's time to take the derivative again.

$$\frac{d^2 y}{dx^2} = \frac{4(y+1) - (4x+1)\left(\dfrac{dy}{dx}\right)}{(y+1)^2}$$

Finally, substitute for $\dfrac{dy}{dx}$.

$$\frac{4(y+1) - (4x+1)\left(\dfrac{4x+1}{y+1}\right)}{(y+1)^2} = \frac{4(y+1)^2 - (4x+1)^2}{(y+1)^3}$$

Try these solved problems without looking at the answers. Then check your work.

PROBLEM 1. Find $\dfrac{dy}{dx}$ if $x^2 + y^2 = 6xy$.

Answer: Differentiate with respect to x.

$$2x + 2y\frac{dy}{dx} = 6x\frac{dy}{dx} + 6y$$

Group all of the $\dfrac{dy}{dx}$ terms on the left and the other terms on the right.

$$2y\frac{dy}{dx} - 6x\frac{dy}{dx} = 6y - 2x$$

Now factor out $\dfrac{dy}{dx}$.

$$\frac{dy}{dx}(2y - 6x) = 6y - 2x$$

Therefore, the first derivative is the following:

$$\frac{dy}{dx} = \frac{6y - 2x}{2y - 6x} = \frac{3y - x}{y - 3x}$$

PROBLEM 2. Find $\dfrac{dy}{dx}$ if $x - \cos y = xy$.

Answer: Differentiate with respect to x.

$$1 + \sin y\frac{dy}{dx} = x\frac{dy}{dx} + y$$

Grouping the terms, you get

$$\sin y\frac{dy}{dx} - x\frac{dy}{dx} = y - 1$$

Now factor out $\dfrac{dy}{dx}$.

$$\frac{dy}{dx}(\sin y - x) = y - 1$$

The derivative is

$$\frac{dy}{dx} = \frac{y - 1}{\sin y - x}$$

PROBLEM 3. Find the derivative of each variable with respect to t of $x^2 + y^2 = z^2$.

Answer: $2x\dfrac{dx}{dt} + 2y\dfrac{dy}{dt} = 2z\dfrac{dz}{dt}$

PROBLEM 4. Find the derivative of each variable with respect to t of $V = \dfrac{1}{3}\pi r^2 h$.

Answer: $\dfrac{dV}{dt} = \dfrac{1}{3}\pi\left(r^2\dfrac{dh}{dt} + 2r\dfrac{dr}{dt}h \right)$

PROBLEM 5. Find $\dfrac{d^2 y}{dx^2}$ if $y^2 = x^2 - 2x$.

Answer: First, take the derivative with respect to x.

$$2y\frac{dy}{dx} = 2x - 2$$

Then, solve for $\dfrac{dy}{dx}$.

$$\frac{dy}{dx} = \frac{2x-2}{2y} = \frac{x-1}{y}$$

The second derivative with respect to x becomes

$$\frac{d^2 y}{dx^2} = \frac{y(1) - (x-1)\dfrac{dy}{dx}}{y^2}$$

Now substitute for $\dfrac{dy}{dx}$ and simplify.

$$\frac{d^2 y}{dx^2} = \frac{y - (x-1)\left(\dfrac{x-1}{y}\right)}{y^2} = \frac{y^2 - (x-1)^2}{y^3}$$

PRACTICE PROBLEM SET 7

Use implicit differentiation to find the following derivatives. The answers are in Chapter 23, starting on page 475.

1. Find $\dfrac{dy}{dx}$ if $x^3 - y^3 = y$.

2. Find $\dfrac{dy}{dx}$ if $x^2 - 16xy + y^2 = 1$.

3. Find $\dfrac{dy}{dx}$ at $(2, 1)$ if $\dfrac{x+y}{x-y} = 3$.

4. Find $\dfrac{dy}{dx}$ if $16x^2 - 16xy + y^2 = 1$ at $(1, 1)$.

5. Find $\dfrac{dy}{dx}$ if $x \sin y + y \sin x = \dfrac{\pi}{2\sqrt{2}}$ at $\left(\dfrac{\pi}{4}, \dfrac{\pi}{4}\right)$.

6. Find $\dfrac{d^2 y}{dx^2}$ if $x^2 + 4y^2 = 1$.

7. Find $\dfrac{d^2 y}{dx^2}$ if $\sin x + 1 = \cos y$.

8. Find $\dfrac{d^2 y}{dx^2}$ if $x^2 - 4x = 2y - 2$.

End of Chapter 7 Drill

The answers are in Chapter 24.

1. Find $\dfrac{dy}{dx}$ if $x^2 + 2y^3 = x^3 - 4y^2$.

    (A) $\dfrac{dy}{dx} = \dfrac{3x^2 + 2x}{6y^2 - 8y}$

    (B) $\dfrac{dy}{dx} = \dfrac{3x^2 - 2x}{6y^2 + 8y}$

    (C) $\dfrac{dy}{dx} = \dfrac{6y^2 + 8y}{3x^2 - 2x}$

    (D) $\dfrac{dy}{dx} = \dfrac{6y^2 - 8y}{3x^2 + 2x}$

2. Find $\dfrac{dy}{dx}$ if $x^2 + 3xy^2 - y^3 = 1$.

    (A) $\dfrac{dy}{dx} = \dfrac{-2x - 3y^2}{6xy - 3y^2}$

    (B) $\dfrac{dy}{dx} = \dfrac{2x + 3y^2}{6xy - 3y^2}$

    (C) $\dfrac{dy}{dx} = 2x + 6xy - 3y^2$

    (D) $\dfrac{dy}{dx} = 2x - 6xy + 3y^2$

3. Find $\dfrac{dy}{dx}$ if $4\sin(x^2) = 3y - y^2$ at $\left(\sqrt{\dfrac{\pi}{6}}, 2\right)$.

    (A) $-4\sqrt{\dfrac{\pi}{6}}$

    (B) $4\sqrt{\dfrac{\pi}{6}}$

    (C) $-4\sqrt{\dfrac{\pi}{2}}$

    (D) $4\sqrt{\dfrac{\pi}{2}}$

4. Find $\dfrac{dy}{dx}$ if $xy^2 - 4x^3y^2 = -3$ at $(1,1)$.

    (A) $-\dfrac{11}{6}$

    (B) $-\dfrac{6}{11}$

    (C) $\dfrac{10}{13}$

    (D) $\dfrac{13}{10}$

5. Find $\dfrac{d^2y}{dx^2}$ if $y = y^3 - x^2$.

    (A) $\dfrac{\left(3y^2 - 1\right)^2 (2) - \left(24x^2y\right)}{\left(3y^2 - 1\right)^3}$

    (B) $\dfrac{\left(3y^2 - 1\right)(2) - \left(24x^2y\right)}{\left(3y^2 - 1\right)^2}$

    (C) $\dfrac{\left(3y^2 - 1\right)(2) - \left(12x^2\right)}{\left(3y^2 - 1\right)^3}$

    (D) $\dfrac{\left(3y^2 - 1\right)^2 (2) - \left(12x^2\right)}{\left(3y^2 - 1\right)^3}$

REFLECT

Respond to the following questions:

- For which topics discussed in this chapter do you feel you have achieved sufficient mastery to answer multiple-choice questions correctly?

- For which topics discussed in this chapter do you feel you have achieved sufficient mastery to answer open-ended questions correctly?

- For which topics discussed in this chapter do you feel you need more work before you can answer multiple-choice questions correctly?

- For which topics discussed in this chapter do you feel you need more work before you can answer open-ended questions correctly?

- What parts of this chapter are you going to re-review?

Unit 2
Differential
Calculus
Applications

Chapter 8
Basic Applications
of the Derivative

EQUATIONS OF TANGENT LINES AND NORMAL LINES

Finding the equation of a line tangent to a certain curve at a certain point is a standard calculus problem. This is because, among other things, the derivative is the slope of a tangent line to a curve at a particular point. Thus, we can find the equation of the tangent line to a curve if we have the equation of the curve and the point at which we want to find the tangent line. Then all we have to do is take the derivative of the equation, plug in the x-coordinate of the point to find the slope, then use the point and the slope to find the equation of the line. Let's take this one step at a time.

Suppose we have a point (x_1, y_1) and a slope m. Then the equation of the line through that point with that slope is

$$(y - y_1) = m(x - x_1)$$

You should remember this formula from algebra. If not, memorize it!

Next, suppose that we have an equation $y = f(x)$, where (x_1, y_1) satisfies that equation. Then $f'(x_1) = m$, and we can plug all of our values into the equation for a line and get the equation of the tangent line. This is much easier to explain with a simple example.

Example 1: Find the equation of the tangent line to the curve $y = 5x^2$ at the point $(3, 45)$.

First of all, notice that the point satisfies the equation: when $x = 3$, $y = 45$. Now, take the derivative of the equation.

$$\frac{dy}{dx} = 10x$$

By the way, the notation for plugging in a point is $\big|_{x=}$. Learn to recognize it!

Now, if you plug in $x = 3$, you'll get the slope of the curve at that point.

$$\frac{dy}{dx}\bigg|_{x=3} = 10(3) = 30$$

Thus, we have the slope and the point, and the equation is
$$(y - 45) = 30(x - 3)$$

It's customary to simplify the equation if it's not too onerous.

$$y = 30x - 45$$

Example 2: Find the equation of the tangent line to $y = x^3 + x^2$ at $(3, 36)$.

The derivative looks like the following:

$$\frac{dy}{dx} = 3x^2 + 2x$$

So, the slope is

$$\left.\frac{dy}{dx}\right|_{x=3} = 3(3)^2 + 2(3) = 33$$

The equation looks like the following:

$$(y - 36) = 33(x - 3), \text{ or } y = 33x - 63$$

Naturally, there are a couple of things that can be done to make the problems harder. First of all, you can be given only the x-coordinate. Second, the equation can be more difficult to differentiate.

In order to find the y-coordinate, all you have to do is plug the x-value into the equation for the curve and solve for y. Remember this: You'll see it again!

Example 3: Find the equation of the tangent line to $y = \dfrac{2x + 5}{x^2 - 3}$ at $x = 1$.

First, find the y-coordinate.

$$y(1) = \frac{2(1) + 5}{1^2 - 3} = \frac{7}{2}$$

Second, take the derivative.

$$\frac{dy}{dx} = \frac{(x^2 - 3)(2) - (2x + 5)(2x)}{(x^2 - 3)^2}$$

You're probably dreading having to simplify this derivative. Don't waste your time! Plug in $x = 1$ right away.

$$\left.\frac{dy}{dx}\right|_{x=1} = \frac{\left(1^2 - 3\right)(2) - \left(2(1) + 5\right)\left(2(1)\right)}{\left(1^2 - 3\right)^2} = \frac{-4 - 14}{4} = -\frac{9}{2}$$

Now, we have a slope and a point, so the equation is

$$y + \frac{7}{2} = -\frac{9}{2}(x - 1), \text{ or } 2y = -9x + 2$$

Sometimes, instead of finding the equation of a tangent line, you will be asked to find the equation of a normal line. A **normal** line is simply the line perpendicular to the tangent line at the same point. You follow the same steps as with the tangent line, but you use the slope that will give you a perpendicular line.

> Remember: the slope of a perpendicular line is just the negative reciprocal of the slope of the tangent line.

Example 4: Find the equation of the line normal to $y = x^5 - x^4 + 1$ at $x = 2$.

First, find the *y*-coordinate.

$$y(2) = 2^5 - 2^4 + 1 = 17$$

Second, take the derivative.

$$\frac{dy}{dx} = 5x^4 - 4x^3$$

Third, find the slope at $x = 2$.

$$\left.\frac{dy}{dx}\right|_{x=2} = 5(2)^4 - 4(2)^3 = 48$$

Fourth, take the negative reciprocal of 48, which is $-\dfrac{1}{48}$.

Finally, the equation becomes

$$y - 17 = -\frac{1}{48}(x - 2)$$

Try these solved problems. Do each problem, covering the answer first, then checking your answer.

PROBLEM 1. Find the equation of the tangent line to the graph of $y = 4 - 3x - x^2$ at the point $(2, -6)$.

Answer: First, take the derivative of the equation.

$$\frac{dy}{dx} = -3 - 2x$$

Now, plug in $x = 2$ to get the slope of the tangent line.

$$\frac{dy}{dx} = -3 - 2(2) = -7$$

Third, plug the slope and the point into the equation for the line.

$$y - (-6) = -7(x - 2)$$

This simplifies to $y = -7x + 8$.

PROBLEM 2. Find the equation of the normal line to the graph of $y = 6 - x - x^2$ at $x = -1$.

Answer: Plug $x = -1$ into the original equation to get the *y*-coordinate.

$$y = 6 + 1 - 1 = 6$$

Once again, take that derivative.

$$\frac{dy}{dx} = -1 - 2x$$

Now plug in $x = -1$ to get the slope of the tangent.

$$\frac{dy}{dx} = -1 - 2(-1) = 1$$

Use the negative reciprocal of the slope in the second step to get the slope of the normal line.

$$m = -1$$

Finally, plug the slope and the point into the equation for the line.

$$y - 6 = -1(x + 1)$$

This simplifies to $y = -x + 5$.

PROBLEM 3. Find the equations of the tangent and normal lines to the graph of $y = \dfrac{10x}{x^2 + 1}$ at the point $(2, 4)$.

Answer: This problem will put your algebra to the test. You have to use the Quotient Rule to take the derivative of this mess.

$$\frac{dy}{dx} = \frac{\left(x^2 + 1\right)(10) - (10x)(2x)}{\left(x^2 + 1\right)^2}$$

Second, plug in $x = 2$ to get the slope of the tangent.

$$\frac{dy}{dx} = \frac{(5)(10) - (20)(4)}{5^2} = -\frac{30}{25} = -\frac{6}{5}$$

Now, plug the slope and the point into the equation for the tangent line.

$$y - 4 = -\frac{6}{5}(x - 2)$$

That simplifies to $6x + 5y = 32$. The equation of the normal line must then be

$$y - 4 = \frac{5}{6}(x - 2)$$

That, in turn, simplifies to $-5x + 6y = 14$.

PROBLEM 4. The curve $y = ax^2 + bx + c$ passes through the point (2, 4) and is tangent to the line $y = x + 1$ at (0, 1). Find a, b, and c.

Answer: The curve passes through (2, 4), so if you plug in $x = 2$, you'll get $y = 4$. Therefore,

$$4 = 4a + 2b + c$$

Second, the curve also passes through the point (0, 1), so $c = 1$.

Because the curve is tangent to the line $y = x + 1$ at (0, 1), they must both have the same slope at that point. The slope of the line is 1. The slope of the curve is the first derivative.

$$\frac{dy}{dx} = 2ax + b$$

$$\left.\frac{dy}{dx}\right|_{x=0} = 2a(0) + b = b$$

At (0, 1), $\frac{dy}{dx} = b$. Therefore, $b = 1$.

Now that you know b and c, plug them back into the equation from the first step and solve for a.

$$4 = 4a + 2 + 1, \text{ and } a = \frac{1}{4}$$

PROBLEM 5. Find the points on the curve $y = 2x^3 - 3x^2 - 12x + 20$ where the tangent is parallel to the x-axis.

Answer: The x-axis is a horizontal line, so it has slope zero. Therefore, you want to know where the derivative of this curve is zero. Take the derivative.

$$\frac{dy}{dx} = 6x^2 - 6x - 12$$

Set it equal to zero and solve for x. Get accustomed to doing this: It's one of the most common questions in differential calculus.

$$\frac{dy}{dx} = 6x^2 - 6x - 12 = 0$$

$$6(x^2 - x - 2) = 0$$

$$6(x - 2)(x + 1) = 0$$

$$x = 2 \text{ or } x = -1$$

Third, find the y-coordinates of these two points.

$$y = 2(8) - 3(4) - 12(2) + 20 = 0$$

$$y = 2(-1) - 3(1) - 12(-1) + 20 = 27$$

Therefore, the points are (2, 0) and (-1, 27).

PRACTICE PROBLEM SET 8

Now try these problems. The answers are in Chapter 23, starting on page 479.

1. Find the equation of the tangent to the graph of $y = 3x^2 - x$ at $x = 1$.

2. Find the equation of the tangent to the graph of $y = x^3 - 3x$ at $x = 3$.

3. Find the equation of the tangent to the graph of $y = \dfrac{1}{\sqrt{x^2 + 7}}$ at $x = 3$.

4. Find the equation of the normal to the graph of $y = \dfrac{x + 3}{x - 3}$ at $x = 4$.

5. Find the equation of the tangent to the graph of $y = 2x^3 - 3x^2 - 12x + 20$ at $x = 2$.

6. Find the equation of the tangent to the graph of $y = \dfrac{x^2 + 4}{x - 6}$ at $x = 5$.

7. Find the equation of the tangent to the graph of $y = \sqrt{x^3} - 15$ at $(4, 7)$.

8. Find the values of x where the tangent to the graph of $y = 2x^3 - 8x$ has a slope equal to the slope of $y - x$.

9. Find the equation of the normal to the graph of $y = \dfrac{3x + 5}{x - 1}$ at $x = 3$.

10. Find the values of x where the normal to the graph of $(x - 9)^2$ is parallel to the y-axis.

11. Find the coordinates where the tangent to the graph of $y = 8 - 3x - x^2$ is parallel to the x-axis.

12. Find the values of a, b, and c where the curves $y = x^2 + ax + b$ and $y = cx + x^2$ have a common tangent line at $(-1, 0)$.

THE MEAN VALUE THEOREM FOR DERIVATIVES

Remember that in order for The Mean Value Theorem for Derivatives to work, the curve must be continuous on the interval <u>and</u> at the endpoints.

If $y = f(x)$ is continuous on the interval $[a, b]$, and is differentiable everywhere on the interval (a, b), then there is at least one number c between a and b such that

$$f'(c) = \frac{f(b) - f(a)}{b - a}$$

In other words, there's some point in the interval where the slope of the tangent line equals the slope of the secant line that connects the endpoints of the interval. (The function has to be continuous at the endpoints of the interval, but it doesn't have to be differentiable at the endpoints. Is this important? Maybe to mathematicians, but probably not to you!) You can see this graphically in the following figure:

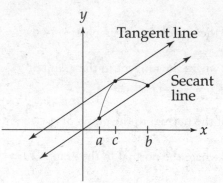

Example 1: Suppose you have the function $f(x) = x^2$, and you're looking at the interval $[1, 3]$. The Mean Value Theorem for derivatives (this is often abbreviated MVTD) states that there is some number c such that

$$f'(c) = \frac{3^2 - 1^2}{3 - 1} = 4$$

Because $f'(x) = 2x$, plug in c for x and solve: $2c = 4$ so $c = 2$. Notice that 2 is in the interval. This is what the MVTD predicted! If you don't get a value for c within the interval, something went wrong; either the function is not continuous and differentiable in the required interval, or you made a mistake.

Example 2: Consider the function $f(x) = x^3 - 12x$ on the interval $[-2, 2]$. The MVTD states that there is a c such that

$$f'(c) = \frac{\left(2^3 - 24\right) - \left((-2)^3 + 24\right)}{2 - (-2)} = -8$$

Then $f'(c) = 3c^2 - 12 = -8$ and $c = \pm\dfrac{2}{\sqrt{3}}$ (which is approximately ±1.155).

Notice that here there are two values of c that satisfy the MVTD. That's allowed. In fact, there can be infinitely many values, depending on the function.

Example 3: Consider the function $f(x) = \dfrac{1}{x}$ on the interval [–2, 2].

Follow the MVTD.

$$f'(c) = \frac{\dfrac{1}{2} - \left(-\dfrac{1}{2}\right)}{2 - (-2)} = \frac{1}{4}$$

Then

$$f'(c) = \frac{-1}{c^2} = \frac{1}{4}$$

There is no value of c that will satisfy this equation! We expected this. Why? Because $f(x)$ is not continuous at $x = 0$, which is in the interval. Suppose the interval had been [1, 3], eliminating the discontinuity. The result would have been

$$f'(c) = \frac{\dfrac{1}{3} - (1)}{3 - 1} = -\frac{1}{3} \text{ and } f'(c) = \frac{-1}{c^2} = -\frac{1}{3}; \; c = \pm\sqrt{3}$$

$c = -\sqrt{3}$ is not in the interval, but $c = \sqrt{3}$ is. The answer is $c = \sqrt{3}$.

Example 4: Consider the function $f(x) = x^2 - x - 12$ on the interval [–3, 4].

Follow the MVTD.

$$f'(c) = \frac{0 - 0}{7} = 0 \text{ and } f'(c) = 2c - 1 = 0 \text{ , so } c = \frac{1}{2}$$

In this last example, you discovered where the derivative of the equation equaled zero. This is going to be the single most common problem you'll encounter in differential calculus. So now, we've got an important tip for you.

> When you don't know what to do, take the derivative of the equation and set it equal to zero!

Remember this advice for the rest of AP Calculus.

ROLLE'S THEOREM

Now let's learn Rolle's Theorem, which is a special case of the MVTD.

> If $y = f(x)$ is continuous on the interval $[a, b]$, and is differentiable everywhere on the interval (a, b), and if $f(a) = f(b) = 0$, then there is at least one number c between a and b such that $f'(c) = 0$.

Graphically, this means that a continuous, differentiable curve has a horizontal tangent between any two points where it crosses the x-axis.

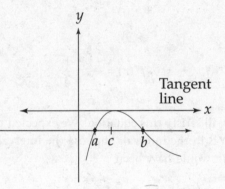

Example 4 was an example of Rolle's Theorem, but let's do another.

Example 5: Consider the function $f(x) = \dfrac{x^2}{2} - 6x$ on the interval $[0, 12]$.

First, show that

$$f(0) = \frac{0}{2} - 6(0) = 0 \ \text{ and } \ f(12) = \frac{144}{2} - 6(12) = 0$$

Then find

$$f'(x) = x - 6, \text{ so } f'(c) = c - 6$$

If you set this equal to zero (remember what we told you!), you get $c = 6$. This value of c falls in the interval, so the theorem holds for this example.

As with the MVTD, you'll run into problems with the theorem when the function is not continuous and differentiable over the interval. This is where you need to look out for a trap set by ETS. Otherwise, just follow what we did here and you won't have any trouble with either Rolle's Theorem or the MVTD. Try these example problems, and cover the responses until you check your work.

PROBLEM 1. Find the values of c that satisfy the MVTD for $f(x) = x^2 + 2x - 1$ on the interval $[0, 1]$.

Answer: First, find $f(0)$ and $f(1)$.

$$f(0) = 0^2 + 2(0) - 1 = -1 \text{ and } f(1) = 1^2 + 2(1) - 1 = 2$$

Then,

$$\frac{2 - (-1)}{1 - 0} = \frac{3}{1} = 3 = f'(c)$$

Next, find $f'(x)$.

$$f'(x) = 2x + 2$$

Thus, $f'(c) = 2c + 2 = 3$, and $c = \dfrac{1}{2}$.

PROBLEM 2. Find the values of c that satisfy the MVTD for $f(x) = x^3 + 1$ on the interval $[1, 2]$.

Answer: Find $f(1) = 1^3 + 1 = 2$ and $f(2) = 2^3 + 1 = 9$. Then,

$$\frac{9 - 2}{2 - 1} = 7 = f'(c)$$

Next, $f'(x) = 3x^2$, so $f'(c) = 3c^2 = 7$ and $c = \pm\sqrt{\dfrac{7}{3}}$.

Notice that there are two answers for c, but only one of them is in the interval. The answer is

$$c = \sqrt{\frac{7}{3}}.$$

PROBLEM 3. Find the values of c that satisfy the MVTD for $f(x) = x + \dfrac{1}{x}$ on the interval $[-4, 4]$.

Answer: First, because the function is not continuous on the interval, there may not be a solution for c. Let's show that this is true. Find $f(-4) = -4 - \dfrac{1}{4} = -\dfrac{17}{4}$ and $f(4) = 4 + \dfrac{1}{4} = \dfrac{17}{4}$.

Then,

$$\frac{\dfrac{17}{4} - \left(-\dfrac{17}{4}\right)}{4 - (-4)} = \frac{17}{16} = f'(c)$$

Next, $f'(x) = 1 - \dfrac{1}{x^2}$. Therefore, $f'(c) = 1 - \dfrac{1}{c^2} = \dfrac{17}{16}$.

There's no solution to this equation.

PROBLEM 4. Find the values of c that satisfy Rolle's Theorem for $f(x) = x^4 - x$ on the interval $[0, 1]$.

Answer: Show that $f(0) = 0^4 - 0 = 0$ and that $f(1) = 1^4 - 1 = 0$.

Next, find $f'(x) = 4x^3 - 1$. By setting $f'(c) = 4c^3 - 1 = 0$ and solving, you'll see that $c = \sqrt[3]{\dfrac{1}{4}}$, which is in the interval.

PRACTICE PROBLEM SET 9

Now try these problems. The answers are in Chapter 23, starting on page 484.

1. Find the values of c that satisfy the MVTD for $f(x) = 3x^2 + 5x - 2$ on the interval $[-1, 1]$.

2. Find the values of c that satisfy the MVTD for $f(x) = x^3 + 24x - 16$ on the interval $[0, 4]$.

3. Find the values of c that satisfy the MVTD for $f(x) = \dfrac{6}{x} - 3$ on the interval $[1, 2]$.

4. Find the values of c that satisfy the MVTD for $f(x) = \dfrac{6}{x} - 3$ on the interval $[-1, 2]$.

5. Find the values of c that satisfy Rolle's Theorem for $f(x) = x^2 - 8x + 12$ on the interval $[2, 6]$.

6. Find the values of c that satisfy Rolle's Theorem for $f(x) = x(1 - x)$ on the interval $[0, 1]$.

7. Find the values of c that satisfy Rolle's Theorem for $f(x) = 1 - \dfrac{1}{x^2}$ on the interval $[-1, 1]$.

8. Find the values of c that satisfy Rolle's Theorem for $f(x) = x^{\frac{2}{3}} - x^{\frac{1}{3}}$ on the interval $[0, 1]$.

End of Chapter 8 Drill

The answers are in Chapter 24.

1. Find the equation of the line tangent to $y = 5x^3 - 20x + 10$ at $x = 2$.

   (A) $y - 10 = 10(x - 2)$
   (B) $y + 10 = 40(x + 2)$
   (C) $y - 10 = 40(x - 2)$
   (D) $y - 10 = 90(x + 2)$

2. Find the equation of the line normal to $y = 4\sec(2x)$ at $x = \dfrac{\pi}{8}$.

   (A) $y + 4\sqrt{2} = 8\sqrt{2}\left(x - \dfrac{\pi}{8}\right)$

   (B) $y - 4\sqrt{2} = -8\sqrt{2}\left(x - \dfrac{\pi}{8}\right)$

   (C) $y + 4\sqrt{2} = -\dfrac{1}{8\sqrt{2}}\left(x + \dfrac{\pi}{8}\right)$

   (D) $y - 4\sqrt{2} = -\dfrac{1}{8\sqrt{2}}\left(x - \dfrac{\pi}{8}\right)$

3. Find the value of c that satisfies the MVTD for $f(x) = x^2 - 6x + 5$ on the interval $[3, 8]$.

   (A) 5
   (B) $\dfrac{11}{2}$
   (C) 11
   (D) There is no value of c.

4. Find the value of c that satisfies the MVTD for $f(x) = \dfrac{2}{x - 4}$ on the interval $[2, 6]$.

   (A) -2
   (B) 6
   (C) 8
   (D) There is no value of c.

5. Find all of the values of c that satisfy Rolle's Theorem for $f(x) = x^4 - 5x^2 + 4$ on the interval $[-3, 3]$.

   (A) $c = 0$

   (B) $c = \pm\sqrt{\dfrac{5}{2}}$

   (C) $c = 0, \pm 2$

   (D) $c = 0, \pm\sqrt{\dfrac{5}{2}}$

REFLECT

Respond to the following questions:

- For which topics discussed in this chapter do you feel you have achieved sufficient mastery to answer multiple-choice questions correctly?

- For which topics discussed in this chapter do you feel you have achieved sufficient mastery to answer open-ended questions correctly?

- For which topics discussed in this chapter do you feel you need more work before you can answer multiple-choice questions correctly?

- For which topics discussed in this chapter do you feel you need more work before you can answer open-ended questions correctly?

- What parts of this chapter are you going to re-review?

Chapter 9
Maxima, Minima, and Curve Stretching

Here's another chapter of material involving more ways to apply the derivative to several other types of problems. This stuff focuses mainly on using the derivative to aid in graphing a function, etc.

APPLIED MAXIMA AND MINIMA PROBLEMS

One of the most common applications of the derivative is to find a maximum or minimum value of a function. These values can be called extreme values, optimal values, or critical points. Each of these problems involves the same, very simple principle.

> A maximum or a minimum of a function occurs at a point where the derivative of a function is zero, or where the derivative fails to exist.

At a point where the first derivative equals zero, the curve has a horizontal tangent line, at which point it could be reaching either a "peak" (maximum) or a "valley" (minimum).

There are a few exceptions to every rule. This rule is no different.

If the derivative of a function is zero at a certain point, it is usually a maximum or minimum—but not always.

There are two different kinds of maxima and minima: relative and absolute. A **relative** or **local** maximum or minimum means that the curve has a horizontal tangent line at that point, but it is not the highest or lowest value that the function attains. In the figure to the right, the two indicated points are relative maxima/minima.

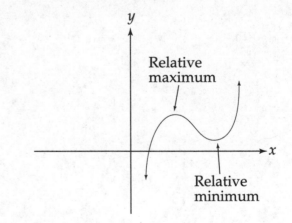

An **absolute** maximum or minimum occurs either at an artificial point or an end point. In the following figure, the two indicated points are absolute maxima/minima. A relative maximum can also be an absolute maximum.

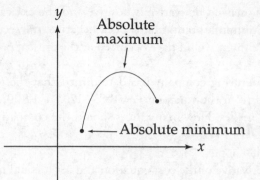

A typical word problem will ask you to find a maximum or a minimum value of a function, as it pertains to a certain situation. Sometimes you're given the equation; other times, you have to figure it out for yourself. Once you have the equation, you find its derivative and set it equal to zero. The values you get are called critical values. That is, if $f'(c) = 0$ or $f'(c)$ does not exist, then c is a critical value. Then, test these values to determine whether each value is a maximum or a minimum. The simplest way to do this is with the second derivative test.

> If a function has a critical value at $x = c$, then that value is a relative maximum if $f''(c) < 0$ and it is a relative minimum if $f''(c) > 0$.

If the second derivative is also zero at $x = c$, then the point is neither a maximum nor a minimum but a point of inflection. More about that later.

It's time to do some examples.

Example 1: Find the minimum value on the curve $y = ax^2$, if $a > 0$.

Take the derivative and set it equal to zero.

$$\frac{dy}{dx} = 2ax = 0$$

The first derivative is equal to zero at $x = 0$. By plugging 0 back into the original equation, we can solve for the y-coordinate of the minimum (the y-coordinate is also 0, so the point is at the origin).

In order to determine if this is a maximum or a minimum, take the second derivative.

$$\frac{d^2y}{dx^2} = 2a$$

Because a is positive, the second derivative is positive and the critical point we obtained from the first derivative is a minimum point. Had a been negative, the second derivative would have been negative and a maximum would have occurred at the critical point.

Example 2: A manufacturing company has determined that the total cost of producing an item can be determined from the equation $C = 8x^2 - 176x + 1,800$, where x is the number of units that the company makes. How many units should the company manufacture in order to minimize the cost?

Once again, take the derivative of the cost equation and set it equal to zero.

$$\frac{dC}{dx} = 16x - 176 = 0$$

$$x = 11$$

This tells us that 11 is a critical point of the equation. Now we need to figure out if this is a maximum or a minimum using the second derivative.

$$\frac{d^2C}{dx^2} = 16$$

Because 16 is always positive, any critical value is going to be a minimum. Therefore, the company should manufacture 11 units in order to minimize its cost.

Example 3: A rocket is fired into the air, and its height in meters at any given time t can be calculated using the formula $h(t) = 1,600 + 196t - 4.9t^2$. Find the maximum height of the rocket and the time at which it occurs.

Take the derivative and set it equal to zero.

$$\frac{dh}{dt} = 196 - 9.8t$$

$$t = 20$$

Now that we know 20 is a critical point of the equation, use the second derivative test.

$$\frac{d^2h}{dt^2} = -9.8$$

This is always negative, so any critical value is a maximum. To determine the maximum height of the rocket, plug $t = 20$ into the equation.

$$h(20) = 1,600 + 196(20) - 4.9(20^2) = 3,560 \text{ meters}$$

> The technique is always the same: (a) take the derivative of the equation; (b) set it equal to zero; and (c) use the second derivative test.

The hardest part of these word problems is when you have to set up the equation yourself. The following is a classic AP problem:

Example 4: Max wants to make a box with no lid from a rectangular sheet of cardboard that is 18 inches by 24 inches. The box is to be made by cutting a square of side x from each corner of the sheet and folding up the sides (see figure below). Find the value of x that maximizes the volume of the box.

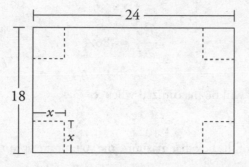

After we cut out the squares of side x and fold up the sides, the dimensions of the box will be:

width: $18 - 2x$

length: $24 - 2x$

depth: x

Using the formula for the volume of a rectangular prism, we can get an equation for the volume in terms of x.

$$V = x(18 - 2x)(24 - 2x)$$

Multiply the terms together (and be careful with your algebra).

$$V = x(18 - 2x)(24 - 2x) = 4x^3 - 84x^2 + 432x$$

Now take the derivative.

$$\frac{dV}{dx} = 12x^2 - 168x + 432$$

Set the derivative equal to zero, and solve for x.

$$12x^2 - 168x + 432 = 0$$

$$x^2 - 14x + 36 = 0$$

$$x = \frac{14 \pm \sqrt{196 - 144}}{2} = 7 \pm \sqrt{13} \approx 3.4, 10.6$$

Common sense tells us that you can't cut out two square pieces that measure 10.6 inches to a side (the sheet's only 18 inches wide!), so the maximizing value has to be 3.4 inches. Here's the second derivative test, just to be sure.

$$\frac{d^2V}{dx^2} = 24x - 168$$

At $x = 3.4$,

$$\frac{d^2V}{dx^2} = -86.4$$

So the volume of the box will be maximized when $x = 3.4$.

Therefore, the dimensions of the box that maximize the volume are approximately: 11.2 in. × 17.2 in. × 3.4 in.

Sometimes, particularly when the domain of a function is restricted, you have to test the endpoints of the interval as well. This is because the highest or lowest value of a function may be at an endpoint of that interval; the critical value you obtained from the derivative might be just a local maximum or minimum. For the purposes of the AP Exam, however, endpoints are considered separate from critical values.

Example 5: Find the absolute maximum and minimum values of $y = x^3 - x$ on the interval $[-3, 3]$.

Take the derivative and set it equal to zero.

$$\frac{dy}{dx} = 3x^2 - 1 = 0$$

Solve for x.

$$x = \pm \frac{1}{\sqrt{3}}$$

Test the critical points.

$$\frac{d^2y}{dx^2} = 6x$$

At $x = \frac{1}{\sqrt{3}}$, we have a minimum. At $x = -\frac{1}{\sqrt{3}}$, we have a maximum.

$$\text{At } x = -\frac{1}{\sqrt{3}}, y = -\frac{1}{3\sqrt{3}} + \frac{1}{\sqrt{3}} = \frac{2}{3\sqrt{3}} \approx 0.385$$

$$\text{At } x = \frac{1}{\sqrt{3}}, y = \frac{1}{3\sqrt{3}} - \frac{1}{\sqrt{3}} = -\frac{2}{3\sqrt{3}} \approx -0.385$$

Now it's time to check the endpoints of the interval.

At $x = -3$, $y = -24$

At $x = 3$, $y = 24$

We can see that the function actually has a *lower* value at $x = -3$ than at its "minimum" when

$x = \dfrac{1}{\sqrt{3}}$. Similarly, the function has a *higher* value at $x = 3$ than at its "maximum" of $x = -\dfrac{1}{\sqrt{3}}$.

This means that the function has a "local minimum" at $x = \dfrac{1}{\sqrt{3}}$, and an "absolute minimum"

when $x = -3$. And, the function has a "local maximum" at $x = -\dfrac{1}{\sqrt{3}}$, and an "absolute maxi-

mum" at $x = 3$.

Example 6: A rectangle is to be inscribed in a semicircle with radius 4, with one side on the semicircle's diameter. What is the largest area this rectangle can have?

Let's look at this on the coordinate axes. The equation for a circle of radius 4, centered at the origin, is $x^2 + y^2 = 16$; a semicircle has the equation $y = \sqrt{16 - x^2}$. Our rectangle can then be expressed as a function of x, where the height is $\sqrt{16 - x^2}$ and the base is $2x$. See the following figure:

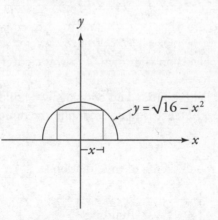

The area of the rectangle is: $A = 2x\sqrt{16 - x^2}$. Let's take the derivative of the area.

$$\frac{dA}{dx} = 2\sqrt{16 - x^2} - \frac{2x^2}{\sqrt{16 - x^2}}$$

The derivative is not defined at $x = \pm 4$. Setting the derivative equal to zero, we get

$$2\sqrt{16 - x^2} - \frac{2x^2}{\sqrt{16 - x^2}} = 0$$

$$2\sqrt{16 - x^2} = \frac{2x^2}{\sqrt{16 - x^2}}$$

$$2\left(16 - x^2\right) = 2x^2$$

$$32 - 2x^2 = 2x^2$$

$$32 = 4x^2$$

$$x = \pm\sqrt{8}$$

If you're wondering why we don't use the negative root, it's because there is no such thing as a negative area.

Note that the domain of this function is $-4 \le x \le 4$, so these numbers serve as endpoints of the interval. Let's compare the critical values and the endpoints.

When $x = -4$, $y = 0$ and the area is 0.

When $x = 4$, $y = 0$ and the area is 0.

When $x = \sqrt{8}$, $y = \sqrt{8}$ and the area is 16.

Thus, the maximum area occurs when $x = \sqrt{8}$ and the area equals 16.

Try some of these solved problems on your own. As always, cover the answers as you work.

PROBLEM 1. A rectangular field, bounded on one side by a building, is to be fenced in on the other three sides. If 3,000 feet of fence is to be used, find the dimensions of the largest field that can be fenced in.

Answer: First, let's make a rough sketch of the situation.

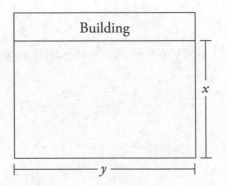

If we call the length of the field y and the width of the field x, the formula for the area of the field becomes

$$A = xy$$

The perimeter of the fencing is equal to the sum of two widths and the length.

$$2x + y = 3,000$$

Now solve this second equation for y:

$$y = 3,000 - 2x$$

When you plug this expression into the formula for the area, you get a formula for A in terms of x.

$$A = x(3,000 - 2x) = 3,000x - 2x^2$$

Next, take the derivative, set it equal to zero, and solve for x.

$$\frac{dA}{dx} = 3,000 - 4x = 0$$

$$x = 750$$

Let's check to make sure it's a maximum. Find the second derivative.

$$\frac{d^2 A}{dx^2} = -4$$

Because we have a negative result, $x = 750$ is a maximum. Finally, if we plug in $x = 750$ and solve for y, we find that $y = 1,500$. The largest field will measure 750 feet by 1,500 feet.

PROBLEM 2. A poster is to contain 100 square inches of picture surrounded by a 4-inch margin at the top and bottom and a 2-inch margin on each side. Find the overall dimensions that will minimize the total area of the poster.

Answer: First, make a sketch.

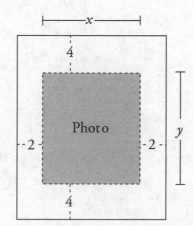

Let the area of the picture be $xy = 100$. The total area of the poster is $A = (x + 4)(y + 8)$. Then, expand the equation.

$$A = xy + 4y + 8x + 32$$

Substitute $xy = 100$ and $y = \dfrac{100}{x}$ into the area equation, and we get

$$A = 132 + \dfrac{400}{x} + 8x$$

Now take the derivative and set it equal to zero.

$$\dfrac{dA}{dx} = 8 - \dfrac{400}{x^2} = 0$$

Solving for x, we find that $x = \sqrt{50}$. Now solve for y by plugging $x = \sqrt{50}$ into the area equation: $y = 2\sqrt{50}$. Then check that these dimensions give us a minimum.

$$\dfrac{d^2 A}{dx^2} = \dfrac{800}{x^3}$$

This is positive when x is positive, so the minimum area occurs when $x = \sqrt{50}$. Thus, the overall dimensions of the poster are $4 + \sqrt{50}$ inches by $8 + 2\sqrt{50}$ inches.

Problem 3. An open-top box with a square bottom and rectangular sides is to have a volume of 256 cubic inches. Find the dimensions that require the minimum amount of material.

Answer: First, make a sketch of the situation.

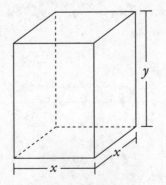

The amount of material necessary to make the box is equal to the surface area.
$$S = x^2 + 4xy$$

The formula for the volume of the box is $x^2 y = 256$.

If we solve the latter equation for y, $y = \dfrac{256}{x^2}$, and plug it into the former equation, we get

$$S = x^2 + 4x\frac{256}{x^2} = x^2 + \frac{1{,}024}{x}$$

Now take the derivative and set it equal to zero.

$$\frac{dS}{dx} = 2x - \frac{1{,}024}{x^2} = 0$$

If we solve this for x, we get $x^3 = 512$ and $x = 8$. Solving for y, we get $y = 4$.

Check that these dimensions give us a minimum.

$$\frac{d^2A}{dx^2} = 2 + \frac{2{,}048}{x^3}$$

This is positive when x is positive, so the minimum surface area occurs when $x = 8$. The dimensions of the box should be 8 inches by 8 inches by 4 inches.

PROBLEM 4. Find the point on the curve $y = \sqrt{x}$ that is a minimum distance from the point $(4, 0)$.

Answer: First, make that sketch.

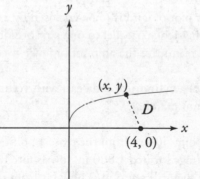

Using the distance formula, we get
$$D^2 = (x - 4)^2 + (y - 0)^2 = x^2 - 8x + 16 + y^2$$

Because $y = \sqrt{x}$,

$$D^2 = x^2 - 8x + 16 + x = x^2 - 7x + 16$$

Next, let $L = D^2$. We can do this because the minimum value of D^2 will occur at the same value of x as the minimum value of D. Therefore, it's simpler to minimize D^2 rather than D (because we won't have to take a square root!).

$$L = x^2 - 7x + 16$$

Now, take the derivative and set it equal to zero.

$$\frac{dL}{dx} = 2x - 7 = 0$$

$$x = \frac{7}{2}$$

Solving for y, we get $y = \sqrt{\frac{7}{2}}$.

Finally, because $\frac{d^2L}{dx^2} = 2$, the point $\left(\frac{7}{2}, \sqrt{\frac{7}{2}} \right)$ is the minimum distance from the point (4, 0).

PRACTICE PROBLEM SET 10

Now try these problems on your own. The answers are in Chapter 23, starting on page 487.

1. A rectangle has its base on the x-axis and its two upper corners on the parabola $y = 12 - x^2$. What is the largest possible area of the rectangle?

2. An open rectangular box is to be made from a 9 × 12 inch piece of tin by cutting squares of side x inches from the corners and folding up the sides. What should x be to maximize the volume of the box?

3. A 384-square-meter plot of land is to be enclosed by a fence and divided into two equal parts by another fence parallel to one pair of sides. What dimensions of the outer rectangle will minimize the amount of fence used?

4. What is the radius of a cylindrical soda can with volume of 512 cubic inches that will use the minimum material?

5. A swimmer is at a point 500 m from the closest point on a straight shoreline. She needs to reach a cottage located 1,800 m downshore from the closest point. If she swims at 4 m/s and she walks at 6 m/s, how far from the cottage should she come ashore so as to arrive at the cottage in the shortest time?

6. Find the closest point on the curve $x^2 + y^2 = 1$ to the point (2, 1).

7. A window consists of an open rectangle topped by a semicircle and is to have a perimeter of 288 inches. Find the radius of the semicircle that will maximize the area of the window.

8. The range of a projectile is $R = \dfrac{v_0^{\,2} \sin 2\theta}{g}$, where v_0 is its initial velocity, g is the acceleration due to gravity and is a constant, and θ is its firing angle. Find the angle that maximizes the projectile's range.

CURVE SKETCHING

Another topic on which students spend a lot of time in calculus is curve sketching. In the old days, whole courses (called "Analytic Geometry") were devoted to the subject, and students had to master a wide variety of techniques to learn how to sketch a curve accurately.

Fortunately (or unfortunately, depending on your point of view), students no longer need to be as good at analytic geometry. There are two reasons for this: (1) The AP Exam tests only a few types of curves; and (2) you can use a graphing calculator. Because of the calculator, you can get an idea of the shape of the curve, and all you need to do is find important points to label the graph. We use calculus to find some of these points.

When it's time to sketch a curve, we'll show you a four-part analysis that'll give you all the information you need.

Step 1: Test the Function

Find where $f(x) = 0$. This tells you the function's x-intercepts (or roots). By setting $x = 0$, we can determine the y-intercepts. Then, find any horizontal and/or vertical asymptotes.

Step 2: Test the First Derivative

Find where $f'(x) = 0$. This tells you the critical points. We can determine whether the curve is rising or falling, as well as where the maxima and minima are. It's also possible to determine if the curve has any points where it's nondifferentiable.

Step 3: Test the Second Derivative

Find where $f''(x) = 0$. This shows you where any points of inflection are. (These are points where the graph of a function changes concavity.) Then we can determine where the graph curves upward and where it curves downward.

Step 4: Test End Behavior

Look at what the general shape of the graph will be, based on the values of y for very large values of $\pm x$. Using this analysis, we can always come up with a sketch of a curve.

And now, the rules.

(1)   When $f'(x) > 0$, the curve is rising; when $f'(x) < 0$, the curve is falling; when $f'(x) = 0$, the curve is at a critical point.

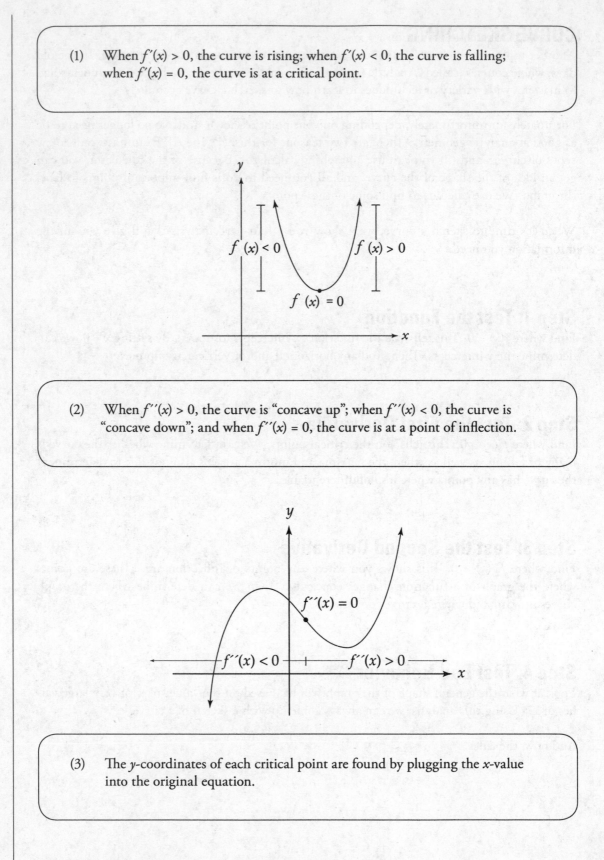

(2)   When $f''(x) > 0$, the curve is "concave up"; when $f''(x) < 0$, the curve is "concave down"; and when $f''(x) = 0$, the curve is at a point of inflection.

(3)   The y-coordinates of each critical point are found by plugging the x-value into the original equation.

As always, this stuff will sink in better if we try a few examples.

Example 1: Sketch the equation $y = x^3 - 12x$.

Step 1: Find the x-intercepts.

$$x^3 - 12x = 0$$

$$x(x^2 - 12) = 0$$

$$x\left(x - \sqrt{12}\right)\left(x + \sqrt{12}\right) = 0$$

$$x = 0, \pm\sqrt{12}$$

The curve has x-intercepts at $\left(\sqrt{12},\ 0\right), \left(-\sqrt{12},\ 0\right)$, and $(0, 0)$.

Next, find the y-intercepts.

$$y = (0)^3 - 12(0) = 0$$

The curve has a y-intercept at $(0, 0)$.

There are no asymptotes, because there's no place where the curve is undefined (you won't have asymptotes for curves that are polynomials).

Step 2: Take the derivative of the function to find the critical points.

$$\frac{dy}{dx} = 3x^2 - 12$$

Set the derivative equal to zero, and solve for x.

$$3x^2 - 12 = 0$$
$$3(x^2 - 4) = 0$$
$$3(x - 2)(x + 2) = 0$$

so $x = 2, -2$.

Next, plug $x = 2, -2$ into the original equation to find the y-coordinates of the critical points.
$$y = (2)^3 - 12(2) = -16$$
$$y = (-2)^3 - 12(-2) = 16$$

Thus, we have critical points at $(2, -16)$ and $(-2, 16)$.

Step 3: Now, take the second derivative to find any points of inflection.

$$\frac{d^2 y}{dx^2} = 6x$$

This equals zero at $x = 0$. We already know that when $x = 0$, $y = 0$, so the curve has a point of inflection at $(0, 0)$.

Now, plug the critical values into the second derivative to determine whether each is a maximum or a minimum. $f''(2) = 6(2) = 12$. This is positive, so the curve has a minimum at $(2, -16)$, and the curve is concave up at that point. $f''(-2) = 6(-2) = -12$. This value is negative, so the curve has a maximum at $(-2, 16)$ and the curve is concave down there.

Armed with this information, we can now plot the graph.

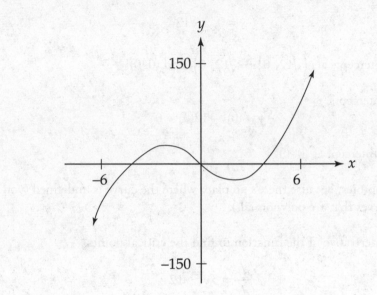

Example 2: Sketch the graph of $y = x^4 + 2x^3 - 2x^2 + 1$.

> **Step 1:** First, let's find the x-intercepts.
> $$x^4 + 2x^3 - 2x^2 + 1 = 0$$

The good news is that if the roots aren't easy to find, ETS won't ask you to find them, or you can find them with your calculator.

> If the equation doesn't factor easily, it's best not to bother to find the function's roots. Convenient, huh?
>
> Next, let's find the y-intercepts.
> $$y = (0)^4 + 2(0)^3 - 2(0)^2 + 1$$

The curve has a y-intercept at $(0, 1)$.

There are no vertical asymptotes because there is no place where the curve is undefined.

> **Step 2:** Now we take the derivative to find the critical points.
>
> $$\frac{dy}{dx} = 4x^3 + 6x^2 - 4x$$

Set the derivative equal to zero.

$$4x^3 + 6x^2 - 4x = 0$$
$$2x(2x^2 + 3x - 2) = 0$$
$$2x(2x - 1)(x + 2) = 0$$

$$x = 0, \frac{1}{2}, -2$$

Next, plug these three values into the original equation to find the y-coordinates of the critical points. We already know that when $x = 0$, $y = 1$.

When $x = \frac{1}{2}$, $y = \left(\frac{1}{2}\right)^4 + 2\left(\frac{1}{2}\right)^3 - 2\left(\frac{1}{2}\right)^2 + 1 = \frac{13}{16}$

When $x = -2$, $y = (-2)^4 + 2(-2)^3 - 2(-2)^2 + 1 = -7$

Thus, we have critical points at $(0, 1)$, $\left(\frac{1}{2}, \frac{13}{16}\right)$, and $(-2, -7)$.

Step 3: Take the second derivative to find any points of inflection.

$$\frac{d^2 y}{dx^2} = 12x^2 + 12x - 4$$

Set this equal to zero.

$$12x^2 + 12x - 4 = 0$$
$$3x^2 + 3x - 1 = 0$$
$$x = \frac{-3 \pm \sqrt{21}}{6} \approx 0.26, -1.26$$

Therefore, the curve has points of inflection at $x = \frac{-3 \pm \sqrt{21}}{6}$.

Now solve for the y-coordinates.

$$(0.26, 0.90) \text{ and } (-1.26, -3.66)$$

We can now plug the critical values into the second derivative to determine whether each is a maximum or a minimum.

$$12(0)^2 + 12(0) - 4 = -4$$

This is negative, so the curve has a maximum at $(0, 1)$; the curve is concave down there.

$$12\left(\frac{1}{2}\right)^2 + 12\left(\frac{1}{2}\right) - 4 = 5$$

This is positive, so the curve has a minimum at $\left(\dfrac{1}{2}, \dfrac{13}{16}\right)$; the curve is concave up there.

$$12(-2)^2 + 12(-2) - 4 = 20$$

This is positive, so the curve has a minimum at $(-2, -7)$ and the curve is also concave up there.

We can now plot the graph.

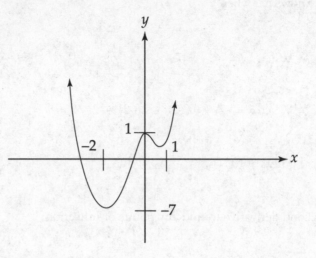

Finding a Cusp

If the derivative of a function approaches ∞ from one side of a point and $-\infty$ from the other, and if the function is continuous at that point, then the curve has a "cusp" at that point. In order to find a cusp, you need to look at points where the first derivative is undefined, as well as where it's zero.

Example 3: Sketch the graph of $y = 2 - x^{\frac{2}{3}}$.

Step 1: Find the x-intercepts.

$$2 - x^{\frac{2}{3}} = 0$$

$$x^{\frac{2}{3}} = 2 \qquad x = \pm 2^{\frac{3}{2}} = \pm 2\sqrt{2}$$

The x-intercepts are at $\left(\pm 2\sqrt{2},\ 0\right)$.

Next, find the y-intercepts.

$$y = 2 - (0)^{\frac{2}{3}} = 2$$

The curve has a y-intercept at $(0, 2)$.

There are no asymptotes because there is no place where the curve is undefined.

Step 2: Now, take the derivative to find the critical points.

$$\frac{dy}{dx} = -\frac{2}{3}x^{-\frac{1}{3}}$$

What's next? You guessed it! Set the derivative equal to zero.

$$-\frac{2}{3}x^{-\frac{1}{3}} = 0$$

There are no values of x for which the equation is zero. But here's the new stuff to deal with: at $x = 0$, the derivative is undefined. If we look at the limit as x approaches 0 from both sides, we can determine whether the graph has a cusp.

$$\lim_{x \to 0^+} -\frac{2}{3}x^{-\frac{1}{3}} = -\infty \quad \text{and} \quad \lim_{x \to 0^-} -\frac{2}{3}x^{-\frac{1}{3}} = \infty$$

Therefore, the curve has a cusp at (0, 2).

There aren't any other critical points. But we can see that when $x < 0$, the derivative is positive (which means that the curve is rising to the left of zero), and when $x > 0$ the derivative is negative (which means that the curve is falling to the right of zero).

Step 3: Now, we take the second derivative to find any points of inflection.

$$\frac{d^2 y}{dx^2} = \frac{2}{9}x^{-\frac{4}{3}}$$

Again, there's no x-value where this is zero. In fact, the second derivative is positive at all values of x except 0. Therefore, the graph is concave up everywhere.

Now it's time to graph this.

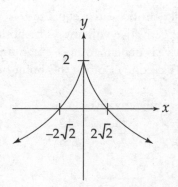

There's one other type of graph you should know about: a rational function. In order to graph a rational function, you need to know how to find that function's asymptotes.

How to Find Asymptotes

A line $y = c$ is a horizontal asymptote of the graph of $y = f(x)$ if

$$\lim_{x \to \infty} f(x) = c \text{ or if } \lim_{x \to -\infty} f(x) = c$$

A line $x = k$ is a vertical asymptote of the graph of $y = f(x)$ if

$$\lim_{x \to k^+} f(x) = \pm\infty \text{ or if } \lim_{x \to k^-} f(x) = \pm\infty$$

Example 4: Sketch the graph of $y = \dfrac{3x}{x+2}$.

Step 1: Find the x-intercepts. A fraction can be equal to zero only when its numerator is equal to zero (provided that the denominator is not also zero there). All we have to do is set $3x = 0$, and you get $x = 0$. Thus, the graph has an x-intercept at $(0, 0)$. Note: This is also the y-intercept.

Next, look for asymptotes. The denominator is undefined at $x = -2$, and if we take the left- and right-hand limits of the function, we see the following:

$$\lim_{x \to -2^+} \frac{3x}{x+2} = -\infty \text{ and } \lim_{x \to -2^-} \frac{3x}{x+2} = \infty$$

The curve has a vertical asymptote at $x = -2$.

If we take $\lim\limits_{x \to \infty} \dfrac{3x}{x+2} = 3$ and $\lim\limits_{x \to -\infty} \dfrac{3x}{x+2} = 3$, the curve has a horizontal asymptote at $y = 3$.

Step 2: Now, take the derivative to figure out the critical points.

$$\frac{dy}{dx} = \frac{(x+2)(3) - (3x)(1)}{(x+2)^2} = \frac{6}{(x+2)^2}$$

There are no values of x that make the derivative equal to zero. Because the numerator is 6 and the denominator is squared, the derivative will always be positive (the curve is always rising). You should note that the derivative is undefined at $x = -2$, but you already know that there's an asymptote at $x = -2$, so you don't need to examine this point further.

Step 3: Now, it's time for the second derivative.

$$\frac{d^2 y}{dx^2} = \frac{-12}{(x+2)^3}$$

This is never equal to zero. The expression is positive when $x < -2$, so the graph is concave up when $x < -2$. The second derivative is negative when $x > -2$, so it's concave down when $x > -2$.

Now plot the graph.

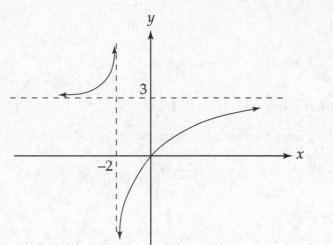

Now it's time to practice some problems. Do each problem, covering the answer first, then check your answer.

PROBLEM 1. Sketch the graph of $y = x^3 - 9x^2 + 24x - 10$. Plot all extrema, points of inflection, and asymptotes.

Answer: Follow the three steps.

First, see if the x-intercepts are easy to find. This is a cubic equation that isn't easily factored. So skip this step.

Next, find the y intercepts by setting $x = 0$.
$$y = (0)^3 - 9(0)^2 + 24(0) - 10 = -10$$

The curve has a y-intercept at $(0, -10)$.

There are no asymptotes, because the curve is a simple polynomial.

Next, find the critical points using the first derivative.
$$\frac{dy}{dx} = 3x^2 - 18x + 24$$

Set the derivative equal to zero and solve for x.

$$3x^2 - 18x + 24 = 0$$
$$3(x^2 - 6x + 8) = 0$$
$$3(x - 4)(x - 2) = 0$$
$$x = 2, 4$$

Plug $x = 2$ and $x = 4$ into the original equation to find the y-coordinates of the critical points.

When $x = 2$, $y = 10$

When $x = 4$, $y = 6$

Thus, we have critical points at $(2, 10)$ and $(4, 6)$.

In our third step, the second derivative indicates any points of inflection.

$$\frac{d^2 y}{dx^2} = 6x - 18$$

This equals zero at $x = 3$.

Next, plug $x = 3$ into the original equation to find the y-coordinates of the point of inflection, which is at $(3, 8)$. Plug the critical values into the second derivative to determine whether each is a maximum or a minimum.

$$6(2) - 18 = -6$$

This is negative, so the curve has a maximum at $(2, 10)$, and the curve is concave down there.

$$6(4) - 18 = 6$$

This is positive, so the curve has a minimum at $(4, 6)$, and the curve is concave up there.

It's graph-plotting time.

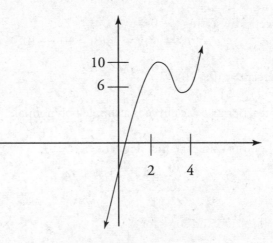

PROBLEM 2. Sketch the graph of $y = 8x^2 - 16x^4$. Plot all extrema, points of inflection, and asymptotes.

Answer: Factor the polynomial.

$$8x^2 \left(1 - 2x^2\right) = 0$$

Solving for x, we get $x = 0$ (a double root), $x = \dfrac{1}{\sqrt{2}}$, and $x = -\dfrac{1}{\sqrt{2}}$.

Find the y-intercepts: when $x = 0$, $y = 0$.

There are no asymptotes because the curve is a simple polynomial.

Find the critical points using the first derivative.

$$\frac{dy}{dx} = 16x - 64x^3$$

Set the derivative equal to zero and solve for x. You get $x = 0$, $x = \dfrac{1}{2}$, and $x = -\dfrac{1}{2}$.

Next, plug $x = 0$, $x = \dfrac{1}{2}$, and $x = -\dfrac{1}{2}$ into the original equation to find the y-coordinates of the critical points.

$$\text{When } x = 0,\ y = 0$$

$$\text{When } x = \frac{1}{2},\ y = 1$$

$$\text{When } x = -\frac{1}{2},\ y = 1$$

Thus, there are critical points at $(0, 0)$, $\left(\dfrac{1}{2}, 1\right)$, and $\left(-\dfrac{1}{2}, 1\right)$.

Take the second derivative to find any points of inflection.

$$\frac{d^2y}{dx^2} = 16 - 192x^2$$

This equals zero at $x = \dfrac{1}{\sqrt{12}}$ and $x = -\dfrac{1}{\sqrt{12}}$.

Next, plug $x = \dfrac{1}{\sqrt{12}}$ and $x = -\dfrac{1}{\sqrt{12}}$ into the original equation to find the y-coordinates of

the points of inflection, which are at $\left(\dfrac{1}{\sqrt{12}}, \dfrac{5}{9}\right)$ and $\left(-\dfrac{1}{\sqrt{12}}, \dfrac{5}{9}\right)$. Now determine whether the

points are maxima or minima.

At $x = 0$, we have a minimum; the curve is concave up there.

At $x = \dfrac{1}{2}$, it's a maximum, and the curve is concave down.

At $x = -\dfrac{1}{2}$, it's also a maximum (still concave down).

Now plot.

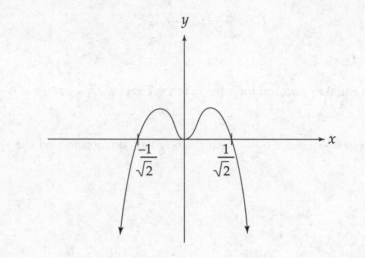

PROBLEM 3. Sketch the graph of $y = \left(\dfrac{x-4}{x+3}\right)^2$. Plot all extrema, points of inflection, and asymptotes.

Answer: This should seem rather routine by now.

Find the x-intercepts by setting the numerator equal to zero; $x = 4$. The graph has an x-intercept at (4, 0). (It's a double root.)

Next, find the y-intercept by plugging in $x = 0$.

$$y = \frac{16}{9}$$

The denominator is undefined at $x = -3$, so there's a vertical asymptote at that point.

Look at the limits.

$$\lim_{x \to \infty} \left(\frac{x-4}{x+3}\right)^2 = 1 \text{ and } \lim_{x \to -\infty} \left(\frac{x-4}{x+3}\right)^2 = 1$$

The curve has a horizontal asymptote at $y = 1$.

It's time for the first derivative.

$$\frac{dy}{dx} = 2\left(\frac{x-4}{x+3}\right)\frac{(x+3)(1)-(x-4)(1)}{(x+3)^2} = \frac{14x-56}{(x+3)^3}$$

The derivative is zero when $x = 4$, and the derivative is undefined at $x = -3$. (There's an asymptote there, so we can ignore the point. If the curve were *defined* at $x = -3$, then it would be a critical point, as you'll see in the next example.)

Now for the second derivative.

$$\frac{d^2 y}{dx^2} = \frac{(x+3)^3(14)-(14x-56)3(x+3)^2}{(x+3)^6} = \frac{-28x+210}{(x+3)^4}$$

This is zero when $x = \frac{15}{2}$. The second derivative is positive (and the graph is concave up) when $x < \frac{15}{2}$, and it's negative (and the graph is concave down) when $x > \frac{15}{2}$.

We can now plug $x = 4$ into the second derivative. It's positive there, so $(4, 0)$ is a minimum.

Your graph should look like the following:

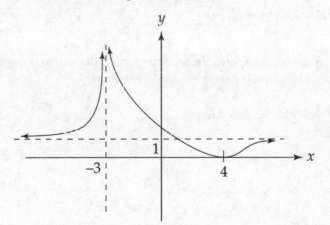

PROBLEM 4. Sketch the graph of $y = (x-4)^{\frac{2}{3}}$. Plot all extrema, points of inflection, and asymptotes.

Answer: By inspection, the x-intercept is at $x = 4$.

Next, find the y-intercepts. When $x = 0$, $y = \sqrt[3]{16} \approx 2.52$.

No asymptotes exist because there's no place where the curve is undefined.

The first derivative is

$$\frac{dy}{dx} = \frac{2}{3}(x-4)^{-\frac{1}{3}}$$

Set it equal to zero.

$$\frac{2}{3}(x-4)^{-\frac{1}{3}} = 0$$

This can never equal zero. But at $x = 4$ the derivative is undefined, so this is a critical point. If you look at the limit as x approaches 4 from both sides, you can see if there's a cusp.

$$\lim_{x \to 4^+} \frac{2}{3}(x-4)^{-\frac{1}{3}} = \infty \quad \text{and} \quad \lim_{x \to 4^-} \frac{2}{3}(x-4)^{-\frac{1}{3}} = -\infty$$

The curve has a cusp at (4, 0).

There were no other critical points. But we can see that when $x > 4$, the derivative is positive and the curve is rising; when $x < 4$ the derivative is negative, and the curve is falling.

The second derivative is

$$\frac{d^2 y}{dx^2} = -\frac{2}{9}(x-4)^{-\frac{4}{3}}$$

No value of x can set this equal to zero. In fact, the second derivative is negative at all values of x except 4. Therefore, the graph is concave down everywhere.

Your graph should look like the following:

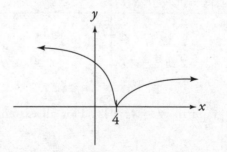

PRACTICE PROBLEM SET 11

It's time for you to try some of these on your own. Sketch each of the graphs below and check the answers in Chapter 23, starting on page 495.

1. $y = x^3 - 9x - 6$

2. $y = -x^3 - 6x^2 - 9x - 4$

3. $y = (x^2 - 4)(9 - x^2)$

4. $y = \dfrac{x - 3}{x + 8}$

5. $y = \dfrac{x^2 - 4}{x - 3}$

6. $y = 3 + x^{\frac{2}{3}}$

7. $y = x^{\frac{2}{3}}\left(3 - 2x^{\frac{1}{3}}\right)$

End of Chapter 9 Drill

The answers are in Chapter 24.

1. A rectangle has its base on the *x*-axis and its two upper corners on the semicircle $y = \sqrt{200 - x^2}$. Find the value of *x* that maximizes the area of the rectangle.

   (A) 0
   (B) 5
   (C) 10
   (D) $\sqrt{200}$

2. An open rectangular box with a square base is to have a volume of 108 in³. Find the dimensions of the box that minimize its surface area (base by height).

   (A) 3 in. by 6 in.
   (B) 6 in. by 6 in.
   (C) 3 in. by 3 in.
   (D) 6 in. by 3 in.

3. Find the coordinates of the maximum and minimum (if they exist) of $y = x^3 - 3x^2 - 24x + 6$.

	Minimum	Maximum
(A)	(–4, –10)	(2, 46)
(B)	(4, –74)	(–2, 34)
(C)	(1, –20)	None
(D)	(1, –20)	(2, 46)

4. At what value(s) of *x* does the curve $y = x^3 - 12x^2 + 120$ have a point (or points) of inflection?

   (A) (4, –8)
   (B) (0, 120)
   (C) (8, –136)
   (D) (–4, –136)

5. On what interval is the curve $y = \sin x + \sqrt{3} \cos x$ decreasing on $[0, \pi]$?

   (A) $\left[0, \dfrac{\pi}{6}\right]$

   (B) $\left[\dfrac{\pi}{6}, \pi\right]$

   (C) $\left[\dfrac{\pi}{2}, 1\right]$

   (D) The function is always increasing.

REFLECT

Respond to the following questions:

- For which topics discussed in this chapter do you feel you have achieved sufficient mastery to answer multiple-choice questions correctly?

- For which topics discussed in this chapter do you feel you have achieved sufficient mastery to answer open-ended questions correctly?

- For which topics discussed in this chapter do you feel you need more work before you can answer multiple-choice questions correctly?

- For which topics discussed in this chapter do you feel you need more work before you can answer open-ended questions correctly?

- What parts of this chapter are you going to re-review?

Chapter 10
Motion

This chapter deals with two different types of word problems that involve motion: related rates and the relationship between velocity and acceleration of a particle. The subject matter might seem arcane, but once you get the hang of them, you'll see that these aren't so hard, either. Besides, the AP Exam tests only a few basic problem types.

RELATED RATES

The idea behind these problems is very simple. In a typical problem, you'll be given an equation relating two or more variables. These variables will change with respect to time, and you'll use derivatives to determine how the rates of change are related. (Hence the name: related rates.) Sounds easy, doesn't it?

Example 1: A circular pool of water is expanding at the rate of 16π in.^{2}/sec. At what rate is the radius expanding when the radius is 4 inches?

Note: The pool is expanding in square inches per second. We've been given the rate that the area is changing, and we need to find the rate of change of the radius. What equation relates the area of a circle to its radius? $A = \pi r^2$.

Step 1: Set up the equation and take the derivative of this equation with respect to t (time).

$$\frac{dA}{dt} = 2\pi r \frac{dr}{dt}$$

In this equation, $\frac{dA}{dt}$ represents the rate at which the area is changing, and $\frac{dr}{dt}$ is the rate at which the radius is changing. The simplest way to explain this is that whenever you have a variable in an equation (r, for example), the derivative with respect to time $\left(\frac{dr}{dt}\right)$ represents the rate at which that variable is increasing or decreasing.

Step 2: Now we can plug in the values for the rate of change of the area and for the radius. (Never plug in the values until after you have taken the derivative or you will get nonsense!)

$$16\pi = 2\pi\,(4)\frac{dr}{dt}$$

Solving for $\frac{dr}{dt}$, we get

$$16\pi = 8\pi\,\frac{dr}{dt} \quad \text{and} \quad \frac{dr}{dt} = 2$$

The radius is changing at a rate of 2 in/sec. It's important to note that this is the rate only when the radius is 4 inches. As the circle gets bigger and bigger, the radius will expand at a slower and slower rate.

Example 2: A 25-foot long ladder is leaning against a wall and sliding toward the floor. If the foot of the ladder is sliding away from the base of the wall at a rate of 15 feet/sec, how fast is the top of the ladder sliding down the wall when the top of the ladder is 7 feet from the ground?

Here's another classic related rates problem. As always, a picture is worth 1,000 words.

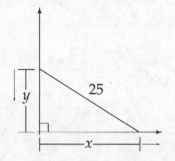

You can see that the ladder forms a right triangle with the wall. Let x stand for the distance from the foot of the ladder to the base of the wall, and let y represent the distance from the top of the ladder to the ground. What's our favorite theorem that deals with right triangles? The Pythagorean Theorem tells us here that $x^2 + y^2 = 25^2$. Now we have an equation that relates the variables to each other.

Now take the derivative of the equation with respect to t.

$$2x\frac{dx}{dt} + 2y\frac{dy}{dt} = 0$$

Just plug in what you know and solve. Because we're looking for the rate at which the vertical distance is changing, we're going to solve for $\frac{dy}{dt}$.

Let's see what we know. We're given the rate at which the ladder is sliding away from the wall: $\frac{dx}{dt} = 15$. The distance from the ladder to the top of the wall is 7 feet ($y = 7$). To find x, use the Pythagorean Theorem. If we plug in $y = 7$ to the equation $x^2 + y^2 = 25^2$, $x = 24$.

Now plug all this information into the derivative equation.

$$2(24)(15) + 2(7)\frac{dy}{dt} = 0$$

$$\frac{dy}{dt} = \frac{-360}{7} \text{ feet/sec}$$

Example 3: A spherical balloon is expanding at a rate of 60π in³/sec. How fast is the surface area of the balloon expanding when the radius of the balloon is 4 in.?

Step 1: You're given the rate at which the volume's expanding, and you know the equation that relates volume to radius. But you have to relate radius to surface area as well, because you have to find the surface area's rate of change. This means that you'll need the equations for volume and surface area of a sphere.

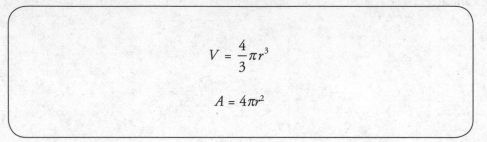

$$V = \frac{4}{3}\pi r^3$$

$$A = 4\pi r^2$$

You're trying to find $\dfrac{dA}{dt}$, but A is given in terms of r, so you have to get $\dfrac{dr}{dt}$ first. Because we know the volume, if we work with the equation that gives us volume in terms of radius, we can find $\dfrac{dr}{dt}$. From there, work with the other equation to find $\dfrac{dA}{dt}$. If we take the derivative of the equation with respect to t, we get $\dfrac{dV}{dt} = 4\pi r^2 \dfrac{dr}{dt}$. Plugging in for $\dfrac{dV}{dt}$ and for r, we get $60\pi = 4\pi(4)^2 \dfrac{dr}{dt}$.

Solving for $\dfrac{dr}{dt}$, we get

$$\frac{dr}{dt} = \frac{15}{16} \text{ in./sec}$$

Step 2: Now we take the derivative of the other equation with respect to t.

$$\frac{dA}{dt} = 8\pi r \frac{dr}{dt}$$

We can plug in for r and $\dfrac{dr}{dt}$ from the previous step to get

$$\frac{dA}{dt} = 8\pi(4)\frac{15}{16} = \frac{480\pi}{16}\frac{\text{in}^2}{\text{sec}} = 30\pi \text{ in.}^2/\text{sec}$$

One final example.

Example 4: An underground conical tank, standing on its vertex, is being filled with water at the rate of 18π ft³/min. If the tank has a height of 30 feet and a radius of 15 feet, how fast is the water level rising when the water is 12 feet deep?

> The volume of a cone is $V = \dfrac{1}{3}\pi r^2 h$. (You'll learn to derive this formula through integration in Chapter 19.)

This "cone" problem is also typical. The key point to getting these right is knowing that the ratio of the height of a right circular cone to its radius is constant. By telling us that the height of the cone is 30 and the radius is 15, we know that at any level, the height of the water will be twice its radius, or $h = 2r$.

You must find the rate at which the water is rising (the height is changing), or $\dfrac{dh}{dt}$. Therefore, you want to eliminate the radius from the volume. By substituting $\dfrac{h}{2} = r$ into the equation for volume, we get

$$V = \frac{1}{3}\pi\left(\frac{h}{2}\right)^2 h = \frac{\pi h^3}{12}$$

Differentiate both sides with respect to t.

$$\frac{dV}{dt} = \frac{\pi}{12}3h^2\frac{dh}{dt}$$

Now we can plug in and solve for $\dfrac{dh}{dt}$.

$$18\pi = \frac{\pi}{12}3(12)^2\frac{dh}{dt}$$

$$\frac{dh}{dt} = \frac{1}{2}\text{ feet/min}$$

In order to solve related rates problems, you have to be good at determining relationships between variables. Once you figure that out, the rest is a piece of cake. Many of these problems involve geometric relationships, so review the formulas for the volumes and areas of cones, spheres, boxes, and other solids. Once you get the hang of setting up the problems, you'll see that these problems follow the same predictable patterns. Look through these sample problems.

PROBLEM 1. A circle is increasing in area at the rate of 16π in.²/sec. How fast is the radius increasing when the radius is 2 in.?

Answer: Use the expression that relates the area of a circle to its radius: $A = \pi r^2$.

Next, take the derivative of the expression with respect to t.

$$\frac{dA}{dt} = 2\pi r\frac{dr}{dt}$$

Now, plug in $\dfrac{dA}{dt} = 16\pi$ and $r = 2$.

$$16\pi = 2\pi(2)\dfrac{dr}{dt}$$

When you solve for $\dfrac{dr}{dt}$, you'll get $\dfrac{dr}{dt} = 4\,\text{in./sec}$.

PROBLEM 2. A rocket is rising vertically at a rate of 5,400 miles per hour. An observer on the ground is standing 20 miles from the rocket's launch point. How fast (in radians per second) is the angle of elevation between the ground and the observer's line of sight of the rocket increasing when the rocket is at an elevation of 40 miles?

Answer: First, draw a picture.

> Notice that velocity is given in miles per hour and the answer asks for radians per second. In situations like this one, you have to be sure to convert the units properly, or you'll get nailed.

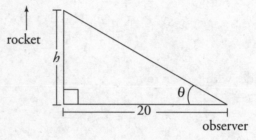

Now, find the equation that relates the angle of elevation to the rocket's altitude.

$$\tan \theta = \dfrac{h}{20}$$

If we take the derivative of both sides of this expression with respect to t, we get

$$\sec^2 \theta \, \dfrac{d\theta}{dt} = \dfrac{1}{20}\dfrac{dh}{dt}$$

We know that $\dfrac{dh}{dt}$ = 5,400 miles per hour, but the problem asks for time in seconds, so we need to convert this number. There are 3,600 seconds in an hour, so $\dfrac{dh}{dt} = \dfrac{3}{2}$ miles per second. Next, we know that $\tan \theta = \dfrac{h}{20}$, so when $h = 40$, $\tan\theta = 2$. Because $1 + \tan^2\theta = \sec^2\theta$, we get $\sec^2\theta = 5$.

Plug in the following information:

$$5\frac{d\theta}{dt} = \frac{1}{20}\left(\frac{3}{2}\right) \text{ and } \frac{d\theta}{dt} = \frac{3}{200} \text{ radians per second}$$

PRACTICE PROBLEM SET 12

Now try these problems on your own. The answers are in Chapter 23, starting on page 504.

1. Oil spilled from a tanker spreads in a circle whose circumference increases at a rate of 40 ft/sec. How fast is the area of the spill increasing when the circumference of the circle is 100π feet?

2. A spherical balloon is inflating at a rate of 27π in.^{3}/sec. How fast is the radius of the balloon increasing when the radius is 3 inches?

3. Cars A and B leave a town at the same time. Car A heads due south at a rate of 80 km/hr and car B heads due west at a rate of 60 km/hr. How fast is the distance between the cars increasing after three hours?

4. The sides of an equilateral triangle are increasing at the rate of 27 in./sec. How fast is the triangle's area increasing when the sides of the triangle are each 18 inches long?

5. An inverted conical container has a diameter of 42 inches and a depth of 15 inches. If water is flowing out of the vertex of the container at a rate of 35π in.^{3}/sec, how fast is the depth of the water dropping when the height is 5 in.?

6. A boat is being pulled toward a dock by a rope attached to its bow through a pulley on the dock 7 feet above the bow. If the rope is hauled in at a rate of 4 ft/sec, how fast is the boat approaching the dock when 25 ft of rope is out?

7. A 6-foot-tall woman is walking at the rate of 4 ft/sec away from a street lamp that is 24 ft tall. How fast is the length of her shadow changing?

8. The minute hand of a clock is 6 inches long. Starting from noon, how fast is the area of the sector swept out by the minute hand increasing in in.^{2}/min at any instant?

POSITION, VELOCITY, AND ACCELERATION

Almost every AP Exam has a question on position, velocity, or acceleration. It's one of the traditional areas of physics where calculus comes in handy. Some of these problems require the use of integral calculus, which we won't talk about until the second half of this book. So, this unit is divided in half; you'll see the other half later.

Please note that these equations are usually functions of time (*t*). Typically, *t* is greater than zero, but it doesn't have to be.

If you have a function that gives you the position of an object (usually called a "particle") at a specified time, then the derivative of that function with respect to time is the velocity of the object, and the second derivative is the acceleration. These are usually represented by the following:

Position: $x(t)$ or sometimes $s(t)$

Velocity: $v(t)$, which is $x'(t)$

Acceleration: $a(t)$, which is $x''(t)$ or $v'(t)$

By the way, speed is the absolute value of velocity.

Example 1: If the position of a particle at a time t is given by the equation $x(t) = t^3 - 11t^2 + 24t$, find the velocity and the acceleration of the particle at time $t = 5$.

First, take the derivative of $x(t)$.

$$x'(t) = 3t^2 - 22t + 24 = v(t)$$

Second, plug in $t = 5$ to find the velocity at that time.
$$v(5) = 3(5^2) - 22(5) + 24 = -11$$

Third, take the derivative of $v(t)$ to find $a(t)$.
$$v'(t) = 6t - 22 = a(t)$$

Finally, plug in $t = 5$ to find the acceleration at that time.
$$a(5) = 6(5) - 22 = 8$$

See the negative velocity? The sign of the velocity is important, because it indicates the direction of the particle. Make sure that you know the following:

When the velocity is negative, the particle is moving to the left.

When the velocity is positive, the particle is moving to the right.

When the velocity and acceleration of the particle have the same signs, the particle's speed is increasing.

When the velocity and acceleration of the particle have opposite signs, the particle's speed is decreasing (or slowing down).

When the velocity is zero and the acceleration is not zero, the particle is momentarily stopped and changing direction.

Example 2: If the position of a particle is given by $x(t) = t^3 - 12t^2 + 36t + 18$, where $t > 0$, find the point at which the particle changes direction.

The derivative is

$$x'(t) = v(t) = 3t^2 - 24t + 36$$

Set it equal to zero and solve for t.

$$x'(t) = 3t^2 - 24t + 36 = 0$$
$$t^2 - 8t + 12 = 0$$
$$(t - 2)(t - 6) = 0$$

So, we know that $t = 2$ or $t = 6$.

You need to check that the acceleration is not 0: $x''(t) = 6t - 24$. This equals 0 at $t = 4$. Therefore, the particle is changing direction at $t = 2$ and $t = 6$.

Example 3: Given the same position function as in Example 2, find the interval of time during which the particle is slowing down.

When $0 < t < 2$ and $t > 6$, the particle's velocity is positive; when $2 < t < 6$, the particle's velocity is negative. You can verify this by graphing the function and seeing when it's above or below the x-axis. Or, try some points in the regions between the roots and outside the roots. Now, we need to determine the same information about the acceleration.

$$a(t) = v'(t) = 6t - 24$$

So the acceleration will be negative when $t < 4$, and positive when $t > 4$.

So we have:

Time	Velocity	Acceleration
$0 < t < 2$	Positive	Negative
$2 < t < 4$	Negative	Negative
$4 < t < 6$	Negative	Positive
$t > 6$	Positive	Positive

Whenever the velocity and acceleration have opposite signs, the particle is slowing down. Here the particle is slowing down during the first two seconds ($0 < t < 2$) and between the fourth and sixth seconds ($4 < t < 6$).

Another typical question you'll be asked is to find the distance a particle has traveled from one time to another. This is the distance that the particle has covered without regard to the sign, not just the displacement. In other words, if the particle had an odometer on it, what would it read? Usually, all you have to do is plug in two times into the position function and find the difference.

Example 4: How far does a particle travel between the eighth and tenth seconds if its position function is $x(t) = t^2 - 6t$?

Find $x(10) - x(8) = (100 - 60) - (64 - 48) = 24$.

Be careful about one very important thing: **if the velocity changes sign during the problem's time interval**, you'll get the wrong answer if you simply follow the method in the paragraph above. For example, suppose we had the same position function as above but we wanted to find the distance that the particle travels from $t = 2$ to $t = 4$.

$$x(4) - x(2) = (-8) - (-8) = 0$$

This is wrong. The particle travels from -8 back to -8, but it hasn't stood still. To fix this problem, divide the time interval into the time when the velocity is negative and the time when the velocity is positive, and add the absolute values of each distance. Here, the velocity is $v(t) = 2t - 6$. The velocity is negative when $t < 3$ and positive when $t > 3$. So we find the absolute value of the distance traveled from $t = 2$ to $t = 3$, and add to that the absolute value of the distance traveled from $t = 3$ to $t = 4$.

Because $x(t) = t^2 - 6t$,

$$\left| x(3) - x(2) \right| + \left| x(4) - x(3) \right| = \left| -9 + 8 \right| + \left| -8 + 9 \right| = 2$$

This is the distance that the particle traveled.

Example 5: Given the position function $x(t) = t^4 - 8t^2$, find the distance that the particle travels from $t = 0$ to $t = 4$.

First, find the first derivative ($v(t) = 4t^3 - 16t$) and set it equal to zero.
$$4t^3 - 16t = 0 \quad 4t(t^2 - 4) = 0 \quad t = 0, 2, -2$$

So we need to divide the time interval into $t = 0$ to $t = 2$ and $t = 2$ to $t = 4$.

$$\left| x(2) - x(0) \right| + \left| x(4) - x(2) \right| = 16 + 144 = 160$$

Here are some solved problems. Do each problem, covering the answer first, then checking your answer.

PROBLEM 1. Find the velocity and acceleration of a particle whose position function is $x(t) = 2t^3 - 21t^2 + 60t + 3$, for $t > 0$.

Answer: Find the first two derivatives.

$$v(t) = 6t^2 - 42t + 60$$
$$a(t) = 12t - 42$$

PROBLEM 2. Given the position function in problem 1, find when the particle's speed is increasing.

Answer: First, set $v(t) = 0$.

$$6t^2 - 42t + 60 = 0$$
$$t^2 - 7t + 10 = 0$$
$$(t - 2)(t - 5) = 0$$
$$t = 2, t = 5$$

You should be able to determine that the velocity is positive from $0 < t < 2$, negative from $2 < t < 5$, and positive again from $t > 5$.

Now, set $a(t) = 0$.

$$12t - 42 = 0$$

$$t = \frac{7}{2}$$

You should be able to determine that the acceleration is negative from $0 < t < \frac{7}{2}$ and positive from $t > \frac{7}{2}$.

The intervals where the velocity and the acceleration have the same sign are $2 < t < \frac{7}{2}$ and $t > 5$.

PROBLEM 3. Given that the position of a particle is found by $x(t) = t^3 - 6t^2 + 1$, $t > 0$, find the distance that the particle travels from $t = 2$ to $t = 5$.

Answer: First, find $v(t)$.

$$v(t) = 3t^2 - 12t$$

Second, set $v(t) = 0$ and find the critical values.

$$3t^2 - 12t = 0 \qquad\qquad 3t(t - 4) = 0 \qquad\qquad t = \{0, 4\}$$

Because the particle changes direction after four seconds, you have to figure out two time intervals separately (from $t = 2$ to $t = 4$ and from $t = 4$ to $t = 5$) and add the absolute values of the distances.

$$\left|x(4) - x(2)\right| + \left|x(5) - x(4)\right| = \left|(-31) - (-15)\right| + \left|(-24) - (-31)\right| = 23$$

PRACTICE PROBLEM SET 13

Now try these problems. The answers are in Chapter 23, starting on page 508.

1. Find the velocity and acceleration of a particle whose position function is $x(t) = t^3 - 9t^2 + 24t$, $t > 0$.

2. Find the velocity and acceleration of a particle whose position function is $x(t) = \sin(2t) + \cos(t)$.

3. If the position function of a particle is $x(t) = \sin\left(\dfrac{t}{2}\right)$, $0 < t < 4\pi$, find when the particle is changing direction.

4. If the position function of a particle is $x(t) = 3t^2 + 2t + 4$, $t > 0$, find the distance that the particle travels from $t = 2$ to $t = 5$.

5. If the position function of a particle is $x(t) = t^2 + 8t$, $t > 0$, find the distance that the particle travels from $t = 0$ to $t = 4$.

6. If the position function of a particle is $x(t) = 2\sin^2 t + 2\cos^2 t$, $t > 0$, find the velocity and acceleration of the particle.

7. If the position function of a particle is $x(t) = t^3 + 8t^2 - 2t + 4$, $t > 0$, find when the particle is changing direction.

8. If the position function of a particle is $x(t) = 2t^3 - 6t^2 + 12t - 18$, $t > 0$, find when the particle is changing direction.

End of Chapter 10 Drill

The answers are in Chapter 24.

1. A 25 ft ladder is sliding down a wall at –6 ft/s. How fast is the bottom of the ladder sliding out when the top of the ladder is 20 ft from the ground?

    (A)   2 ft/s
    (B)   6 ft/s
    (C)   8 ft/s
    (D)   15 ft/s

2. A baseball diamond has right angles at each of the bases, which are 90 ft apart. A baseball player is running from second base to third base at a rate of 30 ft/s. How fast is the distance between the player and home plate changing when the player is halfway from second base to third base?

    (A) $-\dfrac{30}{\sqrt{5}}$ ft/s

    (B)   –30 ft/s

    (C) $-\dfrac{135}{\sqrt{5}}$ ft/s

    (D)   –135 ft/s

3. A spherical balloon is deflating at -24π in.^{3}/s. How fast is the surface area of the balloon decreasing when the radius of the balloon is 6 in. ?

    (A) -6π in.^{2}/s
    (B) -8π in.^{2}/s
    (C) -16π in.^{2}/s
    (D) -24π in.^{2}/s

4. Find where the velocity of a particle is zero if its position function is $x(t)=t\ln t$.

    (A) $-e$

    (B)   1

    (C) $\dfrac{1}{e}$

    (D) e

5. If the position function of a particle is $x(t)=4t^3-12t^2+18$, $t>0$, at what time is the particle changing direction?

    (A) $t=0$

    (B) $t=\dfrac{1}{2}$

    (C) $t=1$

    (D) $t=2$

REFLECT

Respond to the following questions:

*   For which topics discussed in this chapter do you feel you have achieved sufficient mastery to answer multiple-choice questions correctly?

*   For which topics discussed in this chapter do you feel you have achieved sufficient mastery to answer open-ended questions correctly?

*   For which topics discussed in this chapter do you feel you need more work before you can answer multiple-choice questions correctly?

*   For which topics discussed in this chapter do you feel you need more work before you can answer open-ended questions correctly?

*   What parts of this chapter are you going to re-review?

Chapter 11

Derivatives of Exponential and Logarithmic Functions

Log-online-arithms
Remember: if you're not sure logarithms, you can find a focused, downloadable review of Prerequisite Mathematics like this online in your free student tools.

As with trigonometric functions, you'll be expected to remember all of the logarithmic and exponential functions you've studied in the past. Also, this is only part one of our treatment of exponents and logs. Much of what you need to know about these functions requires knowledge of integrals (the second half of the book), so we'll discuss them again later.

THE DERIVATIVE OF ln *x*

When you studied logs in the past, you probably concentrated on common logs (that is, those with a base of 10), and avoided natural logarithms (base *e*) as much as possible. Well, we have bad news for you: most of what you'll see from now on involves natural logs. In fact, common logs almost never show up in calculus. But that's okay. All you have to do is memorize a bunch of rules, and you'll be fine.

Rule No. 1: If $y = \ln x$, then $\dfrac{dy}{dx} = \dfrac{1}{x}$

This rule has a corollary that incorporates the Chain Rule and is actually a more useful rule to memorize.

Rule No. 2: If $y = \ln u$, then $\dfrac{dy}{dx} = \dfrac{1}{u}\dfrac{du}{dx}$

Remember: *u* is a function of *x*, and $\dfrac{du}{dx}$ is its derivative.

You'll see how simple this rule is after we try a few examples.

Example 1: Find the derivative of $f(x) = \ln(x^3)$.

$$f'(x) = \frac{3x^2}{x^3} = \frac{3}{x}$$

If you recall your rules of logarithms, you could have done this another way.

$$\ln(x^3) = 3\ln x$$

Therefore, $f'(x) = 3\left(\dfrac{1}{x}\right) = \dfrac{3}{x}$.

Example 2: Find the derivative of $f(x) = \ln(5x - 3x^6)$.

$$f'(x) = \frac{(5 - 18x^5)}{(5x - 3x^6)}$$

Example 3: Find the derivative of $f(x) = \ln(\cos x)$.

$$f'(x) = \frac{-\sin x}{\cos x} = -\tan x$$

Finding the derivative of a natural logarithm is just a matter of following a simple formula.

THE DERIVATIVE OF e^x

As you'll see in Rule No. 3, the derivative of e^x is probably the easiest thing that you'll ever have to do in calculus.

> **Rule No. 3:** If $y = e^x$, then $\dfrac{dy}{dx} = e^x$

That's not a typo. The derivative is the same as the original function! Incorporating the Chain Rule, we get a good formula for finding the derivative.

> **Rule No. 4:** If $y = e^u$, then $\dfrac{dy}{dx} = e^u \dfrac{du}{dx}$

And you were worried that all of this logarithm and exponential stuff was going to be hard!

Example 4: Find the derivative of $f(x) = e^{3x}$.

$$f'(x) = e^{3x}(3) = 3e^{3x}$$

Example 5: Find the derivative of $f(x) = e^{x^3}$.

$$f'(x) = e^{x^3}\left(3x^2\right) = 3x^2 e^{x^3}$$

Example 6: Find the derivative of $f(x) = e^{\tan x}$.

$$f'(x) = \left(\sec^2 x\right)e^{\tan x}$$

Example 7: Find the second derivative of $f(x) = e^{x^2}$.

$$f'(x) = 2xe^{x^2}$$
$$f''(x) = 2e^{x^2} + 4x^2 e^{x^2}$$

Once again, it's just a matter of following a formula.

THE DERIVATIVE OF $\log_a x$

This derivative is actually a little trickier than the derivative of a natural log. First, if you remember your logarithm rules about change of base, we can rewrite $\log_a x$.

$$\log_a x = \frac{\ln x}{\ln a}$$

Review the unit on Prerequisite Mathematics if this leaves you scratching your head. Anyway, because $\ln a$ is a constant, we can take the derivative.

$$\frac{1}{\ln a}\frac{1}{x}$$

This leads us to our next rule.

> Rule No. 5: If $y = \log_a x$, then $\dfrac{dy}{dx} = \dfrac{1}{x \ln a}$

Once again, incorporating the Chain Rule gives us a more useful formula.

Rule No. 6: If $y = \log_a u$, then $\dfrac{dy}{dx} = \dfrac{1}{u \ln a} \dfrac{du}{dx}$

Note: We refer to the $\log_{10} x$ as $\log x$.

Example 8: Find the derivative of $f(x) = \log_{10} x$.

$$f'(x) = \frac{1}{x \ln 10}$$

Example 9: Find the derivative of $f(x) = \log_8 (x^2 + x)$.

$$f'(x) = \frac{2x + 1}{(x^2 + x) \ln 8}$$

Example 10: Find the derivative of $f(x) = \log_e x$.

$$f'(x) = \frac{1}{x \ln e} = \frac{1}{x}$$

You can expect this result from Rules 1 and 2 involving natural logs.

THE DERIVATIVE OF a^x

You should recall from your precalculus days that we can rewrite a^x as $e^{x \ln a}$. Keep in mind that $\ln a$ is just a constant, which gives us the next rule.

Rule No. 7: If $y = a^x$, then $\dfrac{dy}{dx} = \left(e^{x \ln a} \right) \ln a = a^x (\ln a)$

Given the pattern of this chapter, you can guess what's coming: another rule that incorporates the Chain Rule.

Rule No. 8: If $y = a^u$, then $\dfrac{dy}{dx} = a^u (\ln a) \dfrac{du}{dx}$

And now, some examples.

Example 11: Find the derivative of $f(x) = 3^x$.

$$f'(x) = 3^x \ln 3$$

Example 12: Find the derivative of $f(x) = 8^{4x^5}$.

$$f'(x) = 8^{4x^5} (20x^4) \ln 8$$

Example 13: Find the derivative of $f(x) = \pi^{\sin x}$.

$$f'(x) = \pi^{\sin x} (\cos x) \ln \pi$$

Finally, here's every nasty teacher's favorite exponential derivative.

Example 14: Find the derivative of $f(x) = x^x$.

First, rewrite this as $f(x) = e^{x \ln x}$. Then, take the derivative.

$$f'(x) = e^{x \ln x}\left(\ln x + \frac{x}{x} \right) = e^{x \ln x}(\ln x + 1) = x^x (\ln x + 1)$$

Would you have thought of that? Remember this trick. It might come in handy! Okay. Ready for some practice? Here are some more solved problems. Cover the solutions and get cracking.

PROBLEM 1. Find the derivative of $y = 3\ln(5x^2 + 4x)$.

Answer: Use Rule No. 2.

$$\frac{dy}{dx} = 3\frac{10x + 4}{5x^2 + 4x} = \frac{30x + 12}{5x^2 + 4x}$$

PROBLEM 2. Find the derivative of $f(x) = \ln\left(\sin\left(x^5\right)\right)$.

Answer: Use Rule No. 2 in addition to the Chain Rule.

$$f'(x) = \frac{5x^4 \cos(x^5)}{\sin(x^5)} = 5x^4 \cot(x^5)$$

PROBLEM 3. Find the derivative of $f(x) = e^{3x^7 - 4x^2}$.

Answer: Use Rule No. 4.

$$f'(x) = (21x^6 - 8x)e^{3x^7 - 4x^2}$$

PROBLEM 4. Find the derivative of $f(x) = \log_4(\tan x)$.

Answer: Use Rule No. 6.

$$f'(x) = \frac{1}{\ln 4} \frac{\sec^2 x}{\tan x}$$

PROBLEM 5. Find the derivative of $y = \log_8 \sqrt{\dfrac{x^3}{1+x^2}}$.

Answer: First, use the rules of logarithms to rewrite the equation.

$$y = \frac{1}{2}\left[3\log_8 x - \log_8\left(1 + x^2\right)\right]$$

Now it's much easier to find the derivative.

$$\frac{dy}{dx} = \frac{1}{2}\left[3\frac{1}{x\ln 8}\right] - \frac{1}{2}\left[\frac{2x}{\ln 8 (1+x^2)}\right] = \frac{1}{2\ln 8}\left[\frac{3}{x} - \frac{2x}{(1+x^2)}\right]$$

PROBLEM 6. Find the derivative of $y = 5^{\sqrt{x}}$.

Answer: Use Rule No. 8.

$$\frac{dy}{dx} = 5^{\sqrt{x}}\frac{1}{2\sqrt{x}}\ln 5 = \frac{5^{\sqrt{x}}\ln 5}{2\sqrt{x}}$$

PROBLEM 7. Find the derivative of $y = \dfrac{e^{x^3}}{5^{\cos x}}$.

Answer: Here, you need to use the Quotient Rule and Rules Nos. 4 and 8.

$$\frac{dy}{dx} = \frac{5^{\cos x}\left(3x^2 e^{x^3}\right) - e^{x^3}\left(5^{\cos x}\ln 5(-\sin x)\right)}{\left(5^{\cos x}\right)^2} = 5^{\cos x}e^{x^3}\frac{\left(3x^2\right) + (\sin x \ln 5)}{5^{2\cos x}} = e^{x^3}\frac{3x^2 + \sin x \ln 5}{5^{\cos x}}$$

PRACTICE PROBLEM SET 14

Now find the derivative of each of the following functions. The answers are in Chapter 23, starting on page 511.

1. $f(x) = \ln(x^4 + 8)$

2. $f(x) = \ln(3x\sqrt{3+x})$

3. $f(x) = \ln(\cot x - \csc x)$

4. $f(x) = \ln\left(\dfrac{5x^2}{\sqrt{5+x^2}}\right)$

5. $f(x) = e^{x\cos x}$

6. $f(x) = e^{-3x}\sin 5x$

7. $f(x) = e^{\pi x} - \ln e^{\pi x}$

8. $f(x) = \log_{12}(x^3)$

9. $f(x) = \dfrac{\log_4 x}{e^{4x}}$

10. $f(x) = \log\sqrt{10^{3x}}$

11. $f(x) = e^{3x} - 3^{ex}$

12. $f(x) = 10^{\sin x}$

13. $f(x) = \ln(10^x)$

14. $f(x) = x^5 5^x$

End of Chapter 11 Drill

The answers are in Chapter 24.

1. Find $\dfrac{dy}{dx}$ if $y = \ln(x^3 - 7x + 5)$.

   (A) $\dfrac{1}{x^3 - 7x + 5}$

   (B) $\dfrac{x^3 - 7x + 5}{3x^2 - 7}$

   (C) $\dfrac{3x^2 - 7}{x^3 - 7x + 5}$

   (D) $\dfrac{1}{3x^2 - 7}$

2. Find $\dfrac{dy}{dx}$ if $y = \ln\left(\dfrac{\sqrt{x^2 - 11}}{x^2 + 9}\right)$.

   (A) $\dfrac{x}{x^2 - 11} - \dfrac{2x}{x^2 + 9}$

   (B) $\dfrac{2x}{\sqrt{x^2 - 11}} - \dfrac{2x}{x^2 + 9}$

   (C) $\dfrac{1}{2} \dfrac{x^2 + 9}{x^2 - 11}$

   (D) $\dfrac{x^2 + 9}{\sqrt{x^2 - 11}}$

3. Find $\dfrac{dy}{dx}$ if $y = e^{2\sin x}$.

   (A) $2e^{2\sin x}$

   (B) $e^{2\cos x}$

   (C) $2\sin x e^{2\sin x}$

   (D) $2\cos x e^{2\sin x}$

4. Find $\dfrac{dy}{dx}$ if $y = \log_8(x^4 + \tan x)$.

   (A) $\dfrac{1}{x^4 + \tan x} \dfrac{1}{\ln 8}$

   (B) $\dfrac{4x^3 + \sec^2 x}{x^4 + \tan x} \dfrac{1}{\ln 8}$

   (C) $\dfrac{4x^3 + \sec^2 x}{x^4 + \tan x} \dfrac{1}{\log 8}$

   (D) $\dfrac{1}{x^4 + \tan x} \dfrac{1}{\log 8}$

5. Find $\dfrac{dy}{dx}$ if $y - 7^{x^2 + 1}$.

   (A) $7^{x^2 + 1}(2x \log 7)$

   (B) $7^{x^2 + 1}(2x)$

   (C) $7^{x^2 + 1}(2x \ln 7)$

   (D) $7^{x^2 + 1}(\ln 7)$

REFLECT

Respond to the following questions:

- For which topics discussed in this chapter do you feel you have achieved sufficient mastery to answer multiple-choice questions correctly?

- For which topics discussed in this chapter do you feel you have achieved sufficient mastery to answer open-ended questions correctly?

- For which topics discussed in this chapter do you feel you need more work before you can answer multiple-choice questions correctly?

- For which topics discussed in this chapter do you feel you need more work before you can answer open-ended questions correctly?

- What parts of this chapter are you going to re-review?

Chapter 12
Other Topics
in Differential
Calculus

This chapter is devoted to other topics involving differential calculus that don't fit into a specific category.

THE DERIVATIVE OF AN INVERSE FUNCTION

ETS occasionally asks a question about finding the derivative of an inverse function. To do this, you need to learn only this simple formula.

Suppose we have a function $x = f(y)$ that is defined and differentiable at $y = a$ where $x = c$. Suppose we also know that the $f^{-1}(x)$ exists at $x = c$. Thus, $f(a) = c$ and $f^{-1}(c) = a$. Then, because

$$\frac{dy}{dx} = \frac{1}{\dfrac{dx}{dy}},$$

$$\frac{d}{dx} f^{-1}(x) \bigg|_{x=c} = \frac{1}{\left[\dfrac{d}{dy} f(y)\right]_{y=a}}$$

The short translation of this is: we can find the derivative of a function's inverse at a particular point by taking the reciprocal of the derivative at that point's corresponding y-value. These examples should help clear up any confusion.

Example 1: If $f(x) = x^2$, find a derivative of $f^{-1}(x)$ at $x = 9$.

First, notice that $f(3) = 9$. One of the most confusing parts of finding the derivative of an inverse function is that when you're asked to find the derivative at a value of x, they're *really* asking you for the derivative of the inverse of the function at the value that *corresponds* to $f(x) = 9$. This is because x-values of the inverse correspond to $f(x)$-values of the original function.

The rule is very simple: When you're asked to find the derivative of $f^{-1}(x)$ at $x = c$, you take the reciprocal of the derivative of $f(x)$ at $x = a$, where $f(a) = c$.

We know that $\dfrac{d}{dx} f(x) = 2x$. This means that we're going to plug $x = 3$ into the formula (because $f(3) = 9$). This gives us

$$\frac{1}{2x}\bigg|_{x=3} = \frac{1}{6}$$

We can verify this by finding the inverse of the function first and then taking the derivative. The inverse of the function $f(x) = x^2$ is the function $f^{-1}(x) = \sqrt{x}$. Now we find the derivative and evaluate it at $f(3) = 9$.

$$\frac{d}{dx}\sqrt{x} = \frac{1}{2\sqrt{x}}\bigg|_{x=9} = \frac{1}{6}$$

Remember the rule: Find the value, a, of $f(x)$ that gives you the value of x that the problem asks for. Then plug that value, a, into the reciprocal of the derivative of the *inverse* function.

Example 2: Find a derivative of the inverse of $y = x^3 - 1$ when $y = 7$.

First, we need to find the x-value that corresponds to $y = 7$. A little algebra tells us that this is $x = 2$. Then,

$$\frac{dy}{dx} = 3x^2 \text{ and } \frac{1}{\frac{dy}{dx}} = \frac{1}{3x^2}$$

Therefore, the derivative of the inverse is

$$\frac{1}{3x^2}\bigg|_{x=2} = \frac{1}{12}$$

Verify it: The inverse of the function $y = x^3 - 1$ is the function $y = \sqrt[3]{x+1}$. The derivative of this latter function is

$$\frac{1}{3\sqrt[3]{(x+1)^2}}\bigg|_{x=7} = \frac{1}{12}$$

Let's do one more.

Example 3: Find a derivative of the inverse of $y = x^2 + 4$ when $y = 29$.

At $y = 29$, $x = 5$, the derivative of the function is

$$\frac{dy}{dx} = 2x$$

So, a derivative of the inverse is

$$\frac{1}{2x}\bigg|_{x=5} = \frac{1}{10}$$

Note that $x = -5$ also gives us $y = 29$, so $-\frac{1}{10}$ is also a derivative. It's not that hard, once you get the hang of it.

This is all you'll be required to know involving derivatives of inverses. Naturally, there are ways to create harder problems, but the AP Exam stays away from them and sticks to simpler stuff.

Here are some solved problems. Do each problem, cover the answer first, and then check your answer.

PROBLEM 1. Find a derivative of the inverse of $f(x) = 2x^3 + 5x + 1$ at $y = 8$.

Answer: First, we take the derivative of $f(x)$.

$$f'(x) = 6x^2 + 5$$

A possible value of x is $x = 1$.

Then, we use the formula to find the derivative of the inverse.

$$\frac{1}{f'(1)} = \frac{1}{11}$$

PROBLEM 2. Find a derivative of the inverse of $f(x) = 3x^3 - x + 7$ at $y = 9$.

Answer: First, take the derivative of $f(x)$.

$$f'(x) = 9x^2 - 1$$

A possible value of x is $x = 1$.

Then, use the formula to find the derivative of the inverse.

$$\frac{1}{f'(1)} = \frac{1}{8}$$

PROBLEM 3. Find a derivative of the inverse of $y = \dfrac{8}{x^3}$ at $y = 1$.

Answer: Take the derivative of y.

$$y' = -\frac{24}{x^4}$$

Find the value of x where $y = 1$.

$$1 = \frac{8}{x^3}$$

$$x = 2$$

Use the formula.

$$\frac{1}{\left.\dfrac{dy}{dx}\right|_{x=2}} = \frac{1}{\left.\left(-\dfrac{24}{x^4}\right)\right|_{x=2}} = -\frac{2}{3}$$

Here's one more.

Problem 4. Find a derivative of the inverse of $y = 2x - x^3$ at $y = 1$.

Answer: The derivative of the function is

$$\frac{dy}{dx} = 2 - 3x^2$$

Next, find the value of x where $y = 1$. By inspection, $y = 1$ when $x = 1$.

Then, we use the formula to find the derivative of the inverse.

$$\frac{1}{\left.\dfrac{dy}{dx}\right|_{x=1}} = \frac{1}{\left.\left(2 - 3x^2\right)\right|_{x=1}} = -1$$

PRACTICE PROBLEM SET 15

Find a derivative of the inverse of each of the following functions. The answers are in Chapter 23, starting on page 514.

1. $y = x + \dfrac{1}{x}$ at $y = \dfrac{17}{4}$; where $x > 1$

2. $y = 3x - 5x^3$ at $y = 2$

3. $y = e^x$ at $y = e$

4. $y = x + x^3$ at $y = -2$

5. $y = 4x - x^3$ at $y = 3$

6. $y = \ln x$ at $y = 0$

DERIVATIVES OF PARAMETRIC FUNCTIONS

Although these can seem very difficult, the questions about parametric equations on the AP Exam tend to be very straightforward. As we keep pointing out, don't be intimidated by the difficult topics; the AP Exam tends to keep the questions simple. By contrast, questions on simpler topics tend to be trickier.

What Is a Parametric Function?

Let's use an analogy. Suppose you're driving a car, and you want to determine a function that describes your position on the road. There are two ways that you could arrive at your position. You could figure it out based on how far you've traveled, or you could determine it based on how long you've been traveling. If you let y represent your position and x the distance, you could find your position by $y = f(x)$.

If, on the other hand, you wanted to use the time that you've traveled, you can use two functions: $x = g(t)$, to determine the distance you've traveled, and $y = h(t)$, to determine your position. These latter equations are called "parametric equations." They enable you to define x and y in terms of another variable (usually t), rather than in terms of each other. Parametric equations also follow all of the standard derivative rules.

$$\frac{\dfrac{dy}{dt}}{\dfrac{dx}{dt}} = \frac{dy}{dx}$$

For example, suppose that $x = t^2$ and $y = t^4$. Then,

$$\frac{dx}{dt} = 2t \text{ and } \frac{dy}{dt} = 4t^3$$

$$\frac{dy}{dx} = \frac{4t^3}{2t} = 2t^2$$

Now, because $x = t^2$, $\frac{dy}{dx} = 2x$.

We can verify this by first solving for y in terms of x and then differentiating. The result is

$$y = x^2 \text{ and } \frac{dy}{dx} = 2x$$

You might also be asked to turn an equation in parametric form into an equation in Cartesian form. The simplest thing to do is to solve the equation for t and substitute that equation into the other one. Sometimes the relationship is not so obvious, as in this example.

Example 1: What curve is represented by $x = \cos t$ and $y = \sin t$, where $0 \le t \le 2\pi$?

We know from trigonometry that $\sin^2 t + \cos^2 t = 1$, so we can substitute y for $\sin t$ and x for $\cos t$. If we square both and add them, we get $x^2 + y^2 = 1$. This is the equation of a circle, centered at the origin, with radius 1.

You can test this by picking values of t and finding the coordinates by using the two equations. For example,

$$\text{At } t = \frac{\pi}{2}, \quad x = 0 \text{ and } y = 1; \text{ or}$$

$$\text{At } t = \frac{\pi}{6}, \quad x = \frac{\sqrt{3}}{2} \text{ and } y = \frac{1}{2}$$

Example 2: What curve is represented by $x = t$ and $y = t^2$?

If you substitute x for t, you'll find that $y = x^2$. Thus, this curve is a parabola with vertex at the origin.

Example 3: What curve is represented by $x = a\cos t$ and $y = b\sin t$, where $0 \le t \le 2\pi$?

First, rewrite the two equations.

$$\frac{x}{a} = \cos t \text{ and } \frac{y}{b} = \sin t$$

Now, because we know that $\sin^2 t + \cos^2 t = 1$, we know that

$$\left(\frac{x}{a}\right)^2 + \left(\frac{y}{b}\right)^2 = 1$$

If $a \ne b$, this figure is an ellipse, centered at the origin, with axes of length $2a$ and $2b$. If $a = b$, it's a circle, centered at the origin, with radius a.

Now you know how to convert an equation from parametric form to Cartesian form. You'll also need to know how to work with the parametric equations, even if you can't figure out how to convert them into Cartesian form (sometimes you don't have to). These frequently will be equations of motion.

Example 4: A particle's position in the xy-plane at any time t is given by $x = 2t^2 + 3$ and $y = t^4$.

Find: (a) the x-component of the particle's velocity at time $t = 5$; (b) $\dfrac{dy}{dx}$; and (c) the times at which the x- and y-components of the velocity are the same.

(a) All you do is take the derivative with respect to t.

$$\frac{dx}{dt} = 4t$$

At time $t = 5$, this is 20.

(b) You know $\frac{dx}{dt}$ from part (a), so now compute $\frac{dy}{dt}$. This is $4t^3$. Using the rule, $\frac{dy}{dx} = \frac{\frac{dy}{dt}}{\frac{dx}{dt}}$,

so

$$\frac{4t^3}{4t} = t^2$$

To express the answer in terms of x instead of the parameter, solve for t^2 in terms of x.

$$\frac{x-3}{2} = t^2$$

Then substitute.

$$\frac{dy}{dx} = \frac{x-3}{2}$$

(c) The components of the velocity are the same when $4t^3 = 4t$.

$$4t^3 - 4t = 0$$
$$4t\left(t^2 - 1\right) = 0$$
$$t = 0, \pm 1$$

Now you know all you need to about the basics of parameterized curves. There will be other types of problems involving parametric equations, but you'll see them later in the book.

Try these sample problems, working with an index card, as usual.

PROBLEM 1. Find the Cartesian equation represented by the parametric equations $x = 4\cos t$ and $y = 4\sin t$, $0 \le t \le 2\pi$.

Answer: From Example 1, you know that $\sin^2 t + \cos^2 t = 1$. If you take each equation and rearrange the terms, you get

$$\frac{x}{4} = \cos t \text{ and } \frac{y}{4} = \sin t$$

Next, substitute into $\sin^2 t + \cos^2 t = 1$.

$$\frac{x^2}{16} + \frac{y^2}{16} = 1 \text{, or}$$

$$x^2 + y^2 = 16$$

This is a circle, centered at the origin, with radius 4.

PROBLEM 2. Find an equation of the line tangent to the curve $x = 2\cos t$ and $y = 3\sin t$ at $t = \dfrac{\pi}{4}$.

Answer: You can, of course, eliminate t, obtaining

$$\left(\frac{x}{2}\right)^2 + \left(\frac{y}{3}\right)^2 = 1$$

Proceed as normal, but it's easier to retain the parameter t here and reach the answer. First, find the slope of the tangent line $\dfrac{dy}{dx}$ using the formula $\dfrac{dy}{dx} = \dfrac{\dfrac{dy}{dt}}{\dfrac{dx}{dt}}$.

$$\frac{dy}{dt} = 3\cos t \text{ and } \frac{dx}{dt} = -2\sin t$$

Therefore, $\dfrac{dy}{dx} = \dfrac{3\cos t}{-2\sin t}$.

If we evaluate this at $t = \dfrac{\pi}{4}$,

$$\frac{dy}{dx} = \frac{3\left(\dfrac{1}{\sqrt{2}}\right)}{-2\left(\dfrac{1}{\sqrt{2}}\right)} = -\frac{3}{2}$$

At $t = \dfrac{\pi}{4}$, $x = \dfrac{2}{\sqrt{2}}$ and $y = \dfrac{3}{\sqrt{2}}$. Now you can find the equation of the tangent line.

$$\left(y - \frac{3}{\sqrt{2}}\right) = -\frac{3}{2}\left(x - \frac{2}{\sqrt{2}}\right)$$

This can be rewritten as $3x + 2y - 6\sqrt{2} = 0$.

PROBLEM 3. A particle's position at time t is determined by the equations $x = 3 + 2t^2$ and $y = 4t^4$, $t \geq 0$. Find the x- and y-components of the particle's velocity and the times when these components are equal.

Answer: First, figure out the x- and y-components of the velocity.

$$\frac{dx}{dt} = 4t \text{ and } \frac{dy}{dt} = 16t^3$$

These are equal when $16t^3 = 4t$. Solving for t,

$$t = 0, \pm \frac{1}{2}$$

Throw out the negative value of t, the answer is $t = 0, \dfrac{1}{2}$.

PRACTICE PROBLEM SET 16

Now try these problems. The answers are in Chapter 23, starting on page 517.

1. Find the Cartesian equation of the curve represented by $x = \sec^2 t - 1$ and $y = \tan t$, $-\dfrac{\pi}{2} < t < \dfrac{\pi}{2}$.

2. Find the Cartesian equation of the curve represented by $x = t$ and $y = \sqrt{1 - t^2}$, $-1 < t < 1$.

3. Find the Cartesian equation of the curve represented by $x = 4t + 3$ and $y = 16t^2 - 9$, $-\infty < t < \infty$.

4. Find the equation of the tangent line to $x = t^2 + 4$ and $y = 8t$ at $t = 6$.

5. Find the equation of the tangent line to $x = \sec t$ and $y = \tan t$ at $t = \dfrac{\pi}{4}$.

6. The motion of a particle is given by $x = -2t^2$ and $y = t^3 - 3t + 9$, $t \geq 0$. Find the coordinates of the particle when its instantaneous direction of motion is horizontal.

7. The motion of a particle is given by $x = \ln t$ and $y = t^2 - 4t$. Find the coordinates of the particle when its instantaneous direction of motion is horizontal.

8. The motion of a particle is given by $x = 2\sin t - 1$ and $y = \sin t - \dfrac{t}{2}$, $0 \leq t < 2\pi$.

   Find the times when the horizontal and vertical components of the particle's

   velocity are the same.

L'HÔPITAL'S RULE

L'Hôpital's Rule is a way to find the limit of certain kinds of expressions that are indeterminate forms. If the limit of an expression results in $\frac{0}{0}$ or $\frac{\infty}{\infty}$, the limit is called "indeterminate" and you can use L'Hôpital's Rule to evaluate these expressions.

> If $f(c) - g(c) = 0$, and if $f'(c)$ and $g'(c)$ exist, and if $g'(c) \neq 0$,
>
> then $\lim\limits_{x \to c} \dfrac{f(x)}{g(x)} = \dfrac{f'(c)}{g'(c)}$.

Remember that you can only use L'Hôpital's Rule for indeterminate forms.

Similarly,

> If $f(c) = g(c) = \infty$, and if $f'(c)$ and $g'(c)$ exist, and if $g'(c) \neq 0$,
>
> then $\lim\limits_{x \to c} \dfrac{f(x)}{g(x)} = \dfrac{f'(c)}{g'(c)}$.

In other words, if the limit of the function gives us an undefined expression, like $\frac{0}{0}$ or $\frac{\infty}{\infty}$, L'Hôpital's Rule says we can take the derivative of the top and the derivative of the bottom and see if we get a determinate expression. If not, we can repeat the process.

Example 1: Find $\lim\limits_{x \to 0} \dfrac{\sin x}{x}$.

First, notice that plugging in 0 results in $\frac{0}{0}$, which is indeterminate. Take the derivative of the top and of the bottom.

$$\lim\limits_{x \to 0} \frac{\cos x}{1}$$

The limit equals 1.

Example 2: Find $\lim\limits_{x \to 0} \dfrac{2x - \sin x}{x}$.

If you plug in 0, you get $\dfrac{0-0}{0}$, which is indeterminate. Now, take the derivative of the top and of the bottom.

$$\lim_{x \to 0} \frac{2 - \cos x}{1}$$

This limit also equals 1.

Example 3: Find $\lim\limits_{x \to 0} \dfrac{\sqrt{4+x} - 2}{2x}$.

Again, plugging in 0 is no help; you get $\dfrac{0}{0}$. Take the derivative of the top and bottom.

$$\lim_{x \to 0} \frac{\dfrac{1}{2\sqrt{4+x}}}{2} = \frac{1}{8}$$

Example 4: Find $\lim\limits_{x \to 0} \dfrac{\sqrt{4+x} - 2 - \dfrac{x}{4}}{2x^2}$.

Take the derivative of the top and bottom.

$$\lim_{x \to 0} \frac{\dfrac{1}{2\sqrt{4+x}} - \dfrac{1}{4}}{4x}$$

Now if you take the limit, we still get $\dfrac{0}{0}$. So do it again.

$$\lim_{x \to 0} \frac{-\dfrac{1}{4}(4+x)^{-\frac{3}{2}}}{4} = -\frac{1}{128}$$

Now let's try a couple of $\dfrac{\infty}{\infty}$ forms.

Example 5: Find $\lim\limits_{x \to \infty} \dfrac{5x - 8}{3x + 1}$.

Now, the limit is $\dfrac{\infty}{\infty}$. The derivative of the top and bottom is

$$\lim_{x \to \infty} \frac{5}{3} = \frac{5}{3}$$

Don't you wish that you had learned this back when you first did limits?

Example 6: Find $\lim\limits_{x \to \frac{\pi}{2}} \dfrac{\sec x}{1 + \sec x}$.

The derivative of the top and bottom is

$$\lim_{x \to \frac{\pi}{2}} \frac{\sec x \tan x}{\sec x \tan x} = 1$$

Example 7: Find $\lim\limits_{x \to 0^+} x \cot x$.

Taking this limit results in $(0)(\infty)$, which is also indeterminate. (But you can't use the rule yet!) If you rewrite this expression as $\lim\limits_{x \to 0^+} \dfrac{x}{\tan x}$, it's of the form $\dfrac{0}{0}$, and we can use L'Hôpital's Rule.

$$\lim_{x \to 0^+} \frac{1}{\sec^2 x} = 1$$

That's all that you need to know about L'Hôpital's Rule. Just check to see if the limit results in an indeterminate form. If it does, use the rule until you get a determinate form.

Here are some more examples.

PROBLEM 1. Find $\lim\limits_{x \to 0} \dfrac{\sin 8x}{x}$.

Answer: First, notice that plugging in 0 gives us an indeterminate result: $\dfrac{0}{0}$. Now, take the derivative of the top and of the bottom.

$$\lim_{x \to 0} \frac{8 \cos 8x}{1} = 8$$

PROBLEM 2. Find $\lim\limits_{x \to \infty} x e^{-2x}$.

Answer: First, rewrite this expression as $\dfrac{x}{e^{2x}}$. Notice that when x nears infinity, the expression becomes $\dfrac{\infty}{\infty}$, which is indeterminate.

Take the derivative of the top and bottom.

$$\lim_{x \to \infty} \frac{1}{2e^{2x}} = 0$$

PROBLEM 3. Find $\lim\limits_{x\to\infty}\dfrac{5x^3-4x^2+1}{7x^3+2x-6}$.

Answer: We learned this in Chapter 3, remember? Now we'll use L'Hôpital's Rule. At first glance, the limit is indeterminate: $\dfrac{\infty}{\infty}$. Let's take some derivatives.

$$\frac{15x^2-8x}{21x^2+2}$$

This is still indeterminate, so it's time to take the derivative of the top and bottom again.

$$\frac{30x-8}{42x}$$

It's still indeterminate! If you try it one more time, you'll get a fraction with no variables: $\dfrac{30}{42}$, which can be simplified to $\dfrac{5}{7}$ (as we expected).

PROBLEM 4. Find $\lim\limits_{x\to\frac{\pi}{2}}\dfrac{x-\dfrac{\pi}{2}}{\cos x}$.

Answer: Plugging in $\dfrac{\pi}{2}$ gives you the indeterminate response of $\dfrac{0}{0}$. The derivative is

$$-\frac{1}{\sin x}$$

When we take the limit of this expression, we get –1.

PRACTICE PROBLEM SET 17

Now find these limits using L'Hôpital's Rule. The answers are in Chapter 23, starting on page 519.

1. Find $\lim\limits_{x\to 0}\dfrac{\sin 3x}{\sin 4x}$.

2. Find $\lim\limits_{x\to\pi}\dfrac{x-\pi}{\sin x}$.

3.   Find $\displaystyle\lim_{x\to 0}\frac{x-\sin x}{x^3}$.

4.   Find $\displaystyle\lim_{x\to 0}\frac{e^{3x}-e^{5x}}{x}$.

5.   Find $\displaystyle\lim_{x\to 0}\frac{\tan x-x}{\sin x-x}$.

6.   Find $\displaystyle\lim_{x\to\infty}\frac{x^5}{e^{5x}}$.

7.   Find $\displaystyle\lim_{x\to\infty}\frac{x^5+4x^3-8}{7x^5-3x^2-1}$.

8.   Find $\displaystyle\lim_{x\to 0^+}\frac{\ln(\sin x)}{\ln(\tan x)}$.

9.   Find $\displaystyle\lim_{x\to 0^+}\frac{\cot 2x}{\cot x}$.

10.   Find $\displaystyle\lim_{x\to 0^+}\frac{x}{\ln(x+1)}$.

LINEARIZATION

A differential is a very small quantity that corresponds to a change in a number. We use the symbol Δx to denote a differential. What are differentials used for? The AP Exam mostly wants you to use them to approximate the value of a function or to find the error of an approximation.

Recall the formula for the definition of the derivative.

$$f'(x)=\lim_{h\to 0}\frac{f(x+h)-f(x)}{h}$$

Replace h with Δx, which also stands for a very small increment of x, and get rid of the limit.

$$f'(x) \approx \frac{f(x+\Delta x)-f(x)}{\Delta x}$$

Notice that this is no longer equal to the derivative, but an approximation of it. If Δx is kept small, the approximation remains fairly accurate. Next, rearrange the equation as follows:

$$f(x+\Delta x) \approx f(x)+f'(x)\Delta x$$

This is our formula for differentials. It says that "the value of a function (at x plus a little bit) equals the value of the function (at x) plus the product of the derivative of the function (at x) and the little bit."

Example 1: Use differentials to approximate $\sqrt{9.01}$.

You can start by letting $x = 9$, $\Delta x = +0.01$, $f(x)=\sqrt{x}$. Next, we need to find $f'(x)$.

$$f'(x) = \frac{1}{2\sqrt{x}}$$

Now, plug in to the formula.

$$f(x+\Delta x) \approx f(x)+f'(x)\Delta x$$
$$\sqrt{x+\Delta x} \approx \sqrt{x} + \frac{1}{2\sqrt{x}}\Delta x$$

Now, if we plug in $x = 9$ and $\Delta x = +0.01$,

$$\sqrt{9.01} \approx \sqrt{9} + \frac{1}{2\sqrt{9}}(0.01) \approx 3.001666666$$

If you enter $\sqrt{9.01}$ into your calculator, you get 3.001666204. As you can see, our answer is a pretty good approximation. It's not so good, however, when Δx is too big. How big is too big? Good question.

Example 2: Use differentials to approximate $\sqrt{9.5}$.

Let $x = 9$, $\Delta x = +0.5$, $f(x) = \sqrt{x}$ and plug in to what you found in Example 1.

$$\sqrt{9.5} \approx \sqrt{9} + \frac{1}{2\sqrt{9}}(0.5) \approx 3.083333333$$

However, $\sqrt{9.5}$ equals 3.082207001 on a calculator. This is good to only two decimal places. As the ratio of $\frac{\Delta x}{x}$ grows larger, the approximation gets less accurate, and we start to get away from the actual value.

There's another approximation formula that you'll need to know for the AP Exam. This formula is used to estimate the error in a measurement or to find the effect on a formula when a small change in measurement is made. The formula is

$$dy = f'(x)\, dx$$

Note that this equation is simply a rearrangement of $\frac{dy}{dx} = f'(x)$.

This notation may look a little confusing. It says that the change in a measurement dy, due to a differential dx, is found by multiplying the derivative of the equation for y by the differential. Let's do an example.

Example 3: The radius of a circle is increased from 3 to 3.04. Estimate the change in area.

Let $A = \pi r^2$. Then our formula says that $dA = A'\, dr$, where A' is the derivative of the area with respect to r, and $dr = 0.04$ (the change). First, find the derivative of the area: $A' = 2\pi r$. Now, plug in to the formula.

$$dA = 2\pi r\, dr = 2\pi(3)(0.04) = 0.754$$

The actual change in the area is from 9π to 9.2416π, which is approximately 0.759. As you can see, this approximation formula is pretty accurate.

Here are some sample problems involving this differential formula. Try them out, then check your work against the answers directly beneath.

Problem 1. Use differentials to approximate $(3.98)^4$.

Answer: Let $f(x) = x^4$, $x = 4$, and $\Delta x = -0.02$. Next, find $f'(x)$, which is: $f'(x) = 4x^3$.

Now, plug in to the formula.

$$f(x + \Delta x) \approx f(x) + f'(x)\Delta x$$
$$(x + \Delta x)^4 \approx x^4 + 4x^3\Delta x$$

If you plug in $x = 4$ and $\Delta x = -0.02$, you get

$$(3.98)^4 \approx 4^4 + 4(4)^3(-0.02) \approx 250.88$$

Check $(3.98)^4$ by using your calculator; you should get 250.9182722. Not a bad approximation.

You're probably asking yourself, why can't I just use my calculator every time? Because most math teachers are dedicated to teaching you several complicated ways to calculate things without your calculator.

PROBLEM 2. Use differentials to approximate sin 46°.

Answer: This is a tricky question. The formula doesn't work if you use degrees. Here's why: Let $f(x) = \sin x$, $x = 45°$, and $\Delta x = 1°$. The derivative is $f'(x) = \cos x$.

If you plug this information into the formula, you get $\sin 46° \approx \sin 45° + \cos 45°(1°) = \sqrt{2}$. You should recognize that this is nonsense for two reasons: (1) the sine of any angle is between -1 and 1; and (2) the answer should be close to $\sin 45° = \dfrac{1}{\sqrt{2}}$.

What went wrong? You have to use radians! As we mentioned before, angles in calculus problems are measured in radians, not degrees.

Let $f(x) = \sin x$, $x = \dfrac{\pi}{4}$, and $\Delta x = \dfrac{\pi}{180}$. Now plug in to the formula.

$$\sin\left(\frac{46\pi}{180}\right) \approx \sin\frac{\pi}{4} + \left(\cos\frac{\pi}{4}\right)\left(\frac{\pi}{180}\right) = 0.7194$$

PROBLEM 3. The radius of a sphere is measured to be 4 cm with an error of ± 0.01 cm. Use differentials to approximate the error in the surface area.

Answer: Now it's time for the other differential formula. The formula for the surface area of a sphere is

$$S = 4\pi r^2$$

The formula says that $dS = S'dr$, so first, we find the derivative of the surface area, $(S' = 8\pi r)$ and plug away.

$$dS = 8\pi r\, dr = 8\pi(4)(\pm 0.01) = \pm 1.0053$$

This looks like a big error, but given that the surface area of a sphere with radius 4 is approximately 201 cm², the error is quite small.

PRACTICE PROBLEM SET 18

Use the differential formulas in this chapter to solve these problems. The answers are in Chapter 23, starting on page 525.

1. Approximate $\sqrt{25.02}$.

2. Approximate $\sqrt[3]{63.97}$.

3. Approximate $\tan 61°$.

4. The side of a cube is measured to be 6 in. with an error of ± 0.02 in. Estimate the error in the volume of the cube.

5. When a spherical ball bearing is heated, its radius increases by 0.01 mm. Estimate the change in volume of the ball bearing when the radius is 5 mm.

6. A cylindrical tank is constructed to have a diameter of 5 meters and a height of 20 meters. Find the error in the volume if

    (a) the diameter is exact, but the height is 20.1 meters; and

    (b) the height is exact, but the diameter is 5.1 meters.

LOGARITHMIC DIFFERENTIATION

There's one last topic in differential calculus that you need to know: logarithmic differentiation. It's a very simple and handy technique used to find the derivatives of expressions that involve a lot of algebra. By employing the rules of logarithms, we can find the derivatives of expressions that would otherwise require a messy combination of the Chain Rule, the Product Rule, and the Quotient Rule.

First, let's review a couple of rules of logarithms (remember, when we refer to a logarithm in calculus, we mean the natural log (base e), not the common log).

$$\ln A + \ln B = \ln(AB)$$
$$\ln A - \ln B = \ln\left(\frac{A}{B}\right)$$
$$\ln A^B = B \ln A$$

For example, we can rewrite $\ln(x + 3)^5$ as $5\ln(x + 3)$.

As a quick review exercise, how could you rewrite the following:

$$\ln \frac{x^2}{(x+1)^2}$$

Answer: $2\ln x - 2\ln(x + 1)$.

The other important thing to remember from Chapter 11 is

$$\frac{d}{dx}\ln u = \frac{1}{u}\frac{du}{dx}$$

Bearing these in mind, we can use these rules to find the derivative of a complicated expression without using all that mind-bending algebra. Instead, take the log of both sides of an expression, and then use the rules of logarithms to simplify it before finding the derivative.

Example 1: Find the derivative of $y = \dfrac{x+3}{2x-5}$.

The Quotient Rule works here just fine, but check this out instead. Take the log of both sides.

$$\ln y = \ln \frac{x+3}{2x-5}$$

Using log rules, you can rewrite the expression

$$\ln y = \ln(x + 3) - \ln(2x - 5)$$

Now take the derivative of both sides.

$$\frac{1}{y}\frac{dy}{dx} = \frac{1}{x+3} - \frac{2}{2x-5}$$

Multiply both sides by y, and you get the following:

$$\frac{dy}{dx} = y\left(\frac{1}{x+3} - \frac{2}{2x-5}\right)$$

Finally, substitute for *y*.

$$\frac{dy}{dx} = \left(\frac{x+3}{2x-5}\right)\left(\frac{1}{x+3} - \frac{2}{2x-5}\right)$$

You might be thinking, "Why go through the hassle?" After all, you could have done this with the Quotient Rule, and it would have involved less work. But, as you'll see, this technique comes in very handy when the algebra is especially messy. Also, we don't usually substitute back for the *y* term. Instead, just leave it in the derivative (unless it needs to be defined only in terms of *x*).

Example 2: Find the derivative of *y*, where $y^2 = (x + 4)^3(x - 2)^5$.

Take the log of both sides.

$$\ln y^2 = \ln\left[(x+4)^3(x-2)^5\right]$$

Next, use those log rules to get

$$2\ln y = 3\ln(x + 4) + 5\ln(x - 2)$$

Take the derivative of both sides.

$$\frac{2}{y}\frac{dy}{dx} = \frac{3}{x+4} + \frac{5}{x-2}$$

Finally, multiply both sides by $\frac{y}{2}$.

$$\frac{dy}{dx} = \frac{y}{2}\left[\frac{3}{x+4} + \frac{5}{x-2}\right]$$

That's it. Now, wasn't that easier? Substitute back for *y* only if you have to.

Example 3: Find the derivative of $y = \sqrt{\frac{x+3}{x-5}}\sqrt[3]{\frac{2x+7}{5x-1}}$.

Holy cripes! What a problem! Watch how using logs will save you. Take the log of both sides.

$$\ln y = \ln\left(\sqrt{\frac{x+3}{x-5}}\sqrt[3]{\frac{2x+7}{5x-1}}\right)$$

Now you can simplify using log rules.

$$\ln y = \frac{1}{2}\left[\ln(x+3) - \ln(x-5)\right] + \frac{1}{3}\left[\ln(2x+7) - \ln(5x-1)\right]$$

Take the derivative.

$$\frac{1}{y}\frac{dy}{dx} = \frac{1}{2}\left[\frac{1}{x+3} - \frac{1}{x-5}\right] + \frac{1}{3}\left[\frac{2}{2x+7} - \frac{5}{5x-1}\right]$$

And then multiply by y.

$$\frac{dy}{dx} = \frac{y}{2}\left[\frac{1}{x+3} - \frac{1}{x-5}\right] + \frac{y}{3}\left[\frac{2}{2x+7} - \frac{5}{5x-1}\right]$$

Example 4: Find the derivative of $y = \sqrt{\dfrac{x+3}{5-x}}$.

Take the log of both sides.

$$\ln y = \ln\left(\sqrt{\frac{x+3}{5-x}}\right)$$

Simplify using the log rules.

$$\ln y = \frac{1}{2}\left[\ln(x+3) - \ln(5-x)\right]$$

Take the derivative.

$$\frac{1}{y}\frac{dy}{dx} = \frac{1}{2}\left[\frac{1}{x+3} + \frac{1}{5-x}\right]$$

Multiply by y.

$$\frac{dy}{dx} = \frac{y}{2}\left[\frac{1}{x+3} + \frac{1}{5-x}\right]$$

And you're done.

That's all there is to logarithmic differentiation. It's a really helpful tool for simplifying complicated derivatives. There's generally one question on the AP Exam that uses logarithmic differentiation, and it's usually not that complicated.

Try these solved problems on your own. Cover the answers first, and then check your work.

PROBLEM 1. Use logarithmic differentiation to find $\dfrac{dy}{dx}$ if $y = \dfrac{x^2\sqrt{1-x^3}}{\left(x^2+1\right)^2}$.

Answer: Take the log of both sides.

$$\ln y = \ln\left[\frac{x^2\sqrt{1-x^3}}{\left(x^2+1\right)^2}\right]$$

Next, use the log rules to simplify the expression.

$$\ln y = \ln x^2 + \ln\sqrt{1-x^3} - \ln\left(x^2+1\right)^2$$

$$\ln y = 2\ln x + \frac{1}{2}\ln\left(1-x^3\right) - 2\ln\left(x^2+1\right)$$

Taking the derivative of both sides, you should get

$$\frac{1}{y}\frac{dy}{dx} = \frac{2}{x} - \frac{3x^2}{2\left(1-x^3\right)} - \frac{4x}{x^2+1}$$

Then, get the derivative by itself by multiplying both sides by y.

$$\frac{dy}{dx} = y\left[\frac{2}{x} - \frac{3x^2}{2\left(1-x^3\right)} - \frac{4x}{x^2+1}\right]$$

PROBLEM 2. Use logarithmic differentiation to find $\dfrac{dy}{dx}$ if $y = \dfrac{\left(x^2-5x\right)\cos^2 x}{\left(x^3+1\right)^5}$.

Answer: The four-step process should be second nature by now. Take the log of both sides.

$$\ln y = \ln\left[\frac{\left(x^2-5x\right)\cos^2 x}{\left(x^3+1\right)^5}\right]$$

Simplify the expression using log rules.

$$\ln y = \ln\left(x^2-5x\right) + \ln\left(\cos^2 x\right) - \ln\left(x^3+1\right)^5$$

$$\ln y = \ln\left(x^2-5x\right) + 2\ln\left(\cos x\right) - 5\ln\left(x^3+1\right)$$

Take the derivative of both sides.

$$\frac{1}{y}\frac{dy}{dx} = \frac{2x-5}{x^2-5x} - \frac{2\sin x}{\cos x} - \frac{15x^2}{x^3+1}$$

$$\frac{1}{y}\frac{dy}{dx} = \frac{2x-5}{x^2-5x} - 2\tan x - \frac{15x^2}{x^3+1}$$

Multiply by y.

$$\frac{dy}{dx} = y\left[\frac{2x-5}{x^2-5x} - 2\tan x - \frac{15x^2}{x^3+1}\right]$$

PRACTICE PROBLEM SET 19

Use logarithmic differentiation to find the derivative of each of the following problems. The answers are in Chapter 23, starting on page 527.

1. $y = x\left(\sqrt[4]{1-x^3}\right)$

2. $y = \sqrt{\dfrac{1+x}{1-x}}$

3. $y = \dfrac{\left(x^3+5\right)^{\frac{3}{2}}\sqrt[3]{4-x^2}}{x^4-x^2+6}$

4. $y = \dfrac{\sin x \, \cos x}{\sqrt{x^3-4}}$

5. $y = \dfrac{\left(4x^2-8x\right)^3\left(5-3x^4+7x\right)^4}{\left(x^2+x\right)^3}$

6. $y = \dfrac{x-1}{x\tan x}$

7. $y = \left(x-x^2\right)^2\left(x^3+x^4\right)^3\left(x^6-x^5\right)^4$

8. $y = \sqrt[4]{\dfrac{x(1-x)(1+x)}{\left(x^2-1\right)(5-x)}}$

End of Chapter 12 Drill

The answers are in Chapter 24.

1. Find the derivative of the inverse of $y = 5x^3 + x - 7$ at $y = -7$.

   (A) $-\dfrac{1}{7}$

   (B) $\dfrac{1}{7}$

   (C) 1

   (D) 5131

2. Find the equation of the tangent line to $x = 4\sin^2 t$ and $y = 2\cos t$ at $t = \dfrac{\pi}{3}$.

   (A) $y - 1 = -\dfrac{1}{4\sqrt{3}}(x - 3)$

   (B) $y - \sqrt{3} = -\dfrac{1}{4\sqrt{3}}(x - 1)$

   (C) $y - \sqrt{3} = -\dfrac{1}{2}(x - 1)$

   (D) $y - 1 = -\dfrac{1}{2}(x - 3)$

3. Use L'Hôpital's Rule to find $\displaystyle\lim_{x \to \infty} \dfrac{x^{-\frac{4}{3}}}{\sin\left(\dfrac{1}{x}\right)}$.

   (A) $-\infty$
   (B) 0
   (C) ∞
   (D) The limit does not exist.

4. Approximate $(2.03)^5$.

   (A) 29.6
   (B) 32.03
   (C) 34.4
   (D) 80.03

5. Find the derivative of $y = \sqrt[4]{\dfrac{(8x+1)^3}{(2x-11)^5}}$.

   (A) $y\left(\dfrac{3}{4}\dfrac{8}{8x+1} - \dfrac{5}{4}\dfrac{2}{2x-11}\right)$

   (B) $y\left(\dfrac{3}{4}\dfrac{1}{8x+1} - \dfrac{5}{4}\dfrac{1}{2x-11}\right)$

   (C) $y\left(\dfrac{3}{4}\dfrac{1}{8x+1} + \dfrac{5}{4}\dfrac{1}{2x-11}\right)$

   (D) $y\left(\dfrac{3}{4}\dfrac{8}{8x+1} + \dfrac{5}{4}\dfrac{2}{2x-11}\right)$

REFLECT

Respond to the following questions:

- For which topics discussed in this chapter do you feel you have achieved sufficient mastery to answer multiple-choice questions correctly?

- For which topics discussed in this chapter do you feel you have achieved sufficient mastery to answer open-ended questions correctly?

- For which topics discussed in this chapter do you feel you need more work before you can answer multiple-choice questions correctly?

- For which topics discussed in this chapter do you feel you need more work before you can answer open-ended questions correctly?

- What parts of this chapter are you going to re-review?

Unit 3
Integral
Calculus
Essentials

Chapter 13
The Integral

Welcome to the other half of calculus! This, unfortunately, is the more difficult half, but don't worry. We'll get you through it. In differential calculus, you learned all of the fun things that you can do with the derivative. Now you'll learn to do the reverse: how to take an integral. As you might imagine, there's a bunch of new fun things that you can do with integrals too.

It's also time for a new symbol $\int$, which stands for integration. An integral actually serves several different purposes, but the first, and most basic, is that of the antiderivative.

THE ANTIDERIVATIVE

An antiderivative is a derivative in reverse. Therefore, we're going to reverse some of the rules we learned with derivatives and apply them to integrals. For example, we know that the derivative of x^2 is $2x$. If we are given the derivative of a function and have to figure out the original function, we use antidifferentiation. Thus, the antiderivative of $2x$ is x^2. (Actually, the answer is slightly more complicated than that, but we'll get into that in a few moments.)

Now we need to add some info here to make sure that you get this absolutely correct. First, as far as notation goes, it is traditional to write the antiderivative of a function using its uppercase letter, so the antiderivative of $f(x)$ is $F(x)$, the antiderivative of $g(x)$ is $G(x)$, and so on.

The second idea is very important: each function has more than one antiderivative. In fact, there are an infinite number of antiderivatives of a function. Let's go back to our example to help illustrate this.

Remember that the antiderivative of $2x$ is x^2? Well, consider the fact that if you take the derivative of $x^2 + 1$, you get $2x$. The same is true for $x^2 + 2$, $x^2 - 1$, and so on. In fact, if *any* constant is added to x^2, the derivative is still $2x$ because the derivative of a constant is zero.

Because of this, we write the antiderivative of $2x$ as $x^2 + C$, where C stands for any constant.

Finally, whenever you take the integral (or antiderivative) of a function of x, you always add the term dx (or dy if it's a function of y, etc.) to the integrand (the thing inside the integral). You'll learn why later.

For now, just remember that you must always use the dx symbol, and teachers love to take points off for forgetting the dx. Don't ask why, but they do!

Here is the **Power Rule** for antiderivatives.

$$\text{If } f(x) = x^n, \text{ then } \int f(x)dx = \frac{x^{n+1}}{n+1} + C \text{ (except when } n = -1).$$

Example 1: Find $\int x^3\,dx$. $= \dfrac{x^4}{4} + C$

Using the Power Rule, we get

$$\int x^3\,dx = \frac{x^4}{4} + C$$

Don't forget the constant C, or your teachers will take points off for that too!

Example 2: Find $\int x^{-3}\,dx$. $= \dfrac{x^{-2}}{-2} + C$

The Power Rule works with negative exponents too.

$$\int x^{-3}\,dx = \frac{x^{-2}}{-2} + C$$

Not terribly hard, is it? Now it's time for a few more rules that look remarkably similar to the rules for derivatives that we saw in Chapter 6.

$$\int k f(x)\,dx = k \int f(x)\,dx$$
$$\int \left[f(x) + g(x) \right]\,dx = \int f(x)\,dx + \int g(x)\,dx$$
$$\int k\,dx = kx + C$$

Here are a few more examples to make you an expert.

Example 3: $\int 5\,dx = 5x + C$

Example 4: $\int 7x^3\,dx = \dfrac{7x^4}{4} + C$

Example 5: $\int \left(3x^2 + 2x \right)\,dx = x^3 + x^2 + C$

Example 6: $\int \sqrt{x}\,dx = \dfrac{x^{\frac{3}{2}}}{\frac{3}{2}} + C = \dfrac{2x^{\frac{3}{2}}}{3} + C$

INTEGRALS OF TRIG FUNCTIONS

The integrals of some trigonometric functions follow directly from the derivative formulas in Chapter 8.

We didn't mention the integrals of tangent, cotangent, secant, and cosecant, because you need to know some rules about logarithms to figure them out. We'll get to them in a few chapters. Notice also that each of the answers is divided by a constant. This is to account for the Chain Rule. Let's do some examples.

$$\int \sin ax\ dx = -\frac{\cos ax}{a} + C$$

$$\int \cos ax\ dx = \frac{\sin ax}{a} + C$$

$$\int \sec ax \tan ax\ dx = \frac{\sec ax}{a} + C$$

$$\int \sec^2 ax\ dx = \frac{\tan ax}{a} + C$$

$$\int \csc ax \cot ax\ dx = -\frac{\csc ax}{a} + C$$

$$\int \csc^2 ax\ dx = -\frac{\cot ax}{a} + C$$

Example 7: Check the integral $\int \sin 5x\ dx = -\frac{\cos 5x}{5} + C$ by differentiating the answer.

$$\frac{d}{dx}\left[-\frac{\cos 5x}{5} + C\right] = -\frac{1}{5}(-\sin 5x)(5) = \sin 5x$$

Notice how the constant is accounted for in the answer?

Example 8: $\int \sec^2 3x\ dx = \frac{\tan 3x}{3} + C$

Example 9: $\int \cos \pi x\ dx = \frac{\sin \pi x}{\pi} + C$

Example 10: $\int \sec\left(\frac{x}{2}\right)\tan\left(\frac{x}{2}\right) dx = 2\sec\left(\frac{x}{2}\right) + C$

If you're not sure if you have the correct answer when you take an integral, you can always check by differentiating the answer and seeing if you get what you started with. Try to get in the habit of doing that at the beginning, because it'll help you build confidence in your ability to find integrals properly. You'll see that, although you can differentiate just about any expression that you'll normally encounter, you won't be able to integrate many of the functions you see.

ADDITION AND SUBTRACTION

By using the rules for addition and subtraction, we can integrate most polynomials.

Example 11: Find $\int (x^3 + x^2 - x)\ dx$. $= \frac{x^4}{4} + \frac{x^3}{3} - \frac{x^2}{2} + C$

We can break this into separate integrals, which gives us

$$\int x^3 dx + \int x^2 dx - \int x\ dx$$

Now you can integrate each of these individually.

$$\frac{x^4}{4} + C + \frac{x^3}{3} + C - \frac{x^2}{2} + C$$

You can combine the constants into one constant (it doesn't matter how many C's we use, because their sum is one collective constant whose derivative is zero).

$$\frac{x^4}{4} + \frac{x^3}{3} - \frac{x^2}{2} + C$$

Sometimes you'll be given information about the function you're seeking that will enable you to solve for the constant. Often, this is an "initial value," which is the value of the function when the variable is zero. As we've seen, normally there are an infinite number of solutions for an integral, but when we solve for the constant, there's only one.

Example 12: Find the equation of y where $\dfrac{dy}{dx} = 3x + 5$ and $y = 6$ when $x = 0$.

Let's put this in integral form.

$$y = \int (3x + 5)\, dx$$

Handwritten:
$$dy = 3x + 5\, dx$$
$$y = \int 3x + 5\, dx$$
$$y = \frac{3x^2}{2} + 5x + C$$

Integrating, we get

$$y = \frac{3x^2}{2} + 5x + C$$

Handwritten:
$$6 = \frac{3(0)^2}{2} + 5(0) + C$$

Now we can solve for the constant because we know that $y = 6$ when $x = 0$.

Handwritten: $6 = C$

$$6 = \frac{3(0)^2}{2} + 5(0) + C$$

Therefore, $C = 6$ and the equation is

Handwritten (boxed):
$$y = \frac{3x^2}{2} + 5x + b$$

$$y = \frac{3x^2}{2} + 5x + 6$$

Example 13: Find $f(x)$ if $f'(x) = \sin x - \cos x$ and $f(\pi) = 3$.

Integrate $f'(x)$.

$$f(x) = \int (\sin x - \cos x)\, dx = -\cos x - \sin x + C$$

Handwritten:
$$f(x) = \int \sin x - \cos x\, dx$$
$$f(x) = -\cos x - \sin x + C$$
$$f(\pi) = 3$$

Now solve for the constant.

$$3 = -\cos(\pi) - \sin(\pi) + C$$

$$C = 2$$

Handwritten:
$$3 = -\cos(\pi) - \sin(\pi) + C$$
$$3 = 1 - 0 + C$$
$$C = 2$$

Handwritten (boxed):
$$f(x) = -\cos x - \sin x + 2$$

Therefore, the equation becomes

$$f(x) = -\cos x - \sin x + 2$$

Now we've covered the basics of integration. However, integration is a very sophisticated topic and there are many types of integrals that will cause you trouble. We will need several techniques to learn how to evaluate these integrals. The first and most important is called *u*-substitution, which we will cover in the second half of this chapter.

In the meantime, here are some solved problems. Do each problem, covering the answer first, then check your answer.

PROBLEM 1. Evaluate $\int x^{\frac{3}{5}} dx$. $= \dfrac{x^{8/5}}{8/5} = \dfrac{5x^{8/5}}{8} + C$

Answer: Here's the Power Rule again.

$$\int x^n dx = \frac{x^{n+1}}{n+1} + C$$

Using the rule,

$$\int x^{\frac{3}{5}} dx = \frac{x^{\frac{8}{5}}}{\frac{8}{5}} + C$$

You can rewrite it as $\dfrac{5x^{\frac{8}{5}}}{8} + C$.

PROBLEM 2. Evaluate $\int \left(5x^3 + x^2 - 6x + 4\right) dx$. $= \dfrac{5x^4}{4} + \dfrac{x^3}{3} - 3x^2 + 4x + C$

Answer: We can break this up into several integrals.

$$\int \left(5x^3 + x^2 - 6x + 4\right) dx = \int 5x^3 dx + \int x^2 dx - \int 6x \, dx + \int 4 dx$$

Each of these can be integrated according to the Power Rule.

$$\frac{5x^4}{4} + C + \frac{x^3}{3} + C - \frac{6x^2}{2} + C + 4x + C$$

This can be rewritten as

$$\frac{5x^4}{4} + \frac{x^3}{3} - 3x^2 + 4x + C$$

Notice that we combine the constant terms into one constant term C.

PROBLEM 3. Evaluate $\int (3 - x^2)^2 \, dx$.

Answer: First, expand the integrand.

$$\int (9 - 6x^2 + x^4) \, dx$$

Break this up into several integrals.

$$\int 9 \, dx - \int 6 \, x^2 \, dx + \int x^4 \, dx$$

And integrate according to the Power Rule.

$$9x - 2x^3 + \frac{x^5}{5} + C$$

PROBLEM 4. Evaluate $\int (4 \sin x - 3 \cos x) \, dx$.

Answer: Break this problem into two integrals.

$$4 \int \sin x \, dx - 3 \int \cos x \, dx$$

Each of these trig integrals can be evaluated according to its rule.

$$-4\cos x - 3\sin x + C$$

PROBLEM 5. Evaluate $\int (2 \sec^2 x - 5 \csc^2 x) \, dx$.

Answer: Break the integral in two.

$$2 \int \sec^2 x \, dx - 5 \int \csc^2 x \, dx$$

Each of these trig integrals can be evaluated according to its rule.

$$2\tan x + 5\cot x + C$$

PRACTICE PROBLEM SET 20

Now evaluate the following integrals. The answers are in Chapter 23, starting on page 532.

1. $\displaystyle\int \frac{1}{x^4}\,dx$ $= \int x^{-4}\,dx = \dfrac{x^{-3}}{-3} + C$

2. $\displaystyle\int \frac{5}{\sqrt{x}}\,dx$ $= 5\int x^{-1/2} = 5\left(\dfrac{\sqrt{x}}{1/2} + C\right)$

3. $\displaystyle\int \left(5x^4 - 3x^2 + 2x + 6\right)dx$ $= x^5 - x^3 + x^2 + 6x + C$

4. $\displaystyle\int \left(3x^{-3} - 2x^{-2} + x^4 + 16x^7\right)dx$ $= \dfrac{3x^{-2}}{-2} - \dfrac{2x^{-1}}{-1} + \dfrac{x^5}{5} + 2x^8 + C$

5. $\displaystyle\int x^{\frac{1}{3}}(2+x)\,dx$ $= \int 2x^{1/3} + x^{4/3}\,dx = \dfrac{2x^{4/3}}{4/3} + \dfrac{x^{7/3}}{7/3} + C$

6. $\displaystyle\int \left(x^3 + x\right)^2 dx$

7. $\displaystyle\int \frac{x^6 - 2x^4 + 1}{x^2}\,dx$

8. $\displaystyle\int x(x-1)^3\,dx$

9. $\displaystyle\int \sec x\left(\sec x + \tan x\right)dx$

10. $\displaystyle\int \left(\sec^2 x + x\right)dx$

11. $\displaystyle\int \frac{\cos^3 x + 4}{\cos^2 x}\,dx$

12. $\displaystyle\int \frac{\sin 2x}{\cos x}\,dx$

13. $\displaystyle\int \left(1+\cos^2 x \sec x\right)dx$

14. $\displaystyle\int \frac{1}{\csc x}\,dx$

15. $\displaystyle\int \left(x-\frac{2}{\cos^2 x}\right)dx$

u-SUBSTITUTION

When we discussed differentiation, one of the most important techniques we mastered was the Chain Rule. Now, you'll learn the integration corollary of the Chain Rule (called *u*-substitution), which we use when the integrand is a composite function. All you do is replace the function with *u*, and then you can integrate the simpler function using the Power Rule (as shown below).

$$\int u^n\,du = \frac{u^{n+1}}{n+1} + C$$

Suppose you have to integrate $\displaystyle\int (x-4)^{10}\,dx$. You could expand out this function and integrate each term, but that'll take a while. Instead, you can follow these four steps.

Step 1: Let $u = x - 4$. Then $\dfrac{du}{dx} = 1$ (rearrange this to get $du = dx$).

Step 2: Substitute $u = x - 4$ and $du = dx$ into the integrand.

$$\int u^{10}\,du$$

Step 3: Integrate.

$$\int u^{10}\,du = \frac{u^{11}}{11} + C$$

Step 4: Substitute back for u.

$$\frac{(x-4)^{11}}{11} + C$$

That's u-substitution. The main difficulty you'll have will be picking the appropriate function to set equal to u. The best way to get better is to practice. The object is to pick a function and replace it with u, then take the derivative of u to find du. If we can't replace *all* of the terms in the integrand, we *can't* do the substitution.

Let's do some examples.

Example 1: $\int 10x\left(5x^2 - 3\right)^6 dx =$

Once again, you could expand this out and integrate each term, but that would be difficult. Use u-substitution.

Let $u = 5x^2 - 3$. Then $\frac{du}{dx} = 10x$ and $du = 10x\, dx$. Now you can substitute.

$$\int u^6 du$$

And integrate.

$$\int u^6 du = \frac{u^7}{7} + C$$

Substituting back gives you

$$\frac{\left(5x^2 - 3\right)^7}{7} + C$$

Confirm that this is the integral by differentiating $\dfrac{\left(5x^2 - 3\right)^7}{7} + C$.

$$\frac{d}{dx}\left[\frac{\left(5x^2 - 3\right)^7}{7} + C\right] = \frac{7\left(5x^2 - 3\right)^6}{7}(10x) = \left(5x^2 - 3\right)^6 (10x)$$

Example 2: $\int 2x\sqrt{x^2 - 5}\, dx =$

If $u = x^2 - 5$, then $\frac{du}{dx} = 2x$ and $du = 2x\, dx$. Substitute u into the integrand.

$$\int u^{\frac{1}{2}}\, du$$

Integrate.

$$\int u^{\frac{1}{2}}\,du = \frac{u^{\frac{3}{2}}}{\frac{3}{2}} + C = \frac{2u^{\frac{3}{2}}}{3} + C$$

And substitute back.

$$\frac{2\left(x^2 - 5\right)^{\frac{3}{2}}}{3} + C$$

Note: From now on, we're not going to rearrange $\frac{du}{dx}$; we'll go directly to "$du =$" format. You should be able to do that step on your own.

Example 3: $\displaystyle\int 3\sin(3x - 1)\,dx =$

Let $u = 3x - 1$. Then $du = 3dx$. Substitute the u in the integral.

$$\int \sin u\,du$$

Figure out the integral.

$$\int \sin u\,du = -\cos u + C$$

And throw the x's back in.

$$-\cos(3x - 1) + C$$

So far, this is only the simplest kind of u-substitution; naturally, the process can get worse when the substitution isn't as easy. Usually, you'll have to insert a constant term to put the integrand into a workable form.

Example 4: $\displaystyle\int (5x + 7)^{20}\,dx =$

Let $u = 5x + 7$. Then $du = 5\,dx$. Notice that we can't do the substitution immediately because we need to substitute for dx and we have $5\,dx$. No problem! Because 5 is a constant, just solve for dx.

$$\frac{1}{5}du = dx$$

Now you can substitute.

$$\int (5x + 7)^{20}\,dx = \int u^{20}\left(\frac{1}{5}\right)du$$

Rearrange the integral and solve.

$$\int u^{20}\left(\frac{1}{5}\right)du = \frac{1}{5}\int u^{20}\,du = \frac{1}{5}\frac{u^{21}}{21}+C = \frac{u^{21}}{105}+C$$

And now it's time to substitute back.

$$\frac{u^{21}}{105}+C = \frac{(5x+7)^{21}}{105}+C$$

Example 5: $\int x\cos(3x^2+1)dx =$

Let $u = 3x^2 + 1$. Then $du = 6x\,dx$. We need to substitute for $x\,dx$, so we can rearrange the du term.

$$\frac{1}{6}du = x\,dx$$

Now substitute.

$$\int \frac{1}{6}\cos u\,du$$

Evaluate the integral.

$$\int \frac{1}{6}\cos u\,du = \frac{1}{6}\sin u + C$$

And substitute back.

$$\frac{1}{6}\sin u + C = \frac{1}{6}\sin(3x^2+1)+C$$

Example 6: $\int x\sec^2(x^2)dx =$

Let $u = x^2$. Then $du = 2x\,dx$ and $\frac{1}{2}du = x\,dx$.

Substitute.

$$\int \frac{1}{2}\sec^2 u\,du$$

Evaluate the integral.

$$\int \frac{1}{2}\sec^2 u\,du = \frac{1}{2}\tan u + C$$

Now the original function goes back in.

$$\frac{1}{2}\tan(x^2)+C$$

This is a good technique to master, so practice on the following solved problems. Do each problem, covering the answer first, then check your answer.

PROBLEM 1. Evaluate $\int \sec^2 3x \, dx$.

Answer: Let $u = 3x$ and $du = 3dx$. Then $\frac{1}{3} du = dx$.

Substitute and integrate.

$$\frac{1}{3}\int \sec^2 u \, du = \frac{1}{3}\tan u + C$$

Then substitute back.

$$\frac{1}{3}\tan 3x + C$$

PROBLEM 2. Evaluate $\int \sqrt{5x - 4} \, dx$.

Answer: Let $u = 5x - 4$ and $du = 5dx$. Then $\frac{1}{5} du = dx$.

Substitute and integrate.

$$\frac{1}{5}\int u^{\frac{1}{2}} \, du = \frac{2}{15} u^{\frac{3}{2}} + C$$

Then substitute back.

$$\frac{2}{15}(5x - 4)^{\frac{3}{2}} + C$$

PROBLEM 3. Evaluate $\int x\left(4x^2 - 7\right)^{10} dx$.

Answer: Let $u = 4x^2 - 7$ and $du = 8x \, dx$. Then $\frac{1}{8} du = x \, dx$.

Substitute and integrate.

$$\frac{1}{8}\int u^{10} \, du = \frac{1}{88} u^{11} + C$$

Then substitute back.

$$\frac{1}{88}\left(4x^2 - 7\right)^{11} + C$$

PROBLEM 4. Evaluate $\int \tan \dfrac{x}{3}\ \sec^2 \dfrac{x}{3}\, dx$.

Answer: Let $u = \tan \dfrac{x}{3}$ and $du = \dfrac{1}{3}\sec^2 \dfrac{x}{3}\, dx$. Then $3\, du = \sec^2 \dfrac{x}{3}\, dx$.

Substituting, we get

$$3\int u\ du = \frac{3}{2}u^2 + C$$

Then substitute back.

$$\frac{3}{2}\tan^2 \frac{x}{3} + C$$

PRACTICE PROBLEM SET 21

Now evaluate the following integrals. The answers are in Chapter 23, starting on page 535.

1. $\displaystyle\int \sin 2x \cos 2x\, dx$

2. $\displaystyle\int \frac{3x\, dx}{\sqrt[3]{10 - x^2}}$

3. $\displaystyle\int x^3 \sqrt{5x^4 + 20}\, dx$

4. $\displaystyle\int \left(x^2 + 1\right)\left(x^3 + 3x\right)^{-5} dx$

5. $\displaystyle\int \frac{1}{\sqrt{x}}\sin \sqrt{x}\, dx$

6. $\displaystyle\int x^2 \sec^2 x^3\, dx$

7. $\displaystyle\int \frac{\cos\left(\dfrac{3}{x}\right)}{x^2}\, dx$

8. $\displaystyle\int \sin\left(\sin x\right)\cos x\, dx$

End of Chapter 13 Drill

The answers are in Chapter 24.

1. Evaluate $\int \left(3x^4 - 6x^3 + 2x + 7\right) dx$.

   (A) $12x^3 - 18x^2 + 2 + C$

   (B) $3x^5 - 6x^4 + x^2 + 7x + C$

   (C) $\dfrac{3x^5}{5} - \dfrac{6x^4}{4} + x^2 + 7x + C$

   (D) $\dfrac{3x^4}{4} - \dfrac{6x^3}{3} + \dfrac{x^2}{2} + 7x + C$

2. Evaluate $\int 3\cos x - \dfrac{4}{\sqrt{x}} dx$.

   (A) $3\sin x - 2\sqrt{x} + C$

   (B) $3\sin x - 8\sqrt{x} + C$

   (C) $3\sin x + 2\sqrt{x} + C$

   (D) $-3\sin x + 8\sqrt{x} + C$

3. Evaluate $\int \dfrac{9x^2}{\left(1-x^3\right)^2} dx$.

   (A) $\dfrac{3}{1-x^3} + C$

   (B) $\dfrac{-3}{1-x^3} + C$

   (C) $\dfrac{3}{\left(1-x^3\right)^3} + C$

   (D) $\dfrac{-3}{\left(1-x^3\right)^3} + C$

4. Evaluate $\int \dfrac{\cos\left(\frac{1}{x}\right)}{x^2} dx$.

   (A) $\dfrac{\sin\left(\frac{1}{x}\right)}{x} + C$

   (B) $\dfrac{-\sin\left(\frac{1}{x}\right)}{x} + C$

   (C) $\sin\left(\dfrac{1}{x}\right) + C$

   (D) $-\sin\left(\dfrac{1}{x}\right) + C$

5. Evaluate $\int \cos^3(3x)\sin(3x) dx$.

   (A) $\dfrac{\cos^4(3x)}{12} + C$

   (B) $-\dfrac{\cos^4(3x)}{12} + C$

   (C) $-\dfrac{\sin^4(3x)}{12} + C$

   (D) $\dfrac{\sin^4(3x)}{12} + C$

REFLECT

Respond to the following questions:

- For which topics discussed in this chapter do you feel you have achieved sufficient mastery to answer multiple-choice questions correctly?

- For which topics discussed in this chapter do you feel you have achieved sufficient mastery to answer open-ended questions correctly?

- For which topics discussed in this chapter do you feel you need more work before you can answer multiple-choice questions correctly?

- For which topics discussed in this chapter do you feel you need more work before you can answer open-ended questions correctly?

- What parts of this chapter are you going to re-review?

Chapter 14
Definite Integrals

AREA UNDER A CURVE

It's time to learn one of the most important uses of the integral. We've already discussed how integration can be used to "antidifferentiate" a function; now you'll see that you can also use it to find the area under a curve. First, here's a little background about how to find the area without using integration.

Suppose you have to find the area under the curve $y = x^2 + 2$ from $x = 1$ to $x = 3$. The graph of the curve looks like the following:

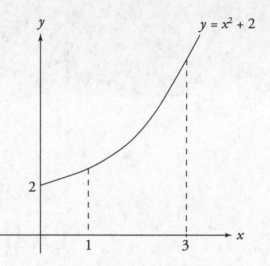

Don't panic yet. Nothing you've learned in geometry thus far has taught you how to find the area of something like this. You have learned how to find the area of a rectangle, though, and we're going to use rectangles to approximate the area between the curve and the x-axis.

Let's divide the region into two rectangles, one from $x = 1$ to $x = 2$ and one from $x = 2$ to $x = 3$, where the top of each rectangle comes just under the curve. It looks like the following:

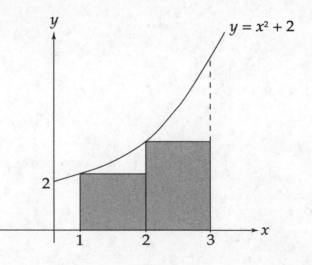

Notice that the width of each rectangle is 1. The height of the left rectangle is found by plugging 1 into the equation $y = x^2 + 2$ (yielding 3); the height of the right rectangle is found by plugging 2 into the same equation (yielding 6). The combined area of the two rectangles is (1)(3) + (1)(6) = 9. So we could say that the area under the curve is approximately 9 square units.

Naturally, this is a pretty rough approximation that significantly underestimates the area. Look at how much of the area we missed by using two rectangles! How do you suppose we could make the approximation better? Divide the region into more, thinner rectangles.

This time, cut up the region into four rectangles, each with a width of $\frac{1}{2}$. It looks like the following:

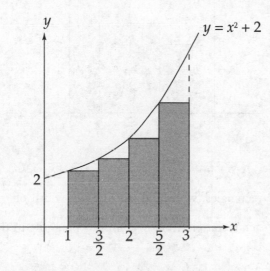

Now find the height of each rectangle the same way as before. Notice that the values we use are the left endpoints of each rectangle. The heights of the rectangles are, respectively,

$$(1)^2 + 2 = 3; \left(\frac{3}{2}\right)^2 + 2 = \frac{17}{4}; \ (2)^2 + 2 = 6 \text{ and } \left(\frac{5}{2}\right)^2 + 2 = \frac{33}{4}$$

Now multiply each height by the width of $\frac{1}{2}$ and add up the areas.

$$\left(\frac{1}{2}\right)(3) + \left(\frac{1}{2}\right)\left(\frac{17}{4}\right) + \left(\frac{1}{2}\right)(6) + \left(\frac{1}{2}\right)\left(\frac{33}{4}\right) = \frac{43}{4}$$

This is a much better approximation of the area, but there's still a lot of space that isn't accounted for. We're still underestimating the area. The rectangles need to be thinner. But before we do that, let's do something else.

Notice how each of the rectangles is inscribed in the region. Suppose we used circumscribed rectangles instead—that is, we could determine the height of each rectangle by the higher of the two *y*-values, not the lower. The following graph shows what the region would look like:

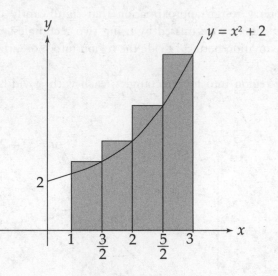

To find the area of the rectangles, we would still use the width of $\frac{1}{2}$, but the heights would change. The heights of each rectangle are now found by plugging in the right endpoint of each rectangle.

$$\left(\frac{3}{2}\right)^2 + 2 = \frac{17}{4}; (2)^2 + 2 = 6; \left(\frac{5}{2}\right)^2 + 2 = \frac{33}{4}; \text{ and } (3)^2 + 2 = 11$$

Once again, multiply each height by the width of $\frac{1}{2}$ and add up the areas.

$$\left(\frac{1}{2}\right)\left(\frac{17}{4}\right) + \left(\frac{1}{2}\right)(6) + \left(\frac{1}{2}\right)\left(\frac{33}{4}\right) + \left(\frac{1}{2}\right)(11) = \frac{59}{4}$$

The area under the curve using four left endpoint rectangles is $\frac{43}{4}$, and the area using four right endpoint rectangles is $\frac{59}{4}$, so why not average the two? This gives us $\frac{51}{4}$, which is a better approximation of the area.

Now that we've found the area using the rectangles a few times, let's turn the method into a formula. Call the left endpoint of the interval *a* and the right endpoint of the interval *b*, and set the number of rectangles we use equal to *n*. So the width of each rectangle is $\frac{b-a}{n}$. The height of the first inscribed rectangle is y_0, the height of the second rectangle is y_1, the height of the third rectangle is y_2, and so on, up to the last rectangle, which is y_{n-1}. If we use the left endpoint of each rectangle, the area under the curve is

$$\left(\frac{b-a}{n}\right)\left[y_0 + y_1 + y_2 + y_3 \ldots + y_{n-1}\right]$$

If we use the right endpoint of each rectangle, then the formula is

$$\left(\frac{b-a}{n}\right)\left[y_1 + y_2 + y_3 \ldots + y_n\right]$$

Now for the fun part. Remember how we said that we could make the approximation better by making more, thinner rectangles? By letting n approach infinity, we create an infinite number of rectangles that are infinitesimally thin. The formula for "left-endpoint" rectangles becomes

$$\lim_{n\to\infty}\left(\frac{b-a}{n}\right)\left[y_0 + y_1 + y_2 + y_3 \ldots + y_{n-1}\right]$$

For "right-endpoint" rectangles, the formula becomes

$$\lim_{n\to\infty}\left(\frac{b-a}{n}\right)\left[y_1 + y_2 + y_3 \ldots + y_n\right]$$

Note: Sometimes these are called "inscribed" and "circumscribed" rectangles, but that restricts the use of the formula. It's more exact to evaluate these rectangles using right and left endpoints.

We could also find the area using the midpoint of each interval. Let's once again use the above example, dividing the region into four rectangles. The region would look like the following:

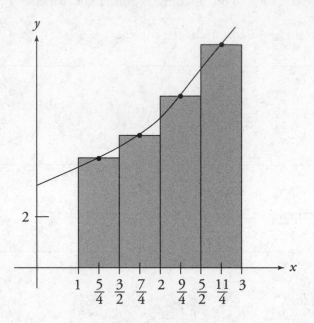

To find the area of the rectangles, we would still use the width of $\frac{1}{2}$, but the heights would now be found by plugging the midpoint of each interval into the equation.

$$\left(\frac{5}{4}\right)^2+2=\frac{57}{16}; \left(\frac{7}{4}\right)^2+2=\frac{81}{16}; \left(\frac{9}{4}\right)^2+2=\frac{113}{16}; \left(\frac{11}{4}\right)^2+2=\frac{153}{16}$$

Once again, multiply each height by the width of $\frac{1}{2}$ and add up the areas.

$$\left(\frac{1}{2}\right)\left(\frac{57}{16}\right)+\left(\frac{1}{2}\right)\left(\frac{81}{16}\right)+\left(\frac{1}{2}\right)\left(\frac{113}{16}\right)+\left(\frac{1}{2}\right)\left(\frac{153}{16}\right)=\frac{404}{32}=\frac{101}{8}=12.625$$

The general formula for approximating the area under a curve using midpoints is

$$\left(\frac{b-a}{n}\right)\left[y_{\frac{1}{2}}+y_{\frac{3}{2}}+y_{\frac{5}{2}}+\ldots+y_{\frac{2n-1}{2}}\right]$$

(Note: The fractional subscript means to evaluate the function at the number halfway between each integral pair of values of n.)

When you take the limit of this infinite sum, you get the integral. (You knew the integral had to show up sometime, didn't you?) Actually, we write the integral as

$$\int_1^3 \left(x^2 + 2\right)dx$$

It is called a **definite integral**, and it means that we're finding the area under the curve $x^2 + 2$ from $x = 1$ to $x = 3$. (We'll discuss how to evaluate this in a moment.) On the AP Exam, you'll only be asked to divide the region into a small number of rectangles, so it won't be very hard. Let's do an example.

Example 1: Approximate the area under the curve $y = x^3$ from $x = 2$ to $x = 3$ using four left-endpoint rectangles.

Draw four rectangles that look like the following:

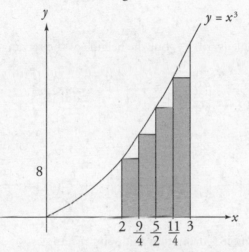

The width of each rectangle is $\dfrac{1}{4}$. The heights of the rectangles are

$$2^3, \left(\frac{9}{4}\right)^3, \left(\frac{5}{2}\right)^3, \text{ and } \left(\frac{11}{4}\right)^3$$

Therefore, the area is

$$\left(\frac{1}{4}\right)\left(2^3\right)+\left(\frac{1}{4}\right)\left(\frac{9}{4}\right)^3+\left(\frac{1}{4}\right)\left(\frac{5}{2}\right)^3+\left(\frac{1}{4}\right)\left(\frac{11}{4}\right)^3 = \frac{893}{64} \approx 13.953$$

Example 2: Repeat Example 1 using four right-endpoint rectangles.

Now draw four rectangles that look like the following:

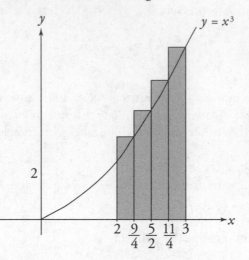

The width of each rectangle is still $\dfrac{1}{4}$, but the heights of the rectangles are now

$$\left(\dfrac{9}{4}\right)^3, \left(\dfrac{5}{2}\right)^3, \left(\dfrac{11}{4}\right)^3, \text{ and } \left(3^3\right)$$

The area is now

$$\left(\dfrac{1}{4}\right)\left(\dfrac{9}{4}\right)^3 + \left(\dfrac{1}{4}\right)\left(\dfrac{5}{2}\right)^3 + \left(\dfrac{1}{4}\right)\left(\dfrac{11}{4}\right)^3 + \left(\dfrac{1}{4}\right)\left(3^3\right) = \dfrac{1197}{64} \approx 18.703$$

Example 3: Repeat Example 1 using four midpoint rectangles.

Now draw four rectangles that look like the following:

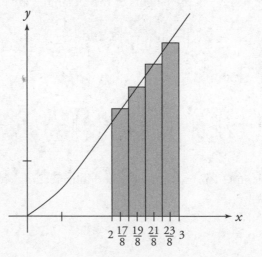

The width of each rectangle is still $\frac{1}{4}$, but the heights of the rectangles are now

$$\left(\frac{17}{8}\right)^3, \left(\frac{19}{8}\right)^3, \left(\frac{21}{8}\right)^3, \text{ and } \left(\frac{23}{8}\right)^3$$

The area is now

$$\left(\frac{1}{4}\right)\left(\frac{17}{8}\right)^3 + \left(\frac{1}{4}\right)\left(\frac{19}{8}\right)^3 + \left(\frac{1}{4}\right)\left(\frac{21}{8}\right)^3 + \left(\frac{1}{4}\right)\left(\frac{23}{8}\right)^3 = \frac{2075}{128} \approx 16.211$$

TABULAR RIEMANN SUMS

Sometimes the AP Exam will ask you to find a Riemann sum, or to approximate an integral (same thing, right?), but won't give you a function to work with. Instead, they will give you a table of values for x and $f(x)$. These are quite simple to evaluate. All you do is use the right-hand or left-hand sum formula, plugging in the appropriate values for $f(x)$. One thing you should watch out for is that sometimes the x-values are not evenly spaced, so make sure that you use the correct values for the widths of the rectangles. Let's do an example.

Example 4: Suppose we are given the following table of values for x and $f(x)$.

x	2	4	6	8	10	12
$f(x)$	10	13	15	14	9	3

Use a *right-hand* Riemann sum with 5 subintervals indicated by the data in the table to approximate $\int_2^{12} f(x)dx$.

Recall that the formula for finding the area under the curve using the right endpoints is $\left(\frac{b-a}{n}\right)[y_1 + y_2 + y_3 + \ldots + y_n]$. Here, the width of each rectangle is 2. We find the height of each rectangle by evaluating $f(x)$ at the appropriate value of x, the right endpoint of each interval on the x-axis. Here, $y_1 = 13$, $y_2 = 15$, $y_3 = 14$, $y_4 = 9$, and $y_5 = 3$. Therefore, we can approximate the integral with $\int_2^{12} f(x)dx = (2)(13) + (2)(15) + (2)(14) + (2)(9) + (2)(3) = 108$.

Let's do another example but this time the values of x will not be evenly spaced on the x-axis.

Example 5: Given the following table of values for x and $f(x)$:

x	0	2	5	11	19	22	23
$f(x)$	4	6	16	18	22	29	50

Use a *left-hand* Riemann sum with 6 subintervals indicated by the data in the table to approximate $\int_0^{23} f(x)dx$.

Recall that the formula for finding the area under the curve using the left endpoints is $\left(\dfrac{b-a}{n}\right)[y_0 + y_1 + y_2 + \dots + y_{n-1}]$. This formula assumes that the x-values are evenly spaced but they aren't here, so we will replace the values of $\left(\dfrac{b-a}{n}\right)$ with the appropriate widths of each rectangle. The width of the first rectangle is $2 - 0 = 2$; the second width is $5 - 2 = 3$; the third is $11 - 5 = 6$; the fourth is $19 - 11 = 8$; the fifth is $22 - 19 = 3$; and the sixth is $23 - 22 = 1$. We find the height of each rectangle by evaluating $f(x)$ at the appropriate value of x, the left endpoint of each interval on the x-axis. Here, $y_0 = 4$, $y_1 = 6$, $y_2 = 16$, $y_3 = 18$, $y_4 = 22$, and $y_5 = 29$. Therefore, we can approximate the integral with $\int_0^{23} f(x)dx = (2)(4) + (3)(6) + (6)(16) + (8)(18) + (3)(22) + (1)(29) = 361$.

That's all there is to approximating the area under a curve using rectangles. Now let's learn how to find the area exactly. In order to evaluate this, you'll need to know. . .

The Fundamental Theorem of Calculus

Before, we said that if you create an infinite number of infinitely thin rectangles, you'll get the area under the curve, which is an integral. For the example above, the integral is

$$\int_2^3 x^3 dx$$

There is a rule for evaluating an integral like this. The rule is called the **Fundamental Theorem of Calculus**, and it says

$$\int_a^b f(x)dx = F(b) - F(a)\text{; where } F(x) \text{ is the antiderivative of } f(x).$$

Using this rule, you can find $\int_2^3 x^3\,dx$ by integrating it, and we get $\dfrac{x^4}{4}$. Now all you do is plug in 3 and 2 and take the difference. We use the following notation to symbolize this

$$\left.\frac{x^4}{4}\right|_2^3$$

Thus, we have

$$\frac{3^4}{4} - \frac{2^4}{4} = \frac{81}{4} - \frac{16}{4} = \frac{65}{4}$$

Because $\dfrac{65}{4} = 16.25$, you can see how close we were with our three earlier approximations.

Example 6: Find $\displaystyle\int_1^3 \left(x^2 + 2\right)dx$.

The Fundamental Theorem of Calculus yields the following:

$$\int_1^3 \left(x^2 + 2\right)dx = \left.\left(\frac{x^3}{3} + 2x\right)\right|_1^3$$

If we evaluate this, we get the following:

$$\left(\frac{3^3}{3} + 2(3)\right) - \left(\frac{1^3}{3} + 2(1)\right) = \frac{38}{3}$$

This is the first function for which we found the approximate area by using inscribed rectangles. Our final estimate, where we averaged the inscribed and circumscribed rectangles, was $\dfrac{51}{4}$, and as you can see, that was very close (off by $\dfrac{1}{12}$). When we used the midpoints, we were off by $\dfrac{1}{24}$.

We're going to do only a few approximations using rectangles, because it's not a big part of the AP Exam. On the other hand, definite integrals are a huge part of the rest of this book.

Example 7: $\displaystyle\int_1^5 \left(x^2 - x\right)dx = \left.\left(\frac{x^3}{3} - \frac{x^2}{2}\right)\right|_1^5 = \left(\frac{125}{3} - \frac{25}{2}\right) - \left(\frac{1}{3} - \frac{1}{2}\right) = \frac{88}{3}$

Example 8: $\displaystyle\int_0^{\frac{\pi}{2}} \sin x\,dx = \left.(-\cos x)\right|_0^{\frac{\pi}{2}} = \left(-\cos\frac{\pi}{2}\right) - (-\cos 0) = 1$

Example 9: $\int_0^{\frac{\pi}{4}} \sec^2 x \, dx = \tan x \Big|_0^{\frac{\pi}{4}} = \tan \frac{\pi}{4} - \tan 0 = 1$

The Trapezoid Rule

There's another approximation method that's even better than the rectangle method. Essentially, all you do is divide the region into trapezoids instead of rectangles. Let's use the problem that we did at the beginning of the chapter.

We get a picture that looks like the following:

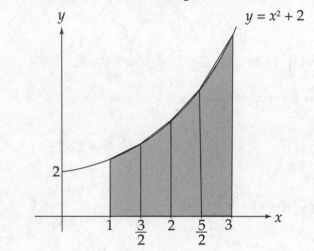

As you should recall from geometry, the formula for the area of a trapezoid is

$$\frac{1}{2}(b_1 + b_2)h$$

(Note: b_1 and b_2 are the two bases of the trapezoid.) Notice that each of the shapes is a trapezoid on its side, so the height of each trapezoid is the length of the interval $\frac{1}{2}$, and the bases are the y-values that correspond to each x-value. We found these earlier in the rectangle example; they are, in order: $3, \frac{17}{4}, 6, \frac{33}{4}$, and 11. We can find the area of each trapezoid and add them up.

$$\frac{1}{2}\left(3 + \frac{17}{4}\right)\left(\frac{1}{2}\right) + \frac{1}{2}\left(\frac{17}{4} + 6\right)\left(\frac{1}{2}\right) + \frac{1}{2}\left(6 + \frac{33}{4}\right)\left(\frac{1}{2}\right) + \frac{1}{2}\left(\frac{33}{4} + 11\right)\left(\frac{1}{2}\right) = \frac{51}{4},$$

or 12.75

Recall that the actual value of the area is $\frac{38}{3}$ or 12.67; the Trapezoid Rule gives a pretty good approximation.

Notice how each trapezoid shares a base with the trapezoid next to it, except for the end ones. This enables us to simplify the formula for using the Trapezoid Rule. Each trapezoid has a height equal to the length of the interval divided by the number of trapezoids we use. If the interval is from $x = a$ to $x = b$, and the number of trapezoids is n, then the height of each trapezoid is $\dfrac{b-a}{n}$. Then our formula is

$$\left(\frac{1}{2}\right)\left(\frac{b-a}{n}\right)\left[y_0 + 2y_1 + 2y_2 + 2y_3 \ldots + 2y_{n-2} + 2y_{n-1} + y_n\right]$$

This is all you need to know about the Trapezoid Rule. Just follow the formula, and you won't have any problems. Let's do one more example.

Example 10: Approximate the area under the curve $y = x^3$ from $x = 2$ to $x = 3$ using four inscribed trapezoids.

Following the rule, the height of each trapezoid is $\dfrac{3-2}{4} = \dfrac{1}{4}$. Thus, the approximate area is

$$\left(\frac{1}{2}\right)\left(\frac{1}{4}\right)\left[2^3 + 2\left(\frac{9}{4}\right)^3 + 2\left(\frac{5}{2}\right)^3 + 2\left(\frac{11}{4}\right)^3 + 3^3\right] = \frac{1{,}045}{64}$$

Compare this answer to the actual value we found earlier—it's pretty close!

PROBLEM 1. Approximate the area under the curve $y = 4 - x^2$ from $x = -1$ to $x = 1$ with $n = 4$ inscribed rectangles.

Answer: Draw four rectangles that look like this.

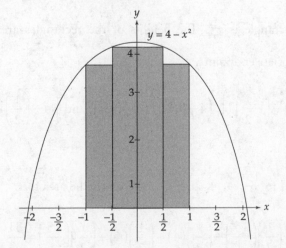

The width of each rectangle is $\frac{1}{2}$. The heights of the rectangles are found by evaluating $y = 4 - x^2$ at the appropriate endpoints.

$$\left(4 - (-1)^2\right), \left(4 - \left(-\frac{1}{2}\right)^2\right), \left(4 - \left(\frac{1}{2}\right)^2\right), \text{ and } \left(4 - (1)^2\right)$$

These can be simplified to $3, \frac{15}{4}, \frac{15}{4}$, and 3. Therefore, the area is

$$\left(\frac{1}{2}\right)(3) + \left(\frac{1}{2}\right)\left(\frac{15}{4}\right) + \left(\frac{1}{2}\right)\left(\frac{15}{4}\right) + \left(\frac{1}{2}\right)(3) = \frac{27}{4}$$

PROBLEM 2. Find the area under the curve $y = 4 - x^2$ from $x = -1$ to $x = 1$ with $n = 4$ circumscribed rectangles.

Answer: Draw four rectangles that look like the following:

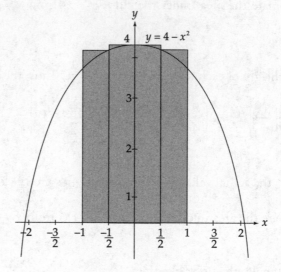

The width of each rectangle is $\frac{1}{2}$. The heights of the rectangles are found by evaluating $y = 4 - x^2$ at the appropriate endpoints.

$$\left(4 - \left(-\frac{1}{2}\right)^2\right), \left(4 - (0)^2\right), \left(4 - (0)^2\right), \text{ and } \left(4 - \left(\frac{1}{2}\right)^2\right)$$

These can be simplified to $\frac{15}{4}, 4, 4$, and $\frac{15}{4}$. Therefore, the area is

$$\left(\frac{1}{2}\right)\left(\frac{15}{4}\right) + \left(\frac{1}{2}\right)(4) + \left(\frac{1}{2}\right)(4) + \left(\frac{1}{2}\right)\left(\frac{15}{4}\right) = \frac{31}{4}$$

PROBLEM 3. Find the area under the curve $y = 4 - x^2$ from $x = -1$ to $x = 1$ using the Trapezoid Rule with $n = 4$.

Answer: Draw four trapezoids that look like the following:

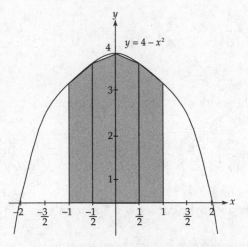

The width of each trapezoid is $\dfrac{1}{2}$. Evaluate the bases of the trapezoid by calculating $y = 4 - x^2$ at the appropriate endpoints. Following the rule, we get that the area is approximately

$$\left(\frac{1}{2}\right)\left(\frac{1}{2}\right)\left[\left(4-(-1)^2\right)+2\left(4-\left(-\frac{1}{2}\right)^2\right)+2\left(4-(0)^2\right)+2\left(4-\left(\frac{1}{2}\right)^2\right)+\left(4-(1)^2\right)\right]=$$
$$\left(\frac{1}{2}\right)\left(\frac{1}{2}\right)\left[3+\frac{15}{2}+8+\frac{15}{2}+3\right]=\frac{29}{4}$$

PROBLEM 4. Find the area under the curve $y = 4 - x^2$ from $x = -1$ to $x = 1$ using the Midpoint Formula with $n = 4$.

Answer: Draw four rectangles that look like the following:

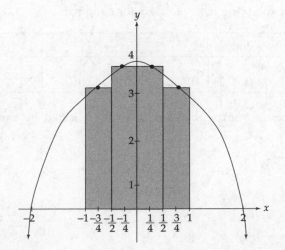

The width of each rectangle is $\dfrac{1}{2}$. The heights of the rectangles are found by evaluating $y = 4 - x^2$ at the appropriate points.

$$4 - \left(-\frac{3}{4}\right)^2 = \frac{55}{16}; \ 4 - \left(-\frac{1}{4}\right)^2 = \frac{63}{16}; \ 4 - \left(\frac{1}{4}\right)^2 = \frac{63}{16}; \ 4 - \left(\frac{3}{4}\right)^2 = \frac{55}{16}$$

Multiply each height by the width of $\dfrac{1}{2}$ and add up the areas.

$$\left(\frac{1}{2}\right)\left(\frac{55}{16}\right) + \left(\frac{1}{2}\right)\left(\frac{63}{16}\right) + \left(\frac{1}{2}\right)\left(\frac{63}{16}\right) + \left(\frac{1}{2}\right)\left(\frac{55}{16}\right) = \frac{59}{8}$$

PROBLEM 5. Find the area under the curve $y = 4 - x^2$ from $x = -1$ to $x = 1$.

Answer: Now we can use the definite integral by evaluating.

$$\int_{-1}^{1} \left(4 - x^2\right) dx$$

This can be rewritten as

$$\left(4x - \frac{x^3}{3}\right)\Bigg|_{-1}^{1}$$

Follow the Fundamental Theorem of Calculus.

$$\left(4 - \frac{1}{3}\right) - \left(4(-1) - \frac{(-1)^3}{3}\right) = \frac{22}{3}$$

PROBLEM 6. Given the following table of values for t and $f(t)$:

t	0	2	4	7	11	13	14
$f(t)$	5	6	10	15	20	26	30

Use a *right-hand* Riemann Sum with 6 subintervals indicated by the data in the table to approximate $\int_{0}^{14} f(t)\, dt$.

Answer: The width of the first rectangle is 2 – 0 = 2; the second width is 4 – 2 = 2; the third is 7 – 4 = 3; the fourth is 11 – 7 = 4; the fifth is 13 – 11 = 2; and the sixth is 14 – 13 = 1. We find the height of each rectangle by evaluating $f(t)$ at the appropriate value of t, the right endpoint of each interval on the t-axis. Here, $y_1 = 6$, $y_2 = 10$, $y_3 = 15$ $y_4 = 20$, $y_5 = 26$, and $y_6 = 30$. Therefore, we can approximate the integral with

$$\int_0^{14} f(t) \, dt = (2)(6) + (2)(10) + (3)(15) + (4)(20) + (2)(26) + (1)(30) = 239$$

PRACTICE PROBLEM SET 22

Here's a great opportunity to practice finding the area beneath a curve and evaluating integrals. The answers are in Chapter 23, starting on page 538.

1. Find the area under the curve $y = 2x - x^2$ from $x = 1$ to $x = 2$ with $n = 4$ left-endpoint rectangles.

2. Find the area under the curve $y = 2x - x^2$ from $x = 1$ to $x = 2$ with $n = 4$ right-endpoint rectangles.

3. Find the area under the curve $y = 2x - x^2$ from $x = 1$ to $x = 2$ using the Trapezoid Rule with $n = 4$.

4. Find the area under the curve $y = 2x - x^2$ from $x = 1$ to $x = 2$ using the Midpoint Formula with $n = 4$.

5. Find the area under the curve $y = 2x - x^2$ from $x = 1$ to $x = 2$.

6. Evaluate $\int_{-\frac{\pi}{2}}^{\frac{\pi}{2}} \cos x \, dx$.

7. Evaluate $\int_1^9 2x\sqrt{x} \, dx$.

8. Evaluate $\int_0^1 \left(x^4 - 5x^3 + 3x^2 - 4x - 6\right) dx$.

9. Evaluate $\int_{-\frac{\pi}{2}}^{\frac{\pi}{2}} \sin x \, dx$.

10. Suppose we are given the following table of values for x and $g(x)$:

x	0	1	3	5	9	14
$g(x)$	10	8	11	17	20	23

Use a left-hand Riemann sum with 5 subintervals indicated by the data in the table to approximate $\int_0^{14} g(x)dx$.

THE MEAN VALUE THEOREM FOR INTEGRALS

As you recall, we did the Mean Value Theorem once before, in Chapter 8, but this time we'll apply it to integrals, not derivatives. In fact, some books refer to it as the "Mean Value Theorem for Integrals" or MVTI. The most important aspect of the MVTI is that it enables you to find the average value of a function. In fact, the AP Exam will often ask you to find the average value of a function, which is just its way of testing your knowledge of the MVTI.

Here's the theorem.

> If $f(x)$ is continuous on a closed interval $[a, b]$, then at some point c in the interval $[a, b]$ the following is true:
>
> $$\int_a^b f(x)dx = f(c)(b-a)$$

This tells you that the area under the curve of $f(x)$ on the interval $[a, b]$ is equal to the value of the function at some value c (between a and b) times the length of the interval. If you look at this graphically, you can see that you're finding the area of a rectangle whose base is the interval and whose height is some value of $f(x)$ that creates a rectangle with the same area as the area under the curve.

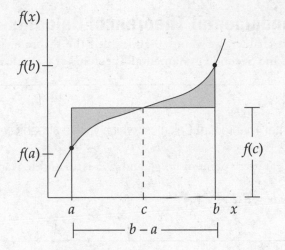

The number $f(c)$ gives us the average value of f on $[a, b]$. Thus, if we rearrange the theorem, we get the formula for finding the average value of $f(x)$ on $[a, b]$.

$$f(c) = \frac{1}{b-a}\int_a^b f(x)dx$$

This is the formula for the average value.

There's all you need to know about finding average values. Try some examples.

Example 1: Find the average value of $f(x) = x^2$ from $x = 2$ to $x = 4$.

Evaluate the integral $\frac{1}{4-2}\int_2^4 x^2 dx$.

$$\frac{1}{4-2}\int_2^4 x^2 dx = \frac{1}{2}\left[\frac{x^3}{3}\Big|_2^4\right] = \frac{1}{2}\left(\frac{64}{3}-\frac{8}{3}\right) = \frac{28}{3}$$

Example 2: Find the average value of $f(x) = \sin x$ on $[0, \pi]$.

Evaluate $\frac{1}{\pi - 0}\int_0^\pi \sin x\ dx$.

$$\frac{1}{\pi - 0}\int_0^\pi \sin x\ dx = \frac{1}{\pi}\left(-\cos x\right)\Big|_0^\pi = \frac{1}{\pi}\left(-\cos\pi + \cos 0\right) = \frac{2}{\pi}$$

The Second Fundamental Theorem of Calculus

As you saw in the last chapter, we've only half-learned the theorem. It has two parts, often referred to as the **First and Second Fundamental Theorems of Calculus.**

The First Fundamental Theorem of Calculus (which you've already seen)

If $f(x)$ is continuous at every point of $[a, b]$, and $F(x)$ is an antiderivative of $f(x)$ on

$[a, b]$, then $\displaystyle\int_a^b f(x)dx = F(b) - F(a)$.

These theorems are sometimes taught in reverse order.

The Second Fundamental Theorem of Calculus

If $f(x)$ is continuous on $[a, b]$, then the derivative of the function
$F(x) = \displaystyle\int_a^x f(t)dt$ is

$$\frac{dF}{dx} = \frac{d}{dx}\int_a^x f(t)dt = f(x)$$

We've already made use of the first theorem in evaluating definite integrals. In fact, we use the first Fundamental Theorem every time we evaluate a definite integral, so we're not going to give you any examples of that here. There is one aspect of the first Fundamental Theorem, however, that involves the area between curves (we'll discuss that in Chapter 18).

But for now, you should know the following:

If we have a point c in the interval $[a, b]$, then

$$\int_a^c f(x)dx + \int_c^b f(x)dx = \int_a^b f(x)dx$$

In other words, we can divide up the region into parts, add them up, and find the area of the region based on the result. We'll get back to this in the chapter on the area between two curves.

The second theorem tells us how to find the derivative of an integral.

Example 3: Find $\dfrac{d}{dx}\displaystyle\int_1^x \cos t\ dt$.

The second Fundamental Theorem says that the derivative of this integral is just cos x.

Example 4: Find $\dfrac{d}{dx}\displaystyle\int_2^x \left(1-t^3\right) dt$.

Here, the theorem says that the derivative of this integral is just $(1-x^3)$.

Isn't this easy? Let's add a couple of nuances. First, the constant term in the limits of integration is a "dummy term." Any constant will give the same answer. For example,

$$\frac{d}{dx}\int_2^x (1-t^3)\,dt = \frac{d}{dx}\int_{-2}^x (1-t^3)\,dt = \frac{d}{dx}\int_\pi^x (1-t^3)\,dt = 1-x^3$$

In other words, all we're concerned with is the variable term.

Second, if the upper limit is a function of x, instead of just plain x, we multiply the answer by the derivative of that term. For example,

$$\frac{d}{dx}\int_2^{x^2} \left(1-t^3\right) dt = \left[1\ \ \left(x^2\right)^3\right](2x) = \left(1-x^6\right)(2x)$$

Example 5: Find $\dfrac{d}{dx}\displaystyle\int_0^{3x^4} \left(t+4t^2\right) dt = \left[3x^4+4(3x^4)^2\right]\left(12x^3\right)$.

Try these solved problems on your own. You know the drill.

PROBLEM 1. Find the average value of $f(x) = \dfrac{1}{x^2}$ on the interval $[1, 3]$.

Answer: According to the Mean Value Theorem, the average value is found by evaluating

$$\frac{1}{3-1}\int_1^3 \frac{dx}{x^2}$$

Your result should be

$$\frac{1}{2}\left(-\frac{1}{x}\right)\Bigg|_1^3 = \frac{1}{2}\left(-\frac{1}{3}+1\right) = \frac{1}{3}$$

PROBLEM 2. Find the average value of $f(x) = \sin x$ on the interval $\left[-\pi, \pi\right]$.

Answer: According to the Mean Value Theorem, the average value is found by evaluating

$$\frac{1}{2\pi}\left(\int_{-\pi}^{\pi}\sin x \; dx\right)$$

Integrating, we get

$$\frac{1}{2\pi}\left(-\cos x\right)\Big|_{-\pi}^{\pi} = \frac{1}{2\pi}\left(-\cos\pi + \cos(-\pi)\right) = 0$$

PROBLEM 3. Find $\dfrac{d}{dx}\displaystyle\int_{1}^{x}\dfrac{dt}{1-\sqrt[3]{t}}$.

Answer: According to the Second Fundamental Theorem of Calculus,

$$\frac{d}{dx}\int_{1}^{x}\frac{dt}{1-\sqrt[3]{t}} = \frac{1}{1-\sqrt[3]{x}}$$

PROBLEM 4. Find $\dfrac{d}{dx}\displaystyle\int_{1}^{x^{2}}\dfrac{t \; dt}{\sin t}$.

Answer: According to the Second Fundamental Theorem of Calculus,

$$\frac{d}{dx}\int_{1}^{x^{2}}\frac{t \; dt}{\sin t} = \frac{x^{2}}{\sin (x^{2})}(2x) = \frac{2x^{3}}{\sin x^{2}}$$

Accumulation Functions

Recently, the AP Exam has been testing problems that deal with **accumulation functions**. These are simply functions of the form $F(x) = \displaystyle\int_{0}^{x} f(t) \; dt$. These are called accumulation functions because the value of the integral is the area under the curve from the constant to the value x, and as x gets bigger, so does the area (it "accumulates"). In these functions, t is a dummy variable that is used as the variable of integration.

Let's do an example.

Example 1: Suppose we have the function $F(x) = \displaystyle\int_{0}^{x} t \; dt$. Let's evaluate this at different values of x. First, let's find $F(1)$. Graphically, we are looking for the area under the curve $y = t$ from $t = 0$ to $t = 1$.

It looks like the following:

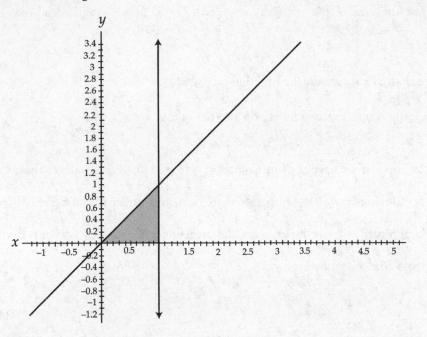

$F(1)$ is just the area of the triangle $A = \frac{1}{2}(1)(1) = \frac{1}{2}$. If we evaluate the integral, we get $F(1) = \int_0^1 t \, dt = \left(\frac{t^2}{2} \right)\Big|_0^1 = \frac{1}{2} - 0 = \frac{1}{2}$. Let's now find $F(2)$. This is the area under the curve $y = t$ from $t = 0$ to $t = 2$.

It looks like the following:

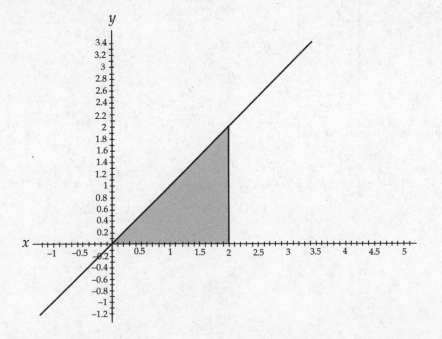

$F(2)$ is just the area of the triangle: $A = \frac{1}{2}(2)(2) = 2$. If we evaluate the integral, we get

$$F(2) = \int_0^2 t \, dt = \left(\frac{t^2}{2}\right)\Big|_0^2 = 2 - 0 = 2.$$

We can see that, as x increases, the function increases.

This was a fairly simple example. Let's do another one.

Example 2: Suppose we have the function $F(x) = \int_0^x \sin t \, dt$. Let's evaluate this as x increases from 0 to π. Obviously $F(0) = 0$ because there is no area under the curve. So, first, let's find $F\left(\frac{\pi}{6}\right)$. Graphically, we are looking for the area under the curve $y = \sin t$ from $t = 0$ to $t = \frac{\pi}{6}$. It looks like the following:

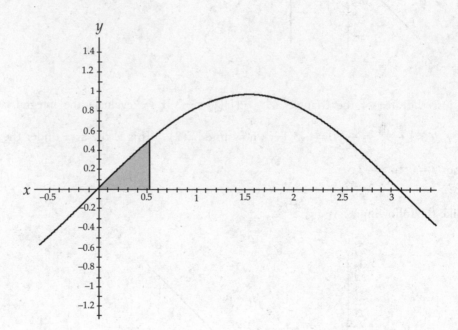

If we evaluate the integral, we get

$$F\left(\frac{\pi}{6}\right) = \int_0^{\frac{\pi}{6}} \sin t \, dt = (-\cos t)\Big|_0^{\frac{\pi}{6}} = -\cos\frac{\pi}{6} + \cos 0 = 1 - \frac{\sqrt{3}}{2} \approx 0.134$$

Now let's find $F\left(\dfrac{\pi}{4}\right)$. It looks like the following:

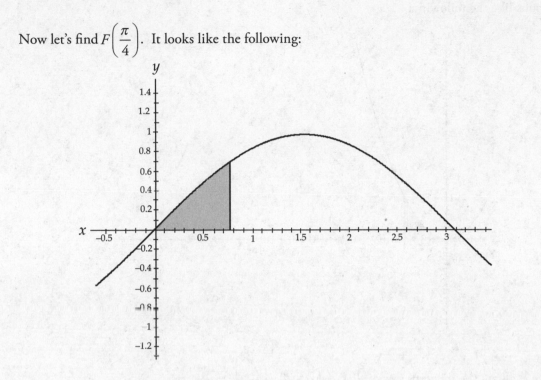

If we evaluate the integral, we get

$$F\left(\frac{\pi}{4}\right) = \int_0^{\frac{\pi}{4}} \sin t \, dt = \left(-\cos t\right)\Big|_0^{\frac{\pi}{4}} = -\cos\frac{\pi}{4} + \cos 0 = 1 - \frac{\sqrt{2}}{2} \approx 0.293$$

Let's make a table of values of the accumulation function for different values of x.

x	0	$\dfrac{\pi}{6}$	$\dfrac{\pi}{4}$	$\dfrac{\pi}{3}$	$\dfrac{\pi}{2}$	π
$F(x)$	0	$1 - \dfrac{\sqrt{3}}{2}$	$1 - \dfrac{\sqrt{2}}{2}$	$\dfrac{1}{2}$	1	2

We can see that the area is accumulating under the curve as x increases from 0 to π. Naturally, the values will shrink as we move from π to 2π because the values of $\sin x$ are negative. But, with accumulation functions, we are usually concerned only with positive areas.

Let's do one more example.

Example 3: Suppose we have the function $F(x) = \int_0^x t^2 \, dt$. Let's evaluate this as x increases from 0 to 4. First, let's find $F(1)$. Graphically, we are looking for the area under the curve $y = t^2$ from $t = 0$ to $t = 1$.

It looks like the following:

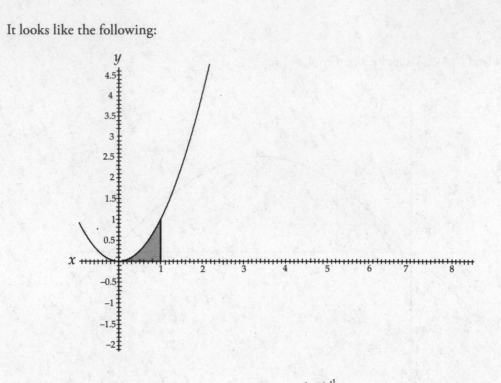

If we evaluate the integral, we get $F(1) = \int_0^1 t^2 \, dt = \left(\dfrac{t^3}{3} \right)\Bigg|_0^1 = \dfrac{1}{3}$.

As in the previous example.

x	1	2	3	4
$F(x)$	$\dfrac{1}{3}$	$\dfrac{8}{3}$	9	$\dfrac{64}{3}$

We can see that the values of $F(x)$ will increase as x increases.

PRACTICE PROBLEM SET 23

Now try these problems. The answers are in Chapter 23, starting on page 543.

1. Find the average value of $f(x) = 4x \cos x^2$ on the interval $\left[0, \sqrt{\dfrac{\pi}{2}}\right]$.

2. Find the average value of $f(x) = \sqrt{x}$ on the interval $[0, 16]$.

3. Find the average value of $f(x) = 2|x|$ on the interval $[-1, 1]$.

4. Find $\dfrac{d}{dx} \displaystyle\int_1^x \sin^2 t \; dt$.

5. Find $\dfrac{d}{dx} \displaystyle\int_1^{3x} \left(t^2 - t\right) dt$.

6. Find $\dfrac{d}{dx} \displaystyle\int_1^x -2\cos t \; dt$.

End of Chapter 14 Drill

The answers are in Chapter 24.

1. Approximate the area under the curve $y = 3 + x^2$ from $x = 1$ to $x = 5$ using $n = 4$ left-endpoint rectangles.

   (A) 21
   (B) 33
   (C) 42
   (D) 66

2. Approximate the area under the curve $y = x^3 + 4$ from $x = 0$ to $x = 8$ using $n = 4$ trapezoids.

   (A) 560
   (B) 1120
   (C) 2240
   (D) 4480

3. Approximate $\int_0^{16} \frac{3}{x} dx$ using the Midpoint Formula with $n = 4$.

   (A) $\frac{22}{35}$

   (B) $\frac{88}{35}$

   (C) $\frac{176}{35}$

   (D) $\frac{352}{35}$

4. Find the average value of $f(x) = \dfrac{x^2}{\left(1 - x^3\right)^2}$ on the interval [2, 4].

   (A) $\frac{2}{189}$

   (B) $\frac{4}{189}$

   (C) $\frac{8}{189}$

   (D) $\frac{16}{189}$

5. Find $\dfrac{d}{dx} \displaystyle\int_2^{x^2} \sin^3 t \, dt$.

   (A) $3x^2 \sin^2 x \cos x$
   (B) $6x \sin^2 x \cos x$
   (C) $x^2 \sin^3 x$
   (D) $2x \sin^3 x$

REFLECT

Respond to the following questions:

- For which topics discussed in this chapter do you feel you have achieved sufficient mastery to answer multiple-choice questions correctly?

- For which topics discussed in this chapter do you feel you have achieved sufficient mastery to answer open-ended questions correctly?

- For which topics discussed in this chapter do you feel you need more work before you can answer multiple-choice questions correctly?

- For which topics discussed in this chapter do you feel you need more work before you can answer open-ended questions correctly?

- What parts of this chapter are you going to re-review?

Chapter 15
Integrals of Exponential, Logarithmic, and Trig Functions

You've learned how to integrate polynomials and some of the trig functions (there are more of them to come), and you have the first technique of integration: *u*-substitution. Now it's time to learn how to integrate some other functions—namely, exponential and logarithmic functions. The first integral is the natural logarithm.

$$\int \frac{du}{u} = \ln|u| + C$$

Notice the absolute value in the logarithm. This ensures that you aren't taking the logarithm of a negative number. If you know that the term you're taking the log of is positive (for example, $x^2 + 1$), we can dispense with the absolute value marks. Let's do some examples.

Example 1: Find $\int \frac{5\,dx}{x+3}$.

Whenever an integrand contains a fraction, check to see if the integral is a logarithm. Usually, the process involves *u*-substitution. Let $u = x + 3$ and $du = dx$. Then,

$$\int \frac{5\,dx}{x+3} = 5\int \frac{du}{u} = 5\ln|u| + C$$

Substituting back, the final result is

$$5\ln|x + 3| + C$$

Example 2: Find $\int \frac{2x\,dx}{x^2 + 1}$.

Let $u = x^2 + 1$ $du = 2x\,dx$ and substitute into the integrand.

$$\int \frac{2x\,dx}{x^2 + 1} = \int \frac{du}{u} = \ln|u| + C$$

Then substitute back.

$$\ln(x^2 + 1) + C$$

MORE INTEGRALS OF TRIG FUNCTIONS

Remember when we started antiderivatives and we didn't do the integral of tangent, cotangent, secant, or cosecant? Well, their time has come.

Example 3: Find $\int \tan x \, dx$.

First, rewrite this integral as

$$\int \frac{\sin x}{\cos x} \, dx$$

Now, we let $u = \cos x$ and $du = -\sin x \, dx$ and substitute.

$$\int -\frac{du}{u}$$

Now, integrate and re-substitute.

$$\int -\frac{du}{u} = -\ln|u| = -\ln|\cos x| + C$$

Thus, $\int \tan x \, dx = -\ln|\cos x| + C$.

Example 4: Find $\int \cot x \, dx$.

Just as before, rewrite this integral in terms of sine and cosine.

$$\int \frac{\cos x}{\sin x} \, dx$$

Now we let $u = \sin x$ and $du = \cos x \, dx$ and substitute.

$$\int \frac{du}{u}$$

Now integrate.

$$\ln|u| + C = \ln|\sin x| + C$$

Therefore, $\int \cot x \; dx = \ln |\sin x| + C$.

This looks a lot like the previous example, doesn't it?

Example 5: Find $\int \sec x \; dx$.

You could rewrite this integral as

$$\int \frac{1}{\cos x} \; dx$$

However, if you try u-substitution at this point, it won't work. So what should you do? You'll probably never guess, so we'll show you: multiply the sec x by $\dfrac{\sec x + \tan x}{\sec x + \tan x}$. This gives you

$$\int \sec x \left(\frac{\sec x + \tan x}{\sec x + \tan x} \right) dx = \int \frac{\sec^2 x + \sec x \tan x}{\sec x + \tan x} \; dx$$

Now you can do u-substitution. Let $u = \sec x + \tan x \; du = (\sec x \tan x + \sec^2 x) \; dx$.

Then rewrite the integral as

$$\int \frac{du}{u}$$

Pretty slick, huh?

The rest goes according to plan as you integrate.

$$\int \frac{du}{u} = \ln |u| + C = \ln |\sec x + \tan x| + C$$

Therefore, $\int \sec x \; dx = \ln |\sec x + \tan x| + C$.

Example 6: Find $\int \csc x \; dx$.

You guessed it! Multiply $\csc x$ by $\dfrac{\csc x + \cot x}{\csc x + \cot x}\, dx$. This gives you

$$\int \csc x \left(\frac{\csc x + \cot x}{\csc x + \cot x} \right) dx = \int \frac{\csc^2 x + \csc x \cot x}{\csc x + \cot x}\, dx$$

Let $u = \csc x + \cot x$ and $du = (-\csc x \cot x - \csc^2 x)\, dx$. And, just as in Example 5, you can rewrite the integral as

$$\int -\frac{du}{u}$$

And integrate.

$$\int -\frac{du}{u} = -\ln |u| + C = -\ln |\csc x + \cot x| + C$$

Therefore, $\displaystyle\int \csc x\, dx = -\ln |\csc x + \cot x| + C$.

As we do more integrals, the natural log will turn up over and over. It's important that you get good at recognizing when integrating requires the use of the natural log.

INTEGRATING e^x AND a^x

Now let's learn how to find the integral of e^x. Remember that $\dfrac{d}{dx} e^x = e^x$? Well, you should be able to predict the following formula:

$$\int e^u\, du = e^u + C$$

As with the natural logarithm, most of these integrals use u-substitution.

Example 7: Find $\displaystyle\int e^{7x}\, dx$.

Let $u = 7x$, $du = 7dx$, and $\frac{1}{7} du = dx$. Then you have

$$\int e^{7x} dx = \frac{1}{7} \int e^u du = \frac{1}{7} e^u + C$$

Substituting back, you get

$$\frac{1}{7} e^{7x} + C$$

In fact, whenever you see $\int e^{kx} dx$, where k is a constant, the integral is

$$\int e^{kx} dx = \frac{1}{k} e^{kx} + C$$

Example 8: Find $\int xe^{3x^2+1} dx$.

Let $u = 3x^2 + 1$, $du = 6x\, dx$, and $\frac{1}{6} du = x\, dx$. The result is

$$\int xe^{3x^2+1} dx = \frac{1}{6} \int e^u\, du = \frac{1}{6} e^u + C$$

Now it's time to put the x's back in.

$$\frac{1}{6} e^{3x^2+1} + C$$

Example 9: Find $\int e^{\sin x} \cos x\, dx$.

Let $u = \sin x$ and $du = \cos x\, dx$. The substitution here couldn't be simpler.

$$\int e^u\, du = e^u + C = e^{\sin x} + C$$

As you can see, these integrals are pretty straightforward. The key is to use u-substitution to transform nasty-looking integrals into simple ones.

There's another type of exponential function whose integral you'll have to find occasionally.

$$\int a^u\, du$$

As you should recall from your rules of logarithms and exponents, the term a^u can be written as $e^{u \ln a}$. Because $\ln a$ is a constant, we can transform $\int a^u \, du$ into $\int e^{u \ln a} \, du$. If you integrate this, you'll get

$$\int e^{u \ln a} \, du = \frac{1}{\ln a} e^{u \ln a} + C$$

Now substituting back a^u for $e^{u \ln a}$,

$$\int a^u \, du = \frac{1}{\ln a} a^u + C$$

Example 10: Find $\int 5^x \, dx$.

Follow the rule we just derived.

$$\int 5^x \, dx = \frac{1}{\ln 5} 5^x + C$$

Because these integrals don't show up too often on the AP Exam, this is the last you'll see of them in this book. You should, however, be able to integrate them using the rule, or by converting them into a form of $\int e^u \, du$.

Try these on your own. Do each problem with the answers covered, and then check your answer.

PROBLEM 1. Evaluate $\int \frac{dx}{3x}$.

Answer: Move the constant term outside of the integral, like this.

$$\frac{1}{3} \int \frac{dx}{x}$$

Now you can integrate.

$$\frac{1}{3} \int \frac{dx}{x} = \frac{1}{3} \ln|x| + C$$

PROBLEM 2. Evaluate $\int \frac{3x^2}{x^3 - 1} \, dx$.

Answer: Let $u = x^3 - 1$ and $du = 3x^2\, dx$, and substitute.

$$\int \frac{du}{u}$$

Now integrate.

$$\ln |u| + C$$

And substitute back.

$$\ln |x^3 - 1| + C$$

PROBLEM 3. Evaluate $\int e^{5x}\, dx$.

Answer: Let $u = 5x$ and $du = 5dx$. Then $\frac{1}{5}\, du = dx$. Substitute in.

$$\frac{1}{5} \int e^u\, du$$

Integrate.

$$\frac{1}{5} e^u + C$$

And substitute back.

$$\frac{1}{5} e^{5x} + C$$

PROBLEM 4. Evaluate $\int 2^{3x}\, dx$.

Answer: Let $u = 3x$ and $du = 3dx$. Then $\frac{1}{3}\, du = dx$. Make the substitution.

$$\frac{1}{3} \int 2^u\, du$$

Integrate according to the rule. Your result should be

$$\frac{1}{3\ln 2} 2^u + C$$

Now get back to the expression as a function of x.

$$\frac{1}{3\ln 2} 2^{3x} + C = \frac{1}{\ln 8} 2^{3x} + C$$

PRACTICE PROBLEM SET 24

Evaluate the following integrals. The answers are in Chapter 23, starting on page 545.

1. $\displaystyle\int \frac{\sec^2 x}{\tan x}\, dx$

2. $\displaystyle\int \frac{\cos x}{1 - \sin x}\, dx$

3. $\displaystyle\int \frac{1}{x \ln x}\, dx$

4. $\displaystyle\int \frac{\sin x - \cos x}{\cos x}\, dx$

5. $\displaystyle\int \frac{dx}{\sqrt{x}\left(1 + 2\sqrt{x}\right)}$

6. $\displaystyle\int \frac{e^x\, dx}{1 + e^x}$

7. $\displaystyle\int x e^{5x^2 - 1}\, dx$

8. $\displaystyle\int \frac{e^x + e^{-x}}{e^x - e^{-x}}\, dx$

9. $\displaystyle\int x 4^{-x^2}\, dx$

10. $\displaystyle\int 7^{\sin x} \cos x\, dx$

End of Chapter 15 Drill

The answers are in Chapter 24.

1. Evaluate $\int \dfrac{x^2}{\left(4x^3+1\right)} dx$.

   (A) $\dfrac{1}{12}\ln\left|4x^3+1\right|+C$

   (B) $\ln\left|4x^3+1\right|+C$

   (C) $\dfrac{-1}{\left(4x^3+1\right)^2}+C$

   (D) $\dfrac{x^3}{\left(4x^3+1\right)^2}+C$

2. Evaluate $\int x \tan x^2\, dx$.

   (A) $\dfrac{1}{2}\sec^2 x^2+C$

   (B) $\sec^2 x^2+C$

   (C) $-\dfrac{1}{2}\ln\left|\cos x^2\right|+C$

   (D) $\dfrac{1}{2}\ln\left|\cos x^2\right|+C$

3. Evaluate $\int x^3 e^{1+x^4} dx$.

   (A) $\dfrac{x^4}{16}e^{1+x^4}+C$

   (B) $\dfrac{x^4}{4}e^{1+x^4}+C$

   (C) $4e^{1+x^4}+C$

   (D) $\dfrac{1}{4}e^{1+x^4}+C$

4. Evaluate $\int x^2 5^{2+x^3} dx$.

   (A) $\dfrac{1}{3\ln 5}5^{2+x^3}+C$

   (B) $\dfrac{1}{3}5^{2+x^3}+C$

   (C) $\dfrac{x^3}{3}5^{2+x^3}+C$

   (D) $\dfrac{x^3}{3\ln 5}5^{2+x^3}+C$

5. Evaluate $\int 4x\sec\left(x^2\right) dx$.

   (A) $\dfrac{1}{2}\ln\left|\sec\left(x^2\right)+\tan\left(x^2\right)\right|+C$

   (B) $2\ln\left|\sec\left(x^2\right)+\tan\left(x^2\right)\right|+C$

   (C) $2\sec\left(x^2\right)\tan\left(x^2\right)+C$

   (D) $\dfrac{1}{2}\sec\left(x^2\right)\tan\left(x^2\right)+C$

REFLECT

Respond to the following questions:

- For which topics discussed in this chapter do you feel you have achieved sufficient mastery to answer multiple-choice questions correctly?

- For which topics discussed in this chapter do you feel you have achieved sufficient mastery to answer open-ended questions correctly?

- For which topics discussed in this chapter do you feel you need more work before you can answer multiple-choice questions correctly?

- For which topics discussed in this chapter do you feel you need more work before you can answer open-ended questions correctly?

- What parts of this chapter are you going to re-review?

Chapter 16
Integration by Parts

This chapter is about one of the most important and powerful techniques of integral calculus. This always shows up on the AP Exam. You'll often find that an integral that otherwise seems to be undoable is simple once you use integration by parts.

THE FORMULA

Remember the Product Rule? It looks like this.

$$\frac{d}{dx}(uv) = u\frac{dv}{dx} + v\frac{du}{dx}$$

If we write this in differential form (i.e., eliminate the dx), we get $d(uv) = u\,dv + v\,du$. Rearranging, we get the following: $u\,dv = d(uv) - v\,du$. Integrating both sides, we obtain the integration by parts formula.

> If you have trouble remembering the order of the integrals, think of the right hand integral as "voodoo."

$$\int u\,dv = uv - \int v\,du$$

This formula allows you to write one integral in terms of another integral. This will often help you turn a difficult integral into an easy one.

Example 1: Find $\int x \sin x\,dx$.

Upon first glance, we've got a problem. If you try u-substitution, neither term is even close to the derivative of the other. Thus, you can't substitute. What do you do? (Hint: look at the title of the chapter.)

Let $u = x$ and $dv = \sin x\,dx$. Then $du = dx$ and $v = -\cos x$. If you plug these parts into the formula, you get

$$\int x \sin x\,dx = -x \cos x + \int \cos x\,dx$$

Now we've turned the difficult left-side integral into an easy right-side integral.

$$\int x \sin x\,dx = -x \cos x + \sin x + C$$

Notice that we let $u = x$ and $dv = \sin x\,dx$. If we had let $u = \sin x$ and $dv = x\,dx$, we would have calculated that

$$du = \cos x\,dx \text{ and } v = \frac{x^2}{2}$$

The formula would have been

$$\int x \sin x \, dx = \frac{x^2 \sin x}{2} - \int \frac{x^2 \cos x}{2} \, dx$$

Not only does this not fix our problem, but it also makes it worse. This leads us to a general rule.

> If one of the terms is a power of x, let that term be u and the other term be dv. (The main exception to this rule is when the other term is $\ln x$.)

Let's assure ourselves that the formula works. If $\int x \sin x \, dx = -x \cos x + \sin x + C$, and you differentiate the right-hand side, the result is

$$-x(-\sin x) + (-1)\cos x + \cos x = x \sin x$$

As you can see, integration by parts is the Product Rule in reverse.

Example 2: Find $\int x^2 \sin x \, dx$.

As with Example 1, we let $u = x^2$ and $dv = \sin x \, dx$. Therefore,

$$du = 2x \, dx \text{ and } v = -\cos x$$

Now plug these into the formula.

$$\int x^2 \sin x \, dx = x^2 \cos x + 2 \int x \cos x \, dx$$

Guess what you have to do now. You have to integrate by parts a second time!

Let $u = x$ and $dv = \cos x \, dx$, so $du = dx$ and $v = \sin x$. Now you have

$$\int x^2 \sin x \, dx = -x^2 \cos x + 2x \sin x - 2 \int \sin x \, dx$$

The final integral becomes

$$\int x^2 \sin x \, dx = -x^2 \cos x + 2x \sin x + 2 \cos x + C$$

Often, you'll have to do integration by parts twice. Good news: the AP Exam never asks you to do it three times.

Example 3: Find $\int \ln x \, dx$.

Let $u = \ln x$ and $dv = dx$. (Yes, we're allowed to do this!) Then $du = \dfrac{1}{x} dx$ and $v = x$. Plugging into the formula, you get

$$\int \ln x \, dx = x \ln x - \int dx$$

Now the right side becomes

$$\int \ln x \, dx = x \ln x - x + C$$

You should add this integral to the others you've memorized or are comfortable with deriving.

$$\int \ln x \, dx = x \ln x - x + C$$

There's one last type of integration by parts technique you need to know for the AP Exam. It requires you to perform two rounds of integration by parts (followed by some simple algebra), and it helps you solve for an unknown integral.

Example 4: Find $\int e^x \cos x \, dx$.

First, we let $u = e^x$ and $dv = \cos x \, dx$. Then, $du = e^x \, dx$ and $v = \sin x$. When you plug this information into the formula, you get

$$\int e^x \cos x \, dx = e^x \sin x - \int e^x \sin x \, dx$$

Now do integration by parts again. This time we let $u = e^x$ and $dv = \sin x \, dx$, so that $du = e^x \, dx$ and $v = -\cos x$. Now the formula looks like the following:

$$\int e^x \cos x \, dx = e^x \sin x - \left(-e^x \cos x + \int e^x \cos x \, dx\right) = e^x \sin x + e^x \cos x - \int e^x \cos x \, dx$$

It looks as if you're back to square one, but notice that the unknown integral is on both the left- and right-hand sides of the equation. If you now add $\int e^x \cos x \, dx$ to both sides (isn't algebra wonderful?), you get

$$2\int e^x \cos x \, dx = e^x \sin x + e^x \cos x$$

Now all you do is divide both sides by 2 and throw in the constant.

$$\int e^x \cos x \, dx = \frac{e^x \sin x + e^x \cos x}{2} + C$$

This is why integration by parts is so powerful. You can evaluate a difficult integral by repeatedly rewriting it in terms of integrals you do know.

The AP Exam always has integration by parts problems, and they're generally just like these. In fact, if you can do all of the examples and problems in this chapter, you should be able to do any integration by parts problem that will appear on the test.

PROBLEM 1. Evaluate $\int x \ln x \, dx$.

Answer: If you let $u = \ln x$ and $dv = x \, dx$, then $du = \dfrac{1}{x} \, dx$ and $v = \dfrac{x^2}{2}$. Notice that this is an exception to our rule about setting the x-term equal to u.

Now you have

$$\int x \ln x \, dx = \left(\frac{x^2}{2} \right) \ln x - \int \frac{x^2}{2} \frac{1}{x} \, dx = \frac{x^2 \ln x}{2} - \frac{1}{2} \int x \, dx$$

Now you can evaluate the right-hand integral.

$$\int x \ln x \, dx = \frac{x^2 \ln x}{2} - \frac{x^2}{4} + C$$

PROBLEM 2. Evaluate $\int x^2 e^x \, dx$.

Answer: Let $u = x^2$ and $dv = e^x \, dx$. Then $du = 2x \, dx$ and $v = e^x$. The formula becomes

$$\int x^2 e^x \, dx = x^2 e^x - 2 \int x e^x \, dx$$

We need to use integration by parts a second time to evaluate the right-hand integral. Let $u = x$ and $dv = e^x \, dx$, so $du = dx$ and $v = e^x$. Now it looks like the following:

$$\int x^2 e^x \, dx = x^2 e^x - 2 \left[x e^x - \int e^x \, dx \right]$$

Now evaluate the right-hand integral.

$$\int x^2 e^x \, dx = x^2 e^x - 2x e^x + 2e^x + C$$

PROBLEM 3. Evaluate $\int \sin x e^{-x} \, dx$.

Answer: Let $u = \sin x$ and $dv = e^{-x}\, dx$. Then $du = \cos x\, dx$ and $v = -e^{-x}$. Now the integral looks like the following:

$$\int \sin x e^{-x}\, dx = -\sin x e^{-x} + \int \cos x e^{-x}\, dx$$

It's time for another round of integration by parts to evaluate the right-hand integral. Let $u = \cos x$ and $dv = e^{-x}\, dx$. Then $du = -\sin x\, dx$ and $v = -e^{-x}$. This gives you

$$\int \sin x e^{-x}\, dx = -\sin x e^{-x} - \cos x e^{-x} - \int \sin x e^{-x}\, dx$$

Now add $\int \sin x e^{-x}\, dx$ to both sides.

$$2\int \sin x e^{-x}\, dx = -\sin x e^{-x} - \cos x e^{-x} + C$$

Therefore,

$$\int \sin x e^{-x}\, dx = \frac{-\sin x e^{-x}}{2} - \frac{\cos x e^{-x}}{2} + C$$

PRACTICE PROBLEM SET 25

Evaluate each of the following integrals. The answers are in Chapter 23, starting on page 548.

1. $\int x \csc^2 x\, dx$

2. $\int x e^{2x}\, dx$

3. $\int \frac{\ln x}{x^2}\, dx$

4. $\int x^2 \cos x\, dx$

5. $\int x \sin 2x\, dx$

6. $\int \ln^2 x\, dx$

7. $\int x \sec^2 x\, dx$

End of Chapter 16 Drill

The answers are in Chapter 24.

1. Evaluate $\int x^2 \ln x \, dx$.

   (A) $\dfrac{x^3}{2} \ln x + \dfrac{x^3}{3} + C$

   (B) $\dfrac{x^3}{2} \ln x + \dfrac{x^3}{9} + C$

   (C) $\dfrac{x^3}{2} \ln x - \dfrac{x^3}{3} + C$

   (D) $\dfrac{x^3}{2} \ln x - \dfrac{x^3}{9} + C$

2. Evaluate $\int x \sec^2 x \, dx$.

   (A) $x \tan x + \ln|\cos x| + C$

   (B) $x \tan x - \ln|\cos x| + C$

   (C) $\dfrac{x^2}{2} \tan x + C$

   (D) $\dfrac{x^2 \sec^3 x}{6} + C$

3. Evaluate $\int x e^{4x} \, dx$.

   (A) $\dfrac{x e^{4x}}{4} \quad \dfrac{e^{4x}}{16} + C$

   (B) $\dfrac{x e^{4x}}{4} + \dfrac{e^{4x}}{16} + C$

   (C) $\dfrac{x e^{4x}}{4} - \dfrac{e^{4x}}{4} + C$

   (D) $\dfrac{x e^{4x}}{4} + \dfrac{e^{4x}}{4} + C$

4. Evaluate $\int x^2 \sin(3x) \, dx$.

   (A) $\dfrac{x^2 \cos(3x)}{3} - \dfrac{2x \sin(3x)}{3} - \dfrac{2\cos(3x)}{3} + C$

   (B) $-\dfrac{x^2 \cos(3x)}{3} + \dfrac{2x \sin(3x)}{3} + \dfrac{2\cos(3x)}{3} + C$

   (C) $-\dfrac{x^2 \cos(3x)}{3} + \dfrac{2x \sin(3x)}{9} + \dfrac{2\cos(3x)}{27} + C$

   (D) $\dfrac{x^2 \cos(3x)}{3} - \dfrac{2x \sin(3x)}{9} - \dfrac{2\cos(3x)}{27} + C$

5. Evaluate $\int e^{\frac{x}{2}} \cos(2x) \, dx$.

   (A) $\dfrac{17}{16} \left[\dfrac{e^{\frac{x}{2}} \sin(2x)}{2} + \dfrac{e^{\frac{x}{2}} \cos(2x)}{8} \right] + C$

   (B) $\left[\dfrac{e^{\frac{x}{2}} \sin(2x)}{2} + \dfrac{e^{\frac{x}{2}} \cos(2x)}{8} \right] + C$

   (C) $\left[\dfrac{e^{\frac{x}{2}} \sin(2x)}{2} - \dfrac{e^{\frac{x}{2}} \cos(2x)}{8} \right] + C$

   (D) $\dfrac{16}{17} \left[\dfrac{e^{\frac{x}{2}} \sin(2x)}{2} + \dfrac{e^{\frac{x}{2}} \cos(2x)}{8} \right] + C$

REFLECT

Respond to the following questions:

- For which topics discussed in this chapter do you feel you have achieved sufficient mastery to answer multiple-choice questions correctly?

- For which topics discussed in this chapter do you feel you have achieved sufficient mastery to answer open-ended questions correctly?

- For which topics discussed in this chapter do you feel you need more work before you can answer multiple-choice questions correctly?

- For which topics discussed in this chapter do you feel you need more work before you can answer open-ended questions correctly?

- What parts of this chapter are you going to re-review?

Chapter 17
Advanced Integrals of Trig Functions

INVERSE TRIG FUNCTIONS

In this unit, you'll learn a broader set of integration techniques. So far, you know how to do only a few types of integrals: polynomials, some trig functions, and logs. Now we'll turn our attention to a set of integrals that result in **inverse trigonometric functions**. But first, we need to go over the derivatives of the inverse trigonometric functions. (You might want to refer back to Chapter 12 on derivatives of inverse functions.)

Suppose you have the equation $\sin y = x$. If you differentiate both sides with respect to x, you get

$$\cos y \frac{dy}{dx} = 1$$

Now divide both sides by $\cos y$.

$$\frac{dy}{dx} = \frac{1}{\cos y}$$

Because $\sin^2 y + \cos^2 y = 1$, we can replace $\cos y$ with $\sqrt{1 - \sin^2 y}$.

$$\frac{dy}{dx} = \frac{1}{\sqrt{1 - \sin^2 y}}$$

Finally, because $x = \sin y$, replace $\sin y$ with x. The derivative equals

$$\frac{1}{\sqrt{1 - x^2}}$$

Now go back to the original equation $\sin y = x$ and solve for y in terms of x: $y = \sin^{-1} x$.

If you differentiate both sides with respect to x, you get

$$\frac{dy}{dx} = \frac{d}{dx} \sin^{-1} x$$

Replace $\frac{dy}{dx}$, and you get the final result.

$$\frac{d}{dx} \sin^{-1} x = \frac{1}{\sqrt{1 - x^2}}$$

This is the derivative of inverse sine. By similar means, you can find the derivatives of all six inverse trig functions. They're not difficult to derive, and they're also not difficult to memorize. The choice is yours. We use u instead of x to account for the Chain Rule.

$$\frac{d}{dx}(\sin^{-1} u) = \frac{1}{\sqrt{1-u^2}}\frac{du}{dx}; \quad -1 < u < 1 \qquad \frac{d}{dx}(\cos^{-1} u) = \frac{-1}{\sqrt{1-u^2}}\frac{du}{dx}; \quad -1 < u < 1$$

$$\frac{d}{dx}(\tan^{-1} u) = \frac{1}{1+u^2}\frac{du}{dx} \qquad\qquad\qquad \frac{d}{dx}(\cot^{-1} u) = \frac{-1}{1+u^2}\frac{du}{dx}$$

$$\frac{d}{dx}(\sec^{-1} u) = \frac{1}{|u|\sqrt{u^2-1}}\frac{du}{dx}; \quad |u| > 1 \qquad \frac{d}{dx}(\csc^{-1} u) = \frac{-1}{|u|\sqrt{u^2-1}}\frac{du}{dx}; \qquad |u| > 1$$

Notice the domain restrictions for inverse sine, cosine, secant, and cosecant.

Example 1: $\dfrac{d}{dx} = \sin^{-1} x^2 = \dfrac{2x}{\sqrt{1-(x^2)^2}} = \dfrac{2x}{\sqrt{1-x^4}}$

Example 2: $\dfrac{d}{dx} = \tan^{-1} 5x = \dfrac{5}{1+(5x)^2} = \dfrac{5}{1+25x^2}$

Example 3: $\dfrac{d}{dx} = \sec^{-1}(x^2 - x) = \dfrac{2x-1}{|x^2 - x|\sqrt{(x^2-x)^2 - 1}} = \dfrac{2x-1}{|x^2 - x|\sqrt{x^4 - 2x^3 + x^2 - 1}}$

Finding the derivatives of inverse trig functions is just a matter of following the formulas. They're very rarely tested on the AP Exam in this form and are not terribly important. In addition, the AP Exam almost always tests only inverse sine and tangent, and then usually only as integrals, not as derivatives.

Therefore, it's time to learn the integrals. The derivative formulas lead directly to the integral formulas, but because the only difference between the derivatives of inverse sine, tangent, and secant and those of inverse cosine, cotangent, and cosecant is the negative sign, only the former functions are generally used.

$$\int \frac{du}{\sqrt{1-u^2}} = \sin^{-1} u + C \qquad \text{(valid for } u^2 < 1\text{)}$$

$$\int \frac{du}{1+u^2} = \tan^{-1} u + C$$

$$\int \frac{du}{u\sqrt{u^2-1}} = \sec^{-1} u + C \qquad \text{(valid for } u^2 > 1\text{)}$$

The first two always show up on the AP Exam, so you should learn to recognize the pattern. Most of the time, the integration requires some sticky algebra to get the integral into the proper form.

Example 4: $\displaystyle\int \frac{x \, dx}{\sqrt{1-x^4}} =$

Don't forget about u-substitution. Let $u = x^2$ and $du = 2x \, dx$. Thus, $\frac{1}{2} du = x \, dx$. Substituting and integrating, we get

$$\frac{1}{2}\int \frac{du}{\sqrt{1-u^2}} = \frac{1}{2}\sin^{-1} u + C$$

Now substitute back.

$$\frac{1}{2}\sin^{-1}(x^2) + C$$

Example 5: $\displaystyle\int \frac{dx}{\sqrt{4-x^2}} =$

Anytime you see an integral where you have the square root of a constant minus a function, try to turn it into an inverse sine. You can do that here with a little simple algebra.

$$\int \frac{dx}{\sqrt{4-x^2}} = \int \frac{dx}{\sqrt{4\left(1-\frac{x^2}{4}\right)}} = \frac{1}{2}\int \frac{dx}{\sqrt{1-\frac{x^2}{4}}}$$

Now use u-substitution: let $u = \dfrac{x}{2}$ and $du = \dfrac{1}{2}\,dx$, so $2du = dx$, and substitute.

$$\frac{1}{2}\int \frac{dx}{\sqrt{1 - \dfrac{x^2}{4}}} = \frac{1}{2}\int \frac{2\,du}{\sqrt{1 - u^2}} = \sin^{-1} u + C$$

When you substitute back, you get

$$\sin^{-1} \frac{x}{2} + C$$

Example 6: $\displaystyle\int \frac{e^x\,dx}{1 + e^{2x}} =$

Again, use u-substitution. Let $u - e^x$ and $du - e^x\,dx$, and substitute.

$$\int \frac{du}{1 + u^2} = \tan^{-1} u + C$$

Then substitute back.

$$\tan^{-1}e^x + C$$

Let's do one last type that's slightly more difficult. For this one, you need to remember how to complete the square.

Example 7: $\displaystyle\int \frac{dx}{x^2 + 4x + 5} =$

Complete the square in the denominator (your algebra teacher warned you about this).

$$x^2 + 4x + 5 = (x + 2)^2 + 1$$

Now rewrite the integral.

$$\int \frac{dx}{1 + (x + 2)^2}$$

Using u-substitution, let $u = x + 2$ and $du = dx$. Then do the substitution.

$$\int \frac{du}{1 + u^2} = \tan^{-1} u + C$$

When you put the function of x back in to the integral, it reads,

$$\tan^{-1}(x + 2) + C$$

Evaluating these integrals involves looking for a particular pattern. If it's there, all you have to do is use algebra and *u*-substitution to make the integrand conform to the pattern. Once that's accomplished, the rest is easy. These integrals will show up again when we do partial fractions. Otherwise, as far as the AP Exam is concerned, this is all you need to know.

Try these solved problems, and don't forget to look for those patterns. Get out your index card, and cover the answers while you work.

PROBLEM 1. Find the derivative of $y = \sin^{-1}\left(\dfrac{x}{2}\right)$.

Answer: The rule is

$$\frac{d}{dx}(\sin^{-1} u) = \frac{1}{\sqrt{1 - u^2}}\frac{du}{dx}$$

The algebra looks like the following:

$$\frac{d}{dx}\left(\sin^{-1}\left(\frac{x}{2}\right)\right) = \frac{1}{\sqrt{1 - \left(\frac{x}{2}\right)^2}}\frac{1}{2} = \frac{1}{2}\left(\frac{1}{\sqrt{1 - \frac{x^2}{4}}}\right) = \frac{1}{2}\left(\frac{1}{\sqrt{\frac{4 - x^2}{4}}}\right) = \frac{1}{2}\left(\frac{2}{\sqrt{4 - x^2}}\right) = \frac{1}{\sqrt{4 - x^2}}$$

PROBLEM 2. Evaluate $\displaystyle\int \frac{dx}{4 + x^2}$.

Answer: First, you need to do a little algebra. Factor 4 out of the denominator.

$$\int \frac{dx}{4\left(1 + \dfrac{x^2}{4}\right)}$$

Next, rewrite the integrand.

$$\frac{1}{4}\int \frac{dx}{\left(1 + \left(\dfrac{x}{2}\right)^2\right)}$$

Now you can use *u*-substitution. Let $u = \dfrac{x}{2}$ and $du = \dfrac{1}{2}dx$, so $2du = dx$.

$$\frac{1}{2}\int \frac{du}{1 + u^2}$$

Now integrate it.

$$\frac{1}{2} \tan^{-1} u + C$$

And re-substitute.

$$\frac{1}{2} \tan^{-1} \frac{x}{2} + C$$

PROBLEM 3. Evaluate $\int \dfrac{dx}{\sqrt{-x^2 + 4x - 3}}$.

Answer: This time, you need to use some algebra by completing the square of the polynomial under the square root sign.

$$-x^2 + 4x - 3 = -(x^2 - 4x + 3) = -\left[(x-2)^2 - 1\right] = \left[1 - (x-2)^2\right]$$

Now rewrite the integrand.

$$\int \frac{dx}{\sqrt{1 - (x-2)^2}}$$

And use *u*-substitution. Let $u = x - 2$ and $du = dx$.

$$\int \frac{du}{\sqrt{1 - u^2}}$$

Now, this looks familiar. Once you integrate, you get

$$\sin^{-1} u + C$$

After you substitute back it becomes

$$\sin^{-1}(x - 2) + C$$

PRACTICE PROBLEM SET 26

Here is some more practice work on derivatives and integrals of inverse trig functions. The answers are in Chapter 23, starting on page 551.

1. Find the derivative of $\dfrac{1}{4}\tan^{-1}\left(\dfrac{x}{4}\right)$.

2. Find the derivative of $\sin^{-1}\left(\dfrac{1}{x}\right)$.

3. Find the derivative of $\tan^{-1}(e^x)$.

4. Evaluate $\displaystyle\int \dfrac{dx}{x\sqrt{x^2-\pi}}$.

5. Evaluate $\displaystyle\int \dfrac{dx}{7+x^2}$.

6. Evaluate $\displaystyle\int \dfrac{dx}{x(1+\ln^2 x)}$.

7. Evaluate $\displaystyle\int \dfrac{\sec^2 x\, dx}{\sqrt{1-\tan^2 x}}$.

8. Evaluate $\displaystyle\int \dfrac{e^{3x}\, dx}{1+e^{6x}}$.

ADVANCED INTEGRALS OF TRIG FUNCTIONS

In Chapter 13, you learned how to find the integrals of some of the trigonometric functions. Now it's time to figure out how to find the integrals of some of the more complicated trig expressions. First, recall some of the basic trigonometric integrals.

$$\int \sin x \, dx = -\cos x + C \qquad\qquad \int \cos x \, dx = \sin x + C$$

$$\int \sec^2 x \, dx = \tan x + C \qquad\qquad \int \csc^2 x \, dx = -\cot x + C$$

$$\int \sec x \tan x \, dx = \sec x + C \qquad\qquad \int \csc x \cot x \, dx = -\csc x + C$$

$$\int \tan x \, dx = -\ln \left| \cos x \right| + C \qquad\qquad \int \cot x \, dx = \ln \left| \sin x \right| + C$$

$$\int \sec x \, dx = \ln \left| \sec x + \tan x \right| + C \qquad\qquad \int \csc x \, dx = -\ln \left| \csc x + \cot x \right| + C$$

Some of these were derived by reversing differentiation, others by u-substitution.

Example 1: Evaluate $\int \sin^2 x \, dx$.

You can start by replacing $\sin^2 x$ with $\dfrac{1 - \cos 2x}{2}$. This comes from taking the Double Angle formula $\cos 2x = 1 - 2\sin^2 x$ and solving for $\sin^2 x$. After you make that replacement, the integral looks like the following:

$$\int \frac{1 - \cos 2x}{2} \, dx = \frac{1}{2} \int dx - \frac{1}{2} \int \cos 2x \, dx = \frac{x}{2} - \frac{\sin 2x}{4} + C$$

Remember the following substitution:

$$\sin^2 x = \frac{1 - \cos 2x}{2}$$

You could also use the following substitution:

$$\cos^2 x = \frac{1 + \cos 2x}{2}$$

(Note: This comes from taking the Double Angle formula $\cos 2x = 2\cos^2 x - 1$ and solving for $\cos^2 x$.) You can also use the substitution to find $\int \cos^2 x \, dx$.

Now let's do some variations.

Example 2: Evaluate $\int \sin^2 x \cos x \, dx$.

Here use some simple u-substitution. Let $u = \sin x$ and $du = \cos x \, dx$ and substitute.

$$\int \sin^2 x \cos x \, dx = \int u^2 \, du = \frac{u^3}{3} + C$$

When you substitute back, you get

$$\frac{\sin^3 x}{3} + C$$

In fact, you can do any integral of the form $\int \sin^2 x \cos x \, dx$ using u-substitution, and the result will be

$$\frac{\sin^{n+1} x}{n+1} + C$$

Similarly,

$$\int \cos^n x \sin x \, dx = -\frac{\cos^{n+1} x}{n+1} + C$$

How about higher powers of sine and cosine?

Example 3: Evaluate $\int \sin^3 x \, dx$.

First, break the integrand into the following:

$$\int (\sin x)(\sin^2 x) \, dx$$

Next, using trig substitution, you get

$$\int (\sin x)(\sin^2 x) \, dx = \int (\sin x)(1 - \cos^2 x) \, dx$$

You can turn the following into two integrals:

$$\int \sin x \, dx - \int (\sin x \cos^2 x) \, dx$$

You can do both of these integrals.

$$-\cos x + \frac{\cos^3 x}{3} + C$$

Example 4: Evaluate $\int \sin^3 x \cos^2 x \, dx$.

Here, break the integrand into

$$\int (\sin x)\left(\sin^2 x\right)\left(\cos^2 x\right) dx$$

Next, using trig substitution, you get

$$\int (\sin x)\left(\sin^2 x\right)\left(\cos^2 x\right) dx = \int (\sin x)\left(1 - \cos^2 x\right)\left(\cos^2 x\right) dx$$

Now you can use u-substitution. Let $u = \cos x$, $du = -\sin x \, dx$, and substitute.

$$-\int \left(1 - u^2\right)u^2 \, du = \int \left(u^4 - u^2\right) du = \frac{u^5}{5} - \frac{u^3}{3} + C$$

Now substitute back, and you're done.

$$\frac{\cos^5 x}{5} - \frac{\cos^3 x}{3} + C$$

As you can see, these integrals all require you to know trig substitutions and u-substitution. The AP Exam doesn't ask too many variations on these integrals, but let's do just a few other types, just in case.

Example 5: Evaluate $\int \tan^2 x \, dx$.

First, use the trigonometric substitution $\tan^2 x = \sec^2 x - 1$.

$$\int \tan^2 x \, dx = \int (\sec^2 x - 1) \, dx$$

Now break this into two integrals that are easy to evaluate.

$$\int \sec^2 x \, dx - \int dx$$

And integrate.

$$\tan x - x + C$$

Example 6: Evaluate $\int \tan^3 x \, dx$.

Recognize what to do here? Right! Break up the integrand!

$$\int (\tan x)(\tan^2 x) \, dx$$

Next, use the trig substitution $\tan^2 x = \sec^2 x - 1$ in the expression.

$$\int (\tan x)(\sec^2 x - 1) \, dx$$

Break this into two integrals.

$$\int \tan x \, \sec^2 x \, dx - \int \tan x \, dx$$

Tackle the first integral using u-substitution. Let $u = \tan x$ and $du = \sec^2 x \, dx$.

$$\int u \, du = \frac{u^2}{2} + C$$

Substituting back gives you

$$\frac{\tan^2 x}{2} + C$$

We've done the second integral before.

$$\int \tan x \, dx = \int \frac{\sin x}{\cos x} \, dx = \ln |\cos x| + C$$

Thus, the integral is

$$\frac{\tan^2 x}{2} + \ln |\cos x| + C$$

Example 7: Evaluate $\int \sec^3 x \, dx$.

Now it's time for integration by parts. Let $u = \sec x$ and $dv = \sec^2 x \, dx$. Therefore, $du = \sec x \tan x \, dx$ and $v = \tan x$. Plug these into the formula, and you get

$$\int \sec^3 x \, dx = \sec x \tan x - \int \tan^2 x \sec x \, dx$$

Next, substitute $\tan^2 x = \sec^2 x - 1$.

$$\int \tan^2 x \sec x \, dx = \int (\sec^2 x - 1)(\sec x) \, dx = \int \sec^3 x \, dx - \int \sec x \, dx$$

So, the integral has become

$$\int \sec^3 x \, dx = \sec x \tan x - \int \sec^3 x \, dx + \int \sec x \, dx$$

Now, add $\int \sec^3 x \, dx$ to both sides.

$$2 \int \sec^3 x \, dx = \sec x \tan x + \int \sec x \, dx$$

Finally, recall that the integral on the right is

$$\int \sec x \, dx = \ln \left| \sec x + \tan x \right| + C$$

The combined integral is

$$2 \int \sec^3 x \, dx = \sec x \tan x + \ln \left| \sec x + \tan x \right| + C$$

When you divide by 2, you get your final answer.

$$\int \sec^3 x \, dx = \frac{\sec x \tan x + \ln \left| \sec x + \tan x \right|}{2} + C$$

Whew! This is about as difficult as the AP Exam ever gets. Believe it or not, trigonometric integrals can get even more complicated, involving higher powers or more difficult combinations of trig functions. Thankfully, the AP Exam tends to limit itself to the simpler ones.

Walk yourself through the step-by-step process by trying some solved problems below. Cover each answer first, then check your answer.

> If you can integrate any power of a trig function up to 4, as well as some of the basic combinations, you'll be able to handle any trigonometric integral that appears on the AP Exam.

PROBLEM 1. $\int \cos^2 x \, dx$

Answer: Use the substitution $\dfrac{1 + \cos 2x}{2}$. Now the integral looks like the following:

$$\int \frac{1 + \cos 2x}{2} \, dx$$

Divide this into two integrals.

$$\int \frac{1}{2} \, dx + \int \frac{\cos 2x}{2} \, dx$$

When you evaluate each one separately, you get

$$\frac{x}{2} + \frac{\sin 2x}{4} + C$$

PROBLEM 2. $\int \cos^3 x \, dx$

Answer: Here, split $\cos^3 x$ into $\cos x \cos^2 x$, and replace the second term with $(1 - \sin^2 x)$.

$$\int \cos x \, (1 - \sin^2 x) \, dx$$

This can be rewritten as

$$\int \cos x \, dx - \int \sin^2 x \cos x \, dx$$

The left-hand integral is $\sin x$. Do the right-hand integral using u-substitution. Let $u = \sin x$ and $du = \cos x \, dx$. You get

$$\int u^2 \, du = \frac{u^3}{3}$$

When you substitute back, the second integral becomes

$$\frac{\sin^3 x}{3} + C$$

Thus, the complete answer is

$$\sin x - \frac{\sin^3 x}{3} + C$$

PROBLEM 3. $\int \tan^2 x \sec^2 x \, dx$

Answer: This won't take long. Use u-substitution, by letting $u = \tan x$ and $du = \sec^2 x \, dx$.

$$\int u^2 \, du = \frac{u^3}{3}$$

The final result when you substitute back is

$$\frac{\tan^3 x}{3} + C$$

PRACTICE PROBLEM SET 27

Evaluate the following integrals. The answers are in Chapter 23, starting on page 554.

1. $\displaystyle\int \sin^4 x\, dx$

2. $\displaystyle\int \cos^4 x\, dx$

3. $\displaystyle\int \cos^4 x \sin x\, dx$

4. $\displaystyle\int \sin^2 x \cos^2 x\, dx$

5. $\displaystyle\int \tan^3 x \sec^2 x\, dx$

6. $\displaystyle\int \tan^5 x\, dx$

7. $\displaystyle\int \cot^2 x \sec x\, dx$

End of Chapter 17 Drill

The answers are in Chapter 24.

1. Evaluate $\int \sin^{19} x \cos^3 x \, dx$.

   (A) $\dfrac{\sin^{20} x}{20} - \dfrac{\cos^4 x}{4} + C$

   (B) $\dfrac{\sin^{20} x}{20} - \dfrac{\sin^{22} x}{22} + C$

   (C) $\dfrac{\sin^{20} x}{20} + \dfrac{\sin^{22} x}{22} + C$

   (D) $\dfrac{\sin^{20} x}{20} + \dfrac{\cos^4 x}{4} + C$

2. Evaluate $\int \cos^3(3x)\sin(3x) \, dx$.

   (A) $\dfrac{\sin^4(3x)}{12}\dfrac{\cos^2(3x)}{3} + C$

   (B) $-\dfrac{\sin^4(3x)}{12}\dfrac{\cos^2(3x)}{3} + C$

   (C) $\dfrac{\cos^4(3x)}{12} + C$

   (D) $-\dfrac{\cos^4(3x)}{12} + C$

3. Evaluate $\int 16\sin^2(2x)\cos^2(2x) \, dx$.

   (A) $2x - \dfrac{\sin 8x}{4} + C$

   (B) $2x + \dfrac{\sin 8x}{4} + C$

   (C) $2x - \dfrac{\sin 2x}{4} + C$

   (D) $2x + \dfrac{\sin 2x}{4} + C$

4. Evaluate $\int \sec^4 x \, dx$.

   (A) $\dfrac{\sec^5 x}{5} + C$

   (B) $-\dfrac{\sec^5 x}{5} + C$

   (C) $\tan x + \dfrac{\tan^3 x}{3} + C$

   (D) $\tan x - \dfrac{\tan^3 x}{3} + C$

5. Evaluate $\int \arctan x \, dx$.

   (A) $\dfrac{1}{|x|\sqrt{x^2-1}} + C$

   (B) $\sec^2 x + C$

   (C) $x \arctan x + \dfrac{1}{2}\ln(1+x^2) + C$

   (D) $x \arctan x - \dfrac{1}{2}\ln(1+x^2) + C$

REFLECT

Respond to the following questions:

- For which topics discussed in this chapter do you feel you have achieved sufficient mastery to answer multiple-choice questions correctly?

- For which topics discussed in this chapter do you feel you have achieved sufficient mastery to answer open-ended questions correctly?

- For which topics discussed in this chapter do you feel you need more work before you can answer multiple-choice questions correctly?

- For which topics discussed in this chapter do you feel you need more work before you can answer open-ended questions correctly?

- What parts of this chapter are you going to re-review?

Unit 4
Integral
Calculus
Applications

Chapter 18
The Area Between
Two Curves

These next two units discuss some of the most difficult topics you'll encounter in AP Calculus. For some reason, students have terrible trouble setting up these problems. Fortunately, the AP Exam asks only relatively simple versions of these problems on the exam.

Unfortunately, this unit and the next are always on the AP Exam. We'll try to make them as simple as possible. You've already learned that if you want to find the area under a curve, you can integrate the function of the curve by using the endpoints as limits. So far, though, we've talked only about the area between a curve and the *x*-axis. What if you have to find the area between two curves?

VERTICAL SLICES

Suppose you wanted to find the area between the curve $y = x$ and the curve $y = x^2$ from $x = 2$ to $x = 4$. First, sketch the curves.

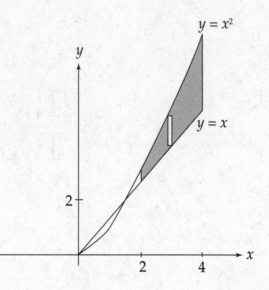

You can find the area by slicing up the region vertically, into a bunch of infinitely thin strips, and adding up the areas of all the strips. The height of each strip is $x^2 - x$, and the width of each strip is dx. Add up all the strips by using the integral.

$$\int_2^4 \left(x^2 - x\right) dx$$

Then, evaluate it.

$$\left(\frac{x^3}{3} - \frac{x^2}{2}\right)\Bigg|_2^4 = \left(\frac{64}{3} - \frac{16}{2}\right) - \left(\frac{8}{3} - \frac{4}{2}\right) = \frac{38}{3}$$

That wasn't so hard, was it? Don't worry. The process gets more complicated, but the idea remains the same. Now let's generalize this and come up with a rule.

If a region is bounded by $f(x)$ above and $g(x)$ below at all points of the interval $[a, b]$, then the area of the region is given by

$$\int_a^b [f(x) - g(x)] dx$$

Example 1: Find the area of the region between the parabola $y = 1 - x^2$ and the line $y = 1 - x$.

First, make a sketch of the region.

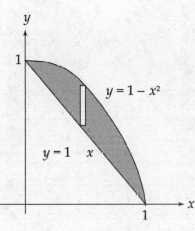

To find the points of intersection of the graphs, set the two equations equal to each other and solve for x.

$$1 - x^2 = 1 - x$$
$$x^2 - x = 0$$
$$x(x - 1) = 0$$
$$x = 0, 1$$

The left-hand edge of the region is $x = 0$ and the right-hand edge is $x = 1$, so the limits of integration are from 0 to 1.

Next, note that the top curve is always $y = 1 - x^2$, and the bottom curve is always $y = 1 - x$. (If the region has a place where the top and bottom curve switch, you need to make two integrals, one for each region. Fortunately, that's not the case here.) Thus, we need to evaluate the following:

$$\int_0^1 \left[(1 - x^2) - (1 - x) \right] dx$$

$$\int_0^1 \left[(1 - x^2) - (1 - x) \right] dx = \int_0^1 (-x^2 + x) dx = \left(-\frac{x^3}{3} + \frac{x^2}{2} \right) \Bigg|_0^1 = \frac{1}{6}$$

Sometimes you're given the endpoints of the region; sometimes you have to find them on your own.

Example 2: Find the area of the region between the curve $y = \sin x$ and the curve $y = \cos x$ from 0 to $\dfrac{\pi}{2}$.

First, sketch the region.

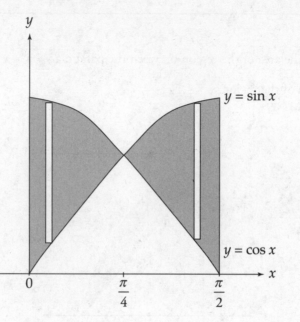

Notice that $\cos x$ is on top between 0 and $\dfrac{\pi}{4}$, then $\sin x$ is on top between $\dfrac{\pi}{4}$ and $\dfrac{\pi}{2}$. The point where they cross is $\dfrac{\pi}{4}$, so you have to divide the area into two integrals: one from 0 to $\dfrac{\pi}{4}$, and the other from $\dfrac{\pi}{4}$ to $\dfrac{\pi}{2}$. In the first region, $\cos x$ is above $\sin x$, so the integral to evaluate is

$$\int_0^{\frac{\pi}{4}} \left(\cos x - \sin x\right)\, dx$$

The integral of the second region is a little different, because $\sin x$ is above $\cos x$.

$$\int_{\frac{\pi}{4}}^{\frac{\pi}{2}} \left(\sin x - \cos x\right) dx$$

If you add the two integrals, you'll get the area of the whole region.

$$\int_0^{\frac{\pi}{4}} (\cos x - \sin x)\,dx = \left(\sin x + \cos x\right)\Big|_0^{\frac{\pi}{4}} = \sqrt{2} - 1$$

$$\int_{\frac{\pi}{4}}^{\frac{\pi}{2}} (\sin x - \cos x)\,dx = \left(-\cos x - \sin x\right)\Big|_{\frac{\pi}{4}}^{\frac{\pi}{2}} = \sqrt{2} - 1$$

Adding these, we get that the area is $2\sqrt{2} - 2$.

HORIZONTAL SLICES

Now for the fun part. We can slice a region vertically when one function is at the top of our section and a different function is at the bottom. But what if the same function is both the top and the bottom of the slice (what we call a double-valued function)? You have to slice the region horizontally.

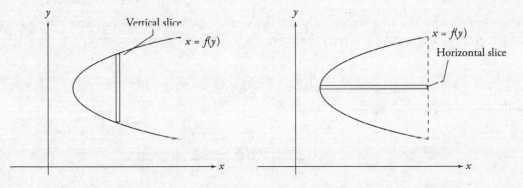

If we were to slice vertically, as in the left-hand picture, we'd have a problem. But if we were to slice horizontally, as in the right-hand picture, we don't have a problem. Instead of integrating an equation $f(x)$ with respect to x, we need to integrate an equation $f(y)$ with respect to y. As a result, our area formula changes a little.

If a region is bounded by $f(y)$ on the right and $g(y)$ on the left at all points of the interval $[c, d]$, then the area of the region is given by

$$\int_c^d \left[f(y) - g(y)\right]dy$$

Example 3: Find the area of the region between the curve $x = y^2$ and the curve $x = y + 6$ from $y = 0$ to $y = 3$.

First, sketch the region.

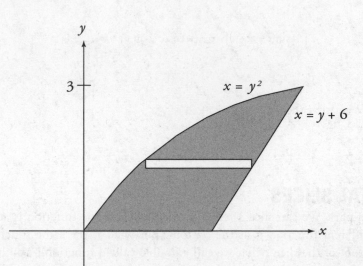

When you slice up the area horizontally, the right end of each section is the curve $x = y + 6$, and the left end of each section is always the curve $x = y^2$. Now set up our integral.

$$\int_0^3 (y + 6 - y^2)\, dy$$

Evaluating this gives us the area.

$$\int_0^3 (y + 6 - y^2)\, dy = \left(\frac{y^2}{2} + 6y - \frac{y^3}{3} \right)\Bigg|_0^3 = \frac{27}{2}$$

Example 4: Find the area between the curve $y = \sqrt{x + 3}$ and the curve $y = \sqrt{3 - x}$ and the x-axis from $x = -3$ to $x = 3$.

First, sketch the curves.

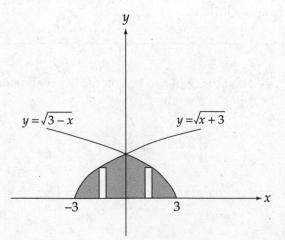

From $x = -3$ to $x = 0$, if you slice the region vertically, the curve $y = \sqrt{x + 3}$ is on top, and the x-axis is on the bottom; from $x = 0$ to $x = 3$, the curve $y = \sqrt{3 - x}$ is on top and the x-axis is on the bottom. Therefore, you can find the area by evaluating two integrals.

$$\int_{-3}^{0} (\sqrt{x + 3} - 0)\, dx \text{ and } \int_{0}^{3} (\sqrt{3 - x} - 0)\, dx$$

Your results should be

$$\frac{2}{3}(x+3)^{\frac{3}{2}}\bigg|_{-3}^{0} + \left(-\frac{2}{3}(3-x)^{\frac{3}{2}} \right)\bigg|_{0}^{3} = 4\sqrt{3}$$

Let's suppose you sliced the region horizontally instead. The curve $y = \sqrt{x + 3}$ is always on the left, and the curve $y = \sqrt{3 - x}$ is always on the right. If you solve each equation for x in terms of y, you save some time by using only one integral instead of two.

The two equations are $x = y^2 - 3$ and $x = 3 - y^2$. We also have to change the limits of integration from x-limits to y-limits. The two curves intersect at $y = \sqrt{3}$, so our limits of integration are from $y = 0$ to $y = \sqrt{3}$. The new integral is

$$\int_{0}^{\sqrt{3}} \left[(3 - y^2) - (y^2 - 3) \right] dy = \int_{0}^{\sqrt{3}} (6 - 2y^2)\, dy = 6y - \frac{2y^3}{3}\bigg|_{0}^{\sqrt{3}} = 4\sqrt{3}$$

You get the same answer no matter which way you integrate (as long as you do it right!). The challenge of area problems is determining which way to integrate and then converting the equation to different terms. Unfortunately, there's no simple rule for how to do this. You have to look at the region and figure out its endpoints, as well as where the curves are with respect to each other.

Once you can do that, then the actual set-up of the integral(s) isn't that hard. Sometimes, evaluating the integrals isn't easy; however, if the integral of an AP question is difficult to evaluate, you'll only be required to set it up, not to evaluate it.

Here are some sample problems. On each, decide the best way to set up the integrals, and then evaluate them. Then check your answer.

PROBLEM 1. Find the area of the region between the curve $y = 3 - x^2$ and the line $y = 1 - x$ from $x = 0$ to $x = 2$.

Answer: First, make a sketch.

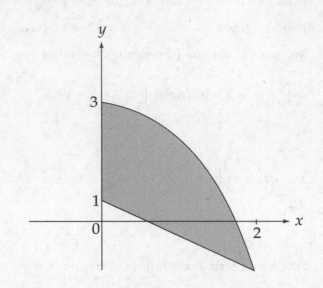

Because the curve $y = 3 - x^2$ is always above $y = 1 - x$ within the interval, you have to evaluate the following integral:

$$\int_0^2 \left[(3 - x^2) - (1 - x) \right] dx = \int_0^2 (2 + x - x^2) \, dx$$

Therefore, the area of the region is

$$\left(2x + \frac{x^2}{2} - \frac{x^3}{3} \right)\Bigg|_0^2 = \frac{10}{3}$$

PROBLEM 2. Find the area between the x-axis and the curve $y = 2 - x^2$ from $x = 0$ to $x = 2$.

Answer: First, sketch the graph over the interval.

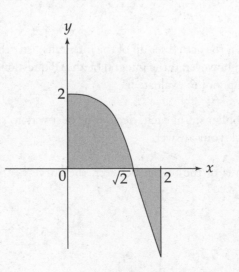

Because the curve crosses the x-axis at $\sqrt{2}$, you have to divide the region into two parts: from $x = 0$ to $x = \sqrt{2}$ and from $x = \sqrt{2}$ to $x = 2$. In the latter region, you'll need to integrate $y = -\left(2 - x^2\right) = x^2 - 2$ to adjust for the region's being below the x-axis. Therefore, we can find the area by evaluating

$$\int_0^{\sqrt{2}} (2 - x^2)\,dx + \int_{\sqrt{2}}^2 (x^2 - 2)\,dx$$

Integrating, we get

$$\left(2x + -\frac{x^3}{3}\right)\Bigg|_0^{\sqrt{2}} + \left(\frac{x^3}{3} - 2x\right)\Bigg|_{\sqrt{2}}^2 = \left(2\sqrt{2} - \frac{2\sqrt{2}}{3}\right) - 0 + \left(\frac{8}{3} - 4\right) - \left(\frac{2\sqrt{2}}{3} - 2\sqrt{2}\right) = \frac{8\sqrt{2} - 4}{3}$$

PROBLEM 3. Find the area of the region between the curve $x = y^2 - 4y$ and the line $x = y$.

Answer: First, sketch the graph over the interval.

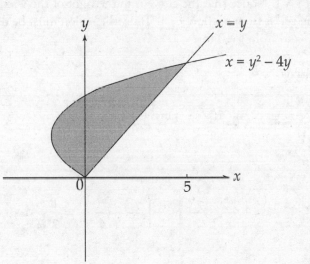

You don't have the endpoints this time, so you need to find where the two curves intersect. If you set them equal to each other, they intersect at $y = 0$ and at $y = 5$. The curve $x = y^2 - 4y$ is always to the left of $x = y$ over the interval we just found, so we can evaluate the following integral:

$$\int_0^5 \left[y - (y^2 - 4y)\right]dy = \int_0^5 (5y - y^2)\,dy$$

The result of the integration should be

$$\left(\frac{5y^2}{2} - \frac{y^3}{3}\right)\Bigg|_0^5 = \frac{125}{6}$$

PROBLEM 4. Find the area between the curve $x = y^3 - y$ and the line $x = 0$ (the y-axis).

Answer: First, sketch the graph over the interval.

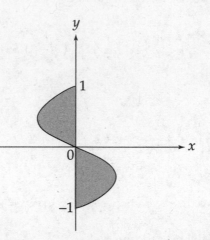

Next, find where the two curves intersect. By setting $y^3 - y = 0$, you'll find that they intersect at $y = -1$, $y = 0$, and $y = 1$. Notice that the curve is to the right of the y-axis from $y = -1$ to $y = 0$ and to the left of the y-axis from $y = 0$ to $y = 1$. Thus, the region must be divided into two parts: from $y = -1$ to $y = 0$ and from $y = 0$ to $y = 1$.

Set up the two integrals.

$$\int_{-1}^{0} (y^3 - y)\,dy + \int_{0}^{1} (y - y^3)\,dy$$

And integrate them.

$$\left(\frac{y^4}{4} - \frac{y^2}{2} \right)\Bigg|_{-1}^{0} + \left(\frac{y^2}{2} - \frac{y^4}{4} \right)\Bigg|_{0}^{1} = \frac{1}{2}$$

PRACTICE PROBLEM SET 28

Find the area of the region between the two curves in each problem, and be sure to sketch each one. (We gave you only endpoints in one of them.) The answers are in Chapter 23, starting on page 558.

1. The curve $y = x^2 - 2$ and the line $y = 2$ $\quad A = \int_{-2}^{2} 2 - (x^2 - 2) \, dx = \int_{-2}^{2} 4 + x^2 = \left. 4x + \frac{x^3}{3} \right|_{-2}^{2} = \frac{32}{3}$

2. The curve $y = x^2$ and the curve $y = 4x - x^2$ $\quad A = \int_{0}^{2} (4x - x^2) - (x^2) \, dx = \int_{0}^{2} 4x - 2x^2 \, dx = \frac{8}{3}$

3. The curve $y = x^2 - 4x - 5$ and the curve $y = 2x - 5$ $\quad A = \int_{0}^{6} 2x - 5 - (x^2 - 4x - 5) = \int_{0}^{6} -x^2 + 6x = 36$

4. The curve $y = x^3$ and the x-axis, from $x = -1$ to $x = 2$

   $A = \int_{-1}^{0} x^3 + \int_{0}^{2} x^3 = \frac{17}{4}$

5. The curve $x = y^2$ and the line $x = y + 2$

6. The curve $x = y^2$ and the curve $x = 3 - 2y^2$

7. The curve $x = y^2 - 4y + 2$ and the line $x = y - 2$

8. The curve $x = y^{\frac{2}{3}}$ and the curve $x = 2 - y^4$

End of Chapter 18 Drill

The answers are in Chapter 24.

1. Find the area between the x-axis and $y = \cos x$ from $x = 0$ to $x = \dfrac{\pi}{2}$.
   - (A) 0
   - (B) 1
   - (C) π
   - (D) 2π

2. Find the area between $y = x^2 + 1$ and $y = 5$.
   - (A) $\dfrac{16}{3}$
   - (B) $\dfrac{32}{3}$
   - (C) 16
   - (D) 32

3. Find the area between $y = 16 - x^2$ and $y = 7$.
   - (A) 18
   - (B) 36
   - (C) 72
   - (D) 108

4. Find the area between the y-axis and $x = 25 - y^2$.
   - (A) $\dfrac{5}{3}$
   - (B) $\dfrac{25}{3}$
   - (C) $\dfrac{125}{3}$
   - (D) $\dfrac{500}{3}$

5. Find the area between the y-axis, $x = \sqrt{y}$ and $x = 4$.
   - (A) $\dfrac{4}{3}$
   - (B) $\dfrac{16}{3}$
   - (C) $\dfrac{64}{3}$
   - (D) $\dfrac{128}{3}$

REFLECT

Respond to the following questions:

- For which topics discussed in this chapter do you feel you have achieved sufficient mastery to answer multiple-choice questions correctly?

- For which topics discussed in this chapter do you feel you have achieved sufficient mastery to answer open-ended questions correctly?

- For which topics discussed in this chapter do you feel you need more work before you can answer multiple-choice questions correctly?

- For which topics discussed in this chapter do you feel you need more work before you can answer open-ended questions correctly?

- What parts of this chapter are you going to re-review?

Chapter 19
The Volume of a
Solid of Revolution

Does the chapter title leave you in a cold sweat? Don't worry. You're not alone. This chapter covers a topic widely seen as one of the most difficult on the AP Exam. There is *always* a volume question on the test. The good news is that you're almost never asked to evaluate the integral— you usually only have to set it up. The difficulty with this chapter, as with Chapter 18, is that there aren't any simple rules to follow. You have to draw the picture and figure it out.

In this chapter, we're going to take the region between two curves, rotate it around a line (usually the *x*- or *y*-axis), and find the volume of the region. There are two methods of doing this: the **washers method** and the **cylindrical shells method**. Sometimes you'll hear the washers method called the **disk method**, but a disk is only a washer without a hole in the middle.

WASHERS AND DISKS

Let's look at the region between the curve $y = \sqrt{x}$ and the *x*-axis (the curve $y = 0$), from $x = 0$ to $x = 1$, and revolve it about the *x*-axis. The picture looks like the following:

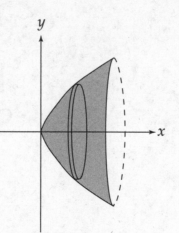

If you slice the resulting solid perpendicular to the *x*-axis, each cross-section of the solid is a circle, or disk (hence the phrase "disk method"). The radii of the disks vary from one value of *x* to the next, but you can find each radius by plugging it into the equation for *y*: each radius is $\sqrt{x}$. Therefore, the area of each disk is

$$\pi\left(\sqrt{x}\right)^2 = \pi x$$

Each disk is infinitesimally thin, so its thickness is *dx*; if you add up the volumes of all the disks, you'll get the entire volume. The way to add these up is by using the integral, with the endpoints of the interval as the limits of integration. Therefore, to find the volume, evaluate the integral.

$$\int_0^1 \pi x \, dx = \frac{\pi x^2}{2}\bigg|_0^1 = \frac{\pi}{2}$$

Now let's generalize this. If you have a region whose area is bounded by the curve $y = f(x)$ and the x-axis on the interval $[a, b]$, each disk has a radius of $f(x)$, and the area of each disk will be

$$\pi \left[f(x) \right]^2$$

To find the volume, evaluate the integral.

$$\pi \int_a^b \left[f(x) \right]^2 \, dx$$

This is the formula for finding the volume using disks.

Example 1: Find the volume of the solid that results when the region between the curve $y = x$ and the x-axis, from $x = 0$ to $x = 1$, is revolved about the x-axis.

As always, sketch the region to get a better look at the problem.

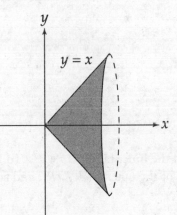

When you slice vertically, the top curve is $y = x$ and the limits of integration are from $x = 0$ to $x = 1$. Using our formula, we evaluate the integral.

$$\pi \int_0^1 x^2 \, dx$$

The result is

$$\pi \int_0^1 x^2 \, dx = \pi \left. \frac{x^3}{3} \right|_0^1 = \frac{\pi}{3}$$

By the way, did you notice that the solid in the problem is a cone with a height and radius of 1?

The formula for the volume of a cone is $\frac{1}{3}\pi r^2 h$, so you should expect to get $\frac{\pi}{3}$.

Now let's figure out how to find the volume of the solid that results when we revolve a region that does not touch the x-axis. Consider the region bounded above by the curve $y = x^3$ and below by the curve $y = x^2$, from $x = 2$ to $x = 4$, which is revolved about the x-axis. Sketch the region first.

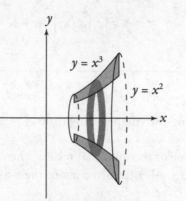

If you slice this region vertically, each cross-section looks like a washer (hence the phrase "washer method").

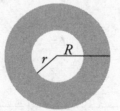

The outer radius is $R = x^3$ and the inner radius is $r = x^2$. To find the area of the region between the two circles, take the area of the outer circle, πR^2, and subtract the area of the inner circle, πr^2.

We can simplify this to

$$\pi R^2 - \pi r^2 = \pi\left(R^2 - r^2\right)$$

Because the outer radius is $R = x^3$ and the inner radius is $r = x^2$, the area of each region is $\pi\left(x^6 - x^4\right)$. You can sum up these regions using the integral.

$$\pi\int_2^4 \left(x^6 - x^4\right)\,dx = \frac{74{,}336\pi}{35}$$

Here's the general idea: In a region whose area is bounded above by the curve $y = f(x)$ and below by the curve $y = g(x)$, on the interval $[a, b]$, then each washer will have an area of

$$\pi\left[f(x)^2 - g(x)^2\right]$$

To find the volume, evaluate the integral.

$$\pi \int_a^b \left[f(x)^2 - g(x)^2 \right] dx$$

This is the formula for finding the volume using washers when the region is rotated around the x-axis.

The formula using disks is just the formula for washers without the second function, $g(x)$.

Example 2: Find the volume of the solid that results when the region bounded by $y = x$ and $y = x^2$, from $x = 0$ to $x = 1$, is revolved about the x-axis.

Sketch it first.

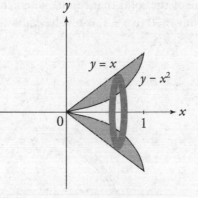

The top curve is $y = x$ and the bottom curve is $y = x^2$ throughout the region. Then our formula tells us that we evaluate the integral.

$$\pi \int_0^1 \left(x^2 - x^4 \right) dx$$

The result is

$$\pi \int_0^1 \left(x^2 - x^4 \right) dx = \pi \left(\frac{x^3}{3} - \frac{x^5}{5} \right) \Bigg|_0^1 = \frac{2\pi}{15}$$

Suppose the region we're interested in is revolved around the y-axis instead of the x-axis. Now, to find the volume, you have to slice the region horizontally instead of vertically. We discussed how to do this in the previous unit on area.

Now, if you have a region whose area is bounded on the right by the curve $x = f(y)$ and on the left by the curve $x = g(y)$, on the interval $[c, d]$, then each washer has an area of

$$\pi\left[f(y)^2 - g(y)^2 \right]$$

To find the volume, evaluate the integral.

$$\pi\int_{c}^{d} \left[f(y)^2 - g(y)^2 \right] dy$$

This is the formula for finding the volume using washers when the region is rotated about the y-axis.

Example 3: Find the volume of the solid that results when the region bounded by the curve $x = y^2$ and the curve $x = y^3$, from $y = 0$ to $y = 1$, is revolved about the y-axis.

Sketch away.

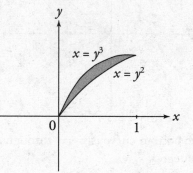

Because $x = y^2$ is always on the outside and $x = y^3$ is always on the inside, you have to evaluate the integral.

$$\pi\int_{0}^{1} \left(y^4 - y^6 \right) dy$$

You should get the following:

$$\pi\int_{0}^{1} \left(y^4 - y^6 \right) dx = \pi\left[\frac{y^5}{5} - \frac{y^7}{7} \right]_{0}^{1} = \frac{2\pi}{35}$$

There's only one more nuance to cover. Sometimes you'll have to revolve the region about a line instead of one of the axes. If so, this will affect the radii of the washers; you'll have to adjust the integral to reflect the shift. Once you draw a picture, it usually isn't too hard to see the difference.

Example 4: Find the volume of the solid that results when the area bounded by the curve $y = x^2$ and the curve $y = 4x$ is revolved about the line $y = -2$. <u>Set up but do not evaluate the integral.</u> (This is how the AP Exam will say it!)

You're not given the limits of integration here, so you need to find where the two curves intersect by setting the equations equal to each other.

$$x^2 = 4x$$
$$x^2 - 4x = 0$$
$$x = 0, 4$$

These will be our limits of integration. Next, sketch the curve.

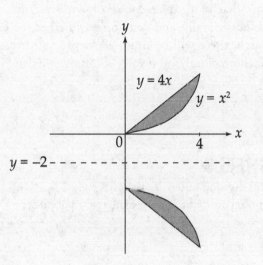

Notice that the distance from the axis of revolution is no longer found by just using each equation. Now, you need to add 2 to each equation to account for the shift in the axis. Thus, the radii are $x^2 + 2$ and $4x + 2$. This means that we need to evaluate the integral.

$$\pi \int_0^4 \left[\left(4x + 2 \right)^2 - \left(x^2 + 2 \right)^2 \right] dx$$

Suppose instead that the region was revolved about the line $x = -2$. Sketch the region again.

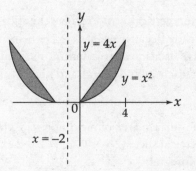

You'll have to slice the region horizontally this time; this means you're going to solve each equation for x in terms of y: $x = \sqrt{y}$ and $x = \dfrac{y}{4}$. We also need to find the y-coordinates of the intersection of the two curves: $y = 0, 16$.

Notice also that, again, each radius is going to be increased by 2 to reflect the shift in the axis of revolution. Thus, we will have to evaluate the integral.

$$\pi \int_0^{16} \left[\left(\sqrt{y} + 2 \right)^2 - \left(\frac{y}{4} + 2 \right)^2 \right] dy$$

Finding the volumes isn't that hard, once you've drawn a picture, figured out whether you need to slice vertically or horizontally, and determined whether the axis of revolution has been shifted. Sometimes, though, there will be times when you want to slice vertically yet revolve around the y-axis (or slice horizontally yet revolve about the x-axis). Here's the method for finding volumes in this way.

CYLINDRICAL SHELLS

Let's examine the region bounded above by the curve $y = 2 - x^2$ and below by the curve $y = x^2$, from $x = 0$ to $x = 1$. Suppose you had to revolve the region about the y-axis instead of the x-axis.

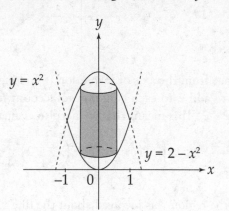

If you slice the region vertically and revolve the slice, you won't get a washer; you'll get a cylinder instead. Because each slice is an infinitesimally thin rectangle, the cylinder's "thickness" is also very, very thin, but real nonetheless. Thus, if you find the surface area of each cylinder and add them up, you'll get the volume of the region.

The formula for the surface area of a cylinder is $2\pi rh$. The height of the cylinder is the length of the vertical slice, $(2 - x^2) - x^2 = 2 - 2x^2$, and the radius of the slice is x. Thus, evaluate the integral:

$$2\pi \int_0^1 x(2 - 2x^2)\, dx$$

Why work in the dark? Just as you spend time practicing formulas in order to memorize them, make sure you actually work on drawing these examples. If you can't visualize the problem, you won't be able to set up the integral.

The math goes like the following:

$$2\pi \int_0^1 x(2 - 2x^2)\,dx = 2\pi \int_0^1 (2x - 2x^3)\,dx = 2\pi\left(x^2 - \frac{x^4}{2}\right)\Bigg|_0^1 = \pi$$

Suppose you tried to slice the region horizontally and use washers. You'd have to convert each equation and find the new limits of integration. Because the region is not bounded by the same pair of curves throughout, you would have to evaluate the region using several integrals. The cylindrical shells method was invented precisely so you can avoid this.

From a general standpoint: If we have a region whose area is bounded above by the curve $y = f(x)$ and below by the curve $y = g(x)$, on the interval $[a, b]$, then each cylinder will have a height of $f(x) - g(x)$, a radius of x, and an area of $2\pi x[f(x) - g(x)]$.

To find the volume, evaluate the integral.

$$2\pi \int_a^b x[(f(x) - g(x)]\,dx$$

This is the formula for finding the volume using cylindrical shells when the region is rotated around the y-axis.

Example 5: Find the volume of the region that results when the region bounded by the curve $y = \sqrt{x}$, the x-axis, and the line $x = 9$ is revolved about the y-axis. <u>Set up but do not evaluate the integral</u>.

Your sketch should look like the following:

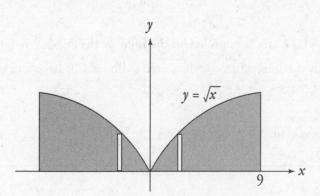

Notice that the limits of integration are from $x = 0$ to $x = 9$, and that each vertical slice is bounded from above by the curve $y = \sqrt{x}$ and from below by the x-axis ($y = 0$). We need to evaluate the integral.

$$2\pi \int_0^9 x(\sqrt{x} - 0)\, dx = 2\pi \int_0^9 x(\sqrt{x})\, dx$$

Example 6: Find the volume that results when the region in Example 5 is revolved about the line $x = -1$. <u>Set up but do not evaluate the integral.</u>

Sketch the figure.

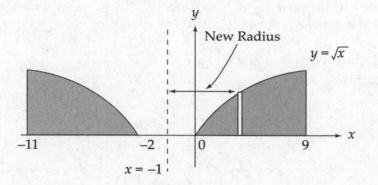

If you slice the region vertically, the height of the shell doesn't change because of the shift in axis of revolution, but you have to add 1 to each radius.

Our integral thus becomes

$$2\pi \int_0^9 (x + 1)(\sqrt{x})\, dx$$

The last formula you need to learn involves slicing the region horizontally and revolving it about the x-axis. As you probably guessed, you'll get a cylindrical shell.

If you have a region whose area is bounded on the right by the curve $x = f(y)$ and on the left by the curve $x = g(y)$, on the interval $[c, d]$, then each cylinder will have a height of $f(y) - g(y)$, a radius of y, and an area of $2\pi y[f(y) - g(y)]$.

To find the volume, evaluate the integral.

$$2\pi \int_c^d y[(f(y) - g(y)]\, dy$$

This is the formula for finding the volume using cylindrical shells when the region is rotated around the *x*-axis.

Example 7: Find the volume of the region that results when the region bounded by the curve $x = y^3$ and the line $x = y$, from $y = 0$ to $y = 1$, is rotated about the *x*-axis. <u>Set up but do not evaluate the integral</u>.

Let your sketch be your guide.

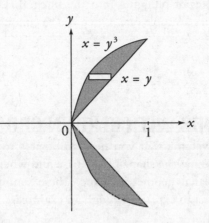

Each horizontal slice is bounded on the right by the curve $x = y$ and on the left by the line $x = y^3$. The integral to evaluate is

$$2\pi \int_0^1 y(y - y^3)\,dy$$

Suppose that you had to revolve this region about the line $y = -1$ instead. Now the region looks like the following:

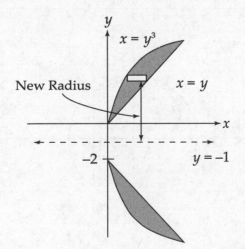

The radius of each cylinder is increased by 1 because of the shift in the axis of revolution, so the integral looks like the following:

$$2\pi \int_0^1 (y+1)(y-y^3)\,dy$$

Wasn't this fun? Volumes of this sort require you to sketch the region carefully and to decide whether it'll be easier to slice the region vertically or horizontally. Once you figure out the slices' boundaries and the limits of integration (and you've adjusted for an axis of revolution, if necessary), it's just a matter of plugging into the integral. Usually, you won't be asked to evaluate the integral unless it's a simple one. Once you've conquered this topic, you're ready for anything.

VOLUMES OF SOLIDS WITH KNOWN CROSS-SECTIONS

There is one other type of volume that you need to be able to find. Sometimes, you will be given an object where you know the shape of the base and where perpendicular cross-sections are all the same regular, planar geometric shape. These sound hard, but are actually quite straightforward. This is easiest to explain through an example.

Example 8: Suppose we are asked to find the volume of a solid whose base is the circle $x^2 + y^2 = 4$, and where cross-sections perpendicular to the x-axis are all squares whose sides lie on the base of the circle. How would we find the volume?

First, make a drawing of the circle.

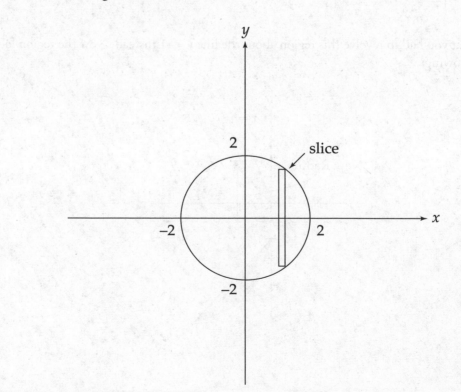

What this problem is telling us is that every time we make a vertical slice, the slice is the length of the base of a square. If we want to find the volume of the solid, all we have to do is integrate the area of the square, from one endpoint of the circle to the other.

The side of the square is the vertical slice whose length is $2y$, which we can find by solving the equation of the circle for y and multiplying by 2. We get $y = \sqrt{4 - x^2}$. Then the length of a side of the square is $2\sqrt{4 - x^2}$. Because the area of a square is $side^2$, we can find the volume with $\int_{-2}^{2}(16 - 4x^2)\,dx$.

Let's perform the integration, although on some problems you will be permitted to find the answer with a calculator.

$$\int_{-2}^{2}(16 - 4x^2)\,dx = \left(16x - \frac{4x^3}{3}\right)\Bigg|_{-2}^{2} = \left(32 - \frac{32}{3}\right) - \left(32 + \frac{32}{3}\right)$$

$$= 64 - \frac{64}{3} = \frac{128}{3}$$

As you can see, the technique is very simple. First, you find the side of the cross-section in terms of y. This will involve a vertical slice. Then, you plug the side into the equation for the area of the cross-section. Then, integrate the area from one endpoint of the base to the other. On the AP Exam, cross-sections will be squares, equilateral triangles, circles, or semi-circles, or maybe isosceles right triangles. So here are some handy formulas to know.

Given the side of an equilateral triangle, the area is $A = (side)^2\,\dfrac{\sqrt{3}}{4}$.

Given the diameter of a semi-circle, the area is $A = (diameter)^2\,\dfrac{\pi}{8}$.

Given the hypotenuse of an isosceles right triangle, the area is $A = \dfrac{(hypotenuse)^2}{4}$.

Example 9: Use the same base as Example 8, except this time the cross-sections are equilateral triangles. We find the side of the triangle just as we did above. It is $2y$, which is $2\sqrt{4 - x^2}$. Now because the area of an equilateral triangle is $(side)^2\,\dfrac{\sqrt{3}}{4}$, we can find the volume by evaluating the integral $\dfrac{\sqrt{3}}{4}\int_{-2}^{2}(4)(4 - x^2)\,dx = \sqrt{3}\int_{-2}^{2}(4 - x^2)\,dx$.

We get $\sqrt{3}\int_{-2}^{2}(4 - x^2)\,dx = \sqrt{3}\left(4x - \dfrac{x^3}{3}\right)\Bigg|_{-2}^{2} = \dfrac{32\sqrt{3}}{3}$.

Example 10: Use the same base as Example 8, except this time the cross-sections are semicircles whose diameters lie on the base. We find the side of the semi-circle just as we did above. It is $2y$, which is $2\sqrt{4 - x^2}$. Now because the area of a semi-circle is $(\textit{diameter})^2\dfrac{\pi}{8}$, we can find the volume by evaluating the integral $\dfrac{\pi}{8}\displaystyle\int_{-2}^{2}(4)(4 - x^2)\,dx = \dfrac{\pi}{2}\displaystyle\int_{-2}^{2}(4 - x^2)\,dx$.

We get $\dfrac{\pi}{2}\displaystyle\int_{-2}^{2}(4 - x^2)\,dx = \dfrac{\pi}{2}\left(4x - \dfrac{x^3}{3}\right)\Bigg|_{-2}^{2} = \dfrac{16\pi}{3}$.

Here are some solved problems. Do each problem, covering the answer first, then check your answer.

PROBLEM 1. Find the volume of the solid that results when the region bounded by the curve $y = 16 - x^2$ and the curve $y = 16 - 4x$ is rotated about the x-axis. Use the washer method and <u>set up but do not evaluate the integral</u>.

Answer: First, sketch the region.

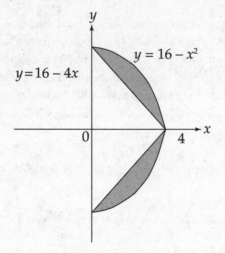

Next, find where the curves intersect by setting the two equations equal to each other.

$$16 - x^2 = 16 - 4x$$
$$x^2 = 4x$$
$$x^2 - 4x = 0$$
$$x = 0,\ 4$$

Slicing vertically, the top curve is always $y = 16 - x^2$ and the bottom is always $y = 16 - 4x$, so the integral looks like the following:

$$\pi\int_{0}^{4}\left[(16 - x^2)^2 - (16 - 4x)^2\right]dx$$

PROBLEM 2. Repeat Problem 1, but revolve the region about the *y*-axis and use the cylindrical shells method. <u>Set up but do not evaluate the integral</u>.

Answer: Sketch the situation.

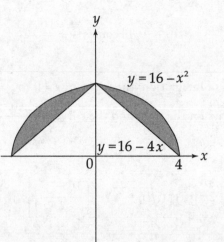

Slicing vertically, the top of each cylinder is $y = 16 - x^2$, the bottom is $y = 16 - 4x$, and the radius is *x*. Therefore, you should set up the following:

$$2\pi \int_0^4 x \left[(16 - x^2) - (16 - 4x) \right] dx$$

PROBLEM 3. Repeat Problem 1, but revolve the region about the *x*-axis and use the cylindrical shells method. <u>Set up but do not evaluate the integral</u>.

Answer: Sketch the situation.

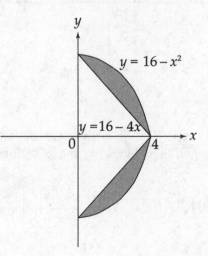

To slice horizontally, you have to solve each equation for x in terms of y and find the limits of integration with respect to y. First, solve for x in terms of y.

$$y = 16 - x^2 \text{ becomes } x = \sqrt{16 - y}$$
$$\text{and}$$
$$y = 16 - 4x \text{ becomes } x = \frac{16 - y}{4}$$

Next, determine the limits of integration: they're from $y = 0$ to $y = 16$. Slicing horizontally, the curve $x = \sqrt{16 - y}$ is always on the right and the curve $x = \frac{16 - y}{4}$ is always on the left. The radius is y, so we evaluate

$$2\pi \int_0^{16} y \left[(\sqrt{16 - y}) - \left(\frac{16 - y}{4} \right) \right] dy$$

PROBLEM 4. Repeat Problem 1, but revolve the region about the y-axis and use the washers method. <u>Set up but do not evaluate the integral</u>.

Answer: Sketch.

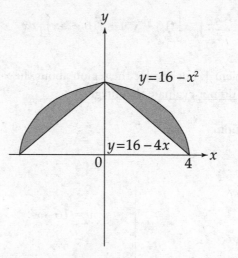

Slicing horizontally, the curve $x = \sqrt{16 - y}$ is always on the right and the curve $x = \frac{16 - y}{4}$ is always on the left. Your integral should look like the following:

$$\pi \int_0^{16} \left[(\sqrt{16 - y})^2 - \left(\frac{16 - y}{4} \right)^2 \right] dy$$

PROBLEM 5. Repeat Problem 1, but revolve the region about the line $y = -3$. You may use either method. <u>Set up but do not evaluate the integral</u>.

Answer: Your sketch should resemble the one below (note that it's not drawn exactly to scale).

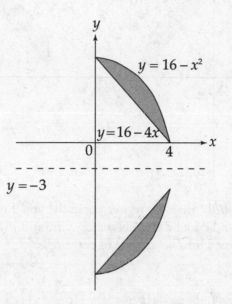

If you were to slice the region vertically, you would use washers. You'll need to add 3 to each radius to adjust for the axis of revolution. The integral to evaluate is

$$\pi \int_0^4 \left[(16 - x^2 + 3)^2 - \left(16 - 4x + 3 \right)^2 \right] dx$$

To slice the region horizontally, use cylindrical shells. The radius of each shell would increase by 3, and you would evaluate

$$2\pi \int_0^{16} (y + 3) \left[(\sqrt{16 - y}) - \left(\frac{16 - y}{4} \right) \right] dy$$

PROBLEM 6. Repeat Problem 1, but revolve the region about the line $x = 8$. You may use either method. <u>Set up but do not evaluate the integral</u>.

Warning! This one is tricky!

Answer: First, sketch the region.

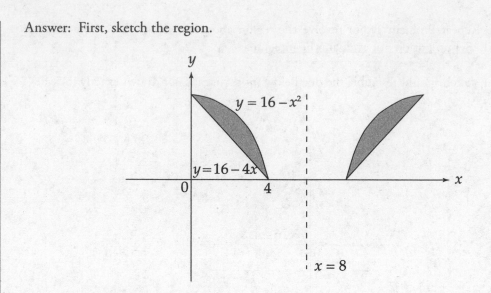

If you choose cylindrical shells, slice the region vertically; you'll need to adjust for the axis of revolution. Each radius can be found by subtracting x from 8. (Not 8 from x. That was the tricky part, in case you missed it.) The integral to evaluate is:

$$2\pi \int_0^4 (8 - x)\left[(16 - x^2) - (16 - 4x) \right] dx$$

If you choose washers, slice the region horizontally. The radius of each washer is found by subtracting each equation from 8. Notice also that the curve $x = \dfrac{16 - y}{4}$ is now the outer radius of the washer, and the curve $x = \sqrt{16 - y}$ is the inner radius. The integral looks like the following:

$$\pi \int_0^{16} \left[\left(8 - \left(\frac{16 - y}{4} \right) \right)^2 - \left(8 - \left(\sqrt{16 - y} \right) \right)^2 \right] dy$$

PROBLEM 7: Find the volume of a solid whose base is the region between the x-axis and the curve $y = 4 - x^2$, and whose cross-sections perpendicular to the x-axis are equilateral triangles with a side that lies on the base.

Answer: The curve $y = 4 - x^2$ intersects the x-axis at $x = -2$ and $x = 2$. The side of the triangle is $4 - x^2$, so all that we have to do is evaluate $\dfrac{\sqrt{3}}{4} \displaystyle\int_{-2}^{2} (4 - x^2)^2 \, dx$.

Expand the integrand to get

$$\frac{\sqrt{3}}{4} \int_{-2}^{2} (16 - 8x^2 + x^4) \, dx$$

Then integrate, which gives you

$$\frac{\sqrt{3}}{4}\left(16x - \frac{8x^3}{3} + \frac{x^5}{5}\right)\Bigg|_{-2}^{2} = \frac{128\sqrt{3}}{15} \approx 14.780$$

PRACTICE PROBLEM SET 29
Calculate the volumes below. The answers are in Chapter 23, starting on page 566.

1. Find the volume of the solid that results when the region bounded by $y = \sqrt{9 - x^2}$ and the x-axis is revolved around the x-axis.

    $V = \pi \int_{-3}^{3} (\sqrt{9-x^2})^2 = 36\pi$

2. Find the volume of the solid that results when the region bounded by $y = \sec x$ and the x-axis from $x = -\frac{\pi}{4}$ to $x = \frac{\pi}{4}$ is revolved around the x-axis.

    $V = \pi \int_{-\pi/4}^{\pi/4} (\sec x)^2 dx = 2\pi$

3. Find the volume of the solid that results when the region bounded by $x = 1 - y^2$ and the y-axis is revolved around the y-axis.

    $V = \pi \int_{-1}^{1} (1-y^2)^2 dy = {}^{16}/_{15}\pi$

4. Find the volume of the solid that results when the region bounded by $x = \sqrt{5}y^2$ and the y-axis from $y = -1$ to $y = 1$ is revolved around the y-axis.

5. Use the method of cylindrical shells to find the volume of the solid that results when the region bounded by $y = x$, $x = 2$, and $y = -\frac{x}{2}$ is revolved around the y-axis.

6. Use the method of cylindrical shells to find the volume of the solid that results when the region bounded by $y = \sqrt{x}$, $y = 2x - 1$, and $x = 0$ is revolved around the y-axis.

7. Use the method of cylindrical shells to find the volume of the solid that results when the region bounded by $y = x^2$, $y = 4$, and $x = 0$ is revolved around the x-axis.

8. Use the method of cylindrical shells to find the volume of the solid that results when the region bounded by $y = 2\sqrt{x}$, $x = 4$, and $y = 0$ is revolved around the y-axis.

9. Find the volume of the solid whose base is the region between the semi-circle $y = \sqrt{16 - x^2}$ and the x-axis, and whose cross-sections perpendicular to the x-axis are squares with a side on the base.

10. Find the volume of the solid whose base is the region between $y = x^2$ and $y = 4$ and whose perpendicular cross-sections are isosceles right triangles with the hypotenuse on the base.

End of Chapter 19 Drill

The answers are in Chapter 24.

1. Find the volume of the solid that results when the region between $y = \dfrac{1}{x}$, $x = 1$, $x = 4$, and $y = 0$ is revolved about the x-axis.

   (A) $\dfrac{3}{4}$

   (B) $\dfrac{3\pi}{4}$

   (C) $\dfrac{63}{64}$

   (D) $\dfrac{63\pi}{64}$

2. Find the volume of the solid that results when the region between $x = \sqrt{1+y}$, $x = 0$, and $y = 3$ is revolved about the y-axis.

   (A) 2π
   (B) 4π
   (C) 8π
   (D) 16π

3. Find the volume of the solid that results when the region between $y = x^2 + 1$, and $y = x + 3$ is revolved about the x-axis.

   (A) $\dfrac{9}{2}$

   (B) $\dfrac{9\pi}{2}$

   (C) $\dfrac{117}{5}$

   (D) $\dfrac{117\pi}{5}$

4. Find the volume of the solid that results when the region between $x = \csc y$, $y = \dfrac{\pi}{4}$, $y = \dfrac{3\pi}{4}$, and $x = 0$ is revolved about the y-axis.

   (A) $\dfrac{\pi}{2}$

   (B) π

   (C) 2π

   (D) 4π

5. Use cylindrical shells to find the volume of the solid that results when the region between $x^2 + y^3 = 4$, $x = 0$, $x = 4$, and $y = 0$ is revolved about the y-axis. **Set up but do not evaluate the integral.**

   (A) $2\pi \displaystyle\int_0^2 x\left(4 - x^2\right)^{\frac{1}{3}} dx - 2\pi \int_2^4 x\left(4 - x^2\right)^{\frac{1}{3}} dx$

   (B) $2\pi \displaystyle\int_0^2 x\left(4 - x^2\right)^{\frac{1}{3}} dx + 2\pi \int_2^4 x\left(4 - x^2\right)^{\frac{1}{3}} dx$

   (C) $\pi \displaystyle\int_0^2 \left(4 - x^2\right)^{\frac{2}{3}} dx - \pi \int_2^4 \left(4 - x^2\right)^{\frac{2}{3}} dx$

   (D) $\pi \displaystyle\int_0^2 \left(4 - x^2\right)^{\frac{2}{3}} dx + \pi \int_2^4 \left(4 - x^2\right)^{\frac{2}{3}} dx$

6. Find the volume of the solid whose base is the region between $y = x^2 + 4$, and $y = 3x + 8$, and whose cross-sections perpendicular to the x-axis are squares with a side on the base.

   (A) $\dfrac{125}{6}$

   (B) $\dfrac{125\pi}{6}$

   (C) $\dfrac{625}{6}$

   (D) $\dfrac{625\pi}{6}$

REFLECT

Respond to the following questions:

- For which topics discussed in this chapter do you feel you have achieved sufficient mastery to answer multiple-choice questions correctly?

- For which topics discussed in this chapter do you feel you have achieved sufficient mastery to answer open-ended questions correctly?

- For which topics discussed in this chapter do you feel you need more work before you can answer multiple-choice questions correctly?

- For which topics discussed in this chapter do you feel you need more work before you can answer open-ended questions correctly?

- What parts of this chapter are you going to re-review?

Chapter 20
Other Applications
of the Integral

LENGTH OF A CURVE

Another way to apply integrals is to find the length of the graph. It's usually a pretty simple problem. All you have to do is use a formula.

Suppose you want to find the length of the graph of a function $y = f(x)$ from $x = a$ to $x = b$. Call this length L. You could divide L into a set of line segments, Δl, add up the lengths of each of them, and you would find L. The formula for the length of each line segment can be found with the Pythagorean Theorem.

$$\Delta l = \sqrt{(\Delta x)^2 + (\Delta y)^2}$$

If we make Δl, Δx, and Δy infinitesimally small, they become dl, dx, and dy. If we then add up all of these line segments, we get

$$L = \int_a^b dl = \int_a^b \sqrt{dx^2 + dy^2}$$

$$= \int_{x_1}^{x_2} \sqrt{1 + \left(\frac{dy}{dx}\right)^2}\, dx$$

$$= \int_{y_1}^{y_2} \sqrt{1 + \left(\frac{dx}{dy}\right)^2}\, dy$$

Here's the rule.

> If the function $f(x)$ is continuous and differentiable on $[a, b]$, then the length of the curve $y = f(x)$ from a to b is
>
> $$L = \int_a^b \sqrt{1 + \left(\frac{dy}{dx}\right)^2}\, dx$$

Example 1: Find the length of the curve $y = x^{\frac{3}{2}}$ from $x = 0$ to $x = 4$.

First, find the first derivative of the function: $\dfrac{dy}{dx} = \dfrac{3}{2} x^{\frac{1}{2}}$.

Next, plug into the formula.

$$L = \int_0^4 \sqrt{1 + \left(\frac{3}{2} x^{\frac{1}{2}}\right)^2}\, dx$$

Now, evaluate the integral.

$$L = \int_0^4 \sqrt{1 + \left(\frac{3}{2} x^{\frac{1}{2}}\right)^2}\, dx = \int_0^4 \sqrt{1 + \frac{9}{4} x}\, dx$$

Using u-substitution, let $u = 1 + \dfrac{9x}{4}$, $du = \dfrac{9}{4}\, dx$, and $\dfrac{4}{9}\, du = dx$. Then substitute.

$$\frac{4}{9} \int u^{\frac{1}{2}}\, du = \frac{8}{27} u^{\frac{3}{2}}$$

Once you substitute back, the result is $\dfrac{8}{27}\left(1 + \dfrac{9}{4} x\right)^{\frac{3}{2}} \Bigg|_0^4 = \dfrac{8}{27}\left(10^{\frac{3}{2}} - 1\right)$.

Sometimes you'll be given the curve as a function of y instead of as a function of x. Then the formula for the length of the curve $x = f(y)$ on the interval $[c, d]$ is

$$L = \int_c^d \sqrt{1 + \left(\frac{dx}{dy}\right)^2}\, dy$$

> As you can see, the formula isn't terribly difficult, but it can often lead to a really ugly integral. Fortunately, the AP Exam will either give you a curve where the integral works out easily, as it did here, or you'll be asked only to set up the integral, not to evaluate it.

Example 2: Find the length of the curve $x = \sin y$, from $y = 0$ to $y = \dfrac{\pi}{2}$. Set up but do not evaluate the integral.

Because $\dfrac{dx}{dy} = \cos y$, we can plug it into the formula: $L = \displaystyle\int_0^{\frac{\pi}{2}} \sqrt{1 + \cos^2 y}\, dy$.

PARAMETRIC FUNCTIONS

Sometimes the curve will be defined parametrically, usually in terms of t (for time). Then the formula for the length of the curve from $t = a$ to $t = b$ is

$$L = \int_a^b \sqrt{\left(\frac{dx}{dt}\right)^2 + \left(\frac{dy}{dt}\right)^2}\, dt$$

Example 3: Find the length of the curve defined by $x = \sin t$ and $y = \cos t$, from $t = 0$ to $t = \pi$.

Take the derivatives of the two t-functions: $\dfrac{dx}{dt} = \cos t$ and $\dfrac{dy}{dt} = -\sin t$. Then use the formula.

$$L = \int_0^\pi \sqrt{\cos^2 t + \sin^2 t}\, dt$$

If you evaluate the integral, you get

$$L = \int_0^\pi \sqrt{1}\, dt = \int_0^\pi dt = \pi$$

That's all there is to finding the length of a curve. As you've seen, many applications of the integral involve simple formulas. All you have to do is to plug in and evaluate the integral. If the integral is a difficult one, you probably just have to set it up.

Try these solved problems. Do each problem, covering the answer first, then check your answer.

PROBLEM 1. Find the length of the curve $y = \dfrac{1}{3}(x^2 + 2)^{\frac{3}{2}}$ from $x = 0$ to $x = 3$.

Answer: First, we find the first derivative $\dfrac{dy}{dx} = \dfrac{1}{3} \cdot \dfrac{3}{2}(x^2 + 2)^{\frac{1}{2}}(2x) = x(x^2 + 2)^{\frac{1}{2}}$.

Next, plug into the formula.

$$L = \int_a^b \sqrt{1 + \left(\frac{dy}{dx}\right)^2}\, dx = \int_0^3 \sqrt{1 + x^2(x^2 + 2)}\, dx$$

Now, you just have to evaluate the integral.

$$L = \int_0^3 \sqrt{1 + x^4 + 2x^2}\, dx = \int_0^3 \sqrt{(x^2 + 1)^2}\, dx = \left(\frac{x^3}{3} + x\right)\Bigg|_0^3 = 12$$

PROBLEM 2. Find the length of the curve $x = \dfrac{y^4}{4} + \dfrac{1}{8y^2}$ from $y = 1$ to $y = 2$.

Answer: Here, you have x in terms of y, so first we find $\dfrac{dx}{dy}$.

$$\frac{dx}{dy} = y^3 - \frac{1}{4y^3}$$

Next, find $\left(\dfrac{dx}{dy}\right)^2$.

$$\left(\frac{dx}{dy}\right)^2 = \left(y^3 - \frac{1}{4y^3}\right)^2 = y^6 - \frac{1}{2} + \frac{1}{16y^6}$$

Plug this into the formula.

$$L = \int_c^d \sqrt{1 + \left(\frac{dx}{dy}\right)^2}\, dy = \int_1^2 \sqrt{1 + \left(y^6 - \frac{1}{2} + \frac{1}{16y^6}\right)}\, dy = \int_1^2 \sqrt{y^6 + \frac{1}{2} + \frac{1}{16y^6}}\, dy$$

And evaluate the integral.

$$L = \int_1^2 \sqrt{y^6 + \frac{1}{2} + \frac{1}{16y^6}}\, dy = \int_1^2 \sqrt{\left(y^3 + \frac{1}{4y^3}\right)^2}\, dy = \left(\frac{y^4}{4} - \frac{1}{8y^2}\right)\Bigg|_1^2 = \frac{123}{32}$$

PROBLEM 3. Find the length of the curve $x = \tan t$ and $y = \sec t$ from $t = 0$ to $t = \dfrac{\pi}{4}$. Set up but do not evaluate the integral.

Answer:

$$L = \int_a^b \sqrt{\left(\frac{dx}{dt}\right)^2 + \left(\frac{dy}{dt}\right)^2}\, dt$$

To use the above formula, you need to determine that $\dfrac{dx}{dt} = \sec^2 t$ and $\dfrac{dy}{dt}$ sec t tan t. Plug these into the formula.

$$L = \int_0^{\frac{\pi}{4}} \sqrt{\sec^4 t + \sec^2 t \, \tan^2 t}\, dt$$

PRACTICE PROBLEM SET 30

Find the length of the following curves between the specified intervals. Evaluate the integrals unless the directions state otherwise. The answers are in Chapter 23, starting on page 570.

1. $y = \dfrac{x^3}{12} + \dfrac{1}{x}$ from $x = 1$ to $x = 2$

2. $y = \tan x$ from $x = -\dfrac{\pi}{6}$ to $x = 0$ (Set up but do not evaluate the integral.)

3. $y = \sqrt{1 - x^2}$ from $x = 0$ to $x = \dfrac{1}{4}$ (Set up but do not evaluate the integral.)

4. $x = \dfrac{y^3}{18} + \dfrac{3}{2y}$ from $y = 2$ to $y = 3$

5. $x = \sqrt{1 - y^2}$ from $y = -\dfrac{1}{2}$ to $y = \dfrac{1}{2}$ (Set up but do not evaluate the integral.)

6. $x = \sin y - y \cos y$ from $y = 0$ to $y = \pi$ (Set up but do not evaluate the integral.)

7. $x = \sqrt{t}$ and $y = \dfrac{1}{t^3}$ from $t = 1$ to $t = 4$ (Set up but do not evaluate the integral.)

8. $x = 3t^2$ and $y = 2t$ from $t = 1$ to $t = 2$ (Set up but do not evaluate the integral.)

THE METHOD OF PARTIAL FRACTIONS

This is the last technique you'll learn to evaluate integrals. There are many, many more types of integrals and techniques to learn; in fact, there are courses in college primarily concerned with integrals and their uses! Fortunately for you, they're not on the AP Exam (and therefore, not in this book). The BC Exam usually has a partial fractions integral or two, and the concept isn't terribly hard.

We use the method of partial fractions to evaluate certain types of integrals that contain rational expressions. First, let's discuss the type of algebra you'll be doing.

If you wanted to add the expressions $\dfrac{3}{x - 1}$ and $\dfrac{5}{x + 2}$, you would do the following:

$$\frac{3}{x - 1} + \frac{5}{x + 2} = \frac{3(x + 2) + 5(x - 1)}{(x - 1)(x + 2)} = \frac{8x + 1}{(x - 1)(x + 2)}$$

Now, suppose you had to do this in reverse; you were given the fraction on the right, and you wanted to determine what two fractions were added to give you that fraction. Another way of asking this is: What constants A and B exist, such that $\dfrac{A}{x-1} + \dfrac{B}{x+2} = \dfrac{8x+1}{(x-1)(x+2)}$?

How do you go about solving for A and B? First, multiply through by $(x-1)(x+2)$ to clear the denominator.

$$A(x+2) + B(x-1) = 8x + 1$$

Next, simplify the left side.

$$Ax + 2A + Bx - B = 8x + 1$$

Now, if you group the terms on the left, you get

$$(A+B)x + (2A - B) = 8x + 1$$

Therefore, $A + B = 8$ and $2A - B = 1$.

If you solve this pair of simultaneous equations, you get $A = 3$ and $B = 5$. Surprised? We hope not.

Why would you need to know this method? Suppose you wanted to find

$$\int \frac{8x+1}{(x-1)(x+2)}\, dx$$

You now know you can rewrite this integral as

$$\int \frac{3}{x-1}\, dx + \int \frac{5}{x-2}\, dx$$

These integrals are easily evaluated.

$$\int \frac{3}{x-1}\, dx + \int \frac{5}{x-2}\, dx = 3\ln|x-1| + 5\ln|x-2| + C$$

Example 1: Evaluate $\displaystyle\int \frac{x+18}{(3x+5)(x+4)}\, dx$.

You need to find A and B such that

$$\frac{A}{3x+5} + \frac{B}{x+4} = \frac{x+18}{(3x+5)(x+4)}$$

First, multiply through by $(3x + 5)(x + 4)$.

$$A(x + 4) + B(3x + 5) = x + 18$$

Next, simplify and group the terms.

$$Ax + 4A + 3Bx + 5B = x + 18$$

$$(A + 3B)x + (4A + 5B) = x + 18$$

You now have two simultaneous equations: $A + 3B = 1$ and $4A + 5B = 18$. If we solve the equations, we get $A = 7$ and $B = -2$. Thus, you can rewrite the integral as

$$\int \frac{7}{3x + 5}\, dx - \int \frac{2}{x + 4}\, dx$$

These are both logarithmic integrals. The solution is

$$\frac{7}{3}\ln|3x + 5| - 2\ln|x + 4| + C$$

There are three main types of partial fractions that appear on the AP Exam. You've just seen the first type: one with two linear factors in the denominator. The second type has a repeated linear term in the denominator.

Example 2: Evaluate $\int \frac{2x + 4}{(x - 1)^2}\, dx$.

Now you need to find two constants A and B, such that

$$\frac{A}{x - 1} + \frac{B}{(x - 1)^2} = \frac{2x + 4}{(x - 1)^2}$$

Multiplying through by $(x - 1)^2$, we get $A(x - 1) + B = 2x + 4$. Now simplify.

$$Ax - A + B = 2x + 4$$
$$Ax + (B - A) = 2x + 4$$

Thus, $A = 2$ and $B - A = 4$, so $B = 6$. Now we can rewrite the integral as

$$\int \frac{2}{x - 1}\, dx + \int \frac{6}{(x - 1)^2}\, dx$$

The solution is

$$2\ln|x - 1| - \frac{6}{x - 1} + C$$

The third type has an irreducible quadratic factor in the denominator.

Example 3: Evaluate $\int \frac{3x + 5}{(x^2 + 1)(x + 2)}\, dx$.

For a quadratic factor, you need to find A, B, and C such that

$$\frac{Ax + B}{x^2 + 1} + \frac{C}{x + 2} = \frac{3x + 5}{(x^2 + 1)(x + 2)}$$

Now you have a term $Ax + B$ over the quadratic term. Whenever you have a quadratic factor, you need to use a linear numerator, not a constant numerator.

Multiply through by $(x^2 + 1)(x + 2)$ and you get
$$(Ax + B)(x + 2) + C(x^2 + 1) = 3x + 5$$

Simplify and group the terms.

$$Ax^2 + 2Ax + Bx + 2B + Cx^2 + C = 3x + 5$$
$$(A + C)x^2 + (2A + B)x + (2B + C) = 3x + 5$$

This gives you three equations.

$$A + C = 0$$
$$2A + B = 3$$
$$2B + C = 5$$

If you solve these three equations, the values are

$$A = \frac{1}{5}, B = \frac{13}{5}, \text{ and } C = -\frac{1}{5}$$

Now you can rewrite the integral.

$$\int \frac{\frac{1}{5}x + \frac{13}{5}}{x^2 + 1}\, dx - \int \frac{\frac{1}{5}}{x + 2}\, dx$$

Break the first integral into two integrals.

$$\frac{1}{5}\int \frac{x}{x^2+1}\,dx + \frac{13}{5}\int \frac{dx}{x^2+1} - \frac{1}{5}\int \frac{1}{x+2}\,dx$$

You can evaluate the first integral using u-substitution; the second integral is an inverse tangent, and the third integral is a natural logarithm.

$$\frac{1}{10}\ln\left|x^2+1\right| + \frac{13}{5}\tan^{-1}(x) - \frac{1}{5}\ln\left|x+2\right| + C$$

Now try these solved problems. Do each problem, covering the answer first, then check your answer.

PROBLEM 1. Evaluate $\displaystyle\int \frac{4}{x^2-6x+5}\,dx$.

Answer: First, factor the denominator.

$$\int \frac{4}{(x-1)(x-5)}\,dx$$

Find A and B such that

$$\frac{A}{x-5} + \frac{B}{x-1} = \frac{4}{(x-5)(x-1)}$$

First, multiply through by $(x-5)(x-1)$.

$$A(x-1) + B(x-5) = 4$$

Next, simplify and group the terms.

$$Ax - A + Bx - 5B = 4$$

$$(A+B)x + (-A-5B) = 4$$

You now have two simultaneous equations: $A + B = 0$ and $-A - 5B = 4$. If you solve the equations, you get $A = 1$ and $B = -1$. Thus, we can rewrite the integral as

$$\int \frac{1}{x-5}\,dx - \int \frac{1}{x-1}\,dx$$

These are both logarithmic integrals. The solution is $\ln|x-5| - \ln|x-1| + C$.

> Notice how all of these integrals contained a natural logarithm term? That's typical of a partial fractions integral.

Handwritten annotations:
$= \dfrac{A}{(x-5)} + \dfrac{B}{(x-1)} \qquad x = 1, 5$

$4 = A(x-1) + B(x-5)$

$4 = B(1-5)$

$-1 = B$

$4 = A(5-1)$

$1 = A$

$\displaystyle\int \frac{1}{x-5} + \int \frac{-1}{x-1}$

$\ln|x-5| - \ln|x-1| + C$

PROBLEM 2. Evaluate $\int \dfrac{(x+1)\,dx}{x(x^2+1)}$.

Answer: You need to find A, B, and C, such that

$$\frac{A}{x} + \frac{Bx+C}{x^2+1} = \frac{x+1}{x(x^2+1)}$$

$x = 0$

$x = 1$

Multiply through by $x(x^2+1)$.

$$A(x^2+1) + (Bx+C)(x) = x+1$$

Simplify and group the terms.

$$Ax^2 + A + Bx^2 + Cx = x+1$$

$$(A+B)x^2 + Cx + A = x+1$$

Now you have three equations.

$$A+B = 0$$
$$C-1$$
$$A = 1$$

You have to solve only for B: because $A = 1$, $B = -1$. Now rewrite the integral as

$$\int \frac{1}{x}\,dx + \int \frac{-x+1}{x^2+1}\,dx$$

Break this into

$$\int \frac{1}{x}\,dx - \int \frac{x\,dx}{x^2+1} + \int \frac{1}{x^2+1}\,dx$$

The first integral is a natural logarithm, the second integral requires u-substitution, and the third integral is an inverse tangent. The answer is

$$\ln|x| - \frac{1}{2}\ln(x^2+1) + \tan^{-1}(x) + C$$

PRACTICE PROBLEM SET 31

Evaluate the following integrals. The answers are in Chapter 23, starting on page 574.

1. $\displaystyle\int \frac{x+4}{(x-1)(x+6)}\,dx$

  (handwritten) $x+4 = A(x-1) + B(x+6)$ $x = 1, -6$
  $1+4 = B(1+6)$ $-6+4 = A(-6-1)$
  $5/7 = B$ $2/7 = A$
  $\frac{2}{7}\int \frac{1}{x+6} + \frac{5}{7}\int \frac{1}{x-1} = \frac{2}{7}\ln|x+6| + \frac{5}{7}\ln|x-1|$

2. $\displaystyle\int \frac{x}{(x-3)(x+1)}\,dx$

  (handwritten) $x = 3, -1$
  $x = A(x-3) + B(x+1)$ $3 = B(3+1)$ $-1 = A(-1-3)$
  $3/4 = B$ $1/4 = A$ $\frac{1}{4}\int \frac{1}{(x+1)} + \frac{3}{4}\int \frac{1}{(x-3)}$
  $1/4 \ln|x+1| + 3/4 \ln|x-3| + C$

3. $\displaystyle\int \frac{1}{x^3 + x^2 - 2x}\,dx$ *(crossed out)*

4. $\displaystyle\int \frac{2x+1}{x^2 - 7x + 12}\,dx$

  (handwritten) $x = 3, 4$
  $2x+1 = A(x-3) + B(x-4)$
  $7 = B(3-4)$ $9 = A(4-3)$ $9\int \frac{1}{x-4} + -7\int \frac{1}{x-3}$
  $B = -7$ $A = 9$
  $9\ln|x-4| - 7\ln|x-3| + C$

5. $\displaystyle\int \frac{2x-1}{(x-1)^2}\,dx$

6. $\displaystyle\int \frac{1}{(x+1)(x^2+1)}\,dx$

7. $\displaystyle\int \frac{x^2 + 3x - 1}{x^3 - 1}\,dx$ *(crossed out)*

IMPROPER INTEGRALS

There's one last type of integral that shows up on the BC Exam. These are improper integrals, which are integrals evaluated over an open interval, rather than a closed one.

> More formally, an integral is improper if: (a) its integrand becomes infinite at one or more points in the interval of integration; or (b) one or both of the limits of integration is infinite.

We integrate an improper integral by evaluating the integral for a constant (*a*) and then taking the limit of the resulting integral as *a* approaches the limit in question. If the limit exists, the integral **converges** and the value of the integral is the limit. If the limit doesn't exist, the integral **diverges** and the integral cannot be evaluated there.

Confusing, huh? Let's clear the air with an example.

Example 1: Evaluate $\int_0^4 \sqrt{\dfrac{4+x}{4-x}}\, dx$.

At $x = 4$, the denominator is zero and the integrand is undefined. Therefore, replace the 4 in the limit of the integral with a and evaluate.

$$\lim_{a \to 4^-} \int_0^a \sqrt{\frac{4+x}{4-x}}\, dx$$

Multiply the numerator and the denominator of the integrand by $\sqrt{4+x}$.

$$\int \sqrt{\frac{4+x}{4-x}}\, \frac{\sqrt{4+x}}{\sqrt{4+x}}\, dx = \int \frac{4+x}{\sqrt{16-x^2}}\, dx = \int \frac{4}{\sqrt{16-x^2}}\, dx + \int \frac{x}{\sqrt{16-x^2}}\, dx$$

Evaluate the first integral by factoring 16 out of the denominator.

$$\int \frac{4\, dx}{4\sqrt{1-\left(\dfrac{x}{4}\right)^2}} = \int \frac{dx}{\sqrt{1-\left(\dfrac{x}{4}\right)^2}}$$

Next, use u-substitution. Let $u = \dfrac{x}{4}$ and $du = \dfrac{1}{4}\, dx$.

$$\int \frac{4\, du}{\sqrt{1-u^2}} = 4\sin^{-1} u = 4\sin^{-1}\left(\frac{x}{4}\right)$$

Evaluate the second integral with u-substitution. Let $u = 16 - x^2$ and $du = -2x\, dx$.

$$-\frac{1}{2}\int u^{-\frac{1}{2}}\, du = -u^{\frac{1}{2}} = -\sqrt{16-x^2}$$

The final result is

$$\int \frac{4}{\sqrt{16-x^2}}\, dx + \int \frac{x\, dx}{\sqrt{16-x^2}} = 4\sin^{-1}\frac{x}{4} - \sqrt{16-x^2}$$

Now you have to evaluate the integral at the limits of integration.

$$4\sin^{-1}\frac{x}{4} - \sqrt{16-x^2}\, \Big|_0^a = \left(4\sin^{-1}\frac{a}{4} - \sqrt{16-a^2}\right) + (4)$$

Finally, take the limit as a approaches 4.

$$\lim_{a \to 4^-}\left(\left(4\sin^{-1}\frac{a}{4} - \sqrt{16 - a^2}\right) + (4)\right) = \left(4\sin^{-1}(1) + (4)\right) = 2\pi + 4$$

This may look complicated, but it was actually a very straightforward process. Because the integrand didn't exist at $x = 4$, replace 4 with a and then integrate. When you evaluated the limits of integration using a instead of 4, you then took the limit as a approaches 4.

Example 2: Evaluate $\displaystyle\int_0^\infty \frac{dx}{1 + x^2}$.

Replace the upper limit of integration with a and take the limit as a approaches infinity.

$$\lim_{a \to \infty}\int_0^a \frac{dx}{1 + x^2}$$

Integrate.

$$\int_0^a \frac{dx}{1 + x^2} = \tan^{-1}x\Big|_0^a = \tan^{-1}a$$

Now, take the limit.

$$\lim_{a \to \infty}\tan^{-1}a = \frac{\pi}{2}$$

Let's do an integral that becomes undefined in the middle of the limits of integration.

Example 3: Evaluate $\displaystyle\int_0^2 \frac{dx}{\sqrt[3]{x - 1}}$.

Note that this integral becomes undefined at $x = 1$. Because of this, you have to divide the integral into two parts: one for $x < 1$ and one for $x > 1$.

$$\lim_{a \to 1^-}\int_0^a \frac{dx}{\sqrt[3]{x - 1}} \quad\text{and}\quad \lim_{b \to 1^+}\int_b^2 \frac{dx}{\sqrt[3]{x - 1}}$$

If either of these limits fails to exist, then the integral from 0 to 2 diverges. Evaluate them both.

$$\lim_{a \to 1^-}\left[\frac{3}{2}(x - 1)^{\frac{2}{3}}\Big|_0^a\right] + \lim_{b \to 1^+}\left[\frac{3}{2}(x - 1)^{\frac{2}{3}}\Big|_b^2\right]$$

$$= \lim_{a \to 1^-}\left[\frac{3}{2}(a - 1)^{\frac{2}{3}} - \frac{3}{2}\right] + \lim_{b \to 1^+}\left[\frac{3}{2} - \frac{3}{2}(b - 1)^{\frac{2}{3}}\right] = -\frac{3}{2} + \frac{3}{2} = 0$$

Because both limits exist and are finite, the integral converges and the value is 0.

There are several other facets of evaluating improper integrals that we won't explore here because they don't appear on the AP Exam. You'll have to do only the most simple of these, usually one with infinity as a limit of integration.

Evaluate these integrals, and check your work.

PROBLEM 1. Evaluate $\int_1^\infty \frac{dx}{x}$.

Answer: First, evaluate the integral.

$$\int_1^a \frac{dx}{x} = \ln x \Big|_1^a = \ln a$$

Next, we evaluate the limit: $\lim_{a\to\infty} \ln a = \infty$.

Therefore, $\int_1^\infty \frac{dx}{x} = \infty$, and the integral diverges.

PROBLEM 2. Evaluate $\int_1^\infty \frac{dx}{x^2}$.

Answer: Evaluate the integral.

$$\int_1^a \frac{dx}{x^2} = -\frac{1}{x}\Big|_1^a = -\frac{1}{a} + 1$$

Next, evaluate the limit.

$$\lim_{a\to\infty}\left(-\frac{1}{a} + 1\right) = 1$$

Therefore, $\int_1^\infty \frac{dx}{x^2} = 1$, and the integral converges.

PROBLEM 3. Evaluate $\int_0^1 \frac{dx}{\sqrt{1-x^2}}$.

Answer: Evaluate the integral.

$$\int_0^a \frac{dx}{\sqrt{1-x^2}} = \sin^{-1} x \Big|_0^a = \sin^{-1}(a) - \sin^{-1}(0) = \sin^{-1}(a)$$

Next, evaluate.

$$\lim_{a \to 1^-} \left(\sin^{-1} a \right) = \frac{\pi}{2}$$

Therefore, $\int_0^1 \frac{dx}{\sqrt{1-x^2}} = \frac{\pi}{2}$, and the integral converges.

PROBLEM 4. Evaluate $\int_{-2}^{2} \frac{dx}{x^2}$.

Answer: The integrand becomes undefined at $x = 0$, so you have to divide the integral into

$$\lim_{a \to 0^-} \int_{-2}^{a} \frac{dx}{x^2} \text{ and } \lim_{b \to 0^+} \int_{b}^{2} \frac{dx}{x^2}$$

Now, evaluating both integrals

$$\lim_{a \to 0^-} \left[-\frac{1}{x} \Big|_{-2}^{a} \right] + \lim_{b \to 0^+} \left[-\frac{1}{x} \Big|_{b}^{2} \right] = \lim_{a \to 0^-} \left[-\frac{1}{a} - \frac{1}{2} \right] + \lim_{b \to 0^+} \left[-\frac{1}{2} + \frac{1}{b} \right] = \infty$$

The integral diverges.

PRACTICE PROBLEM SET 32

Evaluate the following integrals. The answers are in Chapter 23, starting on page 579.

1. $\int_0^1 \frac{dx}{\sqrt{x}}$

2. $\int_{-1}^{1} \frac{dx}{x^{\frac{2}{3}}}$

3. $\int_{-\infty}^{0} e^x \, dx$

4. $\int_1^4 \frac{dx}{1-x}$

5. $\int_0^1 \frac{x+1}{\sqrt{x^2+2x}} \, dx$

6. $\displaystyle\int_{-\infty}^{0}\frac{dx}{\left(2x-1\right)^{3}}$

7. $\displaystyle\int_{0}^{3}\frac{dx}{x-2}$

8. $\displaystyle\int_{-1}^{8}\frac{dx}{\sqrt[3]{x}}$

CALCULUS OF POLAR CURVES

The topic of polar curves usually shows up as only one multiple-choice problem on the BC Exam. However, we will cover the two most typical aspects of the calculus of polar curves. We expect that you had a full treatment of polar coordinates and curves in precalculus, so we won't cover them here.

Slope

The slope of a polar curve $r = f(\theta)$ is $\dfrac{dy}{dx}$. Because x and y are defined parametrically, we need to derive $\dfrac{dy}{dx}$ in terms of r and θ. Remember that in polar coordinates, $x = r\cos\theta = f(\theta)\cos\theta$ and $y = r\sin\theta = f(\theta)\sin\theta$.

Note that $\dfrac{dy}{dx} = \dfrac{\dfrac{dy}{d\theta}}{\dfrac{dx}{d\theta}}$.

We substitute for x and y in the following: $\dfrac{\dfrac{dy}{d\theta}}{\dfrac{dx}{d\theta}} = \dfrac{\dfrac{d}{d\theta}\left[f(\theta)\sin\theta\right]}{\dfrac{d}{d\theta}\left[f(\theta)\cos\theta\right]}$.

Using the Product Rule, we get $\dfrac{dy}{dx} = \dfrac{f'(\theta)\sin\theta + f(\theta)\cos\theta}{f'(\theta)\cos\theta - f(\theta)\sin\theta}$.

This is the formula for finding the slope of a polar curve $r = f(\theta)$.

Example 1: Find the slope of the tangent line to the curve $r = 2 + 4\sin\theta$.

First, let's find $f'(\theta)$: $f'(\theta) = 4\cos\theta$.

This gives us

$$\frac{dy}{dx} = \frac{4\cos\theta\,\sin\theta + (2+4\sin\theta)\cos\theta}{4\cos\theta\,\cos\theta - (2+4\sin\theta)\sin\theta} = \frac{2\cos\theta + 8\sin\theta\cos\theta}{4\cos^2\theta - 4\sin^2\theta - 2\sin\theta} = \frac{2\cos\theta + 4\sin 2\theta}{4\cos 2\theta - 2\sin\theta}$$

Area in Polar Coordinates

We are not going to derive the formula for you. If we want to find the area of a region between the origin and the curve $r = f(\theta)$, $a \le \theta \le b$, we use the formula

$$Area = \int_a^b \frac{1}{2} r^2 \, d\theta$$

Example 2: Find the area of the region in the plane enclosed by the cardioid $r = 4 + 4\sin\theta$.

First, let's graph the curve.

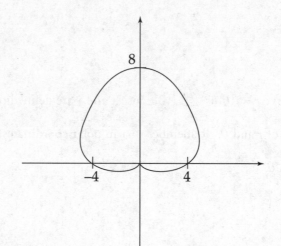

Because r sweeps out the region as θ goes from 0 to 2π, these are our limits of integration.

Plug into the formula for area.

$$A = \int_0^{2\pi} \frac{1}{2} \left(4 + 4\sin\theta\right)^2 d\theta$$

Evaluate the integral.

$$A = \int_0^{2\pi} \frac{1}{2} \left(4 + 4\sin\theta\right)^2 d\theta = \int_0^{2\pi} \left(8 + 16\sin\theta + 8\sin^2\theta\right) d\theta$$

Use the trig identity $\sin^2 \theta = \dfrac{1 - \cos 2\theta}{2}$ to rewrite the integrand.

$$\int_0^{2\pi} \left(8 + 16\sin \theta + 4 - 4\cos 2\theta \right) d\theta = \int_0^{2\pi} \left(12 + 16\sin \theta - 4\cos 2\theta \right) d\theta$$

Evaluate the integral.

$$\int_0^{2\pi} \left(12 + 16\sin \theta - 4\cos 2\theta \right) d\theta = \left(12\theta - 16\cos \theta - 2\sin 2\theta \right)\Big|_0^{2\pi} = (24\pi - 16) - (-16) = 24\pi$$

Example 3: Find the area inside the smaller loop of the limaçon: $r = 1 + 2\cos \theta$.

First, let's graph the curve.

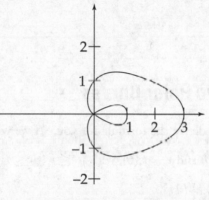

Because in the inner loop, r sweeps out the region as θ goes from $\dfrac{2\pi}{3}$ to $\dfrac{4\pi}{3}$, these are our limits of integration.

Plug into the formula for area.

$$A = \int_{\frac{2\pi}{3}}^{\frac{4\pi}{3}} \frac{1}{2}(1 + 2\cos \theta)^2 \, d\theta$$

Evaluate the integral.

$$A = \int_{\frac{2\pi}{3}}^{\frac{4\pi}{3}} \frac{1}{2}(1 + 2\cos \theta)^2 \, d\theta = \frac{1}{2}\int_{\frac{2\pi}{3}}^{\frac{4\pi}{3}} (1 + 4\cos \theta + 4\cos^2 \theta) \, d\theta$$

Use the trig identity $\cos^2 \theta = \dfrac{1 + \cos 2\theta}{2}$ to rewrite the integrand.

$$\frac{1}{2}\int_{\frac{2\pi}{3}}^{\frac{4\pi}{3}} (1 + 4\cos \theta + 2 + 2\cos \theta)\, d\theta = \frac{1}{2}\int_{\frac{2\pi}{3}}^{\frac{4\pi}{3}} (3 + 4\cos \theta + 2\cos 2\theta)\, d\theta$$

Evaluate the integral.

$$\frac{1}{2}\int_{\frac{2\pi}{3}}^{\frac{4\pi}{3}} (3 + 4\cos \theta + 2\cos 2\theta)\, d\theta = \frac{1}{2}(3\theta + 4\sin \theta + \sin 2\theta)\Big|_{\frac{2\pi}{3}}^{\frac{4\pi}{3}} =$$

$$\frac{1}{2}\left(4\pi - 2\sqrt{3} + \frac{\sqrt{3}}{2}\right) - \frac{1}{2}\left(2\pi + 2\sqrt{3} - \frac{\sqrt{3}}{2}\right) = \pi - \frac{3}{2}\sqrt{3}$$

Area Between Two Polar Curves

Again, we are not going to derive the formula for you. If we want to find the area of a region between the curves $r_1 = f_1(\theta)$ and $r_2 = f_2(\theta)$, with $0 \le r_1(\theta) \le r_2(\theta)$ and $a \le \theta \le b$, we use the formula: $Area = \int_a^b \frac{1}{2}\left(r_2^{\,2} - r_1^{\,2}\right) d\theta$.

Example 4: Find the area of the region inside the circle $r = 4$ and outside $r = 4 - \cos \theta$.

First, let's graph the region.

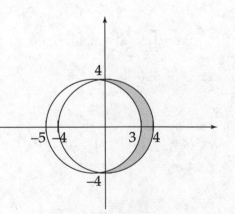

Because r sweeps out the region as θ goes from $-\dfrac{\pi}{2}$ to $\dfrac{\pi}{2}$, these are our limits of integra-

tion. Here, because of symmetry, instead of evaluating $A = \dfrac{1}{2}\displaystyle\int_{-\frac{\pi}{2}}^{\frac{\pi}{2}}\left(r_2^{\,2} - r_1^{\,2}\right)d\theta$, we can evalu-

ate $A = \displaystyle\int_{0}^{\frac{\pi}{2}}\left(r_2^{\,2} - r_1^{\,2}\right)d\theta$.

Plug into the formula for area.

$$A = \int_{0}^{\frac{\pi}{2}}\left[(4)^2 - (4 - \cos\theta)^2\right]d\theta = \int_{0}^{\frac{\pi}{2}}\left[8\cos\theta - \cos^2\theta\right]d\theta$$

Use the trig identity $\cos^2\theta = \dfrac{1 + \cos 2\theta}{2}$ to rewrite the integrand.

$$\int_{0}^{\frac{\pi}{2}}\left[8\cos\theta - \cos^2\theta\right]d\theta = \int_{0}^{\frac{\pi}{2}}\left(8\cos\theta - \frac{1 + \cos 2\theta}{2}\right)d\theta = \int_{0}^{\frac{\pi}{2}}\left(8\cos\theta - \frac{1}{2} - \frac{\cos 2\theta}{2}\right)d\theta$$

Evaluate the integral.

$$\int_{0}^{\frac{\pi}{2}}\left(8\cos\theta - \frac{1}{2} - \frac{\cos 2\theta}{2}\right)d\theta = \left(8\sin\theta - \frac{\theta}{2} - \frac{\sin 2\theta}{4}\right)\Bigg|_{0}^{\frac{\pi}{2}} = 8 - \frac{\pi}{4}$$

PRACTICE PROBLEM SET 33

Do the following problems on your own. The answers are in Chapter 23, starting on page 582.

1. Find the slope of the curve $r = 2\cos 4\theta$.

2. Find the slope of the curve $r = 2 - 3\sin\theta$ at $(2, \pi)$.

3. Find the area inside the limaçon $r = 4 + 2\cos\theta$.

4. Find the area inside one loop of the lemniscate $r^2 = 4\cos 2\theta$.

5. Find the area inside the lemniscate $r^2 = 6\cos 2\theta$ and outside the circle $r = \sqrt{3}$.

End of Chapter 20 Drill

The answers are in Chapter 24.

1. Evaluate $\int \dfrac{8x-7}{x^2-x-2}dx$.

   (A) $8\ln|x-2|+7\ln|x+1|+C$

   (B) $8\ln|x-2|-7\ln|x+1|+C$

   (C) $3\ln|x-2|-5\ln|x+1|+C$

   (D) $3\ln|x-2|+5\ln|x+1|+C$

2. To what value does the integral $\int_{10}^{\infty} \dfrac{e^{-x}dx}{1-e^{-x}}$ converge? Or does it diverge?

   (A) $\ln\left|1-e^{-10}\right|$

   (B) $-\ln\left|1-e^{-10}\right|$

   (C) $-\ln\left|1+e^{-10}\right|$

   (D) The integral diverges.

3. To what value does the integral $\int_{0}^{1} \dfrac{dx}{2x-1}$ converge? Or does it diverge?

   (A) 0

   (B) $\dfrac{1}{2}$

   (C) 1

   (D) The integral diverges.

4. Find the slope of the tangent line to $r=\cos(3\theta)$ at $\theta=\dfrac{3\pi}{4}$.

   (A) -2
   (B) 0
   (C) 2
   (D) The slope is undefined.

5. Find the area of the region inside $r=4+4\cos\theta$ and outside $r=6$.

   (A) $4\sqrt{3}-4\pi$

   (B) $18\sqrt{3}+4\pi$

   (C) $18\sqrt{3}-4\pi$

   (D) $4\sqrt{3}+4\pi$

REFLECT

Respond to the following questions:

- For which topics discussed in this chapter do you feel you have achieved sufficient mastery to answer multiple-choice questions correctly?

- For which topics discussed in this chapter do you feel you have achieved sufficient mastery to answer open-ended questions correctly?

- For which topics discussed in this chapter do you feel you need more work before you can answer multiple-choice questions correctly?

- For which topics discussed in this chapter do you feel you need more work before you can answer open-ended questions correctly?

- What parts of this chapter are you going to re-review?

Chapter 21
Differential
Equations

There are many types of differential equations, but only a very small number of them appear on the AP Exam. There are courses devoted to learning how to solve a wide variety of differential equations, but AP Calculus provides only a very basic introduction to the topic.

SEPARATION OF VARIABLES

If you're given an equation in which the derivative of a function is equal to some other function, you can determine the original function by integrating both sides of the equation and then solving for the constant term.

Example 1: If $\dfrac{dy}{dx} = \dfrac{4x}{y}$ and $y(0) = 5$, find an equation for y in terms of x.

The first step in solving these is to put all of the terms that contain y on the left side of the equals sign and all of the terms that contain x on the right side. We then have $y\,dy = 4x\,dx$. The second step is to integrate both sides.

$$\int y\,dy = \int 4x\,dx$$

And then you integrate.

$$\frac{y^2}{2} = 2x^2 + C$$

You're not done yet. The final step is to solve for the constant by plugging in $x = 0$ and $y = 5$.

$$\frac{5^2}{2} = 2(0^2) + C, \text{ so } C = \frac{25}{2}$$

The solution is $\dfrac{y^2}{2} = 2x^2 + \dfrac{25}{2}$.

That's all there is to it. Separate the variables, integrate both sides, and solve for the constant. Often, the equation will involve a logarithm. Let's do an example.

Example 2: If $\dfrac{dy}{dx} = 3x^2y$ and $y(0) = 2$, find an equation for y in terms of x.

First, put the y terms on the left and the x terms on the right.

$$\frac{dy}{y} = 3x^2\,dx$$

Next, integrate both sides.

$$\int \frac{dy}{y} = \int 3x^2 \, dx$$

The result is $\ln y = x^3 + C$. It's customary to solve this equation for y. You can do this by putting both sides into exponential form.

$$y = e^{x^3 + C}$$

This can be rewritten as $y = e^{x^3} \cdot e^C$ and, because e^C is a constant, the equation becomes

$$y = Ce^{x^3}$$

This is the preferred form of the equation. Now, solve for the constant. Plug in $x = 0$ and $y = 2$, and you get $2 = Ce^0$.

Because $e^0 = 1$, $C = 2$. The solution is $y = 2e^{x^3}$.

This is the typical differential equation that you'll see on the AP Exam. Other common problem types involve position, velocity, and acceleration or exponential growths and decay (Problem 4). We did several problems of this type in Chapter 10, before you knew how to use integrals. In a sample problem, you're given the velocity and acceleration and told to find distance (the reverse of what we did before).

Example 3: If the acceleration of a particle is given by $a(t) = -32$ ft/sec^2, and the velocity of the particle is 64 ft/sec and the height of the particle is 32 ft at time $t - 0$, find: (a) the equation of the particle's velocity at time t; (b) the equation for the particle's height, h, at time t; and (c) the maximum height of the particle.

Part A: Because acceleration is the rate of change of velocity with respect to time, you can write that $\dfrac{dv}{dt} = -32$. Now separate the variables and integrate both sides.

$$\int dv = \int -32 \, dt$$

Integrating this expression, we get $v = -32t + C$. Now we can solve for the constant by plugging in $t = 0$ and $v = 64$. We get $64 = -32(0) + C$ and $C = 64$. Thus, velocity is $v = -32t + 64$.

Part B: Because velocity is the rate of change of displacement with respect to time, you know that

$$\frac{dh}{dt} = -32t + 64$$

Separate the variables and integrate both sides.

$$\int dh = \int \left(-32t + 64 \right) dt$$

Integrate the expression: $h = -16t^2 + 64t + C$. Now solve for the constant by plugging in $t = 0$ and $h = 32$.

$$32 = -16(0^2) + 64(0) + C \text{ and } C = 32$$

Thus, the equation for height is $h = -16t^2 + 64t + 32$.

Part C: In order to find the maximum height, you need to take the derivative of the height with respect to time and set it equal to zero. Notice that the derivative of height with respect to time is the velocity; just set the velocity equal to zero and solve for t.

$$-32t + 64 = 0, \text{ so } t = 2$$

Thus, at time $t = 2$, the height of the particle is a maximum. Now, plug $t = 2$ into the equation for height.

$$h = -16(2)^2 + 64(2) + 32 = 96$$

Therefore, the maximum height of the particle is 96 feet.

Here are some solved problems. Do each problem, covering the answer first, then check your answer.

PROBLEM 1. If $\dfrac{dy}{dx} = \dfrac{3x}{2y}$ and $y(0) = 10$, find an equation for y in terms of x.

Answer: First, separate the variables.

$$2y \, dy = 3x \, dx$$

Then, we take the integral of both sides.

$$\int 2y \, dy = \int 3x \, dx$$

Next, integrate both sides.

$$y^2 = \frac{3x^2}{2} + C$$

Finally, solve for the constant.

$$10^2 = \frac{3(0)^2}{2} + C, \text{ so } C = 100$$

The solution is $y^2 = \dfrac{3x^2}{2} + 100$.

PROBLEM 2. If $\dfrac{dy}{dx} = 4xy^2$ and $y(0) = 1$, find an equation for y in terms of x.

Answer: First, separate the variables: $\dfrac{dy}{y^2} = 4x\,dx$. Then, take the integral of both sides.

$$\int \frac{dy}{y^2} = \int 4x\,dx$$

Next, integrate both sides: $-\dfrac{1}{y} = 2x^2 + C$. You can rewrite this as $y = -\dfrac{1}{2x^2 + C}$.

Finally, solve for the constant.

$$1 = -\frac{1}{2(0)^2 + C} = \frac{-1}{C}, \text{ so } C = -1$$

The solution is $y = -\dfrac{1}{2x^2 - 1}$.

PROBLEM 3. If $\dfrac{dy}{dx} = \dfrac{y^2}{x}$ and $y(1) = \dfrac{1}{3}$, find an equation for y in terms of x.

Answer: This time, separating the variables gives us this: $\dfrac{dy}{y^2} = \dfrac{dx}{x}$.

Then take the integral of both sides: $\displaystyle\int \frac{dy}{y^2} = \int \frac{dx}{x}$.

Next, integrate both sides.

$$-\frac{1}{y} = \ln x + C$$

And rearrange the equation.

$$y = \frac{-1}{\ln x + C}$$

Finally, solve for the constant. $\dfrac{1}{3} = \dfrac{-1}{C}$, so $C = -3$. The solution is $y = \dfrac{-1}{\ln x - 3}$.

PROBLEM 4. A city had a population of 10,000 in 1980 and 13,000 in 1990. Assuming an exponential growth rate, estimate the city's population in 2000.

Answer: The phrase "exponential growth rate" means that $\dfrac{dy}{dt} = ky$, where k is a constant. Take the integral of both sides.

$$\int \frac{dy}{y} = \int k \, dt$$

Then, integrate both sides ($\ln y = kt + C$) and put them in exponential form.

$$y = e^{kt+c} = Ce^{kt}$$

Next, use the information about the population to solve for the constants. If you treat 1980 as $t = 0$ and 1990 as $t = 10$, then

$$10{,}000 = Ce^{k(0)} \text{ and } 13{,}000 = Ce^{k(10)}$$

So $C = 10{,}000$ and $k = \dfrac{1}{10} \ln 1.3 \approx 0.0262$.

The equation for population growth is approximately $y = 10{,}000e^{0.0262t}$. We can estimate that the population in 2000 will be

$$y = 10{,}000e^{0.0262(20)} = 16{,}900$$

EULER'S METHOD

You are going to use your calculator to find an approximate answer to a differential equation. The method is quite simple. First, you need a starting point and an initial slope.

Next, we use increments of h to come up with approximations. Each new approximation will use the following rules:

$$x_n = x_{n-1} + h$$

$$y_n = y_{n-1} + h\left(y'_{n-1}\right)$$

Repeat for $n = 1, 2, 3, \ldots$

This is much easier to understand if we do an example.

Example 1: Use Euler's Method, with $h = 0.2$, to estimate $y(1)$ if $y' = y - 2$ and $y(0) = 4$.

We are given that the curve goes through the point (0, 4). We will call the coordinates of this point $x_0 = 0$ and $y_0 = 4$. The slope is found by $y_0 = 4$ plugging into $y' = y - 2$, so we have an initial slope of $y_0' = 4 - 2 = 2$.

Now we need to find the next set of points.

Step 1: Increase x_0 by h to get x_1.

$$x_1 = 0.2$$

Step 2: Multiply h by y_0' and add to y_0 to get y_1.

$$y_1 = 4 + 0.2(2) = 4.4$$

Step 3: Find y_1' by plugging y_1 into the equation for y'.

$$y_1' = 4.4 - 2 = 2.4$$

Repeat until you get to the desired point (in this case $x = 1$).

Step 1: Increase x_1 by h to get x_2.

$$x_2 = 0.4$$

Step 2: Multiply h by y_1' and add to y_1 to get y_2.

$$y_2 = 4.4 + 0.2(2.4) = 4.88$$

Step 3: Find y_2' by plugging y_2 into the equation for y'.

$$y_2' = 4.88 - 2 = 2.88$$

Step 1: $x_3 = x_2 + h$

$$x_3 = 0.6$$

Step 2: $y_3 = y_2 + h(y_2')$

$$y_3 = 4.88 + 0.2(2.88) = 5.456$$

Step 3: $y_3' = y_3 - 2$

$$y_3' = 5.456 - 2 = 3.456$$

Step 1: $x_4 = x_3 + h$

$$x_4 = 0.8$$

Step 2: $y_4 = y_3 + h(y_3')$

$$y_4 = 5.456 + 0.2(3.456) = 6.1472$$

Step 3: $y_4' = y_4 - 2$

$$y_4' = 6.1472 - 2 = 4.1472$$

Step 1: $x_5 = x_4 + h$

$$x_5 = 1.0$$

Step 2: $y_5 = y_4 + h(y_4')$

$$y_5 = 6.1472 + 0.2(4.1472) = 6.97664$$

We don't need to go any further because we are asked for the value of y when $x = 1$.

The answer is $y = 6.97664$.

Let's do another example.

Example 2: Use Euler's Method, with $h = 0.1$, to estimate $y(0.5)$ if $y' = y - 1$ and $y(0) = 3$.

We start with $x_0 = 0$ and $y_0 = 3$. The slope is found by plugging $y_0 = 3$ into $y' = y - 1$, so we have an initial slope of $y_0' = 3 - 1 = 2$.

Step 1: Increase x_0 by h to get x_1.

$$x_1 = 0.1$$

Step 2: Multiply h by y_0' and add to y_0 to get y_1.

$$y_1 = 3 + 0.1(2) = 3.2$$

Step 3: Find y_1' by plugging y_1 into the equation for y'.

$$y_1' = 3.2 - 1 = 2.2$$

Step 1: $x_2 = x_1 + h$

$$x_2 = 0.2$$

Step 2: $y_2 = y_1 + h(y_1')$

$$y_2 = 3.2 + 0.1(2.2) = 3.42$$

Step 3: $y_2' = y_2 - 1$

$$y_2' = 3.42 - 1 = 2.42$$

Step 1: $x_3 = x_2 + h$

$$x_3 = 0.3 \quad .$$

Step 2: $y_3 = y_2 + h(y_2')$

$$y_3 = 3.42 + 0.1(2.42) = 3.662$$

Step 3: $y_3' = y_3 - 1$

$$y_3' = 3.662 - 1 = 2.662$$

Step 1: $x_4 = x_3 + h$

$$x_4 = 0.4$$

Step 2: $y_4 = y_3 + h(y_3')$

$$y_4 = 3.662 + 0.1(2.662) = 3.9282$$

Step 3: $y_4' = y_4 - 1$

$$y_4' = 3.9282 - 1 = 2.9282$$

Step 1: $x_5 = x_4 + h$

$$x_5 = 0.5$$

Step 2: $y_5 = y_4 + h(y_4')$

$$y_5 = 3.9282 + 0.1(2.9282) = 4.22102$$

The answer is $y = 4.22102$.

Now try these on your own. Do each problem first with the answer covered, and then check your answer.

PROBLEM 1: Use Euler's Method, with $h = 0.2$, to estimate $y(2)$ if $y' = 2y + 1$ and $y(1) = 5$.

We start with $x_0 = 1$ and $y_0 = 5$. The slope is found by plugging $y_0 = 5$ into $y' = 2y + 1$, so we have an initial slope of $y_0' = 2(5) + 1 = 11$.

Step 1: Increase x_0 by h to get x_1.

$$x_1 = 1.2$$

Step 2: Multiply h by y_0' and add to y_0 to get y_1.

$$y_1 = 5 + 0.2(11) = 7.2$$

Step 3: Find y_1' by plugging y_1 into the equation for y'.

$$y_1' = 2(7.2) + 1 = 15.4$$

Step 1: $x_2 = x_1 + h$

$$x_2 = 1.4$$

Step 2: $y_2 = y_1 + h(y_1')$

$$y_2 = 7.2 + 0.2(15.4) = 10.28$$

Step 3: $y_2' = 2y_2 + 1$

$$y_2' = 2(10.28) + 1 = 21.56$$

Step 1: $x_3 = x_2 + h$

$$x_3 = 1.6$$

Step 2: $y_3 = y_2 + h(y_2')$

$$y_3 = 10.28 + 0.2(21.56) = 14.592$$

Step 3: $y_3' = 2y_3 + 1$

$$y_3' = 2(14.592) + 1 = 30.184$$

Step 1: $x_4 = x_3 + h$

$$x_4 = 1.8$$

Step 2: $y_4 = y_3 + h(y_3')$

$$y_4 = 14.592 + 0.2(30.184) = 20.6288$$

Step 3: $y_4' = 2y_4 + 1$

$$y_4' = 2(20.6288) + 1 = 42.2576$$

Step 1: $x_5 = x_4 + h$

$$x_5 = 2$$

Step 2: $y_5 = y_4 + h(y_4')$

$$y_5 = 20.6288 + 0.2(42.2576) = 29.08032$$

The answer is $y = 29.08032$.

PROBLEM 2: Use Euler's Method, with $h = 0.1$, to estimate $y(0.5)$ if $y' = y^2 + 1$ and $y(0) = 0$.

We start with $x_0 = 0$ and $y_0 = 0$. The slope is found by plugging $y_0 = 0$ into $y' = y^2 + 1$, so we have an initial slope of $y_0' = 1$.

Step 1: Increase x_0 by h to get x_1.

$$x_1 = 0.1$$

Step 2: Multiply h by y_0' and add to y_0 to get y_1.

$$y_1 = 0 + 0.1(1) = 0.1$$

Step 3: Find y_1' by plugging y_1 into the equation for y'.

$$y_1' = (0.1)^2 + 1 = 1.01$$

Step 1: $x_2 = x_1 + h$

$$x_2 = 0.2$$

Step 2: $y_2 = y_1 + h(y_1')$

$$y_2 = 0.1 + 0.1(1.01) = 0.201$$

Step 3: $y_2' = (y_2)^2 + 1$

$$y_2' = (0.201)^2 + 1 = 1.040$$

Step 1: $x_3 = x_2 + h$

$$x_3 = 0.3$$

Step 2: $y_3 = y_2 + h(y_2')$

$$y_3' = 0.201 + 0.1(1.040) = 0.305$$

Step 3: $y_3' = (y_3)^2 + 1$

$$y_3' = (0.305)^2 + 1 = 1.093$$

Step 1: $x_4 = x_3 + h$

$$x_4 = 0.4$$

Step 2: $y_4 = y_3 + h(y_3')$

$$y_4 = 0.305 + 0.1(1.093) = 0.414$$

Step 3: $y_4' = (y_4)^2 + 1$

$$y_4' = (0.414)^2 + 1 = 1.171$$

Step 1: $x_5 = x_4 + h$

$$x_5 = 0.5$$

Step 2: $y_5 = y_4 + h(y_4')$

$$y_5 = 0.414 + 0.1(1.171) = 0.531$$

The answer is $y = 0.531$.

SLOPE FIELDS

The idea behind **slope fields**, also known as **direction fields**, is to make a graphical representation of the slope of a function at various points in the plane. We are given a differential equation, but not the equation itself. So how do we do this? Well, it's always easiest to start with an example.

Example 1: Given $\dfrac{dy}{dx} = x$, sketch the slope field of the function.

What does this mean? Look at the equation. It gives us the derivative of the function, which is the slope of the tangent line to the curve at any point x. In other words, the equation tells us that the slope of the curve at any point x is the x-value at that point.

For example, the slope of the curve at $x = 1$ is 1. The slope of the curve at $x = 2$ is 2. The slope of the curve at the origin is 0. The slope of the curve at $x = -1$ is -1. We will now represent these different slopes by drawing small segments of the tangent lines at those points. Let's make a sketch.

See how all of these slopes are independent of the y-values, so for each value of x, the slope is the same vertically, but is different horizontally. Compare this slope field to the next example.

Example 2: Given $\dfrac{dy}{dx} = y$, sketch the slope field of the function.

Here, the slope of the curve at $y = 1$ is 1. The slope of the curve at $y = 2$ is 2. The slope of the curve at the origin is 0. The slope of the curve at $y = -1$ is -1. Let's make a sketch.

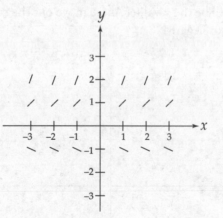

See how all of these slopes are independent of the x-values, so for each value of y, the slope is the same horizontally, but is different vertically.

Now let's do a slightly harder example.

Example 3: Given $\dfrac{dy}{dx} = xy$, sketch the slope field of the function.

Now, we have to think about both the *x*- and *y*-values at each point. Let's calculate a few slopes.

At $(0, 0)$, the slope is $(0)(0) = 0$.

At $(1, 0)$, the slope is $(1)(0) = 0$.

At $(2, 0)$, the slope is $(2)(0) = 0$.

At $(0, 1)$, the slope is $(0)(1) = 0$.

At $(0, 2)$, the slope is $(0)(2) = 0$.

So the slope will be zero at any point on the coordinate axes.

At $(1, 1)$, the slope is $(1)(1) = 1$.

At $(1, 2)$, the slope is $(1)(2) = 2$.

At $(1, -1)$, the slope is $(1)(-1) = -1$.

At $(1, -2)$, the slope is $(1)(-2) = -2$.

So the slope at any point where *x* = 1 will be the *y*-value. Similarly, you should see that the slope at any point where *y* = 1 will be the *x*-value. As we move out the coordinate axes, slopes will get steeper—whether positive or negative.

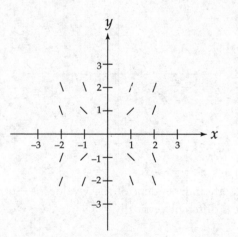

Let's do one more example.

Example 4: Given $\dfrac{dy}{dx} = y - x$, sketch the slope field of the function.

We have to think about both the *x*- and *y*-values at each point. This time, let's make a table of the values of the slope at different points.

	$y = -3$	$y = -2$	$y = -1$	$y = 0$	$y = 1$	$y = 2$	$y = 3$
$x = -3$	0	1	2	3	4	5	6
$x = -2$	-1	0	1	2	3	4	5
$x = -1$	-2	-1	0	1	2	3	4
$x = 0$	-3	-2	-1	0	1	2	3
$x = 1$	-4	-3	-2	-1	0	1	2
$x = 2$	-5	-4	-3	-2	-1	0	1
$x = 3$	-6	-5	-4	-3	-2	-1	0

Now let's make a sketch of the slope field. Notice that the slopes are zero along the line $y = x$ and that the slopes get steeper as we move away from the line in either direction.

That's really all that there is to slope fields. Obviously, there are more complicated slope fields that one could come up with, but on the AP Exam, they will ask you to sketch only the simplest ones.

PRACTICE PROBLEM SET 34

Now try these problems. The answers are in Chapter 23, starting on page 585.

1. If $\dfrac{dy}{dx} = \dfrac{7x^2}{y^3}$ and $y(3) = 2$, find an equation for y in terms of x.

2. If $\dfrac{dy}{dx} = 5x^2 y$ and $y(0) = 6$, find an equation for y in terms of x.

3. If $\dfrac{dy}{dx} = \dfrac{e^x}{y^2}$ and $y(0) = 1$, find an equation for y in terms of x.

4. If $\dfrac{dy}{dx} = \dfrac{y^2}{x^3}$ and $y(1) = 2$, find an equation for y in terms of x.

5. If $\dfrac{dy}{dx} = \dfrac{\sin x}{\cos y}$ and $y(0) = \dfrac{3\pi}{2}$, find an equation for y in terms of x.

6. A colony of bacteria grows exponentially and the colony's population is 4,000 at time $t = 0$ and 6,500 at time $t = 3$. How big is the population at time $t = 10$?

7. A rock is thrown upward with an initial velocity, $v(t)$, of 18 m/s from a height, $h(t)$, of 45 m. If the acceleration of the rock is a constant -9 m/s^2, find the height of the rock at time $t = 4$.

8. A radioactive element decays exponentially in proportion to its mass. One-half of its original amount remains after 5,750 years. If 10,000 grams of the element are present initially, how much will be left after 1,000 years?

9. Use Euler's Method, with $h = 0.25$, to estimate $y(1)$ if $y' = y - x$ and $y(0) = 2$.

10. Use Euler's Method, with $h = 0.2$, to estimate $y(1)$ if $y' = -y$ and $y(0) = 1$.

11. Use Euler's Method, with $h = 0.1$, to estimate $y(0.5)$ if $y' = 4x^3$ and $y(0) = 0$.

12. Sketch the slope field for $\dfrac{dy}{dx} = 2x$.

13. Sketch the slope field for $\dfrac{dy}{dx} = -\dfrac{x}{y}$.

End of Chapter 21 Drill

The answers are in Chapter 24.

1. If $\dfrac{dy}{dx} = -4xy^2$ and $y(0) = 1$, find an equation for y in terms of x.

   (A) $y = \dfrac{-1}{2x^2 - 1}$

   (B) $y = \dfrac{1}{2x^2 - 1}$

   (C) $y = \dfrac{1}{2x^2 + 1}$

   (D) $y = \dfrac{-1}{2x^2 + 1}$

2. If $\sec x \dfrac{dy}{dx} - 4\tan y$ and $y(0) = \dfrac{\pi}{6}$, find an equation for y in terms of x.

   (A) $\sec^2 y = \dfrac{\sqrt{3}}{2} e^{4\sin x}$

   (B) $\sec^2 y = \dfrac{1}{2} e^{4\sin x}$

   (C) $\sin y = \dfrac{\sqrt{3}}{2} e^{4\sin x}$

   (D) $\sin y = \dfrac{1}{2} e^{4\sin x}$

3. Polonium-210 has a half-life of 140 days. It decays exponentially, where rate of decay is proportional to the amount at the time t. If we start with 100 g, how much will remain after 10 weeks?

   (A) 50
   (B) 71
   (C) 90
   (D) 280

4. Use Euler's Method, with $h = 0.25$, to estimate $y(1)$ if $\dfrac{dy}{dx} = x + y$ and $y(0) = 3$.

   (A) 3.25
   (B) 7
   (C) 8.8125
   (D) 10.25

5. Moss is growing on a stone at an exponential rate, where the rate of growth is proportional to the amount at the time t. If there are 16 cm² of moss on June 1, and 60 cm² on July 1 (30 days later), to the nearest square centimeter, how much will there be on September 1 (92 days later)?

   (A) 151
   (B) 921
   (C) 3601
   (D) 9601

REFLECT

Respond to the following questions:

- For which topics discussed in this chapter do you feel you have achieved sufficient mastery to answer multiple-choice questions correctly?

- For which topics discussed in this chapter do you feel you have achieved sufficient mastery to answer open-ended questions correctly?

- For which topics discussed in this chapter do you feel you need more work before you can answer multiple-choice questions correctly?

- For which topics discussed in this chapter do you feel you need more work before you can answer open-ended questions correctly?

- What parts of this chapter are you going to re-review?

Chapter 22
Infinite Series

SEQUENCES AND SERIES

A **sequence** of numbers is an infinite succession of numbers that follow a pattern. For example:

1, 2, 3, 4, ...; or $\frac{1}{2}, \frac{1}{3}, \frac{1}{4}$ ...; or 1, –1, 1, –1 ....

Terms in a sequence are usually denoted with a subscript; the first term of a sequence is a_1, the second term of a sequence is a_2, the nth term of a sequence is a_n, and so on.

Usually, the terms in a sequence are generated by a formula. For example, the sequence $a_n = \frac{n-1}{n}$ beginning with $n = 1$ is: 0, $\frac{1}{2}, \frac{2}{3}, \frac{3}{4}$, ... This is found by plugging in 1 for n, then 2, then 3, and so on. Notice that n is always an integer.

A sequence converges to a number if, as n gets bigger, the terms get closer to a certain number; that number is the limit of the sequence.

Here's the official math jargon.

> A sequence has a limit L if for any $\varepsilon > 0$ there is an associated positive integer N such that $|a_n - L| < \varepsilon$ for all $n \geq N$. If so, the sequence **converges** to L and we write $\lim_{n\to\infty} a_n = L$.

If the sequence has no finite limit, it **diverges**.

For example, the sequence $a_n = \frac{n-1}{n}$ converges, because $\lim_{n\to\infty} = \frac{n-1}{n} = 1$. As n gets bigger, the terms of the sequence get closer and closer to 1. The sequence $a_n = 2^n$ diverges, because $\lim_{n\to\infty} 2^n = \infty$. As n gets bigger, the terms of the sequence get bigger.

Now that we've defined a sequence, we can define a series. A **series** is an expression of the form $a_1 + a_2 + a_3 + a_4 + \dots a_n + \dots$. If the series stops at some final term a_n, the series is finite. If the series continues indefinitely, it's an infinite series. These are the ones that primarily show up on the AP Exam. Because a series is the sum of all the terms of a sequence, we often use sigma notation like the following:

$$\sum_{n=1}^{\infty} a_n$$

The harmonic series, for example, looks like the following:

$$\sum_{n=1}^{\infty} \frac{1}{n} = 1 + \frac{1}{2} + \frac{1}{3} + \frac{1}{4} + \dots$$

A **partial sum** of a series is the sum of the series up to a particular value of n. If the sequence of partial sums of a series converges to a limit L, then the series is said to converge to that limit L, and we write

$$\sum_{n=1}^{\infty} a_n = L$$

If there is no such limit (like the harmonic series, for example), then the series diverges.

Example 1: Does the series $\sum_{n=1}^{\infty} \frac{1}{3^n}$ converge and, if so, to what value?

The nth partial sum of this series is $\frac{1}{3} + \frac{1}{3^2} + \frac{1}{3^3} + \dots \frac{1}{3^n}$. You can figure out the limit using the following method. First, write it out.

$$S_n = \frac{1}{3} + \frac{1}{3^2} + \frac{1}{3^3} + \dots \frac{1}{3^n}$$

Next, multiply through by $\frac{1}{3}$.

$$\frac{1}{3} S_n = \frac{1}{3^2} + \frac{1}{3^3} + \dots \frac{1}{3^{n+1}}$$

If you now subtract the second expression from the first, you get

$$\frac{2}{3} S_n = \frac{1}{3} - \frac{1}{3^{n+1}}$$

Now, multiply through by $\frac{3}{2}$.

$$S_n = \frac{3}{2} \left(\frac{1}{3} - \frac{1}{3^{n+1}} \right)$$

Finally, take the limit.

$$\lim_{n\to\infty} \frac{3}{2}\left(\frac{1}{3} - \frac{1}{3^{n+1}}\right) = \frac{3}{2}\left(\frac{1}{3} - 0\right) = \frac{1}{2}$$

The series converges to $\frac{1}{2}$.

GEOMETRIC SERIES

The problem above contains an example of a series called a geometric series. Generally, this series takes the following form:

$$a + ar + ar^2 + ar^3 + \ldots + ar^{n-1} + \ldots = \sum_{n=1}^{\infty} ar^{n-1}$$

These series often show up on the AP Exam. You're usually asked to determine whether the series converges or diverges, and if it converges, to what limit. The test for convergence of a geometric series is very simple.

If $|r| < 1$ the series converges.

If $|r| \geq 1$, the series diverges.

So, for example, the series $\frac{1}{2} + \frac{1}{2^2} + \frac{1}{2^3} + \ldots$ converges, whereas the series $2 + 2^2 + 2^3 + \ldots$ diverges.

If a geometric series converges, you can always figure out its sum by doing the following:

Let $S_n = a + ar + ar^2 + ar^3 + \ldots ar^{n-1}$

Multiply the expression by r.

$$rS_n = ar + ar^2 + ar^3 + \ldots ar^n$$

Now subtract the second expression from the first.

$$S_n - rS_n = a - ar^n$$

Factor S_n out of the left side.

$$S_n(1 - r) = a - ar^n$$

Divide through by $1 - r$.

$$S_n = \frac{a - ar^n}{(1 - r)} = \frac{a(1 - r^n)}{(1 - r)}$$

Thus, if we want to find the sum of the first n terms of a geometric series, we use the following formula:

$$S_n = \frac{a(1 - r^n)}{(1 - r)}$$

For example, the sum of the first four terms of the series in Example 1 is

$$S_4 = \frac{\frac{1}{3}\left(1 - \left(\frac{1}{3}\right)^4\right)}{\left(1 - \frac{1}{3}\right)} = \frac{40}{81}$$

For an infinite geometric series, if it converges, $\lim\limits_{n \to \infty} r^n = 0$, and its sum is found this way.

$$S = \frac{a}{1 - r}$$

In Example 1, the first term is $a = \frac{1}{3}$ and $r = \frac{1}{3}$. The sum is $\dfrac{\frac{1}{3}}{1 - \frac{1}{3}} = \frac{1}{2}$.

Many students find it is easier to work through the derivation of the sum than to actually memorize the formula. Either way, you should always be able to find the sum of an infinite geometric series.

Example 2: Find the limit to which the series $2 + \dfrac{2}{5} + \dfrac{2}{25} + \dfrac{2}{125} + \ldots$ converges.

The first term is $a = 2$ and $r = \dfrac{1}{5}$, so the limit is $\dfrac{2}{1 - \dfrac{1}{5}} = \dfrac{5}{2}$.

The AP Exam's treatment of series doesn't get very complicated, and the exam usually confines itself to finding the limit of an infinite geometric series. Just remember that if $|r| \geq 1$, the geometric series diverges and it has no limit. No exceptions.

THE RATIO TEST

Sometimes you'll simply be asked to determine whether a series converges. With a geometric series, you already know what to do. If the series isn't a geometric one, you can usually use the Ratio Test.

Let $\sum a_n$ be a series where all of the terms are positive, and suppose that

$$\lim_{n \to \infty} \frac{a_{n+1}}{a_n} = \rho$$

Then

(a)   If $\rho < 1$, the series converges.

(b)   If $\rho > 1$, the series diverges.

(c)   If $\rho = 1$, the test provides insufficient information and the series might converge or diverge.

Example 3: Determine whether the series $\displaystyle\sum_{n=1}^{\infty} \frac{n}{3^n}$ converges.

Use the Ratio Test.

$$\rho = \lim_{n \to \infty} \frac{\dfrac{n+1}{3^{n+1}}}{\dfrac{n}{3^n}} = \lim_{n \to \infty} \frac{n+1}{3^{n+1}} \left(\frac{3^n}{n} \right) = \lim_{n \to \infty} \frac{n+1}{n} \left(\frac{3^n}{3^{n+1}} \right) = \frac{1}{3}$$

Because $\rho < 1$, the series converges. Notice that it does not converge to the limit $\dfrac{1}{3}$. That value simply tells us that the series is a convergent one.

Example 4: Determine whether the series $\sum\limits_{n=1}^{\infty} \dfrac{1}{3n}$ converges.

By the Ratio Test, you discover that

$$\rho = \lim_{n\to\infty} \frac{\dfrac{1}{3(n+1)}}{\dfrac{1}{3n}} = \lim_{n\to\infty} \frac{1}{3(n+1)}\left(\frac{3n}{1}\right) = \lim_{n\to\infty} \frac{3n}{3(n+1)} = 1$$

Because $\rho = 1$, the test is insufficient and we don't know whether the series converges or diverges.

One easy test to use is for a type of series that is called a p series. This is a series of the form $\sum\limits_{n=1}^{\infty} \dfrac{1}{n^p}$. A p series will converge if $p > 1$ and will diverge if $p \leq 1$. You can easily show this using the Ratio Test.

COMPARISON TEST

Another test for convergence is the **Comparison Test** (also known as the Direct Comparison Test).

Let $0 \leq a_n \leq b_n$ for all n.

1) If $\sum\limits_{n=1}^{\infty} b_n$ converges, then $\sum\limits_{n=1}^{\infty} a_n$ converges.

2) If $\sum\limits_{n=1}^{\infty} a_n$ diverges, then $\sum\limits_{n=1}^{\infty} b_n$ diverges.

In other words, if all of the terms of a series are less than those of a convergent series, then it too converges. And, if all of the terms of a series are greater than those of a divergent series, then it too diverges.

Example 5: Does $\sum\limits_{n=1}^{\infty} \dfrac{1}{5 + 2^n}$ converge?

We know that $\dfrac{1}{2^n}$ converges because it is a geometric series with $|r| < 1$.

Compare the series in question with $\frac{1}{2^n}$. Each term of $\frac{1}{5+2^n}$ is less than $\frac{1}{2^n}$, so $\sum\limits_{n=1}^{\infty} \frac{1}{5+2^n}$ converges by comparison.

Example 6: Does $\sum\limits_{n=2}^{\infty} \frac{1}{n-1}$ converge?

We know that $\frac{1}{n}$ diverges because it is the harmonic series.

Compare the series in question with $\frac{1}{n}$. Each term of $\frac{1}{n-1}$ is greater than $\frac{1}{n}$, so $\sum\limits_{n=2}^{\infty} \frac{1}{n-1}$ diverges by comparison.

LIMIT COMPARISON TEST

In addition to the Comparison Test, there is another test for convergence called the *Limit Comparison Test*. This is especially useful for series where a_n is a rational function of n. To perform this test, suppose that $a_n > 0$ and $b_n > 0$ for all $n \geq N$, where N is an integer.

1) If $\lim\limits_{n \to \infty} \frac{a_n}{b_n} = c$ (where $c > 0$), then series a_n and series b_n both converge or both diverge.

2) If $\lim\limits_{n \to \infty} \frac{a_n}{b_n} = 0$, and series b_n converges, then series a_n also converges.

3) $\lim\limits_{n \to \infty} \frac{a_n}{b_n} = \infty$ and series b_n diverges, then series a_n also diverges.

Example 7: Does $\sum\limits_{n=1}^{\infty} \frac{2n}{(n+1)^3}$ converge?

Let's use the Limit Comparison Test. For large n, the series will behave like $\frac{2n}{n^3} = \frac{2}{n^2}$, so let's make $b_n = \frac{2}{n^2}$. We know that the series $\sum\limits_{n=1}^{\infty} \frac{2}{n^2}$ converges.

Let's use the test:

$$\lim_{n\to\infty} \frac{a_n}{b_n} = \lim_{n\to\infty} \frac{\dfrac{2n}{(n+1)^3}}{\dfrac{2}{n^2}} = \lim_{n\to\infty} \frac{2n}{(n+1)^3} \cdot \frac{n^2}{2} = \lim_{n\to\infty} \frac{n^3}{(n+1)^3} = 1$$

This result corresponds to part 1 of the test. Because the ratio is 1 and the series $\sum_{n=1}^{\infty} \dfrac{2}{n^2}$ converges, the series $\sum_{n=1}^{\infty} \dfrac{2n}{(n+1)^3}$ also converges.

Example 8: Does $\sum_{n=1}^{\infty} \dfrac{4n}{(n+1)^2}$ converge?

Let's use the Limit Comparison Test. For large n, the series will behave like $\dfrac{4n}{n^2} = \dfrac{4}{n}$, so let's make $b_n = \dfrac{4}{n}$. We know that the series $\sum_{n=1}^{\infty} \dfrac{4n}{(n-1)^2}$ diverges. Let's use the test:

$$\lim_{n\to\infty} \frac{a_n}{b_n} = \lim_{n\to\infty} \frac{\dfrac{4n}{(n+1)^2}}{\dfrac{4}{n}} = \lim_{n\to\infty} \frac{4n}{(n+1)^2} \cdot \frac{n}{4} = \lim_{n\to\infty} \frac{n^2}{(n+1)^2} = 1$$

This result corresponds to part 1 of the test. Because the ratio is 1 and the series $\sum_{n=1}^{\infty} \dfrac{4}{n}$ diverges, the series $\sum_{n=1}^{\infty} \dfrac{4n}{(n+1)^2}$ also diverges.

Example 9: Does $\sum_{n=1}^{\infty} \dfrac{\ln n}{n^{\frac{5}{2}}}$ converge?

Let's use the Limit Comparison Test. Because $\ln n$ grows more slowly than n^c (where $c > 0$), we can compare the series $\dfrac{\ln n}{n^{\frac{5}{2}}} < \dfrac{n^{\frac{1}{2}}}{n^{\frac{5}{2}}} = \dfrac{1}{n^2}$. We could have used any value for c, so we chose $\dfrac{1}{2}$ to make the numbers work out nicely. So, let's make $b_n = \dfrac{1}{n^2}$. We know that the series $\sum_{n=1}^{\infty} \dfrac{1}{n^2}$ converges.

Let's use the test:

$$\lim_{n\to\infty} \frac{a_n}{b_n} = \lim_{n\to\infty} \frac{\dfrac{\ln n}{n^{\frac{5}{2}}}}{\dfrac{1}{n^2}} = \lim_{n\to\infty} \frac{\ln n}{n^{\frac{5}{2}}} \frac{n^2}{1} = \lim_{n\to\infty} \frac{\ln n}{n^{\frac{1}{2}}} = 1 \text{ (use L'Hôpital's rule).}$$

This result corresponds to part 2 of the test. Because the ratio is 0 and the series $\displaystyle\sum_{n=1}^{\infty} \frac{1}{n^2}$ converges, the series $\displaystyle\sum_{n=1}^{\infty} \frac{\ln n}{n^{\frac{5}{2}}}$ also converges.

CONDITIONAL CONVERGENCE

Sometimes an Alternating Series will converge, but when we look at the absolute value of the terms, it doesn't converge. An alternating series that is not absolutely convergent is called *conditionally convergent*. A good example of this is the series $\displaystyle\sum_{n=1}^{\infty} (-1)^n \frac{1}{n}$, which is called the *alternating harmonic series*. Earlier, we said that the *harmonic* series, $\displaystyle\sum_{n=1}^{\infty} \frac{1}{n}$, diverges. Let's apply the alternating series test to this series to see the difference.

(a) $b_n > 0$ That is, the terms $\frac{1}{n}$ are all positive.

(b) $b_n > b_{n+1}$ for all n. That is, the terms are decreasing.

(c) $b_n \to 0$ as $n \to \infty$.

Therefore, the series converges. Because the series $\displaystyle\sum_{n=1}^{\infty} (-1)^n \frac{1}{n}$ converges but the series $\displaystyle\sum_{n=1}^{\infty} \frac{1}{n}$ diverges, the series $\displaystyle\sum_{n=1}^{\infty} (-1)^n \frac{1}{n}$ converges conditionally.

Example 10: Does $\displaystyle\sum_{n=1}^{\infty} (-1)^n \frac{1}{\sqrt{n}}$ converge absolutely or conditionally?

First, let's use the Alternating Series Test:

(a) $b_n > 0$. That is, the terms $\frac{1}{n}$ are all positive.

(b) $b_n > b_{n+1}$ or all n. That is, the terms are decreasing.

(c) $b_n \to 0$ as $n \to \infty$

Therefore, the series $\displaystyle\sum_{n=1}^{\infty} (-1)^n \frac{1}{\sqrt{n}}$ converges.

Now let's look at the series $\displaystyle\sum_{n=1}^{\infty} \frac{1}{\sqrt{n}}$. This is an example of a p series, $\displaystyle\sum_{n=1}^{\infty} \frac{1}{n^p}$. A p series will converge if $p > 1$ and will diverge if $p \leq 1$. Here, $p = \dfrac{1}{2}$, so the series $\displaystyle\sum_{n=1}^{\infty} \frac{1}{\sqrt{n}}$ diverges. Because the series $\displaystyle\sum_{n=1}^{\infty} (-1)^n \frac{1}{\sqrt{n}}$ converges but the series $\displaystyle\sum_{n=1}^{\infty} \frac{1}{\sqrt{n}}$ diverges, it is conditionally convergent.

ALTERNATING SERIES

The next type of series to concern yourself with is the alternating series. This is a series whose terms are alternately positive and negative. For example,

$$1 - \frac{1}{2} + \frac{1}{3} - \frac{1}{4} + \frac{1}{5} \ldots \frac{(-1)^{n+1}}{n}$$

Usually, you'll be asked to determine whether such a series converges. First, test the following:

The series $\displaystyle\sum_{n=1}^{\infty} (-1)^{n+1} b_n$ converges if all three of the following conditions are satisfied:

(a) $b_n > 0$ (which means that the terms must alternate in sign)
(b) $b_n > b_{n+1}$ for all n
(c) $b_n \to 0$ as $n \to \infty$

Example 11: Determine whether the series $1 - \dfrac{1}{2} + \dfrac{1}{3} - \dfrac{1}{4} \ldots \dfrac{(-1)^{n+1}}{n} + \ldots$ converges.

Let's check the criteria.

(a) Each of the b_n's is positive.
(b) Each b_n is greater than its succeeding b_{n+1}.
(c) The b_n's are tending to zero as n approaches infinity.

Therefore, the series converges. This series is called the alternating harmonic series. Notice that this series converges, whereas the harmonic series diverges.

Usually, you'll be asked to determine whether a series converges "absolutely."

Note that this Absolute Convergence Theorem is not limited to alternating series: it's "absolutely" true for any series.

A series $\sum a_n$ *converges absolutely* if the corresponding series $\sum |a_n|$ converges.

INTEGRAL TEST

Another test that you can use to determine whether a series converges or diverges is the **Integral Test**.

If f is positive, continuous, and decreasing for $x \geq 1$, and $a_n = f(n)$, then

$$\sum_{n=1}^{\infty} a_n \text{ and } \int_1^{\infty} f(x)\, dx$$

Either both converge or both diverge.

In other words, if you want to test convergence for a series with positive terms, evaluate the integral and see if the integral converges.

Example 12: Does $\displaystyle\sum_{n=1}^{\infty} \frac{6n^2}{n^3 + 1}$ converge?

In order to determine whether this converges, we can evaluate the integral, replacing n with x.

Evaluate $\displaystyle\int_1^{\infty} \frac{6x^2}{x^3 + 1}\, dx$.

First, we replace the improper integral with a proper one and a limit. We get $\displaystyle\lim_{a\to\infty} \int_1^a \frac{6x^2}{x^3 + 1}\, dx$.

Use u-substitution. Let $u = x^3 + 1$ and $du = 3x^2\, dx$.

We get

$$\lim_{a\to\infty} 2\int_1^a \frac{du}{u} = \lim_{a\to\infty} 2\left(\ln|u|\right)\Big|_1^a = \lim_{a\to\infty} 2\left(\ln\left|x^3 + 1\right|\right)\Big|_1^a = \lim_{a\to\infty} 2\left(\ln(a^3 + 1) - \ln 1\right) = \infty$$

Therefore, because the integral diverges, the series diverges.

Example 13: Does $\sum_{n=1}^{\infty} e^{-n}$ converge?

In order to determine whether this converges, we can evaluate the integral, replacing n with x.

Evaluate $\int_{1}^{\infty} e^{-x}\,dx$.

First, we replace the improper integral with a proper one and a limit. We get $\lim_{a\to\infty}\int_{1}^{a} e^{-x}\,dx$.

Evaluate the integral. We get $\lim_{a\to\infty}\int_{1}^{a} e^{-x}\,dx = \lim_{a\to\infty}\left(-e^{-x}\right)\Big|_{1}^{a} = \lim_{a\to\infty}\left(-e^{-a}+e^{-1}\right) = \frac{1}{e}$.

Therefore, because the integral converges, the series converges. Note that the series does not necessarily converge to $\frac{1}{e}$.

ERROR BOUND

When summing an alternating series with a finite number of terms, the sum S_n will not be the same as the sum with an infinite number of terms, S. If we want to find the error, we need to evaluate $|S - S_n|$. Sometimes this can be quite difficult to calculate, but we can put a "bound" on the error; that is, a number that the error is less than. This bound will simply be the absolute value of the next term in the alternating series. For example, if we are summing the first ten terms in an alternating series, the error will be less than the absolute value of the eleventh term.

Example 14: Find the sum of the series $\sum_{n=1}^{7}(-1)^{n+1}\frac{1}{n^2}$ and the error bound for this sum and the sum of the infinite series $\sum_{n=1}^{\infty}(-1)^{n+1}\frac{1}{n^2}$.

The sum of the series is simply $1-\frac{1}{4}+\frac{1}{9}-\frac{1}{16}+\frac{1}{25}-\frac{1}{36}+\frac{1}{49}=\frac{48877}{58800}\approx 0.83124$. The error bound of the sum is the absolute value of the next term, namely, $\frac{1}{64}$. In other words, the sum of the infinite series is within $\frac{1}{64}$ of $\frac{48877}{58800}$.

Note that the error bound is unlikely to show up in a multiple choice question. More likely, it will be one part of a written question.

POWER SERIES

The most important type of series, and the one that the AP Exam concentrates on, is the **power series**, which takes the form

$$\sum_{n=0}^{\infty} c_n x^n = c_0 + c_1 x + c_2 x^2 + c_3 x^3 + \ldots + c_n x^n + \ldots.$$

For any power series in x, exactly one of the following is true:

(a) The series converges only for $x = 0$;

(b) The series converges absolutely for all x; or

(c) The series converges absolutely for all x in some open interval $(-R, R)$, and diverges if $x < -R$ or $x > R$. The series may converge or diverge at the endpoints of the interval.

Notice that we're now going from $n = 0$ to $n = \infty$, rather than from $n = 1$ to $n = \infty$. This is because the first term is a constant.

You'll be asked to determine whether a power series converges. The test is to the right.

For case (c), the number R is called the **radius of convergence**. In case (a), the radius of convergence is zero; in case (b) the radius is all x. The set of all values of x in the interval $(-R, R)$ is called the **interval of convergence**.

Example 15: Find the radius and interval of convergence for the power series $\displaystyle\sum_{n=0}^{\infty} x^n$.

This is a geometric series with $r = x$; thus, it converges if $r < 1$.

Therefore, the radius of convergence is $R = 1$, and the interval of convergence is $(-1, 1)$.

Example 16: Find the radius and interval of convergence for the power series $\displaystyle\sum_{n=0}^{\infty} \frac{x^n}{n!}$.

In order to determine convergence, apply the Ratio Test for absolute convergence.

$$\rho = \lim_{n\to\infty}\left[\frac{x^{n+1}}{(n+1)!}\frac{n!}{x^n}\right] = \lim_{n\to\infty}\left|\frac{x}{n+1}\right| = 0$$

Because $\rho = 0$ for all x, the series converges absolutely for all x. Therefore, the radius of convergence is $R = \infty$ and the interval of convergence is $(-\infty, \infty)$.

TAYLOR SERIES AND POLYNOMIALS

The final, and most important, topic in infinite series is the **Taylor series**. The power series and the Taylor series are the most frequent question topics on infinite series on the AP Exam. A Taylor series expansion about a point $x = a$ is a power series expansion that's useful to approximate the function in the neighborhood of the point $x = a$.

If a function f has derivatives of all orders at a, then the Taylor series for f about $x = a$ is

$$\sum_{k=0}^{\infty} \frac{f^{(k)}(a)}{k!}(x-a)^k = f(a) + f'(a)(x-a) + \frac{f''(a)}{2!}(x-a)^2 + ... + \frac{f^{(n)}(a)}{n!}(x-a)^n + ...$$

The special case of the Taylor series for $a = 0$ is called the **MacLaurin series**.

$$\sum_{k=0}^{\infty} \frac{f^{(k)}(0)}{k!}(x)^k = f(0) + f'(0)x + \frac{f''(0)}{2!}x^2 + ... + \frac{f^{(n)}(0)}{n!}x^n + ...$$

Let's do an example to see what one of these looks like.

> The AP Exam usually asks for a Taylor series at $x = 0$. Note that this is the same thing as the MacLaurin series.

Example 17: Find the Taylor series about $a = 1$ generated by $f(x) = \dfrac{1}{x}$.

First, find the derivatives of $\dfrac{1}{x}$ and compute their values at $x = 1$.

$$f(x) = x^{-1} \qquad\qquad f(1) = 1$$

$$f'(x) = -x^{-2} \qquad\qquad f'(1) = -1$$

$$f''(x) = 2x^{-3} \qquad\qquad f''(1) = 2 = 2!$$

$$f'''(x) = -6x^{-4} \qquad\qquad f'''(1) = -6 = -3!$$

$$f^{(4)}(x) = 24x^{-5} \qquad\qquad f^{(4)}(1) = 24 = 4!$$

$$f^{(n)}(x) = (-1)^n \frac{n!}{x^{n+1}} \qquad\qquad f^{(n)}(1) = (-1)^n \, n!$$

Next, plug into the formula to generate the Taylor series.

$$1 - (x-1) + (x-1)^2 + ... = \sum_{k=0}^{\infty} (-1)^k (x-1)^k$$

Example 18: Find the Taylor series about $a = 0$ generated by $f(x) = e^x$.

First, find the derivatives of e^x and compute their values at $x = 0$.

$$f(x) = e^x \qquad\qquad f(0) = 1$$
$$f'(x) = e^x \qquad\qquad f'(0) = 1$$
$$f''(x) = e^x \qquad\qquad f''(0) = 1$$
$$f'''(x) = e^x \qquad\qquad f'''(0) = 1$$
$$f^{(4)}(x) = e^x \qquad\qquad f^{(4)}(0) = 1$$
$$f^{(n)}(x) = e^x \qquad\qquad f^{(n)}(0) = 1$$

Next, plug into the formula to generate the Taylor series.

$$\sum_{k=0}^{\infty} \frac{x^k}{k!} = 1 + x + \frac{x^2}{2!} + \frac{x^3}{3!} + \frac{x^4}{4!} + \dots$$

This is a very important expansion to know. Memorize it and be able to use it on the exam. You're permitted to write the Taylor series expansion for e^x without explaining how to generate it.

That's all you need to know about Taylor series. There are four expansions you should prepare for on the AP Exam. We already discussed the expansion for e^x. You'll also need to know the expansions for $\sin x$, $\cos x$, and $\ln(1 + x)$.

We'll give you the Taylor series expansions here, and you'll derive them in the problems.

Taylor series can also be used to approximate the value of a function at a particular value. All you have to do is to plug that value into the Taylor series expansion for the function. The highest power term that you use is the degree of the Taylor polynomial. This is easier to see with an example.

$$e^x = 1 + x + \frac{x^2}{2!} + \frac{x^3}{3!} + \dots = \sum_{k=0}^{\infty} \frac{x^k}{k!}$$

$$\sin x = x - \frac{x^3}{3!} + \frac{x^5}{5!} - \frac{x^7}{7!} + \dots = \sum_{k=0}^{\infty} (-1)^k \frac{x^{2k+1}}{(2k+1)!}$$

$$\cos x = 1 - \frac{x^2}{2!} + \frac{x^4}{4!} - \frac{x^6}{6!} + \dots = \sum_{k=0}^{\infty} (-1)^k \frac{x^{2k}}{(2k)!}$$

$$\ln(1 + x) = x - \frac{x^2}{2} + \frac{x^3}{3} - \frac{x^4}{4} + \dots = \sum_{k=0}^{\infty} (-1)^k \frac{x^{k+1}}{k+1}$$

Example 19: Use a third degree Taylor polynomial to estimate $e^{0.2}$.

All that we have to do is to plug 0.2 into the Taylor expansion up to the third power term.

We get $e^{0.2} \approx 1 + 0.2 + \frac{0.2^2}{2!} + \frac{0.2^3}{3!} = 1.22133$. If we evaluate $e^{0.2}$ with a calculator, we get 1.22140. As you can see, this is a pretty good approximation.

Sometimes you will be asked to find the "error bound" of a Taylor polynomial. This is also sometimes referred to as the **Lagrange error**. The formula for Lagrange's form of the remainder is $R_n(x, a) \leq f^{(n+1)}(c) \dfrac{(x-a)^{n+1}}{(n+1)!}$, where c is between a and x.

Example 20: Take the first four terms of the MacLaurin series expansion for $\sin x$, namely, $\sin x \approx x - \dfrac{x^3}{3!} + \dfrac{x^5}{5!} - \dfrac{x^7}{7!}$, and estimate the remainder.

Using the formula for the remainder, we get $R_{2n}(x, 0) \dfrac{x^{2n+1}}{(2n+1)!} \sin\left(c + \dfrac{(2n+1)\pi}{2}\right)$. We know that the $|\sin x| \leq 1$, so $R_{2n+1}(x, 0) \leq \dfrac{|x|^{2n+1}}{(2n+1)!}$. Therefore, the remainder is $R(x, 0) \leq \dfrac{x^9}{(9)!}$.

The proof for the rule is quite difficult, so we will simply tell you the rule. If you are finding an nth degree Taylor polynomial, a good approximation to the error bound is the next nonzero term in a decreasing series. For example, if we wanted to find the error in the above example, we know that it's less than the fourth term evaluated at 0.2, which is $\dfrac{0.2^4}{4!} = 0.000066$.

PROBLEM 1. Find the sum of the series $4 + \dfrac{4}{3} + \dfrac{4}{9} + \dfrac{4}{27} + ... + \dfrac{4}{3^6}$.

Answer: This is a geometric series where $a = 4$, $r = \dfrac{1}{3}$ and $n = 7$. According to the formula, the sum is

$$S_n = \frac{a(1-r^n)}{(1-r)}$$

Plugging in, you should get

$$S_7 = \frac{4\left(1 - \left(\dfrac{1}{3}\right)^7\right)}{\left(1 - \dfrac{1}{3}\right)} = 5.997$$

PROBLEM 2. Find the limit to which the series $4 + \dfrac{4}{3} + \dfrac{4}{9} + \dfrac{4}{27} + ...$ converges.

Answer: This is the same series, but now you want the limit at infinity, so use the formula.

$$S = \frac{a}{1-r}$$

The result is

$$S = \frac{4}{1 - \dfrac{1}{3}} = 6$$

PROBLEM 3. Determine whether the series $\displaystyle\sum_{n=1}^{\infty} \frac{n-1}{5^n}$ converges.

Answer: By the Ratio Test,

$$\rho = \lim_{n\to\infty} \frac{\dfrac{n}{5^{n+1}}}{\dfrac{n-1}{5^n}} = \lim_{n\to\infty} \frac{n}{5^{n+1}}\left(\frac{5^n}{n-1}\right) = \lim_{n\to\infty} \frac{n}{n-1}\left(\frac{5^n}{5^{n+1}}\right) = \frac{1}{5}$$

Because $\rho < 1$, the series converges.

PROBLEM 4. Find the Taylor series about $a = 0$ generated by $f(x) = \sin x$.

Answer: First, find the derivatives of $\sin x$ and compute their values at $x = 0$.

$$
\begin{aligned}
f(x) &= \sin x & f(0) &= 0 \\
f'(x) &= \cos x & f'(0) &= 1 \\
f''(x) &= -\sin x & f''(0) &= 0 \\
f'''(x) &= -\cos x & f'''(0) &= -1 \\
f^{(4)}(x) &= \sin x & f^{(4)}(0) &= 0 \\
f^{(5)}(x) &= \cos x & f^{(5)}(0) &= 1
\end{aligned}
$$

As you've seen us do before, plug into the formula to generate the Taylor series.

$$x - \frac{x^3}{3!} + \frac{x^5}{5!} + \ldots = \sum_{k=0}^{\infty} (-1)^k \, \frac{x^{2k+1}}{(2k+1)!}$$

PROBLEM 5. Find the radius and interval of convergence for the series $\displaystyle\sum_{n=1}^{\infty} \frac{x^n}{1+n^2}$.

Answer: In order to determine convergence, apply the Ratio Test for absolute convergence.

$$\rho = \lim_{n\to\infty} \left| \frac{x^{n+1}}{1+(n+1)^2}\left(\frac{1+n^2}{x^n}\right) \right| = \lim_{n\to\infty} \left| x\frac{1+n^2}{\left(1+(n+1)^2\right)} \right| = |x|$$

Thus, if $|x| < 1$, then $\rho < 1$, and the series converges.

If $|x| > 1$, then $\rho > 1$, and the series diverges.

If $x = 1$, then $\displaystyle\sum_{n=0}^{\infty} \frac{x^n}{1 + n^2} = \sum_{n=0}^{\infty} \frac{1}{1 + n^2}$ (which converges).

If $x = -1$, then $\displaystyle\sum_{n=0}^{\infty} \frac{x^n}{1 + n^2} = \sum_{n=0}^{\infty} \frac{(-1)^n}{1 + n^2}$ (which converges).

Therefore, the radius of convergence is 1 and the interval of convergence is $[-1, 1]$.

PRACTICE PROBLEM SET 35

Now try these problems involving the series we've discussed in this chapter. The answers are in Chapter 23, starting on page 594.

1. Find the sum of the series $2 + \dfrac{2}{5} + \dfrac{2}{25} + \ldots + \dfrac{2}{5^4}$.

2. Find the sum of the series $8 + \dfrac{8}{7} + \dfrac{8}{49} + \dfrac{8}{343} + \dfrac{8}{7^4} + \ldots$.

3. Does the series $\displaystyle\sum_{n=1}^{\infty} \frac{5^n}{(n-1)!}$ converge or diverge?

4. Does the series $\displaystyle\sum_{n=1}^{\infty} \frac{5^n}{n^2}$ converge or diverge?

5. Does the series $\displaystyle\sum_{n=1}^{\infty} \frac{n-3}{n^3 - n^2 + 3}$ converge?

6. Does the series $\displaystyle\sum_{n=1}^{\infty} \frac{n^2 + n}{n^3 - n^2 + n - 1}$ converge?

7. Does the series $\displaystyle\sum_{n=1}^{\infty} \frac{2^n}{1 + 4^n}$ converge?

8. Does the series $\displaystyle\sum_{n=1}^{\infty} (-1)^{n+1} \frac{n}{n^3 + 1}$ converge absolutely, converge conditionally, or diverge?

9. Does the series $\displaystyle\sum_{n=1}^{\infty} (-1)^n \frac{1+n}{n^2}$ converge absolutely, converge conditionally, or diverge?

10. Does the series $\displaystyle\sum_{n=1}^{\infty} (-1)^n \frac{3+n}{4+n}$ converge absolutely, converge conditionally, or diverge?

11. Find the Taylor series about $a = 0$ generated by $f(x) = \cos x$.

12. Find the Taylor series about $a = 0$ generated by $f(x) = \ln(1 + x)$.

13. Find the Taylor series about $a = 0$ generated by $f(x) = e^{-x}$.

14. Find the first three nonzero terms of the Taylor series about $a = \dfrac{\pi}{3}$ generated by $f(x) = \sin x$.

15. Find the radius and interval of convergence for the series $\displaystyle\sum_{n=0}^{\infty} 3^n x^n$.

16. Find the radius and interval of convergence for the series $\displaystyle\sum_{n=0}^{\infty} (-1)^n \frac{x^{2n}}{(2n)!}$.

17. Estimate $\cos (0.2)$ using a fourth degree Taylor polynomial about $a = 0$, and find the error bound.

18. Estimate $\ln(1.3)$ using a third degree Taylor polynomial about $a = 0$, and find the error bound.

19. Determine whether $\displaystyle\sum_{n=2}^{\infty} \frac{1}{n \ln n}$ converges.

20. Determine whether $\displaystyle\sum_{n=1}^{\infty} \frac{1}{n \sqrt[4]{n}}$ converges.

21. Determine whether $\displaystyle\sum_{n=2}^{\infty} \frac{\ln n}{n+1}$ converges.

22. Determine whether $\displaystyle\sum_{n=4}^{\infty} \frac{1}{n!}$ converges.

End of Chapter 22 Drill

The answers are in Chapter 24.

1. Find the sum of the series $6 + 2 + \dfrac{2}{3} + \dfrac{2}{9} + \dots$

   (A) 2

   (B) $\dfrac{9}{2}$

   (C) 9

   (D) 18

2. Does the series $\displaystyle\sum_{n-1}^{\infty}\left(\dfrac{9^n}{n!}\right)$ converge or diverge?

3. Does the series $\displaystyle\sum_{n=1}^{\infty}\left(\dfrac{n^3}{1+n^4}\right)$ converge or diverge?

4. Find the first four nonzero terms of the Taylor series about $a = 0$ generated by $f(x) = \sqrt{1+x}$.

   (A) $1 + \dfrac{x}{2} + \dfrac{x^2}{8} + \dfrac{x^3}{16}$

   (B) $1 + \dfrac{x}{2} - \dfrac{x^2}{8} + \dfrac{x^3}{16}$

   (C) $1 - \dfrac{x}{2} + \dfrac{x^2}{8} + \dfrac{x^3}{16}$

   (D) $1 - \dfrac{x}{2} - \dfrac{x^2}{8} - \dfrac{x^3}{16}$

5. Find the radius and interval of convergence for the series $\displaystyle\sum_{n=0}^{\infty}\left(5^n x^n\right)$.

	Radius	Interval
(A)	$\dfrac{1}{5}$	$\left(-\dfrac{1}{5}, \dfrac{1}{5}\right)$
(B)	$\dfrac{1}{5}$	$\left(0, \dfrac{1}{5}\right)$
(C)	$\dfrac{2}{5}$	$\left(-\dfrac{1}{5}, \dfrac{1}{5}\right)$
(D)	$\dfrac{2}{5}$	$\left(0, \dfrac{1}{5}\right)$

REFLECT

Respond to the following questions:

- For which topics discussed in this chapter do you feel you have achieved sufficient mastery to answer multiple-choice questions correctly?

- For which topics discussed in this chapter do you feel you have achieved sufficient mastery to answer open-ended questions correctly?

- For which topics discussed in this chapter do you feel you need more work before you can answer multiple-choice questions correctly?

- For which topics discussed in this chapter do you feel you need more work before you can answer open-ended questions correctly?

- What parts of this chapter are you going to re-review?

Chapter 23
Answers to Practice Problem Sets

PRACTICE PROBLEM SET 1

1. 13

   To find the limit, we simply plug in 8 for x: $\lim\limits_{x \to 8}\left(x^2 - 5x - 11\right) = \left(8^2 - (5)(8) - 11\right) = 13$.

2. $\dfrac{4}{5}$

   To find the limit, we simply plug in 5 for x: $\lim\limits_{x \to 5}\left(\dfrac{x+3}{x^2 - 15}\right) = \dfrac{5+3}{5^2 - 15} = \dfrac{8}{10} = \dfrac{4}{5}$.

3. 4

   If we plug in 3 for x, we get $\dfrac{0}{0}$, which is indeterminate. When this happens, we try to factor the expression in order to get rid of the problem terms. Here we factor the top and get $\lim\limits_{x \to 3}\left(\dfrac{x^2 - 2x - 3}{x - 3}\right) = \lim\limits_{x \to 3}\dfrac{(x-3)(x+1)}{x-3}$. Now we can cancel the term $x - 3$ to get $\lim\limits_{x \to 3}(x+1)$. Notice that we are allowed to cancel the terms because x is not 3 but very close to 3. Now we can plug in 3 for x: $\lim\limits_{x \to 3}(x+1) = 3 + 1 = 4$.

4. $+\infty$

   Here we are finding the limit as x goes to infinity. We divide the top and bottom by the highest power of x in the expression, which is x^4: $\lim\limits_{x \to \infty}\left(\dfrac{x^4 - 8}{10x^2 + 25x + 1}\right) = \lim\limits_{x \to \infty}\left(\dfrac{\dfrac{x^4}{x^4} - \dfrac{8}{x^4}}{\dfrac{10x^2}{x^4} + \dfrac{25x}{x^4} + \dfrac{1}{x^4}}\right)$. Next, simplify the top and bottom: $\lim\limits_{x \to \infty}\left(\dfrac{1 - \dfrac{8}{x^4}}{\dfrac{10}{x^2} + \dfrac{25}{x^3} + \dfrac{1}{x^4}}\right)$. Now, if we take the limit as x goes to infinity, we get $\lim\limits_{x \to \infty}\left(\dfrac{1 - \dfrac{8}{x^4}}{\dfrac{10}{x^2} + \dfrac{25}{x^3} + \dfrac{1}{x^4}}\right) = \dfrac{1 - 0}{0 + 0 + 0} = \infty$.

5. $\dfrac{1}{10}$

Here we are finding the limit as x goes to infinity. We divide the top and bottom by the highest

power of x in the expression, which is x^4: $\displaystyle\lim_{x\to\infty}\left(\dfrac{x^4-8}{10x^4+25x+1}\right)=\lim_{x\to\infty}\left(\dfrac{\dfrac{x^4}{x^4}-\dfrac{8}{x^4}}{\dfrac{10x^4}{x^4}+\dfrac{25x}{x^4}+\dfrac{1}{x^4}}\right)$. Next,

simplify the top and bottom: $\displaystyle\lim_{x\to\infty}\left(\dfrac{1-\dfrac{8}{x^4}}{10+\dfrac{25}{x^3}+\dfrac{1}{x^4}}\right)$. Now, if we take the limit as x goes to infinity,

we get $\displaystyle\lim_{x\to\infty}\left(\dfrac{1-\dfrac{8}{x^4}}{10+\dfrac{25}{x^3}+\dfrac{1}{x^4}}\right)=\dfrac{1-0}{10+0+0}=\dfrac{1}{10}$.

6. $+\infty$

Here we have to think about what happens when we plug in a value that is very close to 6, but a little bit more. The top expression will approach 8. The bottom expression will approach 0, but will be a little bit bigger. Thus, the limit will be $\dfrac{8}{0^+}$, which is $+\infty$.

7. $-\infty$

Here we have to think about what happens when we plug in a value that is very close to 6, but a little bit less. The top expression will approach 8. The bottom expression will approach 0 but will be a little bit less. Thus, the limit will be $\dfrac{8}{0^-}$, which is $-\infty$.

8. The limit *Does Not Exist*.

In order to evaluate the limit as x approaches 6, we find the limit as it approaches 6^+ (from the right) and the limit as it approaches 6^- (from the left). If the two limits approach the same value, or both approach positive infinity or both approach negative infinity, then the limit is that value, or the appropriately signed infinity. If the two limits do not agree, the limit "Does Not Exist." Here, if we look at the solutions to problems 9 and 10, we find that as x approaches 6^+, the limit is $+\infty$, but as x approaches 6^-, the limit is $-\infty$. Because the two limits are *not* the same, the limit *Does Not Exist*.

9. 1

Here we have to think about what happens when we plug in a value that is very close to 0, but a little bit more. The top and bottom expressions will both be positive and the same value, so we get $\displaystyle\lim_{x\to0^+}\dfrac{x}{|x|}=\dfrac{0^+}{0^+}=1$.

10. $+\infty$

Here we have to think about what happens when we plug in a value that is very close to 7, but a little bit more. The top expression will approach 7. The bottom expression will approach 0 but will be a little bit positive. Thus, the limit will be $\dfrac{7}{0^+}$, which is $+\infty$.

11. The limit *Does Not Exist.*

In order to evaluate the limit as x approaches 7, we find the limit as it approaches 7^+ (from the right) and the limit as it approaches 7^- (from the left). If the two limits approach the same value, or both approach positive infinity or both approach negative infinity, then the limit is that value, or the appropriately signed infinity. If the two limits do not agree, the limit "Does Not Exist." Here, if we look at the solutions to problem 14, we see that as x approaches 7^+, the limit is $+\infty$. As x approaches 7^-, the top expression will approach 7. The bottom will approach 0, but will be a little bit negative. Thus, the limit will be $\dfrac{7}{0^-}$, which is $-\infty$. Because the two limits are *not* the same, the limit *Does Not Exist.*

12. (a) 4; (b) 5; (c) The limit *Does Not Exist.*

(a) Notice that $f(x)$ is a piecewise function, which means that we use the function $f(x) = x^2 - 5$ for all values of x less than or equal to 3. Thus, $\lim\limits_{x \to 3^-} f(x) = 3^2 - 5 = 4$.

(b) Here we use the function $f(x) = x + 2$ for all values of x greater than 3. Thus, $\lim\limits_{x \to 3^+} f(x) = 3 + 2 = 5$.

(c) In order to evaluate the limit as x approaches 3, we find the limit as it approaches 3^+ (from the right) and the limit as it approaches 3^- (from the left). If the two limits approach the same value, or both approach positive infinity or both approach negative infinity, then the limit is that value, or the appropriately signed infinity. If the two limits do not agree, the limit "Does Not Exist." Here, if we refer to the solutions in parts (a) and (b), we see that $\lim\limits_{x \to 3^-} f(x) = 4$ and $\lim\limits_{x \to 3^+} f(x) = 5$. Because the two limits are *not* the same, the limit *Does Not Exist.*

13.   (a) 4; (b) 4; (c) 4

(a) Notice that $f(x)$ is a piecewise function, which means that we use the function $f(x) = x^2 - 5$ for all values of x less than or equal to 3. Thus, $\lim\limits_{x \to 3^-} f(x) = 3^2 - 5 = 4$.

(b) Here we use the function $f(x) = x + 1$ for all values of x greater than 3. Thus, $\lim\limits_{x \to 3^+} f(x) = 3 + 1 = 4$.

(c) In order to evaluate the limit as x approaches 3, we find the limit as it approaches 3^+ (from the right) and the limit as it approaches 3^- (from the left). If the two limits approach the same value, or both approach positive infinity or both approach negative infinity, then the limit is that value, or the appropriately signed infinity. If the two limits do not agree, the limit "Does Not Exist." Here, if we refer to the solutions in parts (a) and (b), we see that $\lim\limits_{x \to 3^-} f(x) = 4$ and $\lim\limits_{x \to 3^+} f(x) = 4$. Because the two limits are the same, the limit is 4.

14. $\dfrac{3}{\sqrt{2}}$

Here, if we plug in $\dfrac{\pi}{4}$ for x, we get $\lim\limits_{x \to \frac{\pi}{4}} 3\cos x = 3\cos\dfrac{\pi}{4} = \dfrac{3}{\sqrt{2}}$.

15.   0

Here, if we plug in 0 for x, we get $\lim\limits_{x \to 0} 3\dfrac{x}{\cos x} = 3\dfrac{0}{\cos 0} = 3\dfrac{0}{1} = 0$.

16.   3

Remember Rule No. 1, which says that $\lim\limits_{x \to 0} \dfrac{\sin x}{x} = 1$. If we want to find the limit of its reciprocal, we can write this as $\lim\limits_{x \to 0} \dfrac{1}{\dfrac{\sin x}{x}} = \dfrac{1}{\lim\limits_{x \to 0} \dfrac{\sin x}{x}} = \dfrac{1}{1} = 1$. Here, if we plug in 0 for x, we get $\lim\limits_{x \to 0} \left(3\dfrac{x}{\sin x} \right) = (3)(1) = 3$.

17. $\dfrac{7}{5}$

Here, we can rewrite the expression as $\lim\limits_{x \to 0} \dfrac{\tan 7x}{\sin 5x} = \lim\limits_{x \to 0} \dfrac{\frac{\sin 7x}{\cos 7x}}{\sin 5x} = \lim\limits_{x \to 0} \dfrac{\sin 7x}{(\cos 7x)(\sin 5x)}$. Remem-

ber Rule No. 4, which says that $\lim\limits_{x \to 0} \dfrac{\sin ax}{\sin bx} = \dfrac{a}{b}$. Here, $\lim\limits_{x \to 0} \dfrac{\sin 7x}{(\cos 7x)(\sin 5x)} = \dfrac{7}{(1)(5)} = \dfrac{7}{5}$. If we

want to evaluate the limit the long way, we first divide the numerator and the denominator of the

expression by x: $\lim\limits_{x \to 0} \dfrac{\left(\frac{\sin 7x}{x} \right)}{(\cos 7x)\left(\frac{\sin 5x}{x} \right)}$. Next, we multiply the numerator and the denominator of

the top expression by 7 and the numerator and the denominator of the bottom expression by 5. We

get $\lim\limits_{x \to 0} \dfrac{\left(\frac{7 \sin 7x}{7x} \right)}{(\cos 7x)\left(\frac{5 \sin 5x}{5x} \right)}$. Now, we can evaluate the limit: $\lim\limits_{x \to 0} \dfrac{(7)(1)}{(1)(5)(1)} = \dfrac{7}{5}$.

18. The limit *Does Not Exist*.

The value of sin x oscillates between –1 and 1. Thus, as x approaches infinity, sin x does not approach a specific value. Therefore, the limit *Does Not Exist*.

19. 0

Here, as x approaches infinity, $\dfrac{1}{x}$ approaches 0. Thus, $\lim\limits_{x \to \infty} \sin \dfrac{1}{x} = \sin 0 = 0$.

20. $\dfrac{49}{121}$

We can break up the limit into $\lim\limits_{x \to 0} \dfrac{\sin 7x}{\sin 11x} \dfrac{\sin 7x}{\sin 11x}$. Remember Rule No. 4, which says

that $\lim\limits_{x \to 0} \dfrac{\sin ax}{\sin bx} = \dfrac{a}{b}$. Here, $\lim\limits_{x \to 0} \dfrac{\sin 7x}{\sin 11x} \dfrac{\sin 7x}{\sin 11x} = \dfrac{7}{11} \cdot \dfrac{7}{11} = \dfrac{49}{121}$. If we want to evalu-

ate the limit the long way, we first divide the numerator and the denominator of the expressions by x: $\lim\limits_{x\to 0}\left(\dfrac{\dfrac{\sin 7x}{x}}{\dfrac{\sin 11x}{x}} \dfrac{\dfrac{\sin 7x}{x}}{\dfrac{\sin 11x}{x}}\right)$. Next, we multiply the numerator and the denomi-

nator of the top expression by 7 and the numerator and the denominator of the bottom

expression by 11. We get $\lim\limits_{x\to 0}\left(\dfrac{\dfrac{7\sin 7x}{7x}}{\dfrac{11\sin 11x}{11x}} \dfrac{\dfrac{7\sin 7x}{7x}}{\dfrac{11\sin 11x}{11x}}\right)$. Now, we can evaluate the limit:

$$\lim\limits_{x\to 0}\left(\dfrac{\dfrac{7\sin 7x}{7x}}{\dfrac{11\sin 11x}{11x}} \dfrac{\dfrac{7\sin 7x}{7x}}{\dfrac{11\sin 11x}{11x}}\right) = \dfrac{(7)(1)}{(11)(1)}\dfrac{(7)(1)}{(11)(1)} = \left(\dfrac{7}{11}\right)\left(\dfrac{7}{11}\right) = \dfrac{49}{121}.$$

21.   6

Notice that if we plug in 0 for h, we get $\dfrac{0}{0}$, which is indeterminate. If we expand the expression in

the numerator, we get $\lim\limits_{h\to 0}\dfrac{9+6h+h^2-9}{h}$. This simplifies to $\lim\limits_{h\to 0}\dfrac{6h+h^2}{h}$. Next, factor h out

of the top expression: $\lim\limits_{h\to 0}\dfrac{h(6+h)}{h}$. Now, we can cancel the h and evaluate the limit to get

$$\lim\limits_{h\to 0}\dfrac{h(6+h)}{h} = \lim\limits_{h\to 0}(6+h) = 6+0 = 6.$$

22. $-\dfrac{1}{x^2}$

Notice that if we plug in 0 for h, we get $\dfrac{0}{0}$, which is indeterminate. If we combine

the two expressions on top with a common denominator, we get

$$\lim\limits_{h\to 0}\dfrac{\dfrac{1}{x+h}-\dfrac{1}{x}}{h} = \lim\limits_{h\to 0}\dfrac{\dfrac{x}{x(x+h)}-\dfrac{x+h}{x(x+h)}}{h} = \lim\limits_{h\to 0}\dfrac{\dfrac{x-(x+h)}{x(x+h)}}{h}.$$ We can simplify the top expression,

leaving us with $\lim\limits_{h\to 0}\dfrac{\dfrac{-h}{x(x+h)}}{h}$. Next, simplify the expression into $\lim\limits_{h\to 0}\dfrac{\dfrac{-h}{x(x+h)}}{h} = \lim\limits_{h\to 0}\dfrac{-h}{hx(x+h)}.$

We can cancel the h to get $\lim\limits_{h\to 0}\dfrac{-1}{x(x+h)}$. Now, if we evaluate the limit, we get

$$\lim_{h\to 0}\frac{-1}{x(x+h)}=\frac{-1}{x(x)}=-\frac{1}{x^2}.$$

PRACTICE PROBLEM SET 2

1.   Yes. It satisfies all three conditions.

In order for a function $f(x)$ to be continuous at a point $x = c$, it must fulfill *all three* of the following conditions:

Condition 1: $f(c)$ exists.

Condition 2: $\lim\limits_{x\to c} f(x)$ exists.

Condition 3: $\lim\limits_{x\to c} f(x) = f(c)$

Let's test each condition.

$f(2)=9$, which satisfies condition 1.

$\lim\limits_{x\to 2^-} f(x)=9$ and $\lim\limits_{x\to 2^+} f(x)=9$, so $\lim\limits_{x\to 2} f(x)=9$, which satisfies condition 2.

$\lim\limits_{x\to 2} f(x)=9=f(2)$, which satisfies condition 3. Therefore, $f(x)$ is continuous at $x = 2$.

2.   No. It fails condition 3.

In order for a function $f(x)$ to be continuous at a point $x = c$, it must fulfill *all three* of the following conditions:

Condition 1: $f(c)$ exists.

Condition 2: $\lim\limits_{x\to c} f(x)$ exists.

Condition 3: $\lim\limits_{x\to c} f(x) = f(c)$

Let's test each condition.

$f(3) = 29$, which satisfies condition 1.

$\lim\limits_{x \to 3^-} f(x) = 30$ and $\lim\limits_{x \to 3^+} f(x) = 30$, so $\lim\limits_{x \to 3} f(x) = 30$, which satisfies condition 2.

But $\lim\limits_{x \to 3} f(x) \neq f(3)$. Therefore, $f(x)$ is not continuous at $x = 3$ because it fails condition 3.

3. No. It is discontinuous at any odd integral multiple of $\dfrac{\pi}{2}$.

Recall that $\sec x = \dfrac{1}{\cos x}$. This means that sec x is undefined at any value where

$\cos x = 0$, which are the odd multiples of $\dfrac{\pi}{2}$. Therefore, sec x is not continuous everywhere.

4. No. It is discontinuous at the endpoints of the interval.

Recall that $\sec x = \dfrac{1}{\cos x}$. This means that sec x is undefined at any value where

$\cos x = 0$. Also recall that $\cos\dfrac{\pi}{2} = 0$ and $\cos\left(-\dfrac{\pi}{2}\right) = 0$. Therefore, sec x is not continuous every-

where on the interval $\left[-\dfrac{\pi}{2}, \dfrac{\pi}{2}\right]$.

5. Yes.

Recall that $\sec x = \dfrac{1}{\cos x}$. This means that sec x is undefined at any value where

$\cos x = 0$. Also recall that $\cos\dfrac{\pi}{2} = 0$ and $\cos\left(-\dfrac{\pi}{2}\right) = 0$. Therefore, sec x is continuous everywhere

on the interval $\left(-\dfrac{\pi}{2}, \dfrac{\pi}{2}\right)$ because the interval does not include the endpoints.

6. The function is continuous for $k = \dfrac{9}{16}$.

In order for a function $f(x)$ to be continuous at a point $x = c$, it must fulfill *all three* of the following conditions:

Condition 1: $f(c)$ exists.

Condition 2: $\lim_{x \to c} f(x)$ exists.

Condition 3: $\lim_{x \to c} f(x) = f(c)$

We will need to find a value, or values, of k that enables $f(x)$ to satisfy each condition.

Condition 1: $f(4) = 0$

Condition 2: $\lim_{x \to 4^-} f(x) = 0$ and $\lim_{x \to 4^+} f(x) = 16k - 9$. In order for the limit to exist, the two limits must be the same. If we solve $16k - 9 = 0$, we get $k = \dfrac{9}{16}$.

Condition 3: If we now let $k = \dfrac{9}{16}$, $\lim_{x \to 4} f(x) = 0 = f(4)$. Therefore, the solution is $k = \dfrac{9}{16}$.

7. The removable discontinuity is at $\left(3, \dfrac{11}{5}\right)$.

One of the most common scenarios in which a removable discontinuity occurs is with a rational

expression with common factors in the numerator and denominator. In such a case, the graph will

have a "hole" whose position can be determined by examining the common factor and cancelling

it. If we factor $f(x) = \dfrac{x^2 + 5x - 24}{x^2 - x - 6}$, we get $f(x) = \dfrac{(x+8)(x-3)}{(x+2)(x-3)}$. If we cancel the common fac-

tor, we get $f(x) = \dfrac{(x+8)}{(x+2)}$. Now, if we plug in $x = 3$, we get $f(x) = \dfrac{11}{5}$. Therefore, the removable

discontinuity is at $\left(3, \dfrac{11}{5}\right)$.

8. (a) 0; (b) 0; (c) 1; (d) 1; (e) $f(3)$ *Does Not Exist*; (f) a jump discontinuity at $x = -3$; a removable discontinuity at $x = 3$ and an essential discontinuity at $x = 5$.

(a) If we look at the graph, we can see that $\lim_{x \to -\infty} f(x) = 0$.

(b) If we look at the graph, we can see that $\lim_{x \to \infty} f(x) = 0$.

(c) If we look at the graph, we can see that $\lim_{x \to 3^-} f(x) = 1$.

(d) If we look at the graph, we can see that $\lim_{x \to 3^+} f(x) = 1$.

(e) $f(3)$ Does Not Exist.

(f) There are three discontinuities: (1) a jump discontinuity at $x = -3$; (2) a removable discontinuity at $x = 3$; and (3) an essential discontinuity at $x = 5$.

PRACTICE PROBLEM SET 3

1.　　5

We find the derivative of a function, $f(x)$, using the definition of the derivative, which is

$$f'(x) = \lim_{h \to 0} \frac{f(x+h) - f(x)}{h}.$$ Here, $f(x) = 5x$ and $x = 3$. This means that $f(3) = 5(3) = 15$

and $f(3+h) = 5(3+h) = 15 + 5h$. If we now plug these into the definition of the derivative, we get

$$f'(3) = \lim_{h \to 0} \frac{f(3+h) - f(3)}{h} = \lim_{h \to 0} \frac{15 + 5h - 15}{h}.$$ This simplifies to $f'(3) = \lim_{h \to 0} \frac{5h}{h} = \lim_{h \to 0} 5 = 5$.

If you noticed that the function is simply the equation of a line, then you would have seen that the

derivative is simply the slope of the line, which is 5 everywhere.

2.　　4

We find the derivative of a function, $f(x)$, using the definition of the derivative,

which is $f'(x) = \lim_{h \to 0} \frac{f(x+h) - f(x)}{h}.$ Here, $f(x) = 4x$ and $x = -8$. This means that

$f(-8) = 4(-8) = -32$ and $f(-8+h) = 4(-8+h) = -32 + 4h$. If we now plug these into the defi-

nition of the derivative, we get $f'(-8) = \lim_{h \to 0} \frac{f(-8+h) - f(-8)}{h} = \lim_{h \to 0} \frac{-32 + 4h + 32}{h}.$ This sim-

plifies to $f'(-8) = \lim_{h \to 0} \frac{4h}{h} = \lim_{h \to 0} 4 = 4$.

If you noticed that the function is simply the equation of a line, then you would have seen that the

derivative is simply the slope of the line, which is 4 everywhere.

3.　　−10

We find the derivative of a function, $f(x)$, using the definition of the derivative, which is

$$f'(x) = \lim_{h \to 0} \frac{f(x+h) - f(x)}{h}.$$ Here, $f(x) = 5x^2$ and $x = -1$. This means that $f(-1) = 5(-1)^2 = 5$

and $f(-1+h) = 5(-1+h)^2 = 5(1 - 2h + h^2) = 5 - 10h + 5h^2$. If we now plug these into the definition of the derivative, we get $f'(-1) = \lim\limits_{h \to 0} \dfrac{f(-1+h) - f(-1)}{h} = \lim\limits_{h \to 0} \dfrac{5 - 10h + 5h^2 - 5}{h}$. This

simplifies to $f'(-1) = \lim\limits_{h \to 0} \dfrac{-10h + 5h^2}{h}$. Now we can factor out the h from the numerator and can-

cel it with the h in the denominator: $f'(-1) = \lim\limits_{h \to 0} \dfrac{h(-10 + 5h)}{h} = \lim\limits_{h \to 0}(-10 + 5h)$. Now we take

the limit to get $f'(-1) = \lim\limits_{h \to 0}(-10 + 5h) = -10$.

4. $16x$

We find the derivative of a function, $f(x)$, using the definition of the

derivative, which is $f'(x) = \lim\limits_{h \to 0} \dfrac{f(x+h) - f(x)}{h}$. Here, $f(x) = 8x^2$ and

$f(x + h) = 8(x + h)^2 = 8(x^2 + 2xh + h^2) = 8x^2 + 16xh + 8h^2$. If we now plug these into the defi-

nition of the derivative, we get $f'(x) = \lim\limits_{h \to 0} \dfrac{f(x+h) - f(x)}{h} = \lim\limits_{h \to 0} \dfrac{8x^2 + 16xh + 8h^2 - 8x^2}{h}$. This

simplifies to $f'(x) = \lim\limits_{h \to 0} \dfrac{16xh + 8h^2}{h}$. Now we can factor out the h from the numerator and cancel

it with the h in the denominator: $f'(x) = \lim\limits_{h \to 0} \dfrac{h(16x + 8h)}{h} = \lim\limits_{h \to 0}(16x + 8h)$. Now we take the

limit to get $f'(x) = \lim\limits_{h \to 0}(16x + 8h) = 16x$.

5. $-20x$

We find the derivative of a function, $f(x)$, using the definition of the

derivative, which is $f'(x) = \lim\limits_{h \to 0} \dfrac{f(x+h) - f(x)}{h}$. Here, $f(x) = -10x^2$ and

$f(x + h) = -10(x + h)^2 = -10(x^2 + 2xh + h^2) = -10x^2 - 20xh - 10h^2$.

If we now plug these into the definition of the derivative, we get

$$f'(x) = \lim_{h \to 0} \frac{f(x+h) - f(x)}{h} = \lim_{h \to 0} \frac{-10x^2 - 20xh - 10h^2 + 10x^2}{h}.$$ This simplifies to

$$f'(x) = \lim_{h \to 0} \frac{-20xh - 10h^2}{h}.$$ Now we can factor out the h from the numerator and cancel it with

the h in the denominator: $f'(x) = \lim_{h \to 0} \frac{h(-20x - 10h)}{h} = \lim_{h \to 0}(-20x - 10h)$. Now we take the limit

to get $f'(x) = \lim_{h \to 0}(-20x - 10h) = -20x$.

6. $40a$

We find the derivative of a function, $f(x)$, using the definition of the derivative, which is

$$f'(x) = \lim_{h \to 0} \frac{f(x+h) - f(x)}{h}.$$ Here, $f(x) = 20x^2$ and $x = a$. This means that $f(a) = 20a^2$ and

$f(a + h) = 20(a + h)^2 = 20(a^2 + 2ah + h^2) = 20a^2 + 40ah + 20h^2$. If we now plug these into the

definition of the derivative, we get $f'(a) = \lim_{h \to 0} \frac{f(a+h) - f(a)}{h} = \lim_{h \to 0} \frac{20a^2 + 40ah + 20h^2 - 20a^2}{h}.$

This simplifies to $f'(a) = \lim_{h \to 0} \frac{40ah + 20h^2}{h}$. Now we can factor out the h from the numerator and

cancel it with the h in the denominator: $f'(a) = \lim_{h \to 0} \frac{h(40a + 20h)}{h} = \lim_{h \to 0}(40a + 20h)$. Now we

take the limit to get $f'(a) = \lim_{h \to 0}(40a + 20h) = 40a$.

7. 54

We find the derivative of a function, $f(x)$, using the definition of the

derivative,   which   is $f'(x) = \lim_{h \to 0} \frac{f(x+h) - f(x)}{h}.$ Here, $f(x) = 2x^3$

and $x = -3$.   This   means   that $f(-3) = 2(-3)^3 = -54$ and

$f(-3 + h) = 2(-3 + h)^3 = 2((-3)^3 + 3(-3)^2 h + 3(-3)h^2 + h^3) = -54 + 54h - 18h^2 + 2h^3.$

If we now plug these into the definition of the derivative, we get

$$f'(-3)=\lim_{h\to 0}\frac{f(-3+h)-f(-3)}{h}=\lim_{h\to 0}\frac{-54+54h-18h^2+2h^3+54}{h}.$$ This simplifies to

$$f'(-3)=\lim_{h\to 0}\frac{54h-18h^2+2h^3}{h}.$$ Now we can factor out the h from the numerator and cancel it

with the h in the denominator: $f'(-3)=\lim_{h\to 0}\frac{h\left(54-18h+2h^2\right)}{h}=\lim_{h\to 0}\left(54-18h+2h^2\right)$. Now we

take the limit to get $f'(-3)=\lim_{h\to 0}\left(54-18h+2h^2\right)=54$.

8. $-9x^2$

We find the derivative of a function, $f(x)$, using the definition of the deriva-

tive, which is $f'(x)=\lim_{h\to 0}\frac{f(x+h)-f(x)}{h}$. Here, $f(x)=-3x^3$. This means that

$$f(x+h)=-3(x+h)^3=-3\left(x^3+3x^2h+3xh^2+h^3\right)=-3x^3-9x^2h-9xh^2-3h^3.$$

If we now plug these into the definition of the derivative, we get

$$f'(x)=\lim_{h\to 0}\frac{f(x+h)-f(x)}{h}=\lim_{h\to 0}\frac{-3x^3-9x^2h-9xh^2-3h^3+3x^3}{h}.$$ This simplifies to

$$f'(x)=\lim_{h\to 0}\frac{-9x^2h-9xh^2-3h^3}{h}.$$ Now we can factor out the h from the numerator and cancel it

with the h in the denominator: $f'(x)=\lim_{h\to 0}\frac{h\left(-9x^2-9xh-3h^2\right)}{h}=\lim_{h\to 0}\left(-9x^2-9xh-3h^2\right)$. Now

we take the limit to get $f'(x)=\lim_{h\to 0}\left(-9x^2-9xh-3h^2\right)=-9x^2$.

9. $5x^4$

We find the derivative of a function, $f(x)$, using the definition of the derivative, which is

$f'(x)=\lim_{h\to 0}\frac{f(x+h)-f(x)}{h}$. Here, $f(x)=x^5$.

This means that $f(x+h)=(x+h)^5=\left(x^5+5x^4h+10x^3h^2+10x^2h^3+5xh^4+h^5\right)$.

If we now plug these into the definition of the derivative, we get

$$f'(x) = \lim_{h \to 0} \frac{f(x+h) - f(x)}{h} = \lim_{h \to 0} \frac{x^5 + 5x^4h + 10x^3h^2 + 10x^2h^3 + 5xh^4 + h^5 - x^5}{h}.$$

This simplifies to $f'(x) = \lim_{h \to 0} \dfrac{5x^4h + 10x^3h^2 + 10x^2h^3 + 5xh^4 + h^5}{h}$. Now we can

factor out the h from the numerator and cancel it with the h in the denominator:

$$f'(x) = \lim_{h \to 0} \frac{h\left(5x^4 + 10x^3h + 10x^2h^2 + 5xh^3 + h^4\right)}{h} = \lim_{h \to 0}\left(5x^4 + 10x^3h + 10x^2h^2 + 5xh^3 + h^4\right).$$

Now we take the limit to get $f'(x) = \lim_{h \to 0}\left(5x^4 + 10x^3h + 10x^2h^2 + 5xh^3 + h^4\right) = 5x^4$.

10. $\dfrac{1}{3}$

We find the derivative of a function, $f(x)$, using the definition of the derivative, which is

$f'(x) = \lim_{h \to 0} \dfrac{f(x+h) - f(x)}{h}$. Here, $f(x) = 2\sqrt{x}$ and $x = 9$. This means that $f(9) = 2\sqrt{9} = 6$

and $f(9+h) = 2\sqrt{9+h}$. If we now plug these into the definition of the derivative, we get

$f'(9) = \lim_{h \to 0} \dfrac{f(9+h) - f(9)}{h} = \lim_{h \to 0} \dfrac{2\sqrt{9+h} - 6}{h}$. Notice that if we now take the limit, we

get the indeterminate form $\dfrac{0}{0}$. With polynomials, we merely simplify the expression to elimi-

nate this problem. In any derivative of a square root, we first multiply the top and the bot-

tom of the expression by the conjugate of the numerator and then we can simplify. Here the

conjugate is $2\sqrt{9+h} + 6$. We get $f'(9) = \lim_{h \to 0} \dfrac{2\sqrt{9+h} - 6}{h} \times \dfrac{2\sqrt{9+h} + 6}{2\sqrt{9+h} + 6}$. This simplifies to

$f'(9) = \lim_{h \to 0} \dfrac{4(9+h) - 36}{h\left(2\sqrt{9+h} + 6\right)} = \lim_{h \to 0} \dfrac{36 + 4h - 36}{h\left(2\sqrt{9+h} + 6\right)} = \lim_{h \to 0} \dfrac{4h}{h\left(2\sqrt{9+h} + 6\right)}$. Now we can cancel the

h in the numerator and the denominator to get $f'(9) = \lim_{h \to 0} \dfrac{4}{\left(2\sqrt{9+h} + 6\right)}$. Now we take the limit:

$f'(9) = \lim_{h \to 0} \dfrac{4}{\left(2\sqrt{9+h} + 6\right)} = \dfrac{4}{\left(2\sqrt{9} + 6\right)} = \dfrac{1}{3}$.

11. $\dfrac{5}{4}$

We find the derivative of a function, $f(x)$, using the definition of the derivative, which is

$f'(x) = \lim\limits_{h \to 0} \dfrac{f(x+h)-f(x)}{h}$. Here, $f(x) = 5\sqrt{2x}$ and $x = 8$. This means that $f(8) = 5\sqrt{16} = 20$

and $f(8+h) = 5\sqrt{2(8+h)} = 5\sqrt{16+2h}$. If we now plug these into the definition of the deriva-

tive, we get $f'(8) = \lim\limits_{h \to 0} \dfrac{f(8+h)-f(8)}{h} = \lim\limits_{h \to 0} \dfrac{5\sqrt{16+2h}-20}{h}$. Notice that if we now take the

limit, we get the indeterminate form $\dfrac{0}{0}$. With polynomials, we merely simplify the expression to

eliminate this problem. In any derivative of a square root, we first multiply the top and the bot-

tom of the expression by the conjugate of the numerator and then we simplify. Here the con-

jugate is $5\sqrt{16+2h}+20$. We get $f'(8) = \lim\limits_{h \to 0} \dfrac{5\sqrt{16+2h}-20}{h} \times \dfrac{5\sqrt{16+2h}+20}{5\sqrt{16+2h}+20}$. This simplifies

to $f'(8) = \lim\limits_{h \to 0} \dfrac{25(16+2h)-400}{h\left(5\sqrt{16+2h}+20\right)} = \lim\limits_{h \to 0} \dfrac{400+50h-400}{h\left(5\sqrt{16+2h}+20\right)} = \lim\limits_{h \to 0} \dfrac{50h}{h\left(5\sqrt{16+2h}+20\right)}$. Now we

can cancel the h in the numerator and the denominator to get $f'(8) = \lim\limits_{h \to 0} \dfrac{50}{\left(5\sqrt{16+2h}+20\right)}$.

Now, we take the limit: $f'(8) = \lim\limits_{h \to 0} \dfrac{50}{\left(5\sqrt{16+2h}+20\right)} = \dfrac{50}{(20+20)} = \dfrac{5}{4}$.

12. $\dfrac{1}{2}$

We find the derivative of a function, $f(x)$, using the definition of the derivative,

which is $f'(x) = \lim\limits_{h \to 0} \dfrac{f(x+h)-f(x)}{h}$. Here, $f(x) = \sin x$ and $x = \dfrac{\pi}{3}$. This means that

$f\left(\dfrac{\pi}{3}\right) = \sin\dfrac{\pi}{3} = \dfrac{\sqrt{3}}{2}$ and $f\left(\dfrac{\pi}{3}+h\right) = \sin\left(\dfrac{\pi}{3}+h\right)$. If we now plug these into the definition of the derivative, we get $f'\left(\dfrac{\pi}{3}\right) = \lim\limits_{h\to 0} \dfrac{f\left(\dfrac{\pi}{3}+h\right)-f\left(\dfrac{\pi}{3}\right)}{h} = \lim\limits_{h\to 0} \dfrac{\sin\left(\dfrac{\pi}{3}+h\right)-\dfrac{\sqrt{3}}{2}}{h}$. Notice that if we

now take the limit, we get the indeterminate form $\dfrac{0}{0}$. We cannot eliminate this problem

merely by simplifying the expression the way that we did with a polynomial. Recall that the

trigonometric formula $\sin(A+B)=\sin A\cos B+\cos A\sin B$. Here, we can rewrite the top

expression as $f'\left(\dfrac{\pi}{3}\right) = \lim\limits_{h\to 0}\dfrac{\sin\left(\dfrac{\pi}{3}+h\right)-\dfrac{\sqrt{3}}{2}}{h} = \lim\limits_{h\to 0}\dfrac{\sin\dfrac{\pi}{3}\cos h+\cos\dfrac{\pi}{3}\sin h-\dfrac{\sqrt{3}}{2}}{h}$. We can

break up the limit into $\lim\limits_{h\to 0}\dfrac{\sin\dfrac{\pi}{3}\cos h-\dfrac{\sqrt{3}}{2}}{h}+\dfrac{\cos\dfrac{\pi}{3}\sin h}{h}-\lim\limits_{h\to 0}\dfrac{\dfrac{\sqrt{3}}{2}\cos h-\dfrac{\sqrt{3}}{2}}{h}+\dfrac{\left(\dfrac{1}{2}\right)\sin h}{h}$.

Next, factor $\dfrac{\sqrt{3}}{2}$ out of the top of the left-hand expression: $\lim\limits_{h\to 0}\dfrac{\dfrac{\sqrt{3}}{2}(\cos h-1)}{h}+\dfrac{\left(\dfrac{1}{2}\right)\sin h}{h}$.

Now, we can break this into separate limits: $\lim\limits_{h\to 0}\dfrac{\dfrac{\sqrt{3}}{2}(\cos h-1)}{h}+\lim\limits_{h\to 0}\dfrac{\left(\dfrac{1}{2}\right)\sin h}{h}$.

The left-hand limit is $\lim\limits_{h\to 0}\dfrac{\dfrac{\sqrt{3}}{2}(\cos h-1)}{h}=\dfrac{\sqrt{3}}{2}\lim\limits_{h\to 0}\dfrac{(\cos h-1)}{h}=\dfrac{\sqrt{3}}{2}\cdot 0=0$.

The right-hand limit is $\dfrac{1}{2}\lim\limits_{h\to 0}\dfrac{\sin h}{h}=\dfrac{1}{2}\cdot 1=\dfrac{1}{2}$. Therefore, the limit is $\dfrac{1}{2}$.

13. $2x+1$

We find the derivative of a function, $f(x)$, using the definition of the

derivative, which is $f'(x)=\lim\limits_{h\to 0}\dfrac{f(x+h)-f(x)}{h}$. Here, $f(x)=x^2+x$ and

$f(x+h)=(x+h)^2+(x+h)=x^2+2xh+h^2+x+h$. If we now plug these into the definition of the derivative, we get $f'(x) = \lim_{h \to 0} \dfrac{f(x+h) - f(x)}{h} = \lim_{h \to 0} \dfrac{x^2 + 2xh + h^2 + x + h - (x^2 + x)}{h}$. This

simplifies to $f'(x) = \lim_{h \to 0} \dfrac{2xh + h^2 + h}{h}$. Now we can factor out the h from the numerator and cancel

it with the h in the denominator: $f'(x) = \lim_{h \to 0} \dfrac{2xh + h^2 + h}{h} = \lim_{h \to 0} \dfrac{h(2x + h + 1)}{h} = \lim_{h \to 0} (2x + h + 1)$.

Now we take the limit to get $f'(x) = \lim_{h \to 0} (2x + h + 1) = 2x + 1$.

14. $3x^2 + 3$

We find the derivative of a function, $f(x)$, using the definition of the

derivative,   which   is $f'(x) = \lim_{h \to 0} \dfrac{f(x+h) - f(x)}{h}$.   Here, $f(x) = x^3 + 3x + 2$ and

$f(x+h) = (x+h)^3 + 3(x+h) + 2 = x^3 + 3x^2h + 3xh^2 + h^3 + 3x + 3h + 2$.

If   we   now   plug   these   into   the   definition   of   the   derivative,   we   get

$f'(x) = \lim_{h \to 0} \dfrac{f(x+h) - f(x)}{h} = \lim_{h \to 0} \dfrac{x^3 + 3x^2h + 3xh^2 + h^3 + 3x + 3h + 2 - (x^3 + 3x + 2)}{h}$.

This   simplifies   to $f'(x) = \lim_{h \to 0} \dfrac{3x^2h + 3xh^2 + h^3 + 3h}{h}$.   Now   we   can   fac-

tor   out   the h from   the   numerator   and   cancel   it   with   the h in   the   denominator:

$f'(x) = \lim_{h \to 0} \dfrac{3x^2h + 3xh^2 + h^3 + 3h}{h} \lim_{h \to 0} \dfrac{h(3x^2 + 3xh + h^2 + 3)}{h} = \lim_{h \to 0} (3x^2 + 3xh + h^2 + 3)$.   Now

we take the limit to get $f'(x) = \lim_{h \to 0} (3x^2 + 3xh + h^2 + 3) = 3x^2 + 3$.

15. $-\dfrac{1}{x^2}$

We find the derivative of a function, $f(x)$, using the definition of the derivative, which is

$f'(x) = \lim_{h \to 0} \dfrac{f(x+h) - f(x)}{h}$. Here, $f(x) = \dfrac{1}{x}$ and $f(x+h) = \dfrac{1}{x+h}$. If we now plug these into the definition of the derivative, we get $f'(x) = \lim_{h \to 0} \dfrac{f(x+h) - f(x)}{h} = \lim_{h \to 0} \dfrac{\frac{1}{x+h} - \frac{1}{x}}{h}$.

Notice that if we now take the limit, we get the indeterminate form $\dfrac{0}{0}$. We cannot elimi-

nate this problem merely by simplifying the expression the way that we did with a poly-

nomial. Here we combine the two terms in the numerator of the expression to get

$$f'(x) = \lim_{h \to 0} \frac{\frac{1}{x+h} - \frac{1}{x}}{h} = \lim_{h \to 0} \frac{\frac{x}{x(x+h)} - \frac{x+h}{x(x+h)}}{h} = \lim_{h \to 0} \frac{\frac{-h}{x(x+h)}}{h}.$$ This simplifies to

$$f'(x) = \lim_{h \to 0} \frac{\frac{-h}{x(x+h)}}{h} = \lim_{h \to 0} \frac{-h}{x(x+h)h}.$$ Now we can cancel the factor h in the numerator

and the denominator to get $f'(x) = \lim_{h \to 0} \dfrac{-h}{x(x+h)h} = \lim_{h \to 0} \dfrac{-1}{x(x+h)}$. Now we take the limit:

$$f'(x) = \lim_{h \to 0} \frac{-1}{x(x+h)} = -\frac{1}{x^2}.$$

16. $2ax + b$

We find the derivative of a function, $f(x)$, using the definition of the

derivative, which is $f'(x) = \lim_{h \to 0} \dfrac{f(x+h) - f(x)}{h}$. Here, $f(x) = ax^2 + bx + c$ and

$$f(x+h) = a(x+h)^2 + b(x+h) + c = ax^2 + 2axh + ah^2 + bx + bh + c.$$

If we now plug these into the definition of the derivative, we get

$f'(x) = \lim_{h \to 0} \dfrac{f(x+h) - f(x)}{h} = \lim_{h \to 0} \dfrac{ax^2 + 2axh + ah^2 + bx + bh + c - \left(ax^2 + bx + c\right)}{h}$. This simplifies

to $f'(x) = \lim_{h \to 0} \dfrac{2axh + ah^2 + bh}{h}$. Now we can factor out the h from the numerator and cancel it with the h in the denominator: $f'(x) = \lim_{h \to 0} \dfrac{2axh + ah^2 + bh}{h} = \lim_{h \to 0} \dfrac{h(2ax + ah + b)}{h} = \lim_{h \to 0}(2ax + ah + b)$.

Now we take the limit to get $f'(x) = \lim_{h \to 0}(2ax + ah + b) = 2ax + b$.

PRACTICE PROBLEM SET 4

1. $64x^3 + 16x$

   First, expand $(4x^2 + 1)^2$ to get $16x^4 + 8x^2 + 1$. Now, use the Power Rule to take the derivative of each term. The derivative of $16x^4 = 16(4x^3) = 64x^3$. The derivative of $8x^2 = 8(2x) = 16x$. The derivative of $1 = 0$ (because the derivative of a constant is zero). Therefore, the derivative is $64x^3 + 16x$.

2. $10x^9 + 36x^5 + 18x$

   First, expand $(x^5 + 3x)^2$ to get $x^{10} + 6x^6 + 9x^2$. Now, use the Power Rule to take the derivative of each term. The derivative of $x^{10} = 10x^9$. The derivative of $6x^6 = 6(6x^5) = 36x^5$. The derivative of $9x^2 = 9(2x) = 18x$. Therefore, the derivative is $10x^9 + 36x^5 + 18x$.

3. $77x^6$

   Simply use the Power Rule. The derivative is $11x^7 = 11(7x^6) = 77x^6$.

4. $54x^2 + 12$

   Use the Power Rule to take the derivative of each term. The derivative of $18x^3 = 18(3x^2) = 54x^2$. The derivative of $12x = 12$. The derivative of $11 = 0$ (because the derivative of a constant is zero). Therefore, the derivative is $54x^2 + 12$.

5. $6x^{11}$

   Use the Power Rule to take the derivative of each term. The derivative of $x^{12} = 12x^{11}$. The derivative of $17 = 0$ (because the derivative of a constant is zero). Therefore, the derivative is $\dfrac{1}{2}(12x^{11}) = 6x^{11}$.

6. $\qquad -3x^8 - 2x^2$

Use the Power Rule to take the derivative of each term. The derivative of $x^9 = 9x^8$. The derivative of $2x^3 = 2(3x^2) = 6x^2$. The derivative of $9 = 0$ (because the derivative of a constant is zero). Therefore, the derivative is $-\dfrac{1}{3}(9x^8 + 6x^2) = -3x^8 - 2x^2$.

7. $\qquad 0$

Don't be fooled by the power. π^5 is a constant so the derivative is zero.

8. $\qquad 64x^{-9} + \dfrac{6}{\sqrt{x}}$

Use the Power Rule to take the derivative of each term. The derivative of $-8x^{-8} = -8(-8x^{-9}) = 64x^{-9}$.

The derivative of $12\sqrt{x} = \dfrac{12}{2\sqrt{x}} = \dfrac{6}{\sqrt{x}}$ (remember the shortcut that we showed you on page 129).

Therefore, the derivative is $64x^{-9} + \dfrac{6}{\sqrt{x}}$.

9. $\qquad -42x^{-8} - \dfrac{2}{\sqrt{x}}$

Use the Power Rule to take the derivative of each term. The derivative of $6x^{-7} = 6(-7x^{-8}) = -42x^{-8}$.

The derivative of $4\sqrt{x} = \dfrac{4}{2\sqrt{x}} = \dfrac{2}{\sqrt{x}}$ (remember the shortcut that we showed you on page 129).

Therefore, the derivative is $-42x^{-8} - \dfrac{2}{\sqrt{x}}$.

10. $\qquad \dfrac{-5}{x^6} - \dfrac{8}{x^9}$

Use the Power Rule to take the derivative of each term. The derivative of $x^{-5} = -5x^{-6}$. To find the derivative of $\dfrac{1}{x^8}$, we first rewrite it as x^{-8}. The derivative of $x^{-8} = -8x^{-9}$. Therefore, the derivative is $-5x^{-6} - 8x^{-9} = \dfrac{-5}{x^6} - \dfrac{8}{x^9}$.

11. $216x^2 - 48x + 36$

First, expand $(6x^2 + 3)(12x - 4)$ to get $72x^3 - 24x^2 + 36x - 12$. Now, use the Power Rule to take the derivative of each term. The derivative of $72x^3 = 72(3x^2) = 216x^2$. The derivative of $24x^2 = 24(2x) = 48x$. The derivative of $36x = 36$. The derivative of $12 = 0$ (because the derivative of a constant is zero). Therefore, the derivative is $216x^2 - 48x + 36$.

12. $-6 - 36x^2 + 12x^3 - 5x^4 - 14x^6$

First, expand $(3 - x - 2x^3)(6 + x^4)$ to get $18 - 6x - 12x^3 + 3x^4 - x^5 - 2x^7$. Now, use the Power Rule to take the derivative of each term. The derivative of $18 = 0$ (because the derivative of a constant is zero). The derivative of $6x = 6$. The derivative of $12x^3 = 12(3x^2) = 36x^2$. The derivative of $3x^4 = 3(4x^3) = 12x^3$. The derivative of $x^5 = 5x^4$. The derivative of $2x^7 = 2(7x^6) = 14x^6$. Therefore, the derivative is $-6 - 36x^2 + 12x^3 - 5x^4 - 14x^6$.

13. 0

Don't be fooled by the powers. Each term is a constant so the derivative is zero.

14. $-\dfrac{16}{x^5} + \dfrac{10}{x^6} + \dfrac{36}{x^7}$

First, expand $\left(\dfrac{1}{x} + \dfrac{1}{x^2}\right)\left(\dfrac{4}{x^3} - \dfrac{6}{x^4}\right)$ to get $\dfrac{4}{x^4} - \dfrac{6}{x^5} + \dfrac{4}{x^5} - \dfrac{6}{x^6} = \dfrac{4}{x^4} - \dfrac{2}{x^5} - \dfrac{6}{x^6}$. Next, rewrite the terms as $4x^{-4} - 2x^{-5} - 6x^{-6}$. Now, use the Power Rule to take the derivative of each term. The derivative of $4x^{-4} = 4(-4x^{-5}) = -16x^{-5}$. The derivative of $2x^{-5} = 2(-5x^{-6}) = -10x^{-6}$. The derivative of $6x^{-6} = -36x^{-7}$. Therefore, the derivative is $-16x^{-5} + 10x^{-6} + 36x^{-7} = -\dfrac{16}{x^5} + \dfrac{10}{x^6} + \dfrac{36}{x^7}$.

15. $\dfrac{1}{2\sqrt{x}}$

Use the Power Rule to take the derivative of each term. The derivative of $\sqrt{x} = \dfrac{1}{2\sqrt{x}}$ (remember the shortcut that we showed you on page 129). The derivative of $\dfrac{1}{\sqrt{3}} = 0$ (because the derivative of a constant is zero). Therefore, the derivative is $\dfrac{1}{2\sqrt{x}}$.

16.   0

The derivative of a constant is zero.

17. $3x^2 + 6x + 3$

First, expand $(x+1)^3$ to get $x^3 + 3x^2 + 3x + 1$. Now, use the Power Rule to take the derivative of each term. The derivative of $x^3 = 3x^2$. The derivative of $3x^2 = 3(2x) = 6x$. The derivative of $3x = 3$. The derivative of $1 = 0$ (because the derivative of a constant is zero). Therefore, the derivative is $3x^2 + 6x + 3$.

18. $\dfrac{1}{2\sqrt{x}} + \dfrac{1}{3\sqrt[3]{x^2}} + \dfrac{2}{3\sqrt[3]{x}}$

Use the Power Rule to take the derivative of each term. The derivative of $\sqrt{x} = \dfrac{1}{2\sqrt{x}}$ (remember the shortcut that we showed you on page 129). Rewrite $\sqrt[3]{x}$ as $x^{\frac{1}{3}}$ and $\sqrt[3]{x^2}$ as $x^{\frac{2}{3}}$.

The derivative of $x^{\frac{1}{3}} = \dfrac{1}{3}x^{-\frac{2}{3}}$. The derivative of $x^{\frac{2}{3}} = \dfrac{2}{3}x^{-\frac{1}{3}}$. Therefore, the derivative is

$$\dfrac{1}{2\sqrt{x}} + \dfrac{1}{3}x^{-\frac{2}{3}} + \dfrac{2}{3}x^{-\frac{1}{3}} = \dfrac{1}{2\sqrt{x}} + \dfrac{1}{3\sqrt[3]{x^2}} + \dfrac{2}{3\sqrt[3]{x}}.$$

19. $6x^2 + 6x - 14$

First, expand $x(2x+7)(x-2)$ to get $x(2x^2 + 3x - 14) = 2x^3 + 3x^2 - 14x$. Now, use the Power Rule to take the derivative of each term. The derivative of $2x^3 = 2(3x^2) = 6x^2$. The derivative of $3x^2 = 3(2x) = 6x$. The derivative of $14x = 14$. Therefore, the derivative is $6x^2 + 6x - 14$.

20. $\dfrac{5}{6\sqrt[6]{x}} + \dfrac{7}{10\sqrt[10]{x^3}}$

First, rewrite the terms as $x^{\frac{1}{2}}\left(x^{\frac{1}{3}} + x^{\frac{1}{5}}\right)$. Next, distribute to get: $x^{\frac{1}{2}}\left(x^{\frac{1}{3}} + x^{\frac{1}{5}}\right) = x^{\frac{5}{6}} + x^{\frac{7}{10}}$. Now, use the Power Rule to take the derivative of each term. The derivative of $x^{\frac{5}{6}} = \dfrac{5}{6}x^{-\frac{1}{6}}$. The derivative of

$x^{\frac{7}{10}} = \dfrac{7}{10}x^{-\frac{3}{10}}$. Therefore, the derivative is $\dfrac{5}{6}x^{-\frac{1}{6}} + \dfrac{7}{10}x^{-\frac{3}{10}} = \dfrac{5}{6\sqrt[6]{x}} + \dfrac{7}{10\sqrt[10]{x^3}}$.

PRACTICE PROBLEM SET 5

1. $\dfrac{-80x^9 + 75x^8 + 12x^2 - 6x}{\left(5x^7 + 1\right)^2}$

We find the derivative using the Quotient Rule, which says that if $f(x) = \dfrac{u}{v}$, then

$f'(x) = \dfrac{v\dfrac{du}{dx} - u\dfrac{dv}{dx}}{v^2}$. Here, $f(x) = \dfrac{4x^3 - 3x^2}{5x^7 + 1}$, so $u = 4x^3 - 3x^2$ and $v = 5x^7 + 1$. Using the

Quotient Rule, we get $f'(x) = \dfrac{\left(5x^7 + 1\right)\left(12x^2 - 6x\right) - \left(4x^3 - 3x^2\right)\left(35x^6\right)}{\left(5x^7 + 1\right)^2}$. This can be simplified

to $f'(x) = \dfrac{-80x^9 + 75x^8 + 12x^2 - 6x}{\left(5x^7 + 1\right)^2}$.

2. $3x^2 - 6x - 1$

We find the derivative using the Product Rule, which says that if $f(x) = uv$, then

$f'(x) = u\dfrac{dv}{dx} + v\dfrac{du}{dx}$. Here, $f(x) = \left(x^2 - 4x + 3\right)(x + 1)$, so $u = x^2 - 4x + 3$ and $v = x + 1$.

Using the Product Rule, we get $f'(x) = \left(x^2 - 4x + 3\right)(1) + (x + 1)(2x - 4)$. This can be simplified

to $f'(x) = 3x^2 - 6x - 1$.

3. $\dfrac{16x^2 - 32}{\sqrt{\left(x^2 - 4\right)}}$

We find the derivative using the Chain Rule, which says that if $y = f(g(x))$,

then $y' = \left(\dfrac{df(g(x))}{dg}\right)\left(\dfrac{dg}{dx}\right)$. Here, $f(x) = 8\sqrt{\left(x^4 - 4x^2\right)}$, which can be written as

$f(x) = 8\left(x^4 - 4x^2\right)^{\frac{1}{2}}$. Using the Chain Rule, we get $f'(x) = 8\left(\dfrac{1}{2}\right)\left(x^4 - 4x^2\right)^{-\frac{1}{2}}\left(4x^3 - 8x\right)$.

This can be simplified to $f'(x) = \dfrac{16x^2 - 32}{\sqrt{\left(x^2 - 4\right)}}$.

4. $\dfrac{3x^2 - 3x^4}{\left(x^2 + 1\right)^4}$

Here, we will find the derivative using the Chain Rule. We will also need the Quotient Rule

to take the derivative of the expression inside the parentheses. The Chain Rule says that if

$y = f\left(g(x)\right)$, then $y' = \left(\dfrac{df\left(g(x)\right)}{dg}\right)\left(\dfrac{dg}{dx}\right)$, and the Quotient Rule says that if $f(x) = \dfrac{u}{v}$, then

$f'(x) = \dfrac{v\dfrac{du}{dx} - u\dfrac{dv}{dx}}{v^2}$. We get $f'(x) = 3\left(\dfrac{x}{x^2 + 1}\right)^2 \left(\dfrac{\left(x^2 + 1\right)(1) - x(2x)}{\left(x^2 + 1\right)^2}\right)$. This can be simplified

to $f'(x) = \dfrac{3x^2 - 3x^4}{\left(x^2 + 1\right)^4}$.

5. $\dfrac{1}{4}\left(\dfrac{2x - 5}{5x + 2}\right)^{-\frac{3}{4}}\left(\dfrac{29}{\left(5x + 2\right)^2}\right)$

Here, we will find the derivative using the Chain Rule. We will also need the Quotient Rule

to take the derivative of the expression inside the parentheses. The Chain Rule says that

if $y = f\left(g(x)\right)$, then $y' = \left(\dfrac{df\left(g(x)\right)}{dg}\right)\left(\dfrac{dg}{dx}\right)$, and the Quotient Rule says that if $f(x) = \dfrac{u}{v}$,

then $f'(x) = \dfrac{v\dfrac{du}{dx} - u\dfrac{dv}{dx}}{v^2}$. We get $f'(x) = \dfrac{1}{4}\left(\dfrac{2x - 5}{5x + 2}\right)^{-\frac{3}{4}}\left(\dfrac{(5x + 2)(2) - (2x - 5)(5)}{(5x + 2)^2}\right)$. This

can be simplified to $f'(x) = \dfrac{1}{4}\left(\dfrac{2x - 5}{5x + 2}\right)^{-\frac{3}{4}}\left(\dfrac{29}{\left(5x + 2\right)^2}\right)$.

6. $3x^2 + 1 + \dfrac{1}{x^2} + \dfrac{3}{x^4}$

We have two ways that we could solve this. We could expand the expression first and then take the derivative of each term, or we could find the derivative using the Product Rule. Let's do both methods just to see that they both give us the same answer. First, let's use the Product Rule, which says that if $f(x) = uv$, then $f'(x) = u\dfrac{dv}{dx} + v\dfrac{du}{dx}$. Here,

$f(x) = \left(x + \dfrac{1}{x}\right)\left(x^2 - \dfrac{1}{x^2}\right)$, so $u = \left(x + \dfrac{1}{x}\right)$ and $v = \left(x^2 - \dfrac{1}{x^2}\right)$. Using the Product Rule, we get

$f'(x) = \left(x + \dfrac{1}{x}\right)\left(2x - (-2)x^{-3}\right) + \left(x^2 - \dfrac{1}{x^2}\right)\left(1 - \dfrac{1}{x^2}\right) = \left(x + \dfrac{1}{x}\right)\left(2x + \dfrac{2}{x^3}\right) + \left(x^2 - \dfrac{1}{x^2}\right)\left(1 - \dfrac{1}{x^2}\right)$.

This can be simplified to $f'(x) = \left(2x^2 + \dfrac{2}{x^2} + 2 + \dfrac{2}{x^4}\right) + \left(x^2 - 1 - \dfrac{1}{x^2} + \dfrac{1}{x^4}\right) = 3x^2 + 1 + \dfrac{1}{x^2} + \dfrac{3}{x^4}$.

The other way we could find the derivative is to expand the expression first and then take the derivative.

We get $f(x) = \left(x + \dfrac{1}{x}\right)\left(x^2 - \dfrac{1}{x^2}\right) = x^3 - \dfrac{1}{x} + x - \dfrac{1}{x^3}$. Now we can take the derivative of each term.

We get $f'(x) = 3x^2 + \dfrac{1}{x^2} + 1 - (-3)x^{-4} = 3x^2 + \dfrac{1}{x^2} + 1 + \dfrac{3}{x^4}$. As we can see, the second method is a little quicker, and they both give the same result.

7. $$\frac{4x^3}{(x+1)^5}$$

Here, we will find the derivative using the Chain Rule. We will also need the Quotient Rule

to take the derivative of the expression inside the parentheses. The Chain Rule says that if

$y = f\big(g(x)\big)$, then $y' = \left(\dfrac{df\big(g(x)\big)}{dg}\right)\left(\dfrac{dg}{dx}\right)$, and the Quotient Rule says that if $f(x) = \dfrac{u}{v}$, then

$f'(x) = \dfrac{v\dfrac{du}{dx} - u\dfrac{dv}{dx}}{v^2}$. We get $f'(x) = 4\left(\dfrac{x}{x+1}\right)^3\left(\dfrac{(x+1)(1)-(x)(1)}{(x+1)^2}\right)$. This can be simplified

to $f'(x) = 4\left(\dfrac{x}{x+1}\right)^3\left(\dfrac{1}{(x+1)^2}\right) = \dfrac{4x^3}{(x+1)^5}$.

8. $$100\big(x^2 + x\big)^{99}\big(2x+1\big)$$

Here, we will find the derivative using the Chain Rule. The Chain Rule says that if $y = f\big(g(x)\big)$,

then $y' = \left(\dfrac{df\big(g(x)\big)}{dg}\right)\left(\dfrac{dg}{dx}\right)$. We get $f'(x) = 100\big(x^2 + x\big)^{99}\big(2x+1\big)$.

9. $$\frac{-2x}{\big(x^2+1\big)^{\frac{1}{2}}\big(x^2-1\big)^{\frac{3}{2}}}$$

Here, we will find the derivative using the Chain Rule. We will also need the Quotient Rule

to take the derivative of the expression inside the parentheses. The Chain Rule says that if

$y = f\big(g(x)\big)$, then $y' = \left(\dfrac{df\big(g(x)\big)}{dg}\right)\left(\dfrac{dg}{dx}\right)$, and the Quotient Rule says that if $f(x) = \dfrac{u}{v}$, then

$f'(x) = \dfrac{v\dfrac{du}{dx} - u\dfrac{dv}{dx}}{v^2}$. We get $f'(x) = \dfrac{1}{2}\left(\dfrac{x^2+1}{x^2-1}\right)^{-\frac{1}{2}}\left(\dfrac{(x^2-1)(2x)-(x^2+1)(2x)}{(x^2-1)^2}\right)$. This can be

simplified to $f'(x) = \dfrac{1}{2}\left(\dfrac{x^2+1}{x^2-1}\right)^{-\frac{1}{2}}\left(\dfrac{-4x}{(x-1)^2}\right) = \dfrac{-2x}{(x^2+1)^{\frac{1}{2}}(x^2-1)^{\frac{3}{2}}}$.

10. $\dfrac{9}{64}$

We find the derivative using the Quotient Rule, which says that if $f(x) = \dfrac{u}{v}$, then

$f'(x) = \dfrac{v\dfrac{du}{dx} - u\dfrac{dv}{dx}}{v^2}$. Here, $f(x) = \dfrac{(x+4)(x-8)}{(x+6)(x-6)}$, so $u = (x+4)(x-8)$ and $v = (x+6)(x-6)$.

Before we take the derivative, we can simplify the numerator and denominator of the

expression: $f(x) = \dfrac{(x+4)(x-8)}{(x+6)(x-6)} = \dfrac{x^2-4x-32}{x^2-36}$. Now using the Quotient Rule, we get

$f(x) = \dfrac{(x^2-36)(2x-4)-(x^2-4x-32)(2x)}{(x^2-36)^2}$. Next, we don't simplify. We simply plug in $x = 2$

to get $f(x) = \dfrac{((2)^2-36)(2(2)-4)-((2)^2-4(2)-32)(2(2))}{((2)^2-36)^2} = \dfrac{(-32)(0)-(-36)(4)}{(-32)^2} = \dfrac{9}{64}$.

11. 106

We find the derivative using the Quotient Rule, which says that if $f(x) = \dfrac{u}{v}$,

then $f'(x) = \dfrac{v\dfrac{du}{dx} - u\dfrac{dv}{dx}}{v^2}$. We will also need the Chain Rule to take the derivative of the

expression in the denominator. The Chain Rule says that if $y = f(g(x))$, then

$$y' = \left(\frac{df\left(g\left(x\right)\right)}{dg}\right)\left(\frac{dg}{dx}\right), \text{ here } f\left(x\right) = \frac{x^6 + 4x^3 + 6}{\left(x^4 - 2\right)^2}, \text{ so } u = x^6 + 4x^3 + 6 \text{ and } v = \left(x^4 - 2\right)^2. \text{ We}$$

get $f\left(x\right) = \dfrac{\left(x^4 - 2\right)^2 \left(6x^5 + 12x^2\right) - \left(x^6 + 4x^3 + 6\right)2\left(x^4 - 2\right)\left(4x^3\right)}{\left(x^4 - 2\right)^4}$. Now, we don't simplify.

We simply plug in $x = 1$ to get

$$f\left(x\right) = \frac{\left(\left(1\right)^4 - 2\right)^2 \left(6\left(1\right)^5 + 12\left(1\right)^2\right) - \left(\left(1\right)^6 + 4\left(1\right)^3 + 6\right)2\left(\left(1\right)^4 - 2\right)\left(4\left(1\right)^3\right)}{\left(\left(1\right)^4 - 2\right)^4} =$$

$$\frac{\left(1\right)\left(18\right) - \left(11\right)2\left(-1\right)\left(4\right)}{\left(-1\right)^4} = 106$$

12. $\dfrac{x^2 - 6x + 3}{\left(x - 3\right)^2}$

We find the derivative using the Quotient Rule, which says that if $f\left(x\right) = \dfrac{u}{v}$, then

$f'\left(x\right) = \dfrac{v\dfrac{du}{dx} - u\dfrac{dv}{dx}}{v^2}$. Here, $f\left(x\right) = \dfrac{x^2 - 3}{x - 3}$, so $u = x^2 - 3$ and $v = x - 3$. Using the Quotient

Rule, we get $f'\left(x\right) = \dfrac{\left(x - 3\right)\left(2x\right) - \left(x^2 - 3\right)\left(1\right)}{\left(x - 3\right)^2}$. This can be simplified to $f'\left(x\right) = \dfrac{x^2 - 6x + 3}{\left(x - 3\right)^2}$.

13. 6

We find the derivative using the Product Rule, which says that if

$f\left(x\right) = uv$, then $f'\left(x\right) = u\dfrac{dv}{dx} + v\dfrac{du}{dx}$. Here, $f\left(x\right) = \left(x^4 - x^2\right)\left(2x^3 + x\right)$, so $u = x^4 - x^2$ and

$v = 2x^3 + x$. Using the Product Rule, we get $f'\left(x\right) = \left(x^4 - x^2\right)\left(6x^2 + 1\right) + \left(2x^3 + x\right)\left(4x^3 - 2x\right)$.

Now we don't simplify. We simply plug in $x = 1$ to get

$f'\left(x\right) = \left(\left(1\right)^4 - \left(1\right)^2\right)\left(6\left(1\right)^2 + 1\right) + \left(2\left(1\right)^3 + \left(1\right)\right)\left(4\left(1\right)^3 - 2\left(1\right)\right) = 6$.

14. $-\dfrac{7}{4}$

We find the derivative using the Quotient Rule, which says that if $f(x)=\dfrac{u}{v}$,

then $f'(x)=\dfrac{v\dfrac{du}{dx}-u\dfrac{dv}{dx}}{v^2}$. Here, $f(x)=\dfrac{x^2+2x}{x^4-x^3}$, so $u=x^2+2x$ and $v=x^4-x^3$.

Using the Quotient Rule, we get $f'(x)=\dfrac{\left(x^4-x^3\right)\left(2x+2\right)-\left(x^2+2x\right)\left(4x^3-3x^2\right)}{\left(x^4-x^3\right)^2}$.

Now, we don't simplify. We simply plug in $x=2$ to get

$$f'(x)=\dfrac{\left((2)^4-(2)^3\right)\left(2(2)+2\right)-\left((2)^2+2(2)\right)\left(4(2)^3-3(2)^2\right)}{\left((2)^4-(2)^3\right)^2}=-\dfrac{7}{4}.$$

15. $\dfrac{2x^2+1}{\sqrt{x^2+1}}$

We find the derivative using the Chain Rule, which says that if $y=f\big(g(x)\big)$, then

$y'=\left(\dfrac{df\big(g(x)\big)}{dg}\right)\left(\dfrac{dg}{dx}\right)$. Here, $f(x)=\sqrt{x^4+x^2}=\left(x^4+x^2\right)^{\frac{1}{2}}$. Using the Chain Rule, we get

$f'(x)=\dfrac{1}{2}\left(x^4+x^2\right)^{-\frac{1}{2}}\left(4x^3+2x\right)$. This can be simplified to $f'(x)=\dfrac{2x^2+1}{\sqrt{x^2+1}}$.

16. $-\dfrac{2}{(x-1)^3}$

We find the derivative using the Chain Rule, which says that if $y=y(v)$ and $v=v(x)$, then

$\dfrac{dy}{dx}=\dfrac{dy}{dv}\dfrac{dv}{dx}$. Here, $\dfrac{dy}{du}=2u$ and $\dfrac{du}{dx}=-1(x-1)^{-2}=-\dfrac{1}{(x-1)^2}$. Thus, $\dfrac{dy}{dx}=(2u)\dfrac{-1}{(x-1)^2}$ and

because $u=\dfrac{1}{x-1}$, $\dfrac{dy}{dx}=\left(\dfrac{2}{x-1}\right)\left(\dfrac{-1}{(x-1)^2}\right)=-\dfrac{2}{(x-1)^3}$.

17.   −24

We find the derivative using the Chain Rule, which says that if $y = y(v)$ and $v = v(x)$, then

$\dfrac{dy}{dx} = \dfrac{dy}{dv}\dfrac{dv}{dx}$. Here, $\dfrac{dy}{dt} = \dfrac{(t^2 - 2)(2t) - (t^2 + 2)(2t)}{(t^2 - 2)^2}$ and $\dfrac{dt}{dx} = 3x^2$. Now, we plug $x = 1$ into the

derivative. Note that where $x = 1$, $t = (1)^3 = 1$. We get $\dfrac{dy}{dt} = \dfrac{\left((1)^2 - 2\right)(2(1)) - \left((1)^2 + 2\right)(2(1))}{\left((1)^2 - 2\right)^2} = -8$

and $\dfrac{dt}{dx} = 3(1)^2 = 3$. Thus, $\dfrac{dy}{dx} = \dfrac{dy}{dt}\dfrac{dt}{dx} = (-8)(3) = -24$.

18. $\left(t^3 - 6t^{\frac{5}{2}}\right)\left(5 + \dfrac{1}{2\sqrt{t}}\right) + \left(5t + \sqrt{t}\right)\left(3t^2 - 15t^{\frac{3}{2}}\right)$

Here, the solution will be much simpler if we first substitute $x = \sqrt{t}$ into the expression for y.

We get $y = \left(t^3 - 6t^{\frac{5}{2}}\right)\left(5t + \sqrt{t}\right)$. Now, we find the derivative using the Product Rule, which says

that if $f(x) = uv$, then $f'(x) = u\dfrac{dv}{dx} + v\dfrac{du}{dx}$. Here, $y = \left(t^3 - 6t^{\frac{5}{2}}\right)\left(5t + \sqrt{t}\right)$, so $u = t^3 - 6t^{\frac{5}{2}}$ and

$v = 5t + \sqrt{t}$. Using the Product Rule, we get $\dfrac{dy}{dt} = \left(t^3 - 6t^{\frac{5}{2}}\right)\left(5 + \dfrac{1}{2\sqrt{t}}\right) + \left(5t + \sqrt{t}\right)\left(3t^2 - 15t^{\frac{3}{2}}\right)$.

19.   2

We find the derivative using the Chain Rule, which says that if $y = y(v)$ and $v = v(x)$, then

$\dfrac{dy}{dx} = \dfrac{dy}{dv}\dfrac{dv}{dx}$. Here $\dfrac{dy}{du} = \dfrac{(1 + u^2)(1) - (1 + u)(2u)}{(1 + u^2)^2}$ and $\dfrac{du}{dx} = 2x$. Now, we plug $x = 1$ into the

derivative. Note that, where $x = 1$, $u = 0$. We get $\dfrac{dy}{du} = \dfrac{(1 + 0)(1) - (1 + 0)(0)}{(1 + 0)^2} = 1$ and $\dfrac{du}{dx} = 2$, so

$\dfrac{dy}{dx} = (1)(2) = 2$.

20. $\dfrac{48v^5}{\left(v^2+8\right)^4}$

We find the derivative using the Chain Rule, which says that if $y=y(v)$ and

$v=v(x)$, then $\dfrac{dy}{dx}=\dfrac{dy}{dv}\dfrac{dv}{dx}$, although in this case we have $u(y)$, $y(x)$, and $x(v)$,

so we will find $\dfrac{du}{dv}$ by $\dfrac{du}{dv}=\dfrac{du}{dy}\dfrac{dy}{dx}\dfrac{dx}{dv}$. Here, $\dfrac{du}{dy}=3y^2$, $\dfrac{dy}{dx}=\dfrac{(x+8)(1)-(x)(1)}{(x+8)^2}=\dfrac{8}{(x+8)^2}$,

and $\dfrac{dx}{dv}=2v$. Next, $\dfrac{du}{dv}=\dfrac{du}{dy}\dfrac{dy}{dx}\dfrac{dx}{dv}=\left(3y^2\right)\left(\dfrac{8}{(x+8)^2}\right)(2v)$. Now because $x=v^2$ and

$y=\dfrac{x}{x+8}=\dfrac{v^2}{v^2+8}$, we get $\dfrac{du}{dv}=\left(3\dfrac{\left(v^2\right)^2}{\left(v^2+8\right)^2}\right)\left(\dfrac{8}{\left(v^2+8\right)^2}\right)(2v)=\dfrac{48v^5}{\left(v^2+8\right)^4}$.

PRACTICE PROBLEM SET 6

1. $2\sin x\cos x$ or $\sin 2x$

Recall that $\dfrac{d}{dx}(\sin x)=\cos x$. Here, we use the Chain Rule to find the derivative:

$\dfrac{dy}{dx}=2(\sin x)(\cos x)$. If you recall your trigonometric identities, $2\sin x\cos x=\sin 2x$.

Either answer is acceptable.

2. $-2x\sin\left(x^2\right)$

Recall that $\dfrac{d}{dx}(\cos x)=-\sin x$. Here, we use the Chain Rule to find the derivative:

$\dfrac{dy}{dx}=\left(-\sin\left(x^2\right)\right)(2x)=-2x\sin\left(x^2\right)$.

3. $2\sec^3 x - \sec x$

Recall that $\dfrac{d}{dx}(\tan x) = \sec^2 x$ and that $\dfrac{d}{dx}(\sec x) = \sec x \tan x$. Using the Product Rule, we get

$\dfrac{dy}{dx} = (\tan x)(\sec x \tan x) + (\sec x)(\sec^2 x)$. Using the version of the Pythagorean identity, which

states that $\tan^2 x + 1 = \sec^2 x$, this can be simplified to $\sec^3 x + \sec x \tan^2 x = 2\sec^3 x - \sec x$.

4. $\dfrac{3\cos 3x}{2\sqrt{\sin 3x}}$

Recall that $\dfrac{d}{dx}(\sin x) = \cos x$. Here we use the Chain Rule to find the derivative:

$\dfrac{dy}{dx} = \dfrac{1}{2}(\sin 3x)^{-\frac{1}{2}}(\cos 3x)(3)$. This can be simplified to $\dfrac{dy}{dx} = \dfrac{3\cos 3x}{2\sqrt{\sin 3x}}$.

5. $\dfrac{2\cos x}{(1-\sin x)^2}$

Recall that $\dfrac{d}{dx}(\sin x) = \cos x$. Here, we use the Quotient Rule to find the derivative:

$\dfrac{dy}{dx} = \dfrac{(1-\sin x)(\cos x) - (1+\sin x)(-\cos x)}{(1-\sin x)^2}$. This can be simplified to $\dfrac{2\cos x}{(1-\sin x)^2}$.

6. $-4x\csc^2(x^2)\cot(x^2)$

Recall that $\dfrac{d}{dx}(\csc x) = -\csc x \cot x$. Here we use the Chain Rule to find the derivative:

$\dfrac{dy}{dx} = (2\csc x^2)(-\csc x^2 \cot x^2)(2x) = -4x\csc^2(x^2)\cot(x^2)$.

7. $16\sin 2x$

Recall that $\dfrac{d}{dx}(\sin x) = \cos x$ and that $\dfrac{d}{dx}(\cos x) = -\sin x$. Here we will use the Chain Rule four

times to find the fourth derivative.

The first derivative is $\dfrac{dy}{dx} = (\cos 2x)(2) = 2\cos 2x$.

The second derivative is $\dfrac{d^2 y}{dx^2} = 2(-\sin 2x)(2) = -4\sin 2x$.

The third derivative is $\dfrac{d^3 y}{dx^3} = -4(\cos 2x)(2) = -8\cos 2x$.

And the fourth derivative is $\dfrac{d^4 y}{dx^4} = -8(-\sin 2x)(2) = 16\sin 2x$.

8. $\left[\cos\left(1 + \cos^2 x\right) + \sin\left(1 + \cos^2 x\right)\right]\left(-2\sin x \cos x\right)$

Recall that $\dfrac{d}{dx}(\sin x) = \cos x$ and that $\dfrac{d}{dx}(\cos x) = -\sin x$. Here, we will use

the Chain Rule to find the derivative: $\dfrac{dy}{dt} = \cos t - (-\sin t) = \cos t + \sin t$ and

$\dfrac{dt}{dx} = 2(\cos x)(-\sin x) = -2\sin x \cos x$. Next, because $\dfrac{dy}{dx} = \dfrac{dy}{dt}\dfrac{dt}{dx}$ and $t = 1 + \cos^2 x$, we get

$\dfrac{dy}{dx} = (\cos t + \sin t)(-2\sin x \cos x) = \left[\cos\left(1 + \cos^2 x\right) + \sin\left(1 + \cos^2 x\right)\right]\left(-2\sin x \cos x\right)$.

9. $\dfrac{2\tan x \sec^2 x}{\left(1 - \tan x\right)^3}$

Recall that $\dfrac{d}{dx}(\tan x) = \sec^2 x$. Using the Quotient Rule and the Chain Rule, we get

$\dfrac{dy}{dx} = 2\left(\dfrac{\tan x}{1 - \tan x}\right)\dfrac{\left(1 - \tan x\right)\left(\sec^2 x\right) - \left(\tan x\right)\left(-\sec^2 x\right)}{\left(1 - \tan x\right)^2}$. This simplifies to $\dfrac{2\tan x \sec^2 x}{\left(1 - \tan x\right)^3}$.

10. $-\left[\sin\left(1 + \sin\theta\right)\right]\left(\cos\theta\right)$

Recall that $\dfrac{d}{dx}(\sin x) = \cos x$ and that $\dfrac{d}{dx}(\cos x) = -\sin x$. Using the Chain Rule, we get

$\dfrac{dr}{d\theta} = -\left[\sin\left(1 + \sin\theta\right)\right]\left(\cos\theta\right)$.

11. $\dfrac{\sec\theta\left(\tan\theta-1\right)}{\left(1+\tan\theta\right)^{2}}$

Recall that $\dfrac{d}{dx}\left(\tan x\right)=\sec^{2}x$ and that $\dfrac{d}{dx}\left(\sec x\right)=\sec x\tan x$. Using the Quotient Rule, we

get $\dfrac{dr}{d\theta}=\dfrac{\left(1+\tan\theta\right)\left(\sec\theta\tan\theta\right)-\left(\sec\theta\right)\left(\sec^{2}\theta\right)}{\left(1+\tan\theta\right)^{2}}$. This can be simplified (using trigonometric

identities) to $\dfrac{\sec\theta\left(\tan\theta-1\right)}{\left(1+\tan\theta\right)^{2}}$.

12. $-\dfrac{\left(\dfrac{4}{x^{2}}\right)\left(\csc^{2}\left(\dfrac{2}{x}\right)\right)}{\left(1+\cot\left(\dfrac{2}{x}\right)\right)^{3}}$

Recall that $\dfrac{d}{dx}\left(\cot x\right)=-\csc^{2}x$. Here we use the Chain Rule to find the derivative:

$$\dfrac{dy}{dx}=-2\left(1+\cot\left(\dfrac{2}{x}\right)\right)^{-3}\left(-\csc^{2}\left(\dfrac{2}{x}\right)\right)\left(\dfrac{-2}{x^{2}}\right)=-\dfrac{\left(\dfrac{4}{x^{2}}\right)\left(\csc^{2}\left(\dfrac{2}{x}\right)\right)}{\left(1+\cot\left(\dfrac{2}{x}\right)\right)^{3}}.$$

PRACTICE PROBLEM SET 7

1. $\dfrac{3x^{2}}{1+3y^{2}}$

We take the derivative of each term with respect to x: $\left(3x^{2}\right)\left(\dfrac{dx}{dx}\right)-\left(3y^{2}\right)\left(\dfrac{dy}{dx}\right)=\left(1\right)\left(\dfrac{dy}{dx}\right)$.

Next, because $\dfrac{dx}{dx}=1$, we can eliminate that term and get $\left(3x^{2}\right)-\left(3y^{2}\right)\left(\dfrac{dy}{dx}\right)=\left(\dfrac{dy}{dx}\right)$.

Next, group the terms containing $\dfrac{dy}{dx}$: $\left(3x^{2}\right)=\left(\dfrac{dy}{dx}\right)+\left(3y^{2}\right)\left(\dfrac{dy}{dx}\right)$.

Factor out the term $\dfrac{dy}{dx}$: $\left(3x^{2}\right)=\left(\dfrac{dy}{dx}\right)\left(1+3y^{2}\right)$. Now, we can isolate $\dfrac{dy}{dx}$: $\dfrac{dy}{dx}=\dfrac{3x^{2}}{1+3y^{2}}$.

2. $\dfrac{8y-x}{y-8x}$

We take the derivative of each term with respect to x:

$$(2x)\left(\dfrac{dx}{dx}\right)-16\left[(x)\left(\dfrac{dy}{dx}\right)+(y)\left(\dfrac{dx}{dx}\right)\right]+(2y)\left(\dfrac{dy}{dx}\right)=0.$$

Next, because $\dfrac{dx}{dx}=1$ we can eliminate that term and we can distribute the -16 to get

$$2x-16x\left(\dfrac{dy}{dx}\right)-16y+2y\left(\dfrac{dy}{dx}\right)=0.$$

Next, group the terms containing $\dfrac{dy}{dx}$ on one side of the equals sign and the other terms on the

other side: $-16x\left(\dfrac{dy}{dx}\right)+2y\left(\dfrac{dy}{dx}\right)=16y-2x$.

Factor out the term $\dfrac{dy}{dx}:\left(\dfrac{dy}{dx}\right)(2y-16x)=16y-2x$. Now we can isolate $\dfrac{dy}{dx}:\dfrac{dy}{dx}=\dfrac{16y-2x}{2y-16x}$,

which can be reduced to $\dfrac{dy}{dx}=\dfrac{8y-x}{y-8x}$.

3. $\dfrac{1}{2}$

First, cross-multiply so that we don't have to use the Quotient Rule: $x+y=3x-3y$. Next, simplify

$4y=2x$, which reduces to $y=\dfrac{1}{2}x$. Now, we can take the derivative: $\dfrac{dy}{dx}=\dfrac{1}{2}$. Note that just

because a problem has the x's and y's mixed together doesn't mean that we need to use implicit

differentiation to solve it!

4. $\dfrac{8}{7}$

We take the derivative of each term with respect to x:

$$(32x)\left(\dfrac{dx}{dx}\right)-16\left[(x)\left(\dfrac{dy}{dx}\right)+(y)\left(\dfrac{dx}{dx}\right)\right]+(2y)\left(\dfrac{dy}{dx}\right)=0.$$

Next, because $\dfrac{dx}{dx} = 1$ we can eliminate that term to get $32x - 16x\left(\dfrac{dy}{dx}\right) - 16y + 2y\left(\dfrac{dy}{dx}\right) = 0$.

Next, don't simplify. Plug in (1, 1) for x and y: $32(1) - 16(1)\left(\dfrac{dy}{dx}\right) - 16(1) + 2(1)\left(\dfrac{dy}{dx}\right) = 0$,

which simplifies to $16 - 14\left(\dfrac{dy}{dx}\right) = 0$.

Finally, we can solve for $\dfrac{dy}{dx}$: $\dfrac{dy}{dx} = \dfrac{8}{7}$.

5. -1

We   take   the   derivative   of   each   term   with   respect   to x:

$(x\cos y)\left(\dfrac{dy}{dx}\right) + (\sin y)\left(\dfrac{dx}{dx}\right) + (y\cos x)\left(\dfrac{dx}{dx}\right) + (\sin x)\left(\dfrac{dy}{dx}\right) = 0$.

Next,   because $\dfrac{dx}{dx} = 1$ we   can   eliminate   that   term   to   get

$(x\cos y)\left(\dfrac{dy}{dx}\right) + \sin y + y\cos x + (\sin x)\left(\dfrac{dy}{dx}\right) = 0$. Next, don't simplify. Plug in $\left(\dfrac{\pi}{4}, \dfrac{\pi}{4}\right)$

for x and y: $\left(\dfrac{\pi}{4}\cos\dfrac{\pi}{4}\right)\left(\dfrac{dy}{dx}\right) + \sin\dfrac{\pi}{4} + \dfrac{\pi}{4}\cos\dfrac{\pi}{4} + \left(\sin\dfrac{\pi}{4}\right)\left(\dfrac{dy}{dx}\right) = 0$, which simplifies to

$\left(\dfrac{\pi}{4}\dfrac{1}{\sqrt{2}}\right)\left(\dfrac{dy}{dx}\right) + \dfrac{1}{\sqrt{2}} + \dfrac{\pi}{4}\dfrac{1}{\sqrt{2}} + \dfrac{1}{\sqrt{2}}\left(\dfrac{dy}{dx}\right) = 0$. If we multiply through by $\sqrt{2}$, we get

$\dfrac{\pi}{4}\dfrac{dy}{dx} + 1 + \dfrac{\pi}{4} + \dfrac{dy}{dx} = 0$.

Finally, we can solve for $\dfrac{dy}{dx}$: $\dfrac{dy}{dx} = -1$.

6. $-\dfrac{1}{16y^3}$

We take the derivative of each term with respect to x: $(2x)\left(\dfrac{dx}{dx}\right) + (8y)\left(\dfrac{dy}{dx}\right) = 0$.

Next, because $\dfrac{dx}{dx} = 1$ we can eliminate that term to get $2x + (8y)\left(\dfrac{dy}{dx}\right) = 0$. Next, we can iso-

late $\dfrac{dy}{dx}$: $\dfrac{dy}{dx} = -\dfrac{x}{4y}$. Now, we take the derivative again: $\dfrac{d^2y}{dx^2} = -\dfrac{(4y)\left(\dfrac{dx}{dx}\right) - (x)\left(4\dfrac{dy}{dx}\right)}{16y^2}$. Next, because $\dfrac{dx}{dx} = 1$ and $\dfrac{dy}{dx} = -\dfrac{x}{4y}$, we get $\dfrac{d^2 y}{dx^2} = -\dfrac{4y - 4x\left(-\dfrac{x}{4y}\right)}{16 y^2}$. This can be simplified to

$$\dfrac{d^2 y}{dx^2} = -\dfrac{4y + \dfrac{x^2}{y}}{16 y^2} = -\dfrac{4 y^2 + x^2}{16 y^3} = -\dfrac{1}{16 y^3}.$$

7.　$\dfrac{\sin x \sin^2 y - \cos y \cos^2 x}{\sin^3 y}$

We take the derivative of each term with respect to x: $(\cos x)\left(\dfrac{dx}{dx}\right) = (-\sin y)\left(\dfrac{dy}{dx}\right)$.

Next, because $\dfrac{dx}{dx} = 1$, we can eliminate that term to get

$\cos x = (-\sin y)\left(\dfrac{dy}{dx}\right)$. Next, we can isolate $\dfrac{dy}{dx}$: $\dfrac{dy}{dx} = -\dfrac{\cos x}{\sin y}$. Now, we take the deriva-

tive again: $\dfrac{d^2 y}{dx^2} = \dfrac{(\sin y)(\sin x)\left(\dfrac{dx}{dx}\right) + (\cos x)(\cos y)\left(\dfrac{dy}{dx}\right)}{\sin^2 y}$. Next, because $\dfrac{dx}{dx} = 1$ and

$\dfrac{dy}{dx} = -\dfrac{\cos x}{\sin y}$, we get $\dfrac{d^2 y}{dx^2} = \dfrac{(\sin y)(\sin x) + (\cos x)(\cos y)\left(-\dfrac{\cos x}{\sin y}\right)}{\sin^2 y}$. This can be simplified

to $\dfrac{d^2 y}{dx^2} = \dfrac{\sin x \sin^2 y - \cos y \cos^2 x}{\sin^3 y}$.

8.　1

We can easily isolate y in this equation: $y = \dfrac{1}{2} x^2 - 2x + 1$. We take the derivative: $\dfrac{dy}{dx} = x - 2$. And

we take the derivative again: $\dfrac{d^2 y}{dx^2} = 1$. Note that just because a problem has the x's and y's mixed

together doesn't mean that we need to use implicit differentiation to solve it!

PRACTICE PROBLEM SET 8

1. $y - 2 = 5(x - 1)$

Remember that the equation of a line through a point (x_1, y_1) with slope m is $y - y_1 = m(x - x_1)$. We find the y-coordinate by plugging $x = 1$ into the equation $y = 3x^2 - x$, and we find the slope by plugging $x = 1$ into the derivative of the equation.

First, we find the y-coordinate, y_1: $y = 3(1)^2 - 1 = 2$. This means that the line passes through the point (1, 2).

Next, we take the derivative: $\dfrac{dy}{dx} = 6x - 1$. Now, we can find the slope, m: $\dfrac{dy}{dx}\Big|_{x=1} = 6(1) - 1 = 5$. Finally, we plug in the point $(1, 2)$ and the slope $m = 5$ to get the equation of the tangent line: $y - 2 = 5(x - 1)$.

2. $y - 18 = 24(x - 3)$

Remember that the equation of a line through a point (x_1, y_1) with slope m is $y - y_1 = m(x - x_1)$.

We find the y-coordinate by plugging $x = 3$ into the equation $y = x^3 - 3x$, and we find the slope by plugging $x = 3$ into the derivative of the equation.

First, we find the y-coordinate, y_1: $y = (3)^3 - 3(3) = 18$. This means that the line passes through the point $(3, 18)$.

Next, we take the derivative: $\dfrac{dy}{dx} = 3x^2 - 3$. Now, we can find the slope, m: $\dfrac{dy}{dx}\Big|_{x=3} = 3(3)^2 - 3 = 24$. Finally, we plug in the point $(3, 18)$ and the slope $m = 24$ to get the equation of the tangent line: $y - 18 = 24(x - 3)$.

3. $\quad y - \dfrac{1}{4} = -\dfrac{3}{64}(x-3)$

Remember that the equation of a line through a point (x_1, y_1) with slope m is $y - y_1 = m(x - x_1)$.

We find the y-coordinate by plugging $x = 3$ into the equation $y = \dfrac{1}{\sqrt{x^2 + 7}}$, and we find the slope by plugging $x = 3$ into the derivative of the equation.

First, we find the y-coordinate, y_1: $y = \dfrac{1}{\sqrt{(3)^2 + 7}} = \dfrac{1}{4}$. This means that the line passes through the point $\left(3, \dfrac{1}{4}\right)$.

Next, we take the derivative: $\dfrac{dy}{dx} = -\dfrac{1}{2}\left(x^2 + 7\right)^{-\frac{3}{2}}(2x) = -\dfrac{x}{\left(\sqrt{x^2 + 7}\right)^3}$. Now, we can find the

slope, m: $\left.\dfrac{dy}{dx}\right|_{x=3} = -\dfrac{3}{\left(\sqrt{(3)^2 + 7}\right)^3} = -\dfrac{3}{64}$. Finally, we plug in the point $\left(3, \dfrac{1}{4}\right)$ and the slope

$m = -\dfrac{3}{64}$ to get the equation of the tangent line: $y - \dfrac{1}{4} = -\dfrac{3}{64}(x-3)$.

4. $\quad y - 7 = \dfrac{1}{6}(x-4)$

Remember that the equation of a line through a point (x_1, y_1) with slope m is $y - y_1 = m(x - x_1)$.

We find the y-coordinate by plugging $x = 4$ into the equation $y = \dfrac{x+3}{x-3}$, and we find the slope by plugging $x = 4$ into the derivative of the equation.

First, we find the y-coordinate, y_1: $y = \dfrac{4+3}{4-3} = 7$. This means that the line passes through the point $(4, 7)$.

Next, we take the derivative: $\dfrac{dy}{dx} = \dfrac{(x-3)(1) - (x+3)(1)}{(x-3)^2} = -\dfrac{6}{(x-3)^2}$. Now, we can find the

slope, m: $\left.\dfrac{dy}{dx}\right|_{x=4} = -\dfrac{6}{(4-3)^2} = -6$. However, this is the slope of the *tangent* line. The *normal* line

is perpendicular to the tangent line, so its slope will be the negative reciprocal of the tangent line's slope. In this case, the slope of the normal line is $\dfrac{-1}{-6} = \dfrac{1}{6}$. Finally, we plug in the point $(4, 7)$ and

the slope $m = \dfrac{1}{6}$ to get the equation of the normal line: $y - 7 = \dfrac{1}{6}(x - 4)$.

5. $y = 0$

Remember that the equation of a line through a point (x_1, y_1) with slope m is $y - y_1 = m(x - x_1)$. We find the y-coordinate by plugging $x = 2$ into the equation $y = 2x^3 - 3x^2 - 12x + 20$, and we find the slope by plugging $x = 2$ into the derivative of the equation.

First, we find the y-coordinate, y_1: $y = 2(2)^3 - 3(2)^2 - 12(2) + 20 = 0$. This means that the line passes through the point $(2, 0)$.

Next, we take the derivative: $\dfrac{dy}{dx} = 6x^2 - 6x - 12$. Now, we can find the slope,

m: $\dfrac{dy}{dx}\bigg|_{x=2} = 6(2)^2 - 6(2) - 12 = 0$. Finally, we plug in the point $(2, 0)$ and the slope $m = 0$ to

get the equation of the tangent line: $y - 0 = 0(x - 2)$ or $y = 0$.

6. $y + 29 = -39(x - 5)$

Remember that the equation of a line through a point (x_1, y_1) with slope m is $y - y_1 = m(x - x_1)$.

We find the y-coordinate by plugging $x = 5$ into the equation $y = \dfrac{x^2 + 4}{x - 6}$, and we find the slope by

plugging $x = 5$ into the derivative of the equation.

First, we find the y-coordinate, y_1: $y = \dfrac{(5)^2 + 4}{(5) - 6} = -29$. This means that the line passes through

the point $(5, -29)$.

Next, we take the derivative: $\dfrac{dy}{dx} = \dfrac{(x - 6)(2x) - (x^2 + 4)(1)}{(x - 6)^2} = \dfrac{x^2 - 12x - 4}{(x - 6)^2}$. Now, we can find

the slope, m: $\dfrac{dy}{dx}\bigg|_{x=5} = \dfrac{(5)^2 - 12(5) - 4}{(5 - 6)^2} = -39$. Finally, we plug in the point $(5, -29)$ and the slope

$m = -39$ to get the equation of the tangent line: $y + 29 = -39(x - 5)$.

7. $$y - 7 = \frac{24}{7}(x - 4)$$

Remember that the equation of a line through a point (x_1, y_1) with slope m is $y - y_1 = m(x - x_1)$.

We find the slope by plugging $x = 4$ into the derivative of the equation $y = \sqrt{x^3 - 15}$. First,

we take the derivative: $\frac{dy}{dx} = \frac{1}{2}(x^3 - 15)^{-\frac{1}{2}}(3x^2) = \frac{3x^2}{2\sqrt{x^3 - 15}}$. Now, we can find the slope,

$m: \left.\frac{dy}{dx}\right|_{x=4} = \frac{3(4)^2}{2\sqrt{(4)^3 - 15}} = \frac{24}{7}$. Finally, we plug in the point $(4, 7)$ and the slope $m = \frac{24}{7}$ to get

the equation of the tangent line: $y - 7 = \frac{24}{7}(x - 4)$.

8. $$x = \pm\sqrt{\frac{3}{2}}$$

The slope of the line $y = x$ is 1, so we want to know where the slope of the tangent line is equal

to 1. We find the slope of the tangent line by taking the derivative: $\frac{dy}{dx} = 6x^2 - 8$. Now we set the

derivative equal to 1: $6x^2 - 8 = 1$. If we solve for x, we get $x = \pm\sqrt{\frac{3}{2}}$.

9. $$y - 7 = \frac{1}{2}(x - 3)$$

Remember that the equation of a line through a point (x_1, y_1) with slope m is $y - y_1 = m(x - x_1)$.

We find the y-coordinate by plugging $x = 3$ into the equation $y = \frac{3x + 5}{x - 1}$, and we find the slope by

plugging $x = 3$ into the derivative of the equation.

First, we find the y-coordinate, $y_1: y = \frac{3(3) + 5}{(3) - 1} = 7$. This means that the line passes through the

point $(3, 7)$.

Next, we take the derivative: $\frac{dy}{dx} = \frac{(x - 1)(3) - (3x + 5)(1)}{(x - 1)^2} = \frac{-8}{(x - 1)^2}$. Now, we can find the

slope, $m: \left.\frac{dy}{dx}\right|_{x=3} = \frac{-8}{(3 - 1)^2} = -2$. However, this is the slope of the *tangent* line. The *normal* line is perpendicular to the tangent line, so its slope will be the negative reciprocal of the tangent line's

slope. In this case, the slope of the normal line is $\dfrac{-1}{-2} = \dfrac{1}{2}$. Finally, we plug in the point $(3, 7)$ and

the slope $m = \dfrac{1}{2}$ to get the equation of the normal line: $y - 7 = \dfrac{1}{2}(x - 3)$.

10. $x = 9$

A line that is parallel to the y-axis has an infinite (or undefined) slope. In order to find where the

normal line has an infinite slope, we first take the derivative to find the slope of the tangent line:

$\dfrac{dy}{dx} = 2(x - 9)(1) = 2x - 18$. Next, because the normal line is perpendicular to the tangent line, the

slope of the normal line is the negative reciprocal of the slope of the tangent line: $m = \dfrac{-1}{2x - 18}$. Now,

we need to find where the slope is infinite. This is simply where the denominator of the slope is zero:

$x = 9$.

11. $\left(-\dfrac{3}{2}, \dfrac{41}{4} \right)$

A line that is parallel to the x-axis has a zero slope. In order to find where the tangent line has a

zero slope, we first take the derivative: $\dfrac{dy}{dx} = -3 - 2x$. Now we need to find where the slope is zero.

The derivative $-3 - 2x = 0$ at $x = -\dfrac{3}{2}$. Now we need to find the y-coordinate, which we get by

plugging $x = -\dfrac{3}{2}$ into the equation for y: $8 - 3\left(-\dfrac{3}{2} \right) - \left(-\dfrac{3}{2} \right)^2 = \dfrac{41}{4}$. Therefore, the answer is

$\left(-\dfrac{3}{2}, \dfrac{41}{4} \right)$.

12. $a = 1$, $b = 0$, and $c = 1$.

The two equations will have a common tangent line where they have the same slope, which we find

by taking the derivative of each equation. The derivative of the first equation is: $\dfrac{dy}{dx} = 2x + a$. The

derivative of the second equation is $\dfrac{dy}{dx} = c + 2x$. Setting the two derivatives equal to each other,

we get: $a = c$. Each equation will pass through the point $(-1, 0)$. If we plug $(-1, 0)$ into the first

equation, we get $0 = (-1)^2 + a(-1) + b$, which simplifies to $a - b = 1$. If we plug $(-1, 0)$ into the

second equation, we get $0 = c(-1) + (-1)^2$, which simplifies to $c = 1$. Now we can find the values

for a, b, and c. We get $a = 1$, $b = 0$, and $c = 1$.

PRACTICE PROBLEM SET 9

1. $c = 0$

The Mean Value Theorem says that: If $f(x)$ is continuous on the interval $[a, b]$ and is differentiable

everywhere on the interval (a, b), then there exists at least one number c on the interval (a, b) such

that $f'(c) = \dfrac{f(b) - f(a)}{b - a}$. Here, the function is $f(x) = 3x^2 + 5x - 2$ and the interval is $[-1, 1]$. Thus,

the Mean Value Theorem says that $f'(c) = \dfrac{\left(3(1)^2 + 5(1) - 2\right) - \left(3(-1)^2 + 5(-1) - 2\right)}{(1+1)}$. This simpli-

fies to $f'(c) = 5$. Next, we need to find $f'(c)$. The derivative of $f(x)$ is $f'(x) = 6x + 5$, so $f'(c) = 6c + 5$.

Now, we can solve for c: $6c + 5 = 5$ and $c = 0$. Note that 0 is in the interval $(-1, 1)$, just as we expected.

2. $c = \dfrac{4}{\sqrt{3}}$

The Mean Value Theorem says that: If $f(x)$ is continuous on the interval $[a, b]$ and is differentiable

everywhere on the interval (a, b), then there exists at least one number c on the interval (a, b) such that

$f'(c) = \dfrac{f(b) - f(a)}{b - a}$. Here, the function is $f(x) = x^3 + 24x - 16$ and the interval is $[0, 4]$. Thus, the

Mean Value Theorem says that $f'(c) = \dfrac{\left((4)^3 + 24(4) - 16\right) - \left((0)^3 + 24(0) - 16\right)}{(4 - 0)}$. This simplifies to

$f'(c) = 40$. Next, we need to find $f'(c)$ from the equation. The derivative of $f(x)$ is $f'(x) = 3x^2 + 24$, so

$f'(c) = 3c^2 + 24$. Now, we can solve for c: $3c^2 + 24 = 40$ and $c = \pm\dfrac{4}{\sqrt{3}}$. Note that $\dfrac{4}{\sqrt{3}}$ is in the

interval $(0, 4)$, but $-\dfrac{4}{\sqrt{3}}$ is *not* in the interval. Thus, the answer is only $c = \dfrac{4}{\sqrt{3}}$. It's *very* im-

portant to check that the answers you get for c fall in the given interval when doing Mean Value

Theorem problems.

3. $c = \sqrt{2}$

The Mean Value Theorem says that if $f(x)$ is continuous on the interval $[a, b]$ and is differentiable

everywhere on the interval (a, b), then there exists at least one number c on the interval (a, b) such

that $f'(c) = \dfrac{f(b) - f(a)}{b - a}$. Here, the function is $f(x) = \dfrac{6}{x} - 3$ and the interval is [1, 2]. Thus, the

Mean Value Theorem says that $f'(c) = \dfrac{\left(\dfrac{6}{2} - 3\right) - \left(\dfrac{6}{1} - 3\right)}{(2-1)}$. This simplifies to $f'(c) = -3$. Next, we

need to find $f'(c)$ from the equation. The derivative of $f(x)$ is $f'(x) = -\dfrac{6}{x^2}$, so $f'(c) = -\dfrac{6}{c^2}$. Now,

we can solve for c: $-\dfrac{6}{c^2} = -3$ and $c = \pm\sqrt{2}$. Note that $c = \sqrt{2}$ is in the interval (1, 2), but $-\sqrt{2}$ is

not in the interval. Thus, the answer is only $c = \sqrt{2}$. It's *very* important to check that the answers

you get for c fall in the given interval when doing Mean Value Theorem problems.

4. *No Solution.*

The Mean Value Theorem says that: If $f(x)$ is continuous on the interval $[a, b]$ and is differentiable
everywhere on the interval (a, b), then there exists at least one number c on the interval (a, b)
such that $f'(c) = \dfrac{f(b) - f(a)}{b - a}$. Here, the function is $f(x) = \dfrac{6}{x} - 3$ and the interval is [–1, 2].
Note that the function is *not* continuous on the interval. It has an essential discontinuity (vertical
asymptote) at $x = 0$. Thus, the Mean Value Theorem does not apply on the interval, and there is no
solution.

Suppose that we were to apply the theorem anyway. We would get $f'(c) = \dfrac{\left(\dfrac{6}{2} - 3\right) - \left(\dfrac{6}{-1} - 3\right)}{(2+1)}$.

This simplifies to $f'(c) = 3$. Next, we need to find $f'(c)$ from the equation. The derivative of $f(x)$

is $f'(x) = -\dfrac{6}{x^2}$, so $f'(c) = -\dfrac{6}{c^2}$. Now, we can solve for c: $-\dfrac{6}{c^2} = 3$. This has no real solution.

Therefore, remember that it's *very* important to check that the function is continuous and differentiable everywhere on the given interval (it does not have to be differentiable at the endpoints) when doing Mean Value Theorem problems. If it is not, then the theorem does not apply.

5. $c = 4$

Rolle's Theorem says that if $f(x)$ is continuous on the interval $[a, b]$ and is differentiable everywhere on the interval (a, b), and if $f(a) = f(b) = 0$, then there exists at least one number c on the interval (a, b) such that $f'(c) = 0$. Here, the function is $f(x) = x^2 - 8x + 12$ and the interval is $[2, 6]$. First, we check if the function is equal to zero at both of the endpoints: $f(6) = (6)^2 - 8(6) + 12 = 0$ and $f(2) = (2)^2 - 8(2) + 12 = 0$. Next, we take the derivative to find $f'(c)$: $f'(x) = 2x - 8$, so $f'(c) = 2c - 8$. Now, we can solve for c: $2c - 8 = 0$ and $c = 4$. Note that 4 is in the interval $(2, 6)$, just as we expected.

6. $c = \dfrac{1}{2}$

Rolle's Theorem says that if $f(x)$ is continuous on the interval $[a, b]$ and is differentiable everywhere on the interval (a, b), and if $f(a) = f(b) = 0$, then there exists at least one number c on the interval (a, b) such that $f'(c) = 0$. Here the function is $f(x) = x(1 - x)$ and the interval is $[0, 1]$. First, we check if the function is equal to zero at both of the endpoints: $f(0) = (0)(1 - 0) = 0$ and $f(1) = (1)(1 - 1) = 0$. Next, we take the derivative to find $f'(c)$: $f'(x) = 1 - 2x$, so $f'(c) = 1 - 2c$. Now, we can solve for c: $1 - 2c = 0$ and $c = \dfrac{1}{2}$. Note that $\dfrac{1}{2}$ is in the interval $(0, 1)$, just as we expected.

7.   *No Solution.*

Rolle's Theorem says that if $f(x)$ is continuous on the interval $[a, b]$ and is differentiable everywhere on the interval (a, b), and if $f(a) = f(b) = 0$, then there exists at least one number c on the interval (a, b) such that $f'(c) = 0$. Here, the function is $f(x) = 1 - \dfrac{1}{x^2}$, and the interval is $[-1, 1]$. Note that the function is *not* continuous on the interval. It has an essential discontinuity (vertical asymptote) at $x = 0$. Thus, Rolle's Theorem does not apply on the interval, and there is no solution.

Suppose we were to apply the theorem anyway. First, we check if the function is equal to zero at both of the endpoints: $f(1) = 1 - \dfrac{1}{(1)^2} = 0$ and $f(-1) = 1 - \dfrac{1}{(-1)^2} = 0$. Next, we take the derivative to find $f'(c)$: $f'(x) = \dfrac{2}{x^3}$, so $f'(c) = \dfrac{2}{c^3}$. This has no solution.

Therefore, remember that it's *very* important to check that the function is continuous and differentiable everywhere on the given interval (it does not have to be differentiable at the endpoints) when doing Rolle's Theorem problems. If it is not, then the theorem does not apply.

8.　　$c = \dfrac{1}{8}$

Rolle's Theorem says that if $f(x)$ is continuous on the interval $[a, b]$ and is differentiable everywhere on the interval (a, b), and if $f(a) = f(b) = 0$, then there exists at least one number c on the interval (a, b) such that $f'(c) = 0$. Here, the function is $f(x) = x^{\frac{2}{3}} - x^{\frac{1}{3}}$ and the interval is $[0, 1]$. First, we check if the function is equal to zero at both of the endpoints: $f(0) = (0)^{\frac{2}{3}} - (0)^{\frac{1}{3}} = 0$ and $f(1) = (1)^{\frac{2}{3}} - (1)^{\frac{1}{3}} = 0$.

Next, we take the derivative to find $f'(c)$ $f'(x) = \dfrac{2}{3}x^{-\frac{1}{3}} - \dfrac{1}{3}x^{-\frac{2}{3}} = \dfrac{2}{3\sqrt[3]{x}} - \dfrac{1}{3\sqrt[3]{x^2}}$, so

$f'(c) = \dfrac{2}{3\sqrt[3]{c}} - \dfrac{1}{3\sqrt[3]{c^2}}$. Now, we can solve for c: $\dfrac{2}{3\sqrt[3]{c}} - \dfrac{1}{3\sqrt[3]{c^2}} = 0$ and $c = \dfrac{1}{8}$. Note that $\dfrac{1}{8}$ is in

the interval $(0, 1)$, just as we expected.

PRACTICE PROBLEM SET 10

1.　　The area is 32.

First, let's draw a picture.

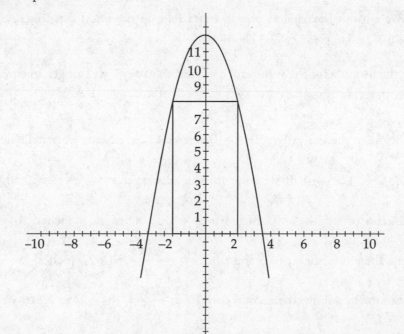

The rectangle can be expressed as a function of x, where the height is $12 - x^2$ and the base is $2x$. The area is $A = 2x(12 - x^2) = 24x - 2x^3$. Now, we take the derivative: $\frac{dA}{dx} = 24 - 6x^2$. Next, we set the derivative equal to zero: $24 - 6x^2 = 0$. If we solve this for x, we get $x = \pm 2$. A negative answer doesn't make any sense in this case, so we use the solution $x = 2$. We can then find the area by plugging in $x = 2$ to get $A = 24(2) - 2(2)^3 = 32$. We can verify that this is a maximum by taking the second derivative: $\frac{d^2A}{dx^2} = -12x$. Next, we plug in $x = 2$: $\frac{d^2A}{dx^2} = -12(2) = -24$. Because the value of the second derivative is negative, according to the second derivative test (see page 183), the area is a maximum at $x = 2$.

2. $x = \dfrac{7 - \sqrt{13}}{2} \approx 1.697$ inches

First, let's draw a picture.

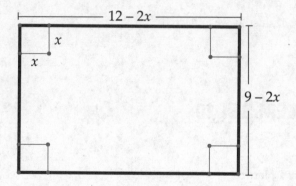

After we cut out the squares of side x and fold up the sides, the dimensions of the box will be: width: $9 - 2x$; length: $12 - 2x$; depth: x.

Using the formula for the volume of a rectangular prism, we can get an equation for the volume of the box in terms of x: $V = x(9 - 2x)(12 - 2x) = 108x - 42x^2 + 4x^3$.

Now, we take the derivative: $\frac{dV}{dx} = 108 - 84x + 12x^2$. Next, we set the derivative equal to zero: $108 - 84x + 12x^2 = 0$. If we solve this for x, we get $x = \dfrac{7 \pm \sqrt{13}}{2} \approx 5.303, 1.697$. We can't cut two squares of length 5.303 inches from a side of length 9 inches, so we can get rid of that answer. Therefore, the answer must be $x = \dfrac{7 - \sqrt{13}}{2} \approx 1.697$ inches. We can verify that this is a maximum by taking the second derivative: $\frac{d^2V}{dx^2} = -84 + 24x$. Next, we plug in $x = 1.697$ to get approximately $\frac{d^2V}{dx^2} = -84 + 24(1.697) = -43.272$. Because the value of the second derivative is negative, according to the second derivative test (see page 183), the volume is a maximum at

$$x = \frac{7 - \sqrt{13}}{2} \approx 1.697 \text{ inches.}$$

3.   16 meters by 24 meters

First, let's draw a picture.

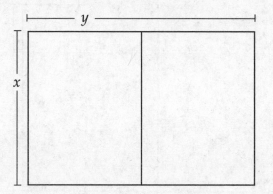

If we call the length of the plot y and the width x, the area of the plot is $A = xy = 384$. The perimeter is $P = 3x + 2y$. So, if we want to minimize the length of the fence, we need to minimize the perimeter of the plot. If we solve the area equation for y, we get $y = \dfrac{384}{x}$. Now we can substitute this for y in the perimeter equation: $P = 3x + 2\left(\dfrac{384}{x}\right) = 3x + \dfrac{768}{x}$. Now we take the derivative of P: $\dfrac{dP}{dx} = 3 - \dfrac{768}{x^2}$. If we solve this for x, we get $x = \pm16$. A negative answer doesn't make any sense in this case, so we use the solution $x = 16$ meters. Now we can solve for y: $y = \dfrac{384}{16} = 24$ meters.

We can verify that this is a minimum by taking the second derivative: $\dfrac{d^2P}{dx^2} = \dfrac{1536}{x^3}$. Next, we plug in $x = 16$ to get $\dfrac{d^2P}{dx^2} = \dfrac{1536}{16^3} = \dfrac{3}{8}$. Because the value of the second derivative is positive, according to the second derivative test (see page 183), the perimeter is a minimum at $x = 16$.

4.   Radius is $\sqrt[3]{\dfrac{256}{\pi}}$ inches.

First, let's draw a picture.

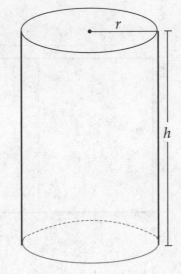

The volume of a cylinder is $V = \pi r^2 h = 512$. The material used for the can is the surface area of the cylinder (don't forget the ends!) $S = 2\pi rh + 2\pi r^2$. If we solve the volume equation for h, we get $h = \dfrac{512}{\pi r^2}$.

Now we can substitute this for h in the surface area equation: $S = 2\pi r\left(\dfrac{512}{\pi r^2}\right) + 2\pi r^2 = \dfrac{1024}{r} + 2\pi r^2$.

Now we take the derivative of S: $\dfrac{dS}{dr} = -\dfrac{1024}{r^2} + 4\pi r$. If we solve this for r, we get $r = \sqrt[3]{\dfrac{256}{\pi}}$

inches. We can verify that this is a minimum by taking the second derivative: $\dfrac{d^2S}{dx^2} = \dfrac{2048}{r^3} + 4\pi$.

Next, we plug in $\sqrt[3]{\dfrac{256}{\pi}}$ and we can see that the value of the second derivative is positive. Therefore,

according to the second derivative test (see page 183), the perimeter is a minimum at $r = \sqrt[3]{\dfrac{256}{\pi}}$.

5.   1,352.786 meters

Let's think about the situation. If the swimmer swims the whole distance to the cottage, she will be traveling the entire time at her slowest speed. If she swims straight to shore first, minimizing her swimming distance, she will be maximizing her running distance. Therefore, there should be a point, somewhere between the cottage and the point on the shore directly opposite her, where the swimmer should come on land to switch from swimming to running to get to the cottage in the shortest time.

Let's draw a picture.

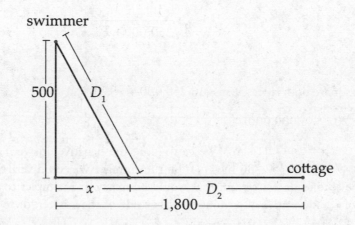

Let x be the distance from the point on the shore directly opposite the swimmer to the point where she comes on land. We have two distances to consider. The first is the diagonal distance that the swimmer swims. This distance, which we'll call D_1 is $D_1 = \sqrt{500^2 + x^2} = \sqrt{250{,}000 + x^2}$.

The second distance, which we'll call D_2, is simply $D_2 = 1{,}800 - x$. Remember that *rate × time = distance*? We'll use this formula to find the total time that the swimmer needs.

The time for the swimmer to travel D_1 is $T_1 = \dfrac{D_1}{4} = \dfrac{\sqrt{250{,}000 + x^2}}{4}$ (because she swims at 4 m/s) and the time for the swimmer to travel D_2 is $T_2 = \dfrac{D_2}{6} = \dfrac{1{,}800 - x}{6} = 300 - \dfrac{x}{6}$.

Therefore, the total time is $T = \dfrac{\sqrt{250{,}000 + x^2}}{4} + 300 - \dfrac{x}{6}$. Now we simply take the derivative:

$$\frac{dT}{dx} = \frac{1}{4}\left(\frac{1}{2}\right)\left(250{,}000 + x^2\right)^{-\frac{1}{2}}(2x) - \frac{1}{6} = \frac{x}{4\sqrt{250{,}000 + x^2}} - \frac{1}{6}.$$ Next, we set this equal to zero and

solve: $\dfrac{x}{4\sqrt{250{,}000 + x^2}} - \dfrac{1}{6} = 0.$ The best way to solve this is to move the $\dfrac{1}{6}$ to the other side of

the equals sign and cross-multiply:

$$\frac{x}{4\sqrt{250{,}000 + x^2}} = \frac{1}{6}$$

$$6x = 4\sqrt{250{,}000 + x^2}$$

Next, we square both sides: $36x^2 = 16(250{,}000 + x^2)$.

Simplify: $20x^2 = 4{,}000{,}000$.

Solve for x: $x = \pm 447.214$ m. (We can ignore the negative answer.) Therefore, she should land $1{,}800 - 447.214 = 1{,}352.786$ meters from the cottage. We could verify that this is a minimum by taking the second derivative but that would be messy. It is simpler to use the calculator to check the sign of the derivative at a point on either side of the answer, or to graph the equation for the time.

6. $\left(\dfrac{2}{\sqrt{5}}, \dfrac{1}{\sqrt{5}}\right)$

First, let's draw a picture.

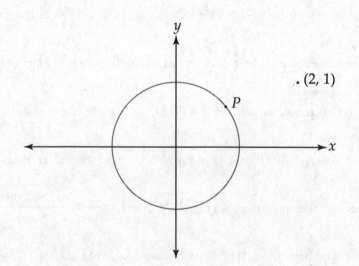

We need to find an expression for the distance from the point P to the point $(2, 1)$ and then minimize the distance. If we call the coordinates of P (x, y), then we can find the distance to $(2, 1)$ using the distance formula: $D^2 = (x - 2)^2 + (y - 1)^2$. Next, just as we did in sample problem 4 (page 185), we can let $L = D^2$ and minimize L: $L = (x - 2)^2 + (y - 1)^2 = x^2 - 4x + 4 + y^2 - 2y + 1$.

Because $x^2 + y^2 = 1$, we can substitute for y to get: $L = x^2 - 4x + 4 + (1 - x^2) - 2\sqrt{1 - x^2} + 1$, which simplifies to $L = -4x + 6 - 2\sqrt{1 - x^2}$. Next, we take the derivative:

$\dfrac{dL}{dx} = -4 - 2\left(\dfrac{1}{2}\right)\left(1 - x^2\right)^{-\frac{1}{2}}(-2x) = -4 + \dfrac{2x}{\sqrt{1 - x^2}}$. Next, we set the derivative equal to zero:

$-4 + \dfrac{2x}{\sqrt{1 - x^2}} = 0$. The best way to solve this is to move the 4 to the other side of the equals sign and cross-multiply.

$$\frac{2x}{\sqrt{1 - x^2}} = 4$$

$$2x = 4\sqrt{1 - x^2}$$

Next, we can simplify and square both sides: $x^2 = 4(1 - x^2)$. Now, we can solve this easily. We get $x = \pm\dfrac{2}{\sqrt{5}}$. Next, we find the y-coordinate: $y = \pm\dfrac{1}{\sqrt{5}}$. There are thus four possible answers but, if we look at the picture, the answer is obviously the point $\left(\dfrac{2}{\sqrt{5}}, \dfrac{1}{\sqrt{5}}\right)$. We could verify that this is a minimum by taking the second derivative, but that will be messy. It is simpler to use the calculator to check the sign of the derivative at a point on either side of the answer, or to graph the equation for the distance.

7. $r = \dfrac{288}{4 + \pi}$ inches

First, let's draw a picture.

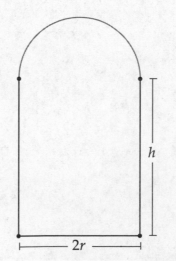

Call the width of the window $2r$. Notice that this is the diameter of the semicircle. Call the height of the window h. The area of the rectangular portion of the window is $2rh$ and the perimeter is $2r + 2h$. The area of the semicircular portion of the window is $\dfrac{\pi r^2}{2}$ and the perimeter is $\dfrac{2\pi r}{2} = \pi r$. Therefore, the area of the window is $A = 2rh + \dfrac{\pi r^2}{2}$, and the perimeter is $2r + 2h + \pi r = 288$. We can use the equation for the perimeter to eliminate a variable from the equation for the area. Let's isolate h: $h = 144 - r - \dfrac{\pi r}{2}$. Now we can substitute for h in the equation for the area: $A = 2r\left(144 - r - \dfrac{\pi r}{2}\right) + \dfrac{\pi r^2}{2} = 288r - 2r^2 - \dfrac{\pi r^2}{2}$. Next, we can take the derivative: $\dfrac{dA}{dr} = 288 - 4r - \pi r$. If we set this equal to zero and solve for r, we get $r = \dfrac{288}{4 + \pi}$ inches. We can verify that this is a maximum by taking the second derivative: $\dfrac{d^2 A}{dr^2} = -4 - \pi$. We can see that the value of the second derivative is negative. Therefore, according to the second derivative test (see page 183), the area is a maximum at $r = \dfrac{288}{4 + \pi}$ inches.

8. $\theta = \dfrac{\pi}{4}$ radians (or 45 degrees)

We simply take the derivative: $\dfrac{dR}{d\theta} = \dfrac{v_0^{\,2}}{g}\left(2\cos 2\theta\right)$. Note that v_0 and g are constants, so the only variable we need to take the derivative with respect to is θ. Now we set the derivative equal to zero: $\dfrac{v_0^{\,2}}{g}\left(2\cos 2\theta\right) = 0$. Although this has an infinite number of solutions, we are interested only in values of θ between 0 and $\dfrac{\pi}{2}$ radians. The value of θ that makes the derivative zero is $\theta = \dfrac{\pi}{4}$ radians (or 45 degrees). We can verify that this is a maximum by taking the second derivative: $\dfrac{d^2 R}{d\theta^2} = \dfrac{v_0^{\,2}}{g}\left(-4\sin 2\theta\right)$. We can see that the value of the second derivative is negative at $\theta = \dfrac{\pi}{4}$. Therefore, according to the second derivative test (see page 183), the range is a maximum at $\theta = \dfrac{\pi}{4}$ radians.

PRACTICE PROBLEM SET 11

1.   Minimum at $\left(\sqrt{3}, -6\sqrt{3} - 6\right)$; Maximum at $\left(-\sqrt{3}, 6\sqrt{3} - 6\right)$; Point of inflection at $(0, -6)$.

First, let's find the y-intercept. We set $x = 0$ to get: $y = (0)^3 - 9(0) - 6 = -6$. Therefore, the y-intercept is $(0, -6)$. Next, we find any critical points using the first derivative. The derivative is $\dfrac{dy}{dx} = 3x^2 - 9$. If we set this equal to zero and solve for x, we get $x = \pm\sqrt{3}$. Plug $x = \sqrt{3}$ and $x = -\sqrt{3}$ into the original equation to find the y-coordinates of the critical points: when $x = \sqrt{3}$, $y = \left(\sqrt{3}\right)^3 - 9\left(\sqrt{3}\right) - 6 = -6\sqrt{3} - 6$.   When $x = -\sqrt{3}$, $y = \left(-\sqrt{3}\right)^3 - 9\left(-\sqrt{3}\right) - 6 = 6\sqrt{3} - 6$. Thus, we have critical points at $\left(\sqrt{3}, -6\sqrt{3} - 6\right)$ and $\left(-\sqrt{3}, 6\sqrt{3} - 6\right)$. Next, we take the second derivative to find any points of inflection. The second derivative is $\dfrac{d^2 y}{dx^2} = 6x$, which is equal to zero at $x = 0$. Note that this is the y-intercept $(0, -6)$, which we already found, so there is a point of inflection at $(0, -6)$. Next, we need to determine if each critical point is maximum, minimum, or something else. If we plug $x = \sqrt{3}$ into the second derivative, the value is obviously positive, so $\left(\sqrt{3}, -6\sqrt{3} - 6\right)$ is a minimum. If we plug $x = -\sqrt{3}$ into the second derivative, the value is obviously negative, so $\left(-\sqrt{3}, 6\sqrt{3} - 6\right)$ is a maximum. Now, we can draw the curve. It looks like the following:

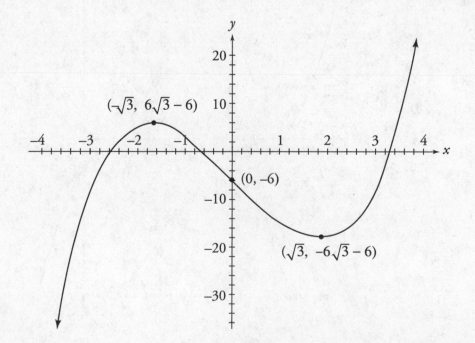

2.   Minimum at (–3, –4); Maximum at (–1, 0); Point of inflection at (–2, –2).

First, let's find the *y*-intercept. We set *x* = 0 to get: $y = -(0)^3 - 6(0)^2 - 9(0) - 4 = -4$. Therefore, the *y*-intercept is (0, –4). Next, we find any critical points using the first derivative. The derivative is $\frac{dy}{dx} = -3x^2 - 12x - 9$. If we set this equal to zero and solve for *x*, we get *x* = –1 and *x* = –3. Plug *x* = –1 and *x* = –3 into the original equation to find the *y*-coordinates of the critical points: when *x* = –1, $y = -(-1)^3 - 6(-1)^2 - 9(-1) - 4 = 0$. When *x* = –3, $y = -(-3)^3 - 6(-3)^2 - 9(-3) - 4 = -4$. Thus, we have critical points at (–1, 0) and (–3, –4). Next, we take the second derivative to find any points of inflection. The second derivative is $\frac{d^2 y}{dx^2} = -6x - 12$, which is equal to zero at *x* = –2. Plug *x* = –2 into the original equation to find the *y*-coordinate: $y = -(-2)^3 - 6(-2)^2 - 9(-2) - 4 = -2$, so there is a point of inflection at (–2, –2). Next, we need to determine if each critical point is a maximum, a minimum, or something else. If we plug *x* = –1 into the second derivative, the value is negative, so (–1, 0) is a maximum. If we plug *x* = –3 into the second derivative, the value is positive, so (–3, –4) is a minimum. Now, we can draw the curve. It looks like the following:

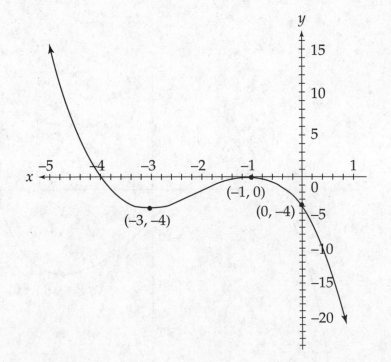

3.   Minimum at $(0, -36)$; Maxima at $\left(\sqrt{\dfrac{13}{2}}, \dfrac{25}{4}\right)$ and $\left(-\sqrt{\dfrac{13}{2}}, \dfrac{25}{4}\right)$; points of inflection at

$\left(\sqrt{\dfrac{13}{6}}, -\dfrac{451}{36}\right)$ and $\left(-\sqrt{\dfrac{13}{6}}, -\dfrac{451}{36}\right)$.

First, we can easily see that the graph has x-intercepts at $x = \pm 2$ and $x = \pm 3$. Next, before we

take the derivative, let's multiply the two terms so we don't have to use the Product Rule. We get

$y = -x^4 + 13x^2 - 36$. Now we can take the derivative: $\dfrac{dy}{dx} = -4x^3 + 26x$. Next, we set the derivative

equal to zero to find the critical points. There are three solutions: $x = 0$, $x = \sqrt{\dfrac{13}{2}}$, and $x = -\sqrt{\dfrac{13}{2}}$.

We plug these values into the original equation to find the y-coordinates of the critical points: when

$x = 0$, $y = -(0)^4 + 13(0)^2 - 36 = -36$. When $x = \sqrt{\dfrac{13}{2}}$, $y = -\left(\sqrt{\dfrac{13}{2}}\right)^4 + 13\left(\sqrt{\dfrac{13}{2}}\right)^2 - 36 = \dfrac{25}{4}$.

When $x = -\sqrt{\dfrac{13}{2}}$, $y = -\left(-\sqrt{\dfrac{13}{2}}\right)^4 + 13\left(-\sqrt{\dfrac{13}{2}}\right)^2 - 36 = \dfrac{25}{4}$.

Thus, we have critical points at $(0, -36)$, $\left(\sqrt{\dfrac{13}{2}}, \dfrac{25}{4}\right)$ and $\left(-\sqrt{\dfrac{13}{2}}, \dfrac{25}{4}\right)$.

Next, we take the second derivative to find any points of inflection. The second derivative is

$\dfrac{d^2 y}{dx^2} = -12x^2 + 26$, which is equal to zero at $x = \sqrt{\dfrac{13}{6}}$ and $x = -\sqrt{\dfrac{13}{6}}$. We plug these values into

the original equation to find the y-coordinates.

When $x = \sqrt{\dfrac{13}{6}}$, then $y = -\left(\sqrt{\dfrac{13}{6}}\right)^4 + 13\left(\sqrt{\dfrac{13}{6}}\right)^2 - 36 = -\dfrac{451}{36}$.

When $x = -\sqrt{\dfrac{13}{6}}$, then $y = -\left(-\sqrt{\dfrac{13}{6}}\right)^4 + 13\left(-\sqrt{\dfrac{13}{6}}\right)^2 - 36 = -\dfrac{451}{36}$. So there are points of

inflection at $\left(\sqrt{\dfrac{13}{6}}, -\dfrac{451}{36}\right)$ and $\left(-\sqrt{\dfrac{13}{6}}, -\dfrac{451}{36}\right)$. Next, we need to determine if each critical

point is a maximum, a minimum, or something else. If we plug $x = 0$ into the second derivative,

the value is positive, so $(0, -36)$ is a minimum. If we plug $x = \sqrt{\dfrac{13}{2}}$ into the second derivative, the

value is negative, so $\left(\sqrt{\dfrac{13}{2}}, \dfrac{25}{4}\right)$ is a maximum. If we plug $x = -\sqrt{\dfrac{13}{2}}$ into the second derivative,

the value is negative, so $\left(-\sqrt{\dfrac{13}{2}}, \dfrac{25}{4}\right)$ is a maximum. Now, we can draw the curve. It looks like

the following:

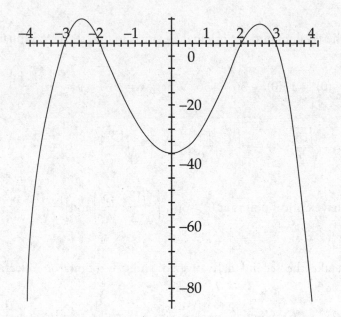

4.   Vertical asymptote at $x = -8$; Horizontal asymptote at $y = 1$; No maxima, minima, or points of inflection.

First, notice that there is an x-intercept at $x = 3$ and that the y-intercept is $\left(0, -\dfrac{3}{8}\right)$. Next, we take

the derivative: $\dfrac{dy}{dx} = \dfrac{(x+8)(1) - (x-3)(1)}{(x+8)^2} = \dfrac{11}{(x+8)^2}$. Next, we set the derivative equal to zero to

find the critical points. There is no solution. Next, we take the second derivative: $\dfrac{d^2 y}{dx^2} = -\dfrac{22}{(x+8)^3}$.

If we set this equal to zero, there is also no solution. Therefore, there are no maxima, minima, or

points of inflection. Note that the first derivative is always positive. This means that the curve is always increasing. Also notice that the second derivative changes sign from positive to negative at

$x = -8$. This means that the curve is concave up for values of x less than $x = -8$ and concave down

for values of x greater than $x = -8$.

Now, we can draw the curve. It looks like the following:

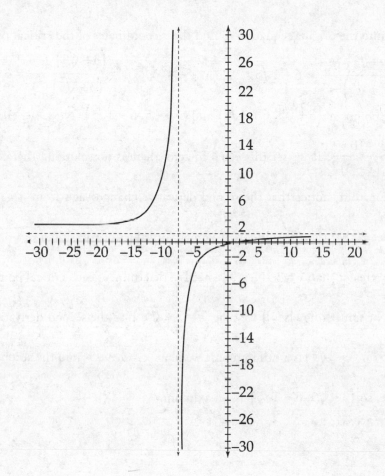

5.   Vertical asymptote at $x = 3$; Oblique asymptote of $y = x + 3$; Maximum at $\left(3+\sqrt{5}, 6+2\sqrt{5}\right)$; Minimum at $\left(3-\sqrt{5}, 6-2\sqrt{5}\right)$; No points of inflection.

First, notice that there are x-intercepts at $x = \pm 2$ and that the y-intercept is $\left(0, \dfrac{4}{3}\right)$. There is a

vertical asymptote at $x = 3$. There is no horizontal asymptote, but notice that the degree of the

numerator of the function is 1 greater than the denominator. This means that there is an oblique

(slant) asymptote. We find this by dividing the denominator into the numerator and looking at the

quotient. We get

$$x - 3\overline{)}\;x + 3 + \dfrac{5}{x-3} \atop x^2 + 0x - 4$$

This means that as $x \to \pm\infty$, the function will behave like the function $y = x + 3$. This means that there is an oblique asymptote of $y = x + 3$. Next, we take the derivative: $\dfrac{dy}{dx} = \dfrac{(x-3)(2x) - (x^2 - 4)(1)}{(x-3)^2} = \dfrac{x^2 - 6x + 4}{(x-3)^2} = 1 - \dfrac{5}{(x-3)^2}$. Next, we set the derivative equal to zero to find the critical points. There are two solutions: $x = 3 + \sqrt{5}$ and $x = 3 - \sqrt{5}$. We plug these values into the original equation to find the y-coordinates of the critical points: when $x = 3 + \sqrt{5}$, $y = \dfrac{\left(3 + \sqrt{5}\right)^2 - 4}{\left(3 + \sqrt{5}\right) - 3} = 6 + 2\sqrt{5}$. When $x = 3 - \sqrt{5}$, $y = \dfrac{\left(3 - \sqrt{5}\right)^2 - 4}{\left(3 - \sqrt{5}\right) - 3} = 6 - 2\sqrt{5}$. Thus, we have critical points at $\left(3 + \sqrt{5}, 6 + 2\sqrt{5}\right)$ and $\left(3 - \sqrt{5}, 6 - 2\sqrt{5}\right)$. Next, we take the second derivative: $\dfrac{d^2 y}{dx^2} = \dfrac{10}{(x-3)^3}$. If we set this equal to zero, there is no solution. Therefore, there is no point of inflection. But, notice that the second derivative changes sign from negative to positive at $x = 3$.

This means that the curve is concave down for values of x less than $x = 3$ and concave up for values of x greater than $x = 3$. Next, we need to determine if each critical point is a maximum, a minimum, or something else. If we plug $x = 3 + \sqrt{5}$ into the second derivative, the value is positive, so $\left(3 + \sqrt{5}, 6 + 2\sqrt{5}\right)$ is a minimum. If we plug $x = 3 - \sqrt{5}$ into the second derivative, the value is positive, so $\left(3 - \sqrt{5}, 6 - 2\sqrt{5}\right)$ is a maximum.

Now, we can draw the curve. It looks like the following:

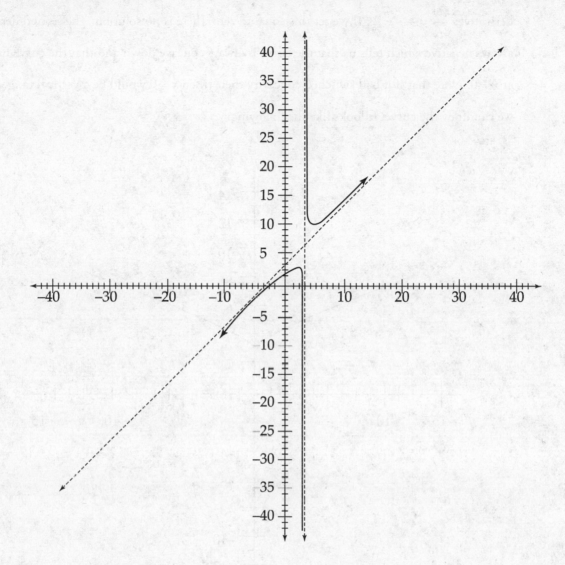

6. No maxima, minima, or points of inflection; Cusp at (0, 3).

First, notice that the function is always positive. There is a *y*-intercept at (0, 3). There is no

x-intercept. Next, we take the derivative: $\dfrac{dy}{dx} = \dfrac{2}{3} x^{-\frac{1}{3}}$. If we set the derivative equal to zero, there

is no solution. But, notice that the derivative is not defined at *x* = 0. This means that the func-

tion has either a vertical tangent or a cusp at *x* = 0. We'll be able to determine which after we take

the second derivative. Notice also that the derivative is negative for *x* < 0 and positive for *x* > 0.

Therefore, the curve is decreasing for $x < 0$ and increasing for $x > 0$. Next, we take the second derivative: $\dfrac{dy}{dx} = -\dfrac{2}{9}x^{-\frac{4}{3}}$. If we set this equal to zero, there is no solution. The second derivative is always negative which tells us that the curve is always concave down and that the curve has a cusp at $x = 0$. Note that if it had switched concavity there, then $x = 0$ would be a vertical tangent. Now, we can draw the curve. It looks like the following:

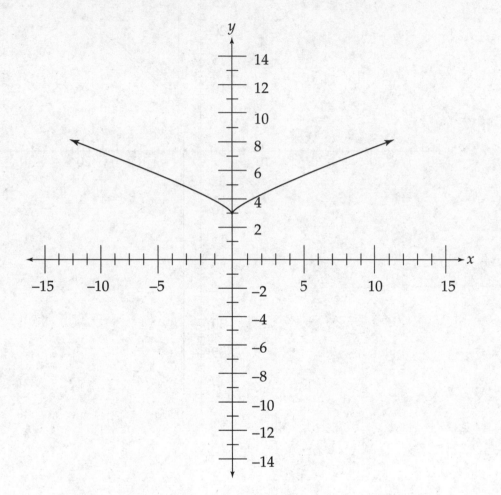

7.   Maximum at (1, 1); No point of inflection; Cusp at (0, 0).

First, we notice that the curve has x-intercepts at $x = 0$ and $x = \dfrac{27}{8}$ and that there is a y-intercept at $(0, 0)$. Before we take the derivative, let's expand the expression. That way, we won't have to use the Product Rule. We get $y = 3x^{\frac{2}{3}} - 2x$. Next, we take the derivative: $\dfrac{dy}{dx} = 2x^{-\frac{1}{3}} - 2$. Next, we set the derivative equal to zero to find the critical points. There is one solution: $x = 1$. We plug this value into the original equation to find the y-coordinate of the critical point: when $x = 1$,

$y = 3(1)^{\frac{2}{3}} - 2(1) = 1$. Thus, we have a critical point at (1, 1). *But*, notice that the derivative is not defined at $x = 0$. This means that the function has either a vertical tangent or a cusp at $x = 0$. We'll be able to determine which after we take the second derivative. Notice also that the derivative is negative for $x < 0$ and for $x > 1$ and positive for $0 < x < 1$. Therefore, the curve is decreasing for $x < 0$ and for $x > 1$, and increasing for $0 < x < 1$. Next, we take the second derivative: $\dfrac{d^2y}{dx^2} = -\dfrac{2}{3}x^{-\frac{4}{3}}$. If we set this equal to zero, there is no solution. Therefore, there is no point of inflection. The second derivative is always negative which tells us that the curve is always concave down and that the curve has a cusp at $x = 0$. Note that if it had switched concavity there, then $x = 0$ would be a vertical tangent. Next, we need to determine if each critical point is a maximum, a minimum, or something else. If we plug $x = 1$ into the second derivative, the value is negative, so (1, 1) is a maximum. Now, we can draw the curve. It looks like the following:

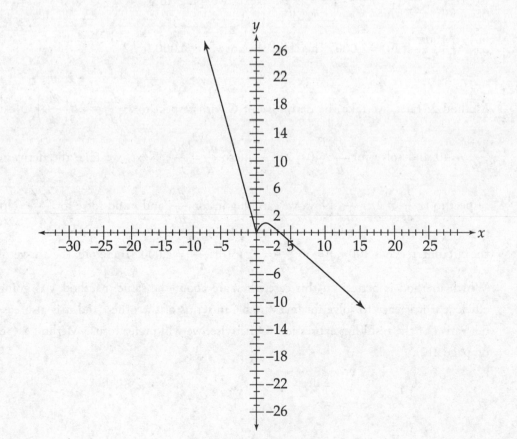

PRACTICE PROBLEM SET 12

1.   2,000 ft²/s

We are given the rate at which the circumference is increasing, $\dfrac{dC}{dt} = 40$, and are looking for the

rate at which the area is increasing, $\dfrac{dA}{dt}$. Thus, we need to find a way to relate the area of a circle

to its circumference. Recall that the circumference of a circle is $C = 2\pi r$, and the area is $A = \pi r^2$.

We could find C in terms of r and then plug it into the equation for A, or we could work with the

equations separately and then relate them. Let's do both and compare.

Method 1: First, we find C in terms of r: $r = \dfrac{C}{2\pi}$. Now, we plug this in for r in the equa-

tion for A: $A = \pi \left(\dfrac{C}{2\pi}\right)^2 = \dfrac{C^2}{4\pi}$. Next, we take the derivative of the equation with respect

to t: $\dfrac{dA}{dt} = \dfrac{1}{4\pi}(2C)\dfrac{dC}{dt} = \left(\dfrac{C}{2\pi}\right)\dfrac{dC}{dt}$. Next, we plug in $C = 100\pi$ and $\dfrac{dC}{dt} = 40$ and solve:

$\dfrac{dA}{dt} = \left(\dfrac{100\pi}{2\pi}\right)(40) = 2,000$. Therefore, the answer is 2,000 ft²/s.

Method 2: First, we take the derivative of C with respect to t: $\dfrac{dC}{dt} = 2\pi\left(\dfrac{dr}{dt}\right)$. Next, we plug in

$\dfrac{dC}{dt} = 40$ and solve for $\dfrac{dr}{dt}$: $40 = 2\pi\left(\dfrac{dr}{dt}\right)$, so $\dfrac{dr}{dt} = \dfrac{20}{\pi}$. Next, we take the derivative of A with

respect to t: $\dfrac{dA}{dt} = 2\pi r\dfrac{dr}{dt}$. Now, we can plug in for $\dfrac{dr}{dt}$ and r and solve for $\dfrac{dA}{dt}$. Note that when

the circumference is 100π, $r = 50$: $\dfrac{dA}{dt} = 2\pi(50)\left(\dfrac{20}{\pi}\right) = 2,000$. Therefore, the answer is 2,000 ft²/s.

Which method is better? In this case, they are about the same. Method 1 is going to be more

efficient if it is easy to solve for one variable in terms of the other, and it is also easy to take the

derivative of the resulting expression. Otherwise, we will prefer to use Method 2 (See Example 3

on page 212).

2. $\dfrac{3}{4}$ in./sec

We are given the rate at which the volume is increasing, $\dfrac{dV}{dt} = 27\pi$ and are looking for the rate at which the radius is increasing, $\dfrac{dr}{dt}$. Thus, we need to find a way to relate the volume of a sphere to its radius. Recall that the volume of a sphere is $V = \dfrac{4}{3}\pi r^3$. All we have to do is take the derivative of the equation with respect to t: $\dfrac{dV}{dt} = \dfrac{4}{3}\pi \left(3r^2\right)\left(\dfrac{dr}{dt}\right) = 4\pi r^2 \dfrac{dr}{dt}$. Now we substitute $\dfrac{dV}{dt} = 27\pi$ and $r = 3$: $27\pi = 4\pi(3)^2 \dfrac{dr}{dt}$. If we solve for $\dfrac{dr}{dt}$, we get $\dfrac{dr}{dt} = \dfrac{3}{4}$ in./sec.

3.   100 km/hr

We are given the rate at which Car A is moving south, $\dfrac{dA}{dt} = 80$, and the rate at which Car B is moving west, $\dfrac{dB}{dt} = 60$, and are looking for the rate at which the distance between them is increasing, which we'll call $\dfrac{dC}{dt}$. Note that the directions south and west are at right angles to each other. Thus, the distance that Car A is from the starting point, which we'll call A, and the distance that Car B is from the starting point, which we'll call B, are the legs of a right triangle, with C as the hypotenuse. We can relate the three distances using the Pythagorean Theorem. Here, because A and B are the legs and C is the hypotenuse, $A^2 + B^2 = C^2$. Now we take the derivative of the equation with respect to t: $2A\dfrac{dA}{dt} + 2B\dfrac{dB}{dt} = 2C\dfrac{dC}{dt}$. This simplifies to $A\dfrac{dA}{dt} + B\dfrac{dB}{dt} = C\dfrac{dC}{dt}$. We know that Car A has been driving for 3 hours at 80 km/hr and Car B has been driving for 3 hours at 60 km/hr, so $A = 240$ and $B = 180$. Using the Pythagorean Theorem, $240^2 + 180^2 = C^2$, so $C = 300$. Now we can substitute into our derivative equation: $(240)(80) + (180)(60) = (300)\dfrac{dC}{dt}$. If we solve for C, we get $C = 100$ km/hr.

4. $243\sqrt{3}$ in.²/sec

We are given the rate at which the sides of the triangle are increasing, $\dfrac{ds}{dt} = 27$, and are looking for the rate at which the area is increasing, $\dfrac{dA}{dt}$. Thus, we need to find a way to relate the area of an equilateral triangle to the length of a side. We know that the area of an equilateral triangle, in terms of its sides, is $A = \dfrac{s^2\sqrt{3}}{4}$. (If you don't know this formula, memorize it! It will

come in very handy in future math problems.) Now we take the derivative of this equation with

respect to t: $\dfrac{dA}{dt} = \dfrac{\sqrt{3}}{4}(2s)\left(\dfrac{ds}{dt}\right)$. Next, we plug $\dfrac{ds}{dt} = 27$ and $s = 18$ into the derivative and we get

$\dfrac{dA}{dt} = \dfrac{\sqrt{3}}{4}(36)(27) = 243\sqrt{3}$ in.²/sec.

5. $-\dfrac{5}{7}$ in./sec

We are given the rate at which the water is flowing out of the container, $\dfrac{dV}{dt} = -35\pi$ (Why is it

negative?), and are looking for the rate at which the depth of the water is dropping, $\dfrac{dh}{dt}$. Thus,

we need to find a way to relate the volume of a cone to its height. We know that the volume of a

cone is $V = \dfrac{1}{3}\pi r^2 h$. Notice that we have a problem. We have a third variable, r, in the equation.

We cannot treat it as a constant the way we did in problem 4 because as the volume of a cone

changes, both its height and radius change. But, we also know that in any cone, the ratio of the

radius to the height is a constant. Here, when the radius is 21 (because the diameter is 42), the

height is 15. Thus, $\dfrac{r}{h} = \dfrac{21}{15} = \dfrac{7}{5}$. We can now isolate r in this equation: $r = \dfrac{7h}{5}$. Now we can plug

it into the volume formula to get rid of r: $V = \dfrac{1}{3}\pi\left(\dfrac{7h}{5}\right)^2 h$. This simplifies to $V = \dfrac{49\pi}{75}h^3$. Now

we can take the derivative of this equation with respect to t: $\dfrac{dV}{dt} = \dfrac{49\pi}{75}(3h^2)\dfrac{dh}{dt}$. Next, we plug

$\dfrac{dV}{dt} = -35\pi$ and $h = 5$ into the derivative to get $-35\pi = \dfrac{49\pi}{75}\left(3(5)^2\right)\dfrac{dh}{dt}$. Now we can solve for

$\dfrac{dh}{dt}$: $\dfrac{dh}{dt} = -\dfrac{5}{7}$ in./sec.

6. $-\dfrac{25}{6}$ ft/sec

We are given the rate at which the length of the rope, R, is changing, $\dfrac{dR}{dt} = -4$, and are looking

for the rate at which the boat, B, is approaching the dock, $\dfrac{dB}{dt}$. The key to this problem is to realize that the vertical distance from the dock to the bow, the distance from the boat to the dock, and the length of the rope form a right triangle.

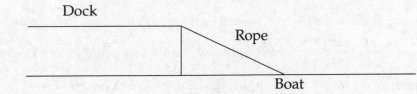

The vertical distance from the dock to the bow is always 7, so, using the Pythagorean Theorem, we get $7^2 + B^2 = R^2$. Now we can take the derivative of this equation with respect to t: $2B\dfrac{dB}{dt} = 2R\dfrac{dR}{dt}$, which simplifies to $B\dfrac{dB}{dt} = R\dfrac{dR}{dt}$. We know that $R = 25$ and can use the Pythagorean Theorem to find B: $7^2 + B^2 = 25^2$, so $B = 24$. Now we plug $\dfrac{dR}{dt} = -4$, $R = 25$, and $B = 24$ into the derivative and we get $24\dfrac{dB}{dt} = 25(-4)$. Now we can solve for $\dfrac{dB}{dt}$: $\dfrac{dB}{dt} = -\dfrac{25}{6}$ ft/sec.

7. $\dfrac{4}{3}$ ft/sec

We are given the rate at which the woman is walking away from the street lamp and are looking for the rate at which the length of her shadow is changing. It helps to draw a picture of the situation.

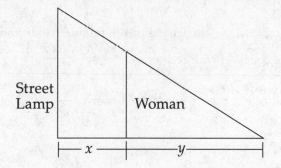

If we label the distance between the woman and the street lamp, x, and the length of the woman's shadow, y, we can use similar triangles to get $\dfrac{y}{6} = \dfrac{x+y}{24}$. We can cross-multiply and simplify.

$$24y = 6x + 6y$$

$$3y = x$$

Next, we take the derivative of the equation with respect to t: $3\dfrac{dy}{dt} = \dfrac{dx}{dt}$. Now, we can plug $\dfrac{dx}{dt} = 4$ into the derivative and solve: $\dfrac{dy}{dt} = \dfrac{4}{3}$ ft/sec.

8. $\dfrac{3\pi}{5}$ in.^{2}/min

We know that it takes 60 minutes for the minute hand of a clock to make one complete revolution, so the rate at which the angle, θ, formed by the minute hand and noon increasing, in terms of radians/min, is $\dfrac{d\theta}{dt} = \dfrac{2\pi}{60} = \dfrac{\pi}{30}$. Next, we know that the area of the sector of a circle is proportional to its central angle, so $\dfrac{\theta}{2\pi} = \dfrac{S}{\pi r^2}$, which can be simplified to $S = \dfrac{1}{2} r^2 \theta$.

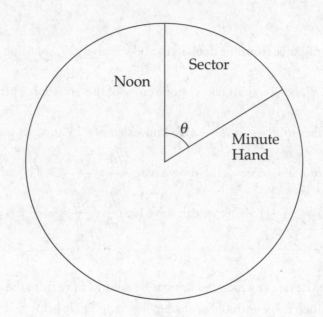

Note that the radius is 6 (the length of the minute hand) and is a constant. Next, we take the derivative of the equation with respect to t: $\dfrac{dS}{dt} = \dfrac{1}{2} r^2 \dfrac{d\theta}{dt}$.

Finally, we substitute $\dfrac{d\theta}{dt} = \dfrac{\pi}{30}$ into the derivative: $\dfrac{dS}{dt} = \dfrac{1}{2} (6)^2 \left(\dfrac{\pi}{30} \right) = \dfrac{3\pi}{5}$ in.^{2}/sec.

PRACTICE PROBLEM SET 13

1. $v(t) = 3t^2 - 18t + 24 \,; a(t) = 6t - 18$

In order to find the velocity function of the particle, we simply take the derivative of the position function with respect to t: $\dfrac{dx}{dt} = v(t) = 3t^2 - 18t + 24$. In order to find the acceleration function of the particle, we simply take the derivative of the velocity function with respect to t: $\dfrac{dv}{dt} = 6t - 18$.

2. $v(t) = 2\cos(2t) - \sin(t)$; $a(t) = -4\sin(2t) - \cos(t)$

In order to find the velocity function of the particle, we simply take the derivative of the position function with respect to t: $\dfrac{dx}{dt} = v(t) = 2\cos(2t) - \sin(t)$. In order to find the acceleration function of the particle, we simply take the derivative of the velocity function with respect to t: $\dfrac{dv}{dt} = a(t) = -4\sin(2t) - \cos(t)$.

3. $t = \pi$ and $t = 3\pi$

In order to find where the particle is changing direction, we need to find where the velocity of the particle changes sign. The velocity function of the particle is the derivative of the position function: $\dfrac{dx}{dt} = v(t) = \dfrac{1}{2}\cos\left(\dfrac{t}{2}\right)$. Next, we set the velocity equal to zero. The solutions are $t = \pi$ and $t = 3\pi$. Actually, there are an infinite number of solutions but remember that we are restricted to $0 < t < 4\pi$. Next, we check the sign of the velocity on the intervals $0 < t < \pi$, $\pi < t < 3\pi$, and $3\pi < t < 4\pi$. When $0 < t < \pi$, the velocity is positive, so the particle is moving to the right. When $\pi < t < 3\pi$, the velocity is negative, so the particle is moving to the left. When $3\pi < t < 4\pi$, the velocity is positive, so the particle is moving to the right. Therefore, the particle is changing direction at $t = \pi$ and $t = 3\pi$.

4.   The distance is 69.

In order to find the distance that the particle travels, we need to look at the position of the particle at $t = 2$ and at $t = 5$. We also need to see if the particle changes direction anywhere on the interval between the two times. If so, we will need to look at the particle's position at those "turning points" as well. The way to find out if the particle is changing direction is to look at the velocity of the particle, which we find by taking the derivative of the position function. We get $\dfrac{dx}{dt} = v(t) = 6t + 2$. If we set the velocity equal to zero, we get $t = -\dfrac{1}{3}$, which is not in the time interval. This means that the velocity doesn't change signs, and thus the particle does not change direction. Now we look at the position of the particle on the interval. At $t = 2$, the particle's position is $x = 3(2)^2 + 2(2) + 4 = 20$. At $t = 5$, the particle's position is: $x = 3(5)^2 + 2(5) + 4 = 89$. Therefore, the particle travels a distance of 69.

5.   The distance is 48.

In order to find the distance that the particle travels, we need to look at the position of the particle at $t = 0$ and at $t = 4$. We also need to see if the particle changes direction anywhere on the interval between the two times. If so, we will need to look at the particle's position at those "turning points" as well. The way to find out if the particle is changing direction is to look at the velocity of the particle, which we find by taking the derivative of the position function. We get $\dfrac{dx}{dt} = v(t) = 2t + 8$. If we set the velocity equal to zero, we get $t = -4$, which is not in the time interval. This means that the velocity doesn't change signs, and thus the particle does not change direction. Now we look at the position of the particle on the interval. At $t = 0$, the particle's position is: $x = (0)^2 + 8(0) = 0$. At $t = 4$, the particle's position is $x = (4)^2 + 8(4) = 48$. Therefore, the particle travels a distance of 48.

6.   Velocity is 0; acceleration is 0.

This should not be a surprise because $2\sin^2 t + 2\cos^2 t = 2$, so the position is a constant. This means that the particle is not moving and thus has a velocity and acceleration of 0.

7. $t = \dfrac{-8 + \sqrt{70}}{3} \approx 0.122$

In order to find where the particle is changing direction, we need to find where the velocity of the particle changes sign. The velocity function of the particle is the derivative of the position function: $\dfrac{dx}{dt} = v(t) = 3t^2 + 16t - 2$. Next, we set the velocity equal to zero. The solutions are $t = \dfrac{-8 + \sqrt{70}}{3}$ and $t = \dfrac{-8 - \sqrt{70}}{3}$. We can eliminate the second solution because it is negative and we are restricted to $t > 0$. Next, we check the sign of the velocity on the intervals $0 < t < \dfrac{-8 + \sqrt{70}}{3}$ and $t > \dfrac{-8 + \sqrt{70}}{3}$. When $0 < t < \dfrac{-8 + \sqrt{70}}{3}$, the velocity is negative, so the particle is moving to the left. When $t > \dfrac{-8 + \sqrt{70}}{3}$, the velocity is positive, so the particle is moving to the right. Therefore, the particle is changing direction at $t = \dfrac{-8 + \sqrt{70}}{3}$.

8.   The velocity is never 0, which means that it never changes signs and thus the particle does not change direction.

In order to find where the particle is changing direction, we need to find where the velocity of the particle changes signs. The velocity function of the particle is the derivative of the position function: $\dfrac{dx}{dt} = v(t) = 6t^2 - 12t + 12$. Next, we set the velocity equal to zero. There are no real solutions. If we try a few values, we can see that the velocity is always positive. Therefore, the particle does not change direction.

PRACTICE PROBLEM SET 14

1. $f'(x) = \dfrac{4x^3}{x^4 + 8}$

    The rule for finding the derivative of $y = \ln u$ is $\dfrac{dy}{dx} = \dfrac{1}{u}\dfrac{du}{dx}$, where u is a function of x. Here, $u = x^4 + 8$. Therefore, the derivative is $f'(x) = \dfrac{1}{x^4 + 8}\left(4x^3\right) = \dfrac{4x^3}{x^4 + 8}$.

2. $f'(x) = \dfrac{1}{x} + \dfrac{1}{6 + 2x}$

    The rule for finding the derivative of $y = \ln u$ is $\dfrac{dy}{dx} = \dfrac{1}{u}\dfrac{du}{dx}$, where u is a function of x. Before we find the derivative, we can use the laws of logarithms to expand the logarithm. This way, we won't have to use the Product Rule. We get $\ln\left(3x\sqrt{3 + x}\right) = \ln 3 + \ln x + \ln\sqrt{3 + x} = \ln 3 + \ln x + \dfrac{1}{2}\ln(3 + x)$.

    Now we can find the derivative: $f'(x) = 0 + \dfrac{1}{x} + \dfrac{1}{2}\dfrac{1}{3 + x} = \dfrac{1}{x} + \dfrac{1}{6 + 2x}$.

3. $f'(x) - \csc x$

    The rule for finding the derivative of $y = \ln u$ is $\dfrac{dy}{dx} = \dfrac{1}{u}\dfrac{du}{dx}$, where u is a function of x.

    Here, $u = \cot x - \csc x$. Therefore, the derivative is $f'(x) = \dfrac{1}{\cot x - \csc x}\left(-\csc^2 x + \csc x \cot x\right)$.

    This can be simplified to $f'(x) = \dfrac{(-\csc x + \cot x)(\csc x)}{\cot x - \csc x} = \csc x$.

4. $f'(x) = \dfrac{2}{x} - \dfrac{x}{5 + x^2}$

    The rule for finding the derivative of $y = \ln u$ is $\dfrac{dy}{dx} = \dfrac{1}{u}\dfrac{du}{dx}$, where u is a function of x. Before we find the derivative, we can use the laws of logarithms to expand the logarithm. This way, we won't have to use the Product Rule or the Quotient Rule. We get

$\ln\left(\dfrac{5x^2}{\sqrt{5+x^2}}\right) = \ln 5 + \ln x^2 - \ln\sqrt{5+x^2} = \ln 5 + 2\ln x - \dfrac{1}{2}\ln(5+x^2)$. Now we can find the

derivative: $f'(x) = 0 + 2\dfrac{1}{x} - \dfrac{1}{2}\dfrac{1}{5+x^2}(2x) = \dfrac{2}{x} - \dfrac{x}{5+x^2}$.

5. $f'(x) = e^{x\cos x}\left(\cos x - x\sin x\right)$

The rule for finding the derivative of $y = e^u$ is $\dfrac{dy}{dx} = e^u\dfrac{du}{dx}$, where u is a function of x. Here we

will use the Product Rule to find the derivative of the exponent: $f'(x) = e^{x\cos x}\left(\cos x - x\sin x\right)$.

6. $f'(x) = -3e^{-3x}\sin 5x + 5e^{-3x}\cos 5x$

The rule for finding the derivative of $y = e^u$ is $\dfrac{dy}{dx} = e^u\dfrac{du}{dx}$, where u is a function of x. Here we

will use the Product Rule to find the derivative: $f'(x) = e^{-3x}(-3)\sin 5x + e^{-3x}(5\cos 5x)$, which we

can rearrange to $f'(x) = -3e^{-3x}\sin 5x + 5e^{-3x}\cos 5x$.

7. $f'(x) = \pi e^{\pi x} - \pi$

The rule for finding the derivative of $y = e^u$ is $\dfrac{dy}{dx} = e^u\dfrac{du}{dx}$, where u is a function of x and the rule

for finding the derivative of $y = \ln u$ is $\dfrac{dy}{dx} = \dfrac{1}{u}\dfrac{du}{dx}$, where u is a function of x.

We get $f'(x) = e^{\pi x}(\pi) - \dfrac{1}{e^{\pi x}}\pi e^{\pi x}$. This can be simplified to $f'(x) = \pi e^{\pi x} - \pi$. You might have

noticed that $\ln e^{\pi x} = \pi x$, which would have made the derivative a little easier.

8. $f'(x) = \dfrac{3}{x\ln 12}$

The rule for finding the derivative of $y = \log_a u$ is $\dfrac{dy}{dx} = \dfrac{1}{u\ln a}\dfrac{du}{dx}$, where u is a function of x.

Before we find the derivative, we can use the laws of logarithms to expand the logarithm. We get

$f(x) = \log_{12} x^3 = 3\log_{12} x$. Now we can find the derivative: $f'(x) = \dfrac{3}{x\ln 12}$.

9. $f'(x) = \dfrac{1}{xe^{4x}\ln 4} - \dfrac{4\log_4 x}{e^{4x}}$

The rule for finding the derivative of $y = \log_a u$ is $\dfrac{dy}{dx} = \dfrac{1}{u\ln a}\dfrac{du}{dx}$, and the rule for finding the

derivative of $y = e^u$ is $\dfrac{dy}{dx} = e^u\dfrac{du}{dx}$, where u is a function of x. Here we will use the Quotient

Rule to find the derivative: $f'(x) = \dfrac{\left(e^{4x}\right)\left(\dfrac{1}{x \ln 4}\right) - \left(\log_4 x\right)\left(e^{4x}\right)(4)}{\left(e^{4x}\right)^2}$. This can be simplified to

$f'(x) = \dfrac{1}{xe^{4x} \ln 4} - \dfrac{4 \log_4 x}{e^{4x}}$.

10. $f'(x) = \dfrac{3}{2}$

The rule for finding the derivative of $y = \log_a u$ is $\dfrac{dy}{dx} = \dfrac{1}{u \ln a} \dfrac{du}{dx}$, and the rule for finding the

derivative of $y = a^u$ is $\dfrac{dy}{dx} = a^u \left(\ln a\right) \dfrac{du}{dx}$, where u is a function of x. Before we find the derivative,

we can use the laws of logarithms to simplify the logarithm. We get $f(x) = \dfrac{1}{2} \log 10^{3x}$. Now if

you are alert, you will remember that $\log 10^{3x} = 3x$, so this simplifies to $f(x) = \dfrac{1}{2} 3x = \dfrac{3}{2} x$. The

derivative is simply $f'(x) = \dfrac{3}{2}$.

11. $f'(x) = 3e^{3x} - 3^{ex}(e)\left(\ln 3\right)$

The rule for finding the derivative of $y = e^u$ is $\dfrac{dy}{dx} = e^u \dfrac{du}{dx}$, and the rule for finding the derivative

of $y = a^u$ is $\dfrac{dy}{dx} = a^u \left(\ln a\right) \dfrac{du}{dx}$, where u is a function of x. We get $f'(x) = 3e^{3x} - 3^{ex}(e)(\ln 3)$.

12. $f'(x) = 10^{\sin x}\left(\cos x\right)\left(\ln 10\right)$

The rule for finding the derivative of $y = a^u$ is $\dfrac{dy}{dx} = a^u \left(\ln a\right) \dfrac{du}{dx}$, where u is a function of x. We get

$f'(x) = 10^{\sin x}\left(\cos x\right)(\ln 10)$.

13. $f'(x) = \ln 10$

Before we find the derivative, we can use the laws of logarithms to simplify the logarithm. We get

$f(x) = \ln\left(10^x\right) = x \ln 10$. Now the derivative is simply $f'(x) = \ln 10$.

14. $f'(x) = x^4 5^x\left(5 + x \ln 5\right)$

The rule for finding the derivative of $y = a^u$ is $\dfrac{dy}{dx} = a^u \left(\ln a\right) \dfrac{du}{dx}$, where u is a function of x. Here

we will use the Product Rule to find the derivative: $f'(x) = x^5\left(5^x \ln 5\right) + \left(5^x\right)\left(5x^4\right)$, which sim-

plifies to $f'(x) = x^4 5^x\left(5 + x \ln 5\right)$.

PRACTICE PROBLEM SET 15

1. $\dfrac{16}{15}$

    First, we take the derivative of y: $\dfrac{dy}{dx} = 1 - \dfrac{1}{x^2}$. Next, we find the value of x where $y = \dfrac{17}{4}$:

    $\dfrac{17}{4} = x + \dfrac{1}{x}$.

    With a little algebra, you should get $x = 4$ or $x = \dfrac{1}{4}$. Because $x > 1$, we can ignore the second

    answer. Or, if you are permitted, use your calculator.

    Now we can use the formula for the derivative of the inverse of $f(x)$ (see page 234):

    $$\dfrac{d}{dx} f^{-1}(x) \bigg|_{x=c} = \dfrac{1}{\left[\dfrac{d}{dy} f(y) \right]_{y=a}}, \text{ where } f(a) = c. \text{ This formula means that we find the derivative}$$

    of the inverse of a function at a value a by taking the reciprocal of the derivative and plugging in

    the value of x that makes y equal to a.

    $$\dfrac{1}{\dfrac{dy}{dx}\bigg|_{x=4}} = \dfrac{1}{1 - \dfrac{1}{x^2}\bigg|_{x=4}} = \dfrac{16}{15}$$

2. $-\dfrac{1}{12}$

    First, we take the derivative of y: $\dfrac{dy}{dx} = 3 - 15x^2$. Next, we find the value of x where $y = 2$:
    $2 = 3x - 5x^3$.

    With a little algebra, you should get $x = -1$. Or, if you are permitted, use your calculator.

Now we can use the formula for the derivative of the inverse of $f(x)$ (see page 234):

$$\frac{d}{dx}f^{-1}(x)\bigg|_{x=c} = \frac{1}{\left[\dfrac{d}{dy}f(y)\right]_{y=a}}$$, where $f(a) = c$ This formula means that we find the derivative

of the inverse of a function at a value a by taking the reciprocal of the derivative and plugging in

the value of x that makes y equal to a.

$$\frac{1}{\dfrac{dy}{dx}\bigg|_{x=-1}} = \frac{1}{3-15x^2}\bigg|_{x=-1} = -\frac{1}{12}$$

3. $\dfrac{1}{e}$

First, we take the derivative of y: $\dfrac{dy}{dx} = e^x$. Next, we find the value of x where $y = e : e = e^x$. It should be obvious that $x = 1$.

Now we can use the formula for the derivative of the inverse of $f(x)$ (see page 234):

$$\frac{d}{dx}f^{-1}(x)\bigg|_{x=c} = \frac{1}{\left[\dfrac{d}{dy}f(y)\right]_{y=a}}$$, where $f(a) = c$. This formula means that we find the derivative

of the inverse of a function at a value a by taking the reciprocal of the derivative and plugging in

the value of x that makes y equal to a.

$$\frac{1}{\dfrac{dy}{dx}\bigg|_{x=1}} = \frac{1}{e^x}\bigg|_{x=1} = \frac{1}{e}$$

4. $\dfrac{1}{4}$

First, we take the derivative of y: $\dfrac{dy}{dx} = 1 + 3x^2$. Next, we find the value of x where $y = -2$:

$-2 = x + x^3$. You should be able to tell by inspection that $x = -1$ is a solution. Or, if you are permit-

ted, use your calculator. Remember that the AP Exam won't give you a problem where it is very

difficult to solve for the inverse value of y. If the algebra looks difficult, look for an obvious solu-

tion, such as $x = 0$ or $x = 1$ or $x = -1$.

Now we can use the formula for the derivative of the inverse of $f(x)$: $\left. \dfrac{d}{dx} f^{-1}(x) \right|_{x=c} = \dfrac{1}{\left[\dfrac{d}{dy} f(y) \right]_{y=a}}$,

where $f(a) = c$. This formula means that we find the derivative of the inverse of a function at a

value a by taking the reciprocal of the derivative and plugging in the value of x that makes y equal

to a: $\left. \dfrac{1}{\dfrac{dy}{dx} \right|_{x=-1}} = \left. \dfrac{1}{1+3x^2} \right|_{x=-1} = \dfrac{1}{4}$.

5.   1

First, we take the derivative of y: $\dfrac{dy}{dx} = 4 - 3x^2$. Next, we find the value of x where $y = 3$:

$3 = 4x - x^3$. You should be able to tell by inspection that $x = 1$ is a solution. Or, if you are permitted,

use your calculator. Remember that the AP Exam won't give you a problem where it is very difficult

to solve for the inverse value of y. If the algebra looks difficult, look for an obvious solution, such as

$x = 0$ or $x = 1$ or $x = -1$.

Now we can use the formula for the derivative of the inverse of $f(x)$: $\left. \dfrac{d}{dx} f^{-1}(x) \right|_{x=c} = \dfrac{1}{\left[\dfrac{d}{dy} f(y) \right]_{y=a}}$,

where $f(a) = c$. This formula means that we find the derivative of the inverse of a function at a

value a by taking the reciprocal of the derivative and plugging in the value of x that makes y equal

to a: $\left. \dfrac{1}{\dfrac{dy}{dx} \right|_{x=1}} = \left. \dfrac{1}{4-3x^2} \right|_{x=1} = 1$.

6.   1

First, we take the derivative of y: $\dfrac{dy}{dx} = \dfrac{1}{x}$. Next, we find the value of x where $y = 0$: $\ln x = 0$. You

should know that $x = 1$ is the solution. Now we can use the formula for the derivative of the inverse of $f(x)$: $\dfrac{d}{dx}f^{-1}(x)\bigg|_{x=c} = \dfrac{1}{\left[\dfrac{d}{dy}f(y)\right]_{y=a}}$, where $f(a)=c$. This formula means that we find the

derivative of the inverse of a function at a value a by taking the reciprocal of the derivative and

plugging in the value of x that makes y equal to a.

$$\dfrac{1}{\dfrac{dy}{dx}\bigg|_{x=1}} = \dfrac{1}{\dfrac{1}{x}\bigg|_{x=1}} = 1$$

PRACTICE PROBLEM SET 16

1. $x = y^2$

Recall the trigonometric identity $1 + \tan^2\theta = \sec^2\theta$. We can rewrite this identity as $\tan^2\theta = \sec^2\theta - 1$. Here, we are given that $x = \sec^2 t - 1$ and $y = \tan t$. Using this identity and plugging in our parametric equations, we get the equation $x = y^2$.

2. $y = \sqrt{1-x^2}$

Here, we are given that $x = t$ and $y = \sqrt{1-t^2}$. We can simply substitute x for t in the equation for y to get $y = \sqrt{1-x^2}$.

3. $y = x^2 - 6x$

Here, we are given that $x = 4t + 3$ and $y = 16t^2 - 9$. If we take the equation for x and isolate $4t$, we get $x - 3 = 4t$. Next, because $(4t)^2 = 16t^2$, we can substitute into the equation for y. We get $y = (x-3)^2 - 9$. This can be simplified to $y = x^2 - 6x$.

4. $y - 48 = \dfrac{2}{3}(x - 40)$

We could first come up with an equation for y in terms of x by eliminating the parameter, but

this is often more time-consuming than necessary. Instead, we first find the slope of the tangent

line using the formula $\dfrac{dy}{dx} = \dfrac{\dfrac{dy}{dt}}{\dfrac{dx}{dt}}$. Taking the derivative of each equation with respect to t, we get

$\dfrac{dy}{dt}=8$ and $\dfrac{dx}{dt}=2t$. Therefore, $\dfrac{dy}{dx}=\dfrac{8}{2t}=\dfrac{4}{t}$. If we evaluate this at $t=6$, we get $\dfrac{dy}{dx}=\dfrac{4}{6}=\dfrac{2}{3}$.

At $t=6$, we know that $y=48$ and $x=40$. Now we can find the equation of the tangent line:

$y-48=\dfrac{2}{3}(x-40).$

5. $y-1=\sqrt{2}\left(x-\sqrt{2}\right)$

We could first come up with an equation for y in terms of x by eliminating the parameter, but

this is often more time consuming than necessary. Instead, we first find the slope of the tangent

line using the formula $\dfrac{dy}{dx}=\dfrac{\frac{dy}{dt}}{\frac{dx}{dt}}$. Taking the derivative of each equation with respect to t, we

get $\dfrac{dy}{dt}=\sec^2 t$ and $\dfrac{dx}{dt}=\sec t\tan t$. Therefore, $\dfrac{dy}{dx}=\dfrac{\sec^2 t}{\sec t\tan t}=\dfrac{\sec t}{\tan t}$. If we evaluate this at

$t=\dfrac{\pi}{4}$, we get $\dfrac{dy}{dx}=\dfrac{\sec\frac{\pi}{4}}{\tan\frac{\pi}{4}}=\dfrac{\sqrt{2}}{1}=\sqrt{2}$.

At $t=\dfrac{\pi}{4}$, we know that $y=\tan\dfrac{\pi}{4}=1$ and $x=\sec\dfrac{\pi}{4}=\sqrt{2}$. Now we can find the equation of the

tangent line: $y-1=\sqrt{2}\left(x-\sqrt{2}\right)$.

6. $\left(-2,7\right)$

The particle's direction will be horizontal when it is moving only in the x direction. In other words,

when $\dfrac{dy}{dt}=0$. We take the derivative of y with respect to t: $\dfrac{dy}{dt}=3t^2-3$. If we set this equal to

zero and solve, we get $3t^2-3=0$ and $t=\pm 1$. We can ignore the negative solution because t must

be positive. Now, in order to find the coordinates of the particle, we simply plug $t=1$ into the

equations of the particle's position. We get $x=-2(1)^2=-2$ and $y=(1)^3-3(1)+9=7$. There-

fore, the particle's position is $\left(-2,7\right)$.

7. $\left(\ln 2, -4\right)$

The particle's direction will be horizontal when it is moving only in the x direction. In other words, when $\dfrac{dy}{dt} = 0$. We take the derivative of y with respect to t: $\dfrac{dy}{dt} = 2t - 4$. If we set this equal to zero and solve, we get $2t - 4 = 0$ and $t = 2$. Now, in order to find the coordinates of the particle, we simply plug $t = 2$ into the equations of the particle's position. We get $x = \ln 2$ and $y = (2)^2 - 4(2) = -4$. Therefore, the particle's position is $\left(\ln 2, -4\right)$.

8. $t = \dfrac{2\pi}{3}, \dfrac{4\pi}{3}$

We find the horizontal and vertical components of the particle's velocity by taking the derivative of each parametric equation with respect to t. We get $x = 2\cos t$ and $y = \cos t - \dfrac{1}{2}$. If we set these equal to each other, we get $\cos t - \dfrac{1}{2} = 2\cos t$, which can be simplified to $\cos t = -\dfrac{1}{2}$. If we solve this, we get $t = \dfrac{2\pi}{3}, \dfrac{4\pi}{3}$. Note that there are actually an infinite number of solutions, but we are restricted to $0 \le t \le 2\pi$.

PRACTICE PROBLEM SET 17

1. $\dfrac{3}{4}$

Recall L'Hôpital's Rule: If $f(c) = g(c) = 0$, or if $f(c) = g(c) = \infty$, and if $f'(c)$ and $g'(c)$ exist, and if $g'(c) \ne 0$, then $\lim\limits_{x \to c} \dfrac{f(x)}{g(x)} = \lim\limits_{x \to c} \dfrac{f'(x)}{g'(x)}$. Here, $f(x) = \sin 3x$ and $g(x) = \sin 4x$, and $\sin 0 = 0$. This means that we can use L'Hôpital's Rule to find the limit. We take the derivative of the numerator and the denominator: $\lim\limits_{x \to 0} \dfrac{\sin 3x}{\sin 4x} = \lim\limits_{x \to 0} \dfrac{3\cos 3x}{4\cos 4x}$. If we take the new limit, we get $\lim\limits_{x \to 0} \dfrac{3\cos 3x}{4\cos 4x} = \dfrac{3\cos 0}{4\cos 0} = \dfrac{(3)(1)}{(4)(1)} = \dfrac{3}{4}$.

2. -1

Recall L'Hôpital's Rule: If $f(c) = g(c) = 0$, or if $f(c) = g(c) = \infty$, and if $f'(c)$ and $g'(c)$ exist, and if $g'(c) \ne 0$, then $\lim\limits_{x \to c} \dfrac{f(x)}{g(x)} = \lim\limits_{x \to c} \dfrac{f'(x)}{g'(x)}$. Here, $f(x) = x - \pi$ and $g(x) = \sin x$. We can see that $x - \pi = 0$ when $x = \pi$, and that $\sin \pi = 0$. This means that we can use L'Hôpital's Rule to find

the limit. We take the derivative of the numerator and the denominator: $\lim_{x \to \pi} \dfrac{x - \pi}{\sin x} = \lim_{x \to \pi} \dfrac{1}{\cos x}$. If

we take the new limit, we get $\lim_{x \to \pi} \dfrac{1}{\cos x} = \dfrac{1}{\cos \pi} = \dfrac{1}{-1} = -1$.

3. $\dfrac{1}{6}$

Recall L'Hôpital's Rule: If $f(c) = g(c) = 0$, or if $f(c) = g(c) = \infty$, and if $f'(c)$ and $g'(c)$ exist, and

if $g'(c) \neq 0$, then $\lim_{x \to c} \dfrac{f(x)}{g(x)} = \lim_{x \to c} \dfrac{f'(x)}{g'(x)}$. Here, $f(x) = x - \sin x$ and $g(x) = x^3$. We can see that

$x - \sin x = 0$ when $x = 0$, and that $0^3 = 0$. This means that we can use L'Hôpital's Rule to find the

limit. We take the derivative of the numerator and the denominator: $\lim_{x \to 0} \dfrac{x - \sin x}{x^3} = \lim_{x \to 0} \dfrac{1 - \cos x}{3x^2}$.

If we take the new limit, we get $\lim_{x \to 0} \dfrac{1 - \cos x}{3x^2} = \dfrac{1 - \cos 0}{3(0)^2} = \dfrac{1 - 1}{0} = \dfrac{0}{0}$. But this is still indetermi-

nate, so what do we do? Use L'Hôpital's Rule again! We take the derivative of the numerator and

the denominator: $\lim_{x \to 0} \dfrac{1 - \cos x}{3x^2} = \lim_{x \to 0} \dfrac{\sin x}{6x}$. If we take the limit, we get $\lim_{x \to 0} \dfrac{\sin x}{6x} = \dfrac{\sin 0}{6(0)} = \dfrac{0}{0}$.

We need to use L'Hôpital's Rule one more time. We get $\lim_{x \to 0} \dfrac{\sin x}{6x} = \lim_{x \to 0} \dfrac{\cos x}{6}$. Now, if we take

the limit, we get $\lim_{x \to 0} \dfrac{\cos x}{6} = \dfrac{\cos 0}{6} = \dfrac{1}{6}$. Notice that if L'Hôpital's Rule results in an indeterminate

form, we can use the rule again and again (but not infinitely often).

4. -2

Recall L'Hôpital's Rule: If $f(c) = g(c) = 0$, or if $f(c) = g(c) = \infty$, and if $f'(c)$ and $g'(c)$

exist, and if $g'(c) \neq 0$, then $\lim_{x \to c} \dfrac{f(x)}{g(x)} = \lim_{x \to c} \dfrac{f'(x)}{g'(x)}$. Here, $f(x) = e^{3x} - e^{5x}$ and $g(x) = x$.

We can see that $e^{3x} - e^{5x} = 0$ when $x = 0$, and that the denominator is obviously zero at $x = 0$.

This means that we can use L'Hôpital's Rule to find the limit. We take the derivative of the

numerator and the denominator: $\lim\limits_{x\to0}\dfrac{e^{3x}-e^{5x}}{x}=\lim\limits_{x\to0}\dfrac{3e^{3x}-5e^{5x}}{1}$. If we take the new limit, we get

$$\lim\limits_{x\to0}\dfrac{3e^{3x}-5e^{5x}}{1}=\dfrac{3e^{0}-5e^{0}}{1}=\dfrac{3(1)-5(1)}{1}=-2.$$

5. -2

Recall L'Hôpital's Rule: If $f(c)=g(c)=0$, or if $f(c)=g(c)=\infty$, and if $f'(c)$ and

$g'(c)$ exist, and if $g'(c)\ne0$, then $\lim\limits_{x\to c}\dfrac{f(x)}{g(x)}=\lim\limits_{x\to c}\dfrac{f'(x)}{g'(x)}$. Here, $f(x)=\tan x-x$ and

$g(x)=\sin x-x$. We can see that $\tan x-x=0$ and $\sin x-x=0$ when $x=0$. This means

that we can use L'Hôpital's Rule to find the limit. We take the derivative of the numera-

tor and the denominator: $\lim\limits_{x\to0}\dfrac{\tan x-x}{\sin x-x}=\lim\limits_{x\to0}\dfrac{\sec^{2}x-1}{\cos x-1}$. If we take the new limit, we get

$\lim\limits_{x\to0}\dfrac{\sec^{2}x-1}{\cos x-1}=\dfrac{\sec^{2}(0)-1}{\cos(0)-1}=\dfrac{1-1}{1-1}=\dfrac{0}{0}$. But this is still indeterminate, so what do we do?

Use L'Hôpital's Rule again! We take the derivative of the numerator and the denomina-

tor: $\lim\limits_{x\to0}\dfrac{\sec^{2}x-1}{\cos x-1}=\lim\limits_{x\to0}\dfrac{2\sec x(\sec x\tan x)}{-\sin x}=\lim\limits_{x\to0}\dfrac{2\sec^{2}x\tan x}{-\sin x}$. If we take the limit, we get

$\lim\limits_{x\to0}\dfrac{2\sec^{2}(0)\tan(0)}{-\sin(0)}=\dfrac{0}{0}$. At this point you might be getting nervous as the derivatives start to

get messier. Let's try using trigonometric identities to simplify the limit. If we rewrite the numera-

tor in terms of $\sin x$ and $\cos x$, we get $\dfrac{2\sec^{2}x\tan x}{-\sin x}=\dfrac{2\left(\dfrac{1}{\cos^{2}x}\right)\left(\dfrac{\sin x}{\cos x}\right)}{-\sin x}$. We can simplify this to $\dfrac{2\left(\dfrac{1}{\cos^2 x}\right)\left(\dfrac{\sin x}{\cos x}\right)}{-\sin x} = -2\dfrac{\dfrac{\sin x}{\cos^3 x}}{\sin x}$. This simplifies to $-2\dfrac{\sin x}{\cos^3 x}\dfrac{1}{\sin x} = \dfrac{-2}{\cos^3 x}$. Now, if we take

the limit, we get $\lim\limits_{x\to 0}\dfrac{-2}{\cos^3 x} = \dfrac{-2}{\cos^3(0)} = \dfrac{-2}{1^3} = -2$.

Note that we could have used trigonometric identities on either of the first two limits as well. Remember that when you have a limit that is an indeterminate form, you can sometimes use algebra or trigonometric identities (or both) to simplify the limit. Sometimes this will get rid of the problem.

6.   0

Recall L'Hôpital's Rule: If $f(c) = g(c) = 0$, or if $f(c) = g(c) = \infty$, and if $f'(c)$ and $g'(c)$

exist, and if $g'(c) \neq 0$, then $\lim\limits_{x\to c}\dfrac{f(x)}{g(x)} = \lim\limits_{x\to c}\dfrac{f'(x)}{g'(x)}$. Here, $f(x) = x^5$ and $g(x) = e^{5x}$. We

can see that $x^5 = \infty$ when $x = \infty$, and that $e^\infty = \infty$. This means that we can use L'Hôpital's

Rule to find the limit. We take the derivative of the numerator and the denominator:

$\lim\limits_{x\to\infty}\dfrac{x^5}{e^{5x}} = \lim\limits_{x\to\infty}\dfrac{5x^4}{5e^{5x}} = \lim\limits_{x\to\infty}\dfrac{x^4}{e^{5x}}$. If we take the new limit, we get $\lim\limits_{x\to 0}\dfrac{x^4}{e^{5x}} = \dfrac{\infty}{\infty}$. But this is still

indeterminate, so what do we do? Use L'Hôpital's Rule again! We take the derivative of the

numerator and the denominator: $\lim\limits_{x\to\infty}\dfrac{x^4}{e^{5x}} = \lim\limits_{x\to\infty}\dfrac{4x^3}{5e^{5x}}$. We are going to need to use L'Hôpital's

Rule again. In fact, we can see that each time we use the rule we are reducing the power of the

x in the numerator and that we are going to need to keep doing so until the x term is gone. Let's

take the derivative: $\lim\limits_{x\to\infty}\dfrac{4x^3}{5e^{5x}} = \lim\limits_{x\to\infty}\dfrac{12x^2}{25e^{5x}}$. And again: $\lim\limits_{x\to\infty}\dfrac{12x^2}{25e^{5x}} = \lim\limits_{x\to\infty}\dfrac{24x}{125e^{5x}}$. And one more

time: $\lim\limits_{x\to\infty}\dfrac{24x}{125e^{5x}} = \lim\limits_{x\to\infty}\dfrac{24}{625e^{5x}}$. Now, if we take the limit, we get $\lim\limits_{x\to\infty}\dfrac{24}{625e^{5x}} = 0$. Notice that

if L'Hôpital's Rule results in an indeterminate form, we can use the rule again and again (but not

infinitely often).

7. $\dfrac{1}{7}$

Recall L'Hôpital's Rule: If $f(c) = g(c) = 0$, or if $f(c) = g(c) = \infty$, and if $f'(c)$ and $g'(c)$

exist, and if $g'(c) \neq 0$, then $\lim\limits_{x \to c} \dfrac{f(x)}{g(x)} = \lim\limits_{x \to c} \dfrac{f'(x)}{g'(x)}$. Here, $f(x) = x^5 + 4x^3 - 8$ and

$g(x) = 7x^5 - 3x^2 - 1$. We can see that $x^5 + 4x^3 - 8 = \infty$ and $7x^5 - 3x^2 - 1 = \infty$ when $x = \infty$. This

means that we can use L'Hôpital's Rule to find the limit. We take the derivative of the numera-

tor and the denominator: $\lim\limits_{x \to \infty} \dfrac{x^5 + 4x^3 - 8}{7x^5 - 3x^2 - 1} = \lim\limits_{x \to \infty} \dfrac{5x^4 + 12x^2}{35x^4 - 6x}$. If we take the new limit, we get

$\lim\limits_{x \to \infty} \dfrac{5x^4 + 12x^2}{35x^4 - 6x} = \dfrac{\infty}{\infty}$. But this is still indeterminate, so what do we do? Use L'Hôpital's Rule again!

We take the derivative of the numerator and the denominator: $\lim\limits_{x \to \infty} \dfrac{5x^4 + 12x^2}{35x^4 - 6x} = \lim\limits_{x \to \infty} \dfrac{20x^3 + 24x}{140x^3 - 6}$.

We are going to need to use L'Hôpital's Rule again. In fact, we can see that each time we use

the rule we are reducing the power of the x terms in the numerator and denominator and that

we are going to need to keep doing so until the x terms are gone. Let's take the derivative:

$\lim\limits_{x \to \infty} \dfrac{20x^3 + 24x}{140x^3 - 6} = \lim\limits_{x \to \infty} \dfrac{60x^2 + 24}{420x^2}$. And again, $\lim\limits_{x \to \infty} \dfrac{60x^2 + 24}{420x^2} = \lim\limits_{x \to \infty} \dfrac{120x}{840x}$. Now we can take the

limit: $\lim\limits_{x \to \infty} \dfrac{120x}{840x} = \lim\limits_{x \to \infty} \dfrac{1}{7} = \dfrac{1}{7}$. Notice that if L'Hôpital's Rule results in an indeterminate form, we

can use the rule again and again (but not infinitely often).

8.   1

Recall L'Hôpital's Rule: If $f(c) = g(c) = 0$, or if $f(c) = g(c) = \infty$, and if $f'(c)$ and $g'(c)$

exist, and if $g'(c) \neq 0$, then $\lim\limits_{x \to c} \dfrac{f(x)}{g(x)} = \lim\limits_{x \to c} \dfrac{f'(x)}{g'(x)}$. Here, $f(x) = \ln(\sin x)$ and $g(x) = \ln(\tan x)$, and both of these approach infinity as x approaches 0 from the right. This means that we can use L'Hôpital's Rule to find the limit. We take the derivative of the numerator and the denominator: $\lim\limits_{x\to 0^+} \dfrac{\ln(\sin x)}{\ln(\tan x)} = \lim\limits_{x\to 0^+} \dfrac{\dfrac{\cos x}{\sin x}}{\dfrac{\sec^2 x}{\tan x}}$. Let's use trigonometric identities to simplify the

limit: $\dfrac{\dfrac{\cos x}{\sin x}}{\dfrac{\sec^2 x}{\tan x}} = \dfrac{\cos x}{\sin x}\dfrac{\tan x}{\sec^2 x} = \dfrac{\cos x}{\sin x}\dfrac{\dfrac{\sin x}{\cos x}}{\sec^2 x} = \dfrac{1}{\sec^2 x}$. Now, if we take the new limit, we get

$\lim\limits_{x\to 0^+} \dfrac{1}{\sec^2 x} = 1$.

9. $\dfrac{1}{2}$

Recall L'Hôpital's Rule: If $f(c) = g(c) = 0$, or if $f(c) = g(c) = \infty$, and if $f'(c)$ and $g'(c)$ exist, and if $g'(c) \neq 0$, then $\lim\limits_{x\to c}\dfrac{f(x)}{g(x)} = \lim\limits_{x\to c}\dfrac{f'(x)}{g'(x)}$. Here, $f(x)$ = cot $2x$ and $g(x)$ = cot x, and both of these approach infinity as x approaches 0 from the right. This means that we can use L'Hôpital's Rule to find the limit. We take the derivative of the

numerator and the denominator: $\lim\limits_{x\to 0^+}\dfrac{\cot 2x}{\cot x} = \lim\limits_{x\to 0^+}\dfrac{-2\cot 2x \csc 2x}{-\cot x \csc x}$. This seems to be worse than what we started with. Instead, let's use trigonometric identities to simplify the

limit: $\dfrac{\cot 2x}{\cot x} = \dfrac{\dfrac{\cos 2x}{\sin 2x}}{\dfrac{\cos x}{\sin x}} = \dfrac{\cos 2x}{\sin 2x}\dfrac{\sin x}{\cos x} = \dfrac{\cos^2 x - \sin^2 x}{2\sin x \cos x}\dfrac{\sin x}{\cos x}$. Notice that our prob-

lem is with sin x as x approaches zero (because it becomes zero), not with cos x. As long as we are multiplying the numerator and denominator by sin x, we are going to get an indeter-

minate form. So, thanks to trigonometric identities, we can eliminate the problem term:

$\dfrac{\cos^2 x - \sin^2 x}{2\sin x \cos x} \dfrac{\sin x}{\cos x} = \dfrac{\cos^2 x - \sin^2 x}{2\cos^2 x}$. If we take the limit of this expression, it is not indeter-

minate. We get $\lim\limits_{x \to 0^+} \dfrac{\cos^2 x - \sin^2 x}{2\cos^2 x} = \dfrac{\cos^2 0 - \sin^2 0}{2\cos^2 0} = \dfrac{1^2 - 0}{2(1^2)} = \dfrac{1}{2}$. Notice that we didn't need to use

L'Hôpital's Rule here. You should bear in mind that just because a limit is indeterminate does not

mean that the best way to evaluate it is with L'Hôpital's Rule.

10.   1

Recall L'Hôpital's Rule: If $f(c) = g(c) = 0$, or if $f(c) = g(c) = \infty$, and if $f'(c)$ and $g'(c)$

exist, and if $g'(c) \neq 0$, then $\lim\limits_{x \to c} \dfrac{f(x)}{g(x)} = \lim\limits_{x \to c} \dfrac{f'(x)}{g'(x)}$. Here, $f(x) = x$ and $g(x) = \ln(x+1)$,

and both of these approach zero as x approaches 0 from the right. This means that we can use

L'Hôpital's Rule to find the limit. We take the derivative of the numerator and the denominator:

$\lim\limits_{x \to 0^+} \dfrac{1}{\dfrac{1}{x+1}} = \lim\limits_{x \to 0^+} (x+1)$. Now, if we take the new limit, we get $\lim\limits_{x \to 0^+} (x+1) = 1$.

PRACTICE PROBLEM SET 18

1.   5.002

Recall the differential formula that we use for approximating the value of a function:

$f(x + \Delta x) \approx f(x) + f'(x)\Delta x$. Here, we want to approximate the value of $\sqrt{25.02}$, so we'll use

$f(x) = \sqrt{x}$ with $x = 25$ and $\Delta x = 0.02$. First, we need to find $f'(x)$: $f'(x) = \dfrac{1}{2\sqrt{x}}$. Now, we plug into the formula: $f(x+\Delta x) \approx \sqrt{x} + \dfrac{1}{2\sqrt{x}} \Delta x$. If we plug in $x = 25$ and $\Delta x = 0.02$, we get

$\sqrt{25+0.02} \approx \sqrt{25} + \dfrac{1}{2\sqrt{25}}(0.02)$. If we evaluate this, we get $\sqrt{25.02} \approx 5 + \dfrac{1}{10}(0.02) = 5.002$.

2.   3.999375

Recall the differential formula that we use for approximating the value of a function: $f(x + \Delta x) \approx f(x) + f'(x)\Delta x$. Here, we want to approximate the value of $\sqrt{63.97}$, so we'll use $f(x) = \sqrt[3]{x}$ with $x = 64$ and $\Delta x = -0.03$. First, we need to find $f'(x)$: $f'(x) = \dfrac{1}{3} x^{-\frac{2}{3}} = \dfrac{1}{3\sqrt[3]{x^2}}$. Now, we plug into the formula: $f(x + \Delta x) \approx \sqrt[3]{x} + \dfrac{1}{3\sqrt[3]{x^2}} \Delta x$. If we plug in $x = 64$ and $\Delta x = -0.03$, we get $\sqrt[3]{64-0.03} \approx \sqrt[3]{64} + \dfrac{1}{3\sqrt[3]{64^2}}(-0.03)$. If we evaluate this, we get $\sqrt[3]{63.97} \approx 4 + \dfrac{1}{48}(-0.03) = 3.999375$.

3.   1.802

Recall the differential formula that we use for approximating the value of a function: $f(x + \Delta x) \approx f(x) + f'(x)\Delta x$. Here, we want to approximate the value of $\tan 61°$. Be careful! Whenever we work with trigonometric functions, it is *very* important to work with radians, *not* degrees! Remember that $60° = \dfrac{\pi}{3}$ radians and $1° = \dfrac{\pi}{180}$ radians, so we'll use $f(x) = \tan x$ with $x = \dfrac{\pi}{3}$ and $\Delta x = \dfrac{\pi}{180}$. First, we need to find $f'(x)$: $f'(x) = \sec^2 x$. Now, we plug into the formula: $f(x + \Delta x) \approx \tan x + \sec^2 x \, \Delta x$. If we plug in $x = \dfrac{\pi}{3}$ and $\Delta x = \dfrac{\pi}{180}$, we get $\tan\left(\dfrac{\pi}{3} + \dfrac{\pi}{180}\right) \approx \tan\left(\dfrac{\pi}{3}\right) + \sec^2\left(\dfrac{\pi}{3}\right)\left(\dfrac{\pi}{180}\right)$. If we evaluate this, we get $\tan\left(\dfrac{61\pi}{180}\right) \approx \sqrt{3} + (2)^2\left(\dfrac{\pi}{180}\right) \approx 1.802$.

4.   ±2.16 in.³

Recall the formula that we use when we want to approximate the error in a measurement: $dy = f'(x)\,dx$. Here, we want to approximate the error in the volume of a cube when we know that it has a side of length 6 in. with an error of ±0.02 in., where $V(x) = x^3$ (the volume of a cube of side x)

with $dx = \pm0.02$. We find the derivative of the volume: $V'(x) = 3x^2$. Now we can plug into the formula: $dV = 3x^2 dx$. If we plug in $x = 6$ and $dx = \pm0.02$, we get $dV = 3(6)^2(\pm0.02) = \pm2.16$.

5.　　$\pi \approx 3.142$ mm^3

Recall the formula that we use when we want to approximate the error in a measurement: $dy = f'(x)\,dx$. Here, we want to approximate the increase in the volume of a sphere when we know that it has a radius of length 5 mm with an increase of 0.01 mm, where $V(r) = \frac{4}{3}\pi r^3$ (the volume of a sphere of radius r) with $dr = 0.01$. We find the derivative of the volume: $V'(r) = \frac{4}{3}\pi\left(3r^2\right) = 4\pi r^2$. Now we can plug into the formula: $dV = 4\pi r^2\,dr$. If we plug in $r = 5$ and $dr = 0.01$, we get

$dV = 4\pi(5)^2(0.01) = \pi \approx 3.142$.

6.　　(a)　1.963 m^3;　(b)　15.708 m^3

(a) Recall the formula that we use when we want to approximate the error in a measurement: $dy = f'(x)\,dx$. Here, we want to approximate the error in the volume of a cylinder when we know that it has a diameter of length 5 m (which means that its radius is 2.5 m) and its height is 20 m, with an error in the height of 0.1 m, where $V = \pi r^2 h$ with $dh = 0.1$. Note that the radius is exact, so when we take the derivative we will treat only the height as a variable. We find the derivative of the volume: $V' = \pi r^2$. Now we can plug into the formula: $dV = \pi r^2\,dh$. If we plug in $r = 2.5$, $h = 20$, and $dh = 0.1$, we get $dV = \pi(2.5)^2(0.1) = 0.625\pi \approx 1.963$.

(b) Here, we want to approximate the error in the volume of a cylinder when we know that it has a diameter of length 5 m (which means that its radius is 2.5 m) and its height is 20 m, with an error in the diameter of 0.1 m (which means that the error in the radius is 0.05 m), where $V = \pi r^2 h$ with $dr = 0.05$. Note that the height is exact, so when we take the derivative we will treat only the radius as a variable. We find the derivative of the volume: $V' = 2\pi rh$. Now we can plug into the formula: $dV = 2\pi rh\,dr$. If we plug in $r = 2.5$, $h = 20$, and $dr = 0.05$, we get $dV = 2\pi(2.5)(20)(0.05) = 5\pi \approx 15.708$.

PRACTICE PROBLEM SET 19

1.　　$y\left[\dfrac{1}{x} - \dfrac{3x^2}{4\left(1 - x^3\right)}\right]$

Anytime we are presented with finding the derivative of a complex expression, we look to see if we can use logarithmic differentiation to simplify the problem. Often, doing so means that we can avoid a messy expression involving the Product, Quotient, or Chain Rules. We always do the same four steps.

Step One: Take the log of both sides: $\ln y = \ln\left(x\sqrt[4]{1-x^3}\right)$.

Step Two: Simplify the expression using log rules.

$$\ln y = \ln x + \ln\left(\sqrt[4]{1-x^3}\right)$$

$$\ln y = \ln x + \frac{1}{4}\ln\left(1-x^3\right)$$

Step Three: Take the derivative of both sides: $\dfrac{1}{y}\dfrac{dy}{dx} = \dfrac{1}{x} + \dfrac{1}{4}\dfrac{-3x^2}{1-x^3} = \dfrac{1}{x} - \dfrac{3x^2}{4\left(1-x^3\right)}$.

Step Four: Multiply both sides by y: $\dfrac{dy}{dx} = y\left[\dfrac{1}{x} - \dfrac{-3x^2}{4\left(1-x^3\right)}\right]$.

2. $\dfrac{y}{2}\left[\dfrac{1}{1+x} + \dfrac{1}{1-x}\right]$

Anytime we are presented with finding the derivative of a complex expression, we look to see if we can use logarithmic differentiation to simplify the problem. Often, doing so means that we can avoid a messy expression involving the Product, Quotient, or Chain Rules. We always do the same four steps.

Step One: Take the log of both sides: $\ln y = \ln\sqrt{\dfrac{1+x}{1-x}}$.

Step Two: Simplify the expression using log rules.

$$\ln y = \frac{1}{2}\ln\left(\frac{1+x}{1-x}\right)$$

$$\ln y = \frac{1}{2}\left[\ln(1+x) - \ln(1-x)\right]$$

Step Three: Take the derivative of both sides: $\dfrac{1}{y}\dfrac{dy}{dx} = \dfrac{1}{2}\left[\dfrac{1}{1+x} - \dfrac{1}{1-x}\right]$.

Step Four: Multiply both sides by y: $\dfrac{dy}{dx} = \dfrac{y}{2}\left[\dfrac{1}{1+x} - \dfrac{1}{1-x}\right]$.

3.
$$y\left[\frac{9x^2}{2(x^3+5)}-\frac{2x}{3(4-x^2)}-\frac{4x^3-2x}{(x^4-x^2+6)}\right]$$

Anytime we are presented with finding the derivative of a complex expression, we look to see if we can use logarithmic differentiation to simplify the problem. Often, doing so means that we can avoid a messy expression involving the Product, Quotient, or Chain Rules. We always do the same four steps.

Step One: Take the log of both sides: $\ln y = \ln\dfrac{(x^3+5)^{\frac{3}{2}}\left(\sqrt[3]{4-x^2}\right)}{(x^4-x^2+6)}$.

Step Two: Simplify the expression using log rules.

$$\ln y = \ln\left(x^3+5\right)^{\frac{3}{2}}+\ln\left(\sqrt[3]{4-x^2}\right)-\ln\left(x^4-x^2+6\right)$$

$$\ln y = \frac{3}{2}\ln\left(x^3+5\right)+\frac{1}{3}\ln\left(4-x^2\right)-\ln\left(x^4-x^2+6\right)$$

Step Three: Take the derivative of both sides.

$$\frac{1}{y}\frac{dy}{dx}=\frac{3}{2}\left(\frac{3x^2}{x^3+5}\right)+\frac{1}{3}\left(\frac{-2x}{4-x^2}\right)-\frac{4x^3-2x}{x^4-x^2+6}=\frac{9x^2}{2(x^3+5)}-\frac{2x}{3(4-x^2)}-\frac{4x^3-2x}{(x^4-x^2+6)}$$

Step Four: Multiply both sides by y: $\dfrac{dy}{dx}=y\left[\dfrac{9x^2}{2(x^3+5)}-\dfrac{2x}{3(4-x^2)}-\dfrac{4x^3-2x}{(x^4-x^2+6)}\right]$.

4.
$$y\left[\cot x-\tan x-\frac{3x^2}{2(x^3-4)}\right]$$

Anytime we are presented with finding the derivative of a complex expression, we look to see if we can use logarithmic differentiation to simplify the problem. Often, doing so means that we can avoid a messy expression involving the Product, Quotient, or Chain Rules. We always do the same four steps.

Step One: Take the log of both sides: $\ln y = \ln\dfrac{\sin x\cos x}{\sqrt{x^3-4}}$.

Step Two: Simplify the expression using log rules.

$$\ln y = \ln \sin x + \ln \cos x - \ln \sqrt{x^3 - 4}$$

$$\ln y = \ln \sin x + \ln \cos x - \frac{1}{2}\ln\left(x^3 - 4\right)$$

Step Three: Take the derivative of both sides.

$$\frac{1}{y}\frac{dy}{dx} = \frac{\cos x}{\sin x} + \frac{-\sin x}{\cos x} - \frac{1}{2}\frac{3x^2}{x^3 - 4} = \cot x - \tan x - \frac{3x^2}{2\left(x^3 - 4\right)}$$

Step Four: Multiply both sides by y: $\dfrac{dy}{dx} = y\left[\cot x - \tan x - \dfrac{3x^2}{2\left(x^3 - 4\right)}\right]$.

5. $$y\left[\frac{6(x-1)}{\left(x^2 - 2x\right)} + \frac{4\left(-12x^3 + 7\right)}{\left(5 - 3x^4 + 7x\right)} - \frac{3(2x+1)}{\left(x^2 + x\right)}\right]$$

Anytime we are presented with finding the derivative of a complex expression, we look to see if we can use logarithmic differentiation to simplify the problem. Often, doing so means that we can avoid a messy expression involving the Product, Quotient, or Chain Rules. We always do the same four steps.

Step One: Take the log of both sides: $\ln y = \ln \dfrac{\left(4x^2 - 8x\right)^3\left(5 - 3x^4 + 7x\right)^4}{\left(x^2 + x\right)^3}$.

Step Two: Simplify the expression using log rules.

$$\ln y = \ln(4x^2 - 8x)^3 + \ln(5 - 3x^4 + 7x)^4 - \ln(x^2 + x)^3$$

$$\ln y = 3\ln(4x^2 - 8x) + 4\ln(5 - 3x^4 + 7x) - 3\ln(x^2 + x)$$

Step Three: Take the derivative of both sides.

$$\frac{1}{y}\frac{dy}{dx} = 3\frac{8x-8}{4x^2-8x} + 4\frac{-12x^3+7}{5-3x^4+7x} - 3\frac{2x+1}{x^2+x} = \frac{3(2x-2)}{x^2-2x} + \frac{4\left(-12x^3+7\right)}{\left(5-3x^4+7x\right)} - \frac{3(2x+1)}{\left(x^2+x\right)}$$

Step Four: Multiply both sides by y: $\dfrac{dy}{dx} = y\left[\dfrac{6(x-1)}{\left(x^2-2x\right)} + \dfrac{4\left(-12x^3+7\right)}{\left(5-3x^4+7x\right)} - \dfrac{3(2x+1)}{\left(x^2+x\right)}\right]$.

6. $$y\left[\frac{1}{x-1}-\frac{1}{x}-\sec x \csc x\right]$$

Anytime we are presented with finding the derivative of a complex expression, we look to see if we can use logarithmic differentiation to simplify the problem. Often, doing so means that we can avoid a messy expression involving the Product, Quotient, or Chain Rules. We always do the same four steps.

Step One: Take the log of both sides: $\ln y = \ln\dfrac{x-1}{x \tan x}$.

Step Two: Simplify the expression using log rules.

$$\ln y = \ln(x-1) - \ln x - \ln \tan x$$

Step Three: Take the derivative of both sides: $\dfrac{1}{y}\dfrac{dy}{dx} = \dfrac{1}{x-1} - \dfrac{1}{x} - \dfrac{\sec^2 x}{\tan x} = \dfrac{1}{x-1} - \dfrac{1}{x} - \sec x \csc x.$

Step Four: Multiply both sides by y: $\dfrac{dy}{dx} = y\left[\dfrac{1}{x-1} - \dfrac{1}{x} - \sec x \csc x\right].$

7. $$y\left[\frac{2(1-2x)}{\left(x-x^2\right)}+\frac{3\left(3x^2+4x^3\right)}{\left(x^3+x^4\right)}+\frac{4\left(6x^5-5x^4\right)}{\left(x^6-x^5\right)}\right]$$

Anytime we are presented with finding the derivative of a complex expression, we look to see if we can use logarithmic differentiation to simplify the problem. Often, doing so means that we can avoid a messy expression involving the Product, Quotient, or Chain Rules. We always do the same four steps.

Step One: Take the log of both sides: $\ln y = \ln\left[\left(x-x^2\right)^2\left(x^3+x^4\right)^3\left(x^6-x^5\right)^4\right]$.

Step Two: Simplify the expression using log rules.

$$\ln y = \ln(x-x^2)^2 + \ln(x^3+x^4)^3 + \ln(x^6-x^5)^4$$

$$\ln y = 2\ln(x-x^2) + 3\ln(x^3+x^4) + 4\ln(x^6-x^5)$$

Step Three: Take the derivative of both sides: $\dfrac{1}{y}\dfrac{dy}{dx} = 2\left(\dfrac{1-2x}{x-x^2}\right) + 3\left(\dfrac{3x^2+4x^3}{x^3+x^4}\right) + 4\left(\dfrac{6x^5-5x^4}{x^6-x^5}\right).$

Step Four: Multiply both sides by y: $\dfrac{dy}{dx} = y\left[\dfrac{2(1-2x)}{\left(x-x^2\right)} + \dfrac{3\left(3x^2+4x^3\right)}{\left(x^3+x^4\right)} + \dfrac{4\left(6x^5-5x^4\right)}{\left(x^6-x^5\right)}\right].$

8. $\dfrac{y}{4}\left[\dfrac{1}{x}-\dfrac{1}{1-x}+\dfrac{1}{1+x}-\dfrac{2x}{x^2-1}+\dfrac{1}{5-x}\right]$

Anytime we are presented with finding the derivative of a complex expression, we look to see if we can use logarithmic differentiation to simplify the problem. Often, doing so means that we can avoid a messy expression involving the Product, Quotient, or Chain Rules. We always do the same four steps.

Step One: Take the log of both sides: $\ln y=\ln\sqrt[4]{\dfrac{x(1-x)(1+x)}{(x^2-1)(5-x)}}$.

Step Two: Simplify the expression using log rules.

$$\ln y=\dfrac{1}{4}\ln\dfrac{x(1-x)(1+x)}{(x^2-1)(5-x)}$$

$$\ln y=\dfrac{1}{4}\left[\ln x+\ln(1-x)+\ln(1+x)-\ln(x^2-1)-\ln(5-x)\right]$$

Step Three: Take the derivative of both sides: $\dfrac{1}{y}\dfrac{dy}{dx}=\dfrac{1}{4}\left[\dfrac{1}{x}-\dfrac{1}{1-x}+\dfrac{1}{1+x}-\dfrac{2x}{x^2-1}+\dfrac{1}{5-x}\right]$.

Step Four: Multiply both sides by y: $\dfrac{dy}{dx}=\dfrac{y}{4}\left[\dfrac{1}{x}-\dfrac{1}{1-x}+\dfrac{1}{1+x}-\dfrac{2x}{x^2-1}+\dfrac{1}{5-x}\right]$.

PRACTICE PROBLEM SET 20

1. $-\dfrac{1}{3x^3}+C$

Here we will use the Power Rule, which says that $\int x^n dx=\dfrac{x^{n+1}}{n+1}+C$. The integral is

$\int\dfrac{1}{x^4}dx=\int x^{-4}dx=\dfrac{x^{-3}}{-3}+C=-\dfrac{1}{3x^3}+C$.

2. $10\sqrt{x}+C$

Here we will use the Power Rule, which says that $\int x^n dx=\dfrac{x^{n+1}}{n+1}+C$. The integral is

$\int\dfrac{5}{\sqrt{x}}dx=5\int x^{-\frac{1}{2}}dx=5\dfrac{x^{\frac{1}{2}}}{\frac{1}{2}}+C=10\sqrt{x}+C$.

3. $x^5 - x^3 + x^2 + 6x + C$

Here we will use the Power Rule, which says that $\int x^n \, dx = \dfrac{x^{n+1}}{n+1} + C$. The integral is

$$\int \left(5x^4 - 3x^2 + 2x + 6\right) dx = 5\frac{x^5}{5} - 3\frac{x^3}{3} + 2\frac{x^2}{2} + 6x + C = x^5 - x^3 + x^2 + 6x + C.$$

4. $-\dfrac{3}{2x^2} + \dfrac{2}{x} + \dfrac{x^5}{5} + 2x^8 + C$

Here we will use the Power Rule, which says that $\int x^n \, dx = \dfrac{x^{n+1}}{n+1} + C$. The integral is

$$\int \left(3x^{-3} - 2x^{-2} + x^4 + 16x^7\right) dx = 3\frac{x^{-2}}{-2} - 2\frac{x^{-1}}{-1} + \frac{x^5}{5} + 16\frac{x^8}{8} + C = -\frac{3}{2x^2} + \frac{2}{x} + \frac{x^5}{5} + 2x^8 + C.$$

5. $\dfrac{3x^{\frac{4}{3}}}{2} + \dfrac{3x^{\frac{7}{3}}}{7} + C$

Here we will use the Power Rule, which says that $\int x^n \, dx = \dfrac{x^{n+1}}{n+1} + C$. First, let's simplify

the integrand: $\int x^{\frac{1}{3}}(2 + x) \, dx = \int \left(2x^{\frac{1}{3}} + x^{\frac{4}{3}}\right) dx$. Now, we can evaluate the integral:

$$\int \left(2x^{\frac{1}{3}} + x^{\frac{4}{3}}\right) dx = 2\frac{x^{\frac{4}{3}}}{\frac{4}{3}} + \frac{x^{\frac{7}{3}}}{\frac{7}{3}} + C = \frac{3x^{\frac{4}{3}}}{2} + \frac{3x^{\frac{7}{3}}}{7} + C.$$

6. $\dfrac{x^7}{7} + \dfrac{2x^5}{5} + \dfrac{x^3}{3} + C$

Here we will use the Power Rule, which says that $\int x^n \, dx = \dfrac{x^{n+1}}{n+1} + C$. First, let's simplify the integrand: $\int (x^3 + x)^2 \, dx = \int (x^6 + 2x^4 + x^2) \, dx$. Now, we can evaluate the integral:

$$\int \left(x^6 + 2x^4 + x^2\right) dx = \frac{x^7}{7} + \frac{2x^5}{5} + \frac{x^3}{3} + C.$$

7. $\dfrac{x^5}{5} - \dfrac{2x^3}{3} - \dfrac{1}{x} + C$

Here we will use the Power Rule, which says that $\int x^n \, dx = \dfrac{x^{n+1}}{n+1} + C$. First,

let's simplify the integrand: $\int \dfrac{x^6 - 2x^4 + 1}{x^2} \, dx = \int \left(\dfrac{x^6}{x^2} - 2\dfrac{x^4}{x^2} + \dfrac{1}{x^2}\right) dx = \int \left(x^4 - 2x^2 + x^{-2}\right) dx$.

Now, we can evaluate the integral: $\int \left(x^4 - 2x^2 + x^{-2}\right) dx = \dfrac{x^5}{5} - \dfrac{2x^3}{3} + \dfrac{x^{-1}}{-1} + C = \dfrac{x^5}{5} - \dfrac{2x^3}{3} - \dfrac{1}{x} + C.$

8. $\dfrac{x^5}{5} - \dfrac{3x^4}{4} + x^3 - \dfrac{x^2}{2} + C$

Here we will use the Power Rule, which says that $\displaystyle\int x^n\, dx = \dfrac{x^{n+1}}{n+1} + C$.

First, let's simplify the integrand: $\displaystyle\int x(x-1)^3\, dx = \int x(x^3 - 3x^2 + 3x - 1)\, dx = \int (x^4 - 3x^3 + 3x^2 - x)\, dx$.

Now, we can evaluate the integral:

$$\int \left(x^4 - 3x^3 + 3x^2 - x\right)\, dx = \dfrac{x^5}{5} - \dfrac{3x^4}{4} + \dfrac{3x^3}{3} - \dfrac{x^2}{2} + C = \dfrac{x^5}{5} - \dfrac{3x^4}{4} + x^3 - \dfrac{x^2}{2} + C$$

9. $\tan x + \sec x + C$

Here we will use the Rules for the Integrals of Trig Functions, namely, $\displaystyle\int \sec^2 x\, dx = \tan x + C$ and $\displaystyle\int (\sec x \tan x)\, dx = \sec x + C$. First, let's expand the integrand: $\displaystyle\int \sec x\,(\sec x + \tan x)\, dx = \int (\sec^2 x + \sec x \tan x)\, dx$. We get $\displaystyle\int (\sec^2 x + \sec x \tan x)\, dx = \tan x + \sec x + C$.

10. $\tan x + \dfrac{x^2}{2} + C$

Here we will use the Rules for the Integrals of Trig Functions, namely, $\displaystyle\int \sec^2 x\, dx = \tan x + C$ and the Power Rule, which says that $\displaystyle\int x^n\, dx = \dfrac{x^{n+1}}{n+1} + C$. We get $\displaystyle\int \left(\sec^2 x + x\right)\, dx = \tan x + \dfrac{x^2}{2} + C$.

11. $\sin x + 4 \tan x + C$

Here we will use the Rules for the Integrals of Trig Functions, namely, $\displaystyle\int \sec^2 x\, dx = \tan x + C$ and $\displaystyle\int \cos x\, dx = \sin x + C$.

First, we need to rewrite the integrand, using trig identities:

$\displaystyle\int \dfrac{\cos^3 x + 4}{\cos^2 x}\, dx = \int \left(\dfrac{\cos^3 x}{\cos^2 x} + \dfrac{4}{\cos^2 x}\right) dx = \int \left(\cos x + 4\sec^2 x\right) dx$. Now, we can evaluate the integral: $\displaystyle\int (\cos x + 4\sec^2 x)\, dx = \sin x + 4\tan x + C$.

12. $-2\cos x + C$

Here we will use the Rules for the Integrals of Trig Functions, namely, $\displaystyle\int \sin x\, dx = -\cos x + C$. First, we need to rewrite the integrand, using trig identities: $\displaystyle\int \dfrac{\sin 2x}{\cos x}\, dx = \int \dfrac{2\sin x \cos x}{\cos x}\, dx = \int 2\sin x\, dx$. Now, we can evaluate the integral: $\displaystyle\int 2\sin x\, dx = -2\cos x + C$.

13. $x + \sin x + C$

Here we will use the Rules for the Integrals of Trig Functions, namely, $\int \cos x\, dx = \sin x + C$. First, we need to rewrite the integrand, using trig identities: $\int \left(1 + \cos^2 x \sec x\right) dx = \int \left(1 + \dfrac{\cos^2 x}{\cos x}\right) dx = \int \left(1 + \cos x\right) dx$. Now, we can evaluate the integral: $\int (1 + \cos x)\, dx = x + \sin x + C$.

14. $-\cos x + C$

Here we will use the Rules for the Integrals of Trig Functions, namely, $\int \sin x\, dx = -\cos x + C$. First, we need to rewrite the integrand, using trig identities: $\int \dfrac{1}{\csc x}\, dx = \int \sin x\, dx$. Now, we can evaluate the integral: $\int \sin x\, dx = -\cos x + C$.

15. $\dfrac{x^2}{2} - 2\tan x + C$

Here we will use the Rules for the Integrals of Trig Functions, namely, $\int \sec^2 x\, dx = \tan x + C$. First, we need to rewrite the integrand, using trig identities: $\int \left(x - \dfrac{2}{\cos^2 x}\right) dx = \int \left(x - 2\sec^2 x\right) dx$. Now, we can evaluate the integral: $\int \left(x - 2\sec^2 x\right) dx = \dfrac{x^2}{2} - 2\tan x + C$.

PRACTICE PROBLEM SET 21

1. $\dfrac{\sin^2 2x}{4} + C$

If we let $u = \sin 2x$, then $du = 2\cos 2x\, dx$. We need to substitute for $\cos 2x\, dx$, so we can divide the du term by 2: $\dfrac{du}{2} = \cos 2x\, dx$. Next we can substitute into the integral: $\int \sin 2x \cos 2x\, dx = \dfrac{1}{2} \int u\, du$. Now we can integrate: $\dfrac{1}{2} \int u\, du = \dfrac{1}{2}\left(\dfrac{u^2}{2}\right) + C = \dfrac{u^2}{4} + C$. Last, we substitute back and get $\dfrac{\sin^2 2x}{4} + C$.

2. $-\dfrac{9}{4}\left(10-x^2\right)^{\frac{2}{3}}+C$

First, pull the constant out of the integrand: $\displaystyle\int\frac{3x\,dx}{\sqrt[3]{10-x^2}}\,dx=3\int\frac{x\,dx}{\sqrt[3]{10-x^2}}\,dx$. If we let $u=10-x^2$,

then $du=-2x\,dx$. We need to substitute for $x\,dx$, so we can divide the du term by -2: $-\dfrac{du}{2}=x\,dx$.

Next we can substitute into the integral: $3\displaystyle\int\frac{x\,dx}{\sqrt[3]{10-x^2}}\,dx=-\frac{3}{2}\int u^{-\frac{1}{3}}\,du$. Now we can integrate:

$-\dfrac{3}{2}\displaystyle\int u^{-\frac{1}{3}}\,du=-\frac{3}{2}\frac{u^{\frac{2}{3}}}{\frac{2}{3}}+C=-\frac{9}{4}u^{\frac{2}{3}}+C$. Last, we substitute back and get $-\dfrac{9}{4}\left(10-x^2\right)^{\frac{2}{3}}+C$.

3. $\dfrac{1}{30}\left(5x^4+20\right)^{\frac{3}{2}}+C$

If we let $u=5x^4+20$, then $du=20x^3\,dx$. We need to substitute for $x^3\,dx$, so we can divide the du

term by 20: $\dfrac{du}{20}=x^3\,dx$. Next $\dfrac{1}{20}\displaystyle\int u^{\frac{1}{2}}\,du=\frac{1}{20}\frac{u^{\frac{3}{2}}}{\frac{3}{2}}+C=\frac{1}{30}u^{\frac{3}{2}}+C$. Last, we substitute back and

get $\dfrac{1}{30}\left(5x^4+20\right)^{\frac{3}{2}}+C$.

4. $-\dfrac{1}{12\left(x^3+3x\right)^4}+C$

If we let $u=x^3+3x$, then $du=(3x^2+3)\,dx$. We need to substitute for $(x^2+1)\,dx$, so we can

divide the du term by 3: $\dfrac{du}{3}=\left(x^2+1\right)dx$. Now we can substitute into the integral:

$\displaystyle\int\left(x^2+1\right)\left(x^3+3x\right)^{-5}dx=\frac{1}{3}\int u^{-5}\,du$. Next we can integrate: $\dfrac{1}{3}\displaystyle\int u^{-5}\,du=\frac{1}{3}\frac{u^{-4}}{-4}+C=-\frac{1}{12}\frac{1}{u^4}+C$.

Last, we substitute back and get $-\dfrac{1}{12\left(x^3+3x\right)^4}+C$.

5. $-2\cos\sqrt{x}+C$

If we let $u=\sqrt{x}$, then $du=\dfrac{1}{2\sqrt{x}}\,dx$. We need to substitute for $\dfrac{1}{\sqrt{x}}\,dx$, so we can multiply the

du term by 2: $2\,du=\dfrac{1}{\sqrt{x}}\,dx$. Next we can substitute into the integral: $\displaystyle\int\frac{1}{\sqrt{x}}\sin\sqrt{x}\,dx=2\int\sin u\,du$.

Now we can integrate: $2\displaystyle\int\sin u\,du=-2\cos u+C$. Last, we substitute back and get $-2\cos\sqrt{x}+C$.

6. $\dfrac{1}{3}\tan\left(x^3\right)+C$

If we let $u = x^3$, then $du = 3x^2\, dx$. We need to substitute for $x^2\, dx$, so we can divide the du term by

3: $\dfrac{du}{3} = x^2\, dx$. Next we can substitute into the integral: $\displaystyle\int x^2 \sec^2\left(x^3\right)dx = \dfrac{1}{3}\int \sec^2 u\, du$. Now we

can integrate: $\dfrac{1}{3}\displaystyle\int \sec^2 u\, du = \dfrac{1}{3}\tan u + C$. Last, we substitute back and get $\dfrac{1}{3}\tan\left(x^3\right)+C$.

7. $-\dfrac{1}{3}\sin\left(\dfrac{3}{x}\right)+C$

If we let $u = \dfrac{3}{x}$, then $du = -\dfrac{3}{x^2}\, dx$. We need to substitute for $\dfrac{1}{x^2}\, dx$, so we can divide the du

term by -3: $-\dfrac{du}{3} = \dfrac{1}{x^2}\, dx$. Next we can substitute into the integral: $\displaystyle\int \dfrac{\cos\left(\dfrac{3}{x}\right)}{x^2}\, dx = -\dfrac{1}{3}\int \cos u\, du$.

Now we can integrate: $-\dfrac{1}{3}\displaystyle\int \cos u\, du = -\dfrac{1}{3}\sin u + C$. Last, we substitute back and get

$-\dfrac{1}{3}\sin\left(\dfrac{3}{x}\right)+C$.

8. $-\cos(\sin x) + C$

If we let $u = \sin x$, then $du = \cos x\, dx$. Next, we can substitute into the integral: $\int\sin(\sin x)\cos x\, dx = \int\sin u\, du$. Now we can integrate: $\int\sin u\, du = -\cos u + C$. Last, we substitute back and get $-\cos(\sin x) + C$.

PRACTICE PROBLEM SET 22

1. $\dfrac{25}{32}$

First, let's draw a picture.

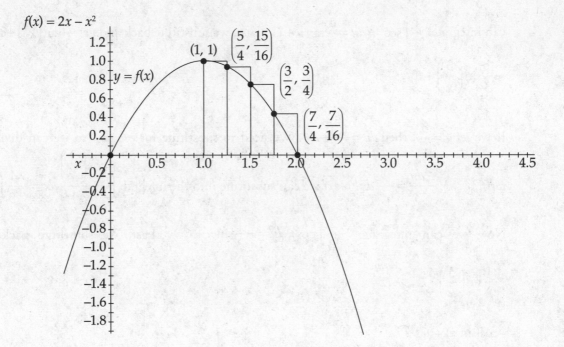

The width of each rectangle is found by taking the difference between the endpoints and dividing by n. Here the width of each rectangle is $\dfrac{2-1}{4} = \dfrac{1}{4}$.

We find the heights of the rectangles by evaluating $y = 2x - x^2$ at the appropriate endpoints.

$y(1) = 2(1) - (1)^2 = 1$; $y\left(\dfrac{5}{4}\right) = 2\left(\dfrac{5}{4}\right) - \left(\dfrac{5}{4}\right)^2 = \dfrac{15}{16}$; $y\left(\dfrac{3}{2}\right) = 2\left(\dfrac{3}{2}\right) - \left(\dfrac{3}{2}\right)^2 = \dfrac{3}{4}$

and $y\left(\dfrac{7}{4}\right) = 2\left(\dfrac{7}{4}\right) - \left(\dfrac{7}{4}\right)^2 = \dfrac{7}{16}$.

Therefore, the area is $\left(\dfrac{1}{4}\right)\left(1 + \dfrac{15}{16} + \dfrac{3}{4} + \dfrac{7}{16}\right) = \dfrac{25}{32}$.

2. $\dfrac{17}{32}$

First, let's draw a picture.

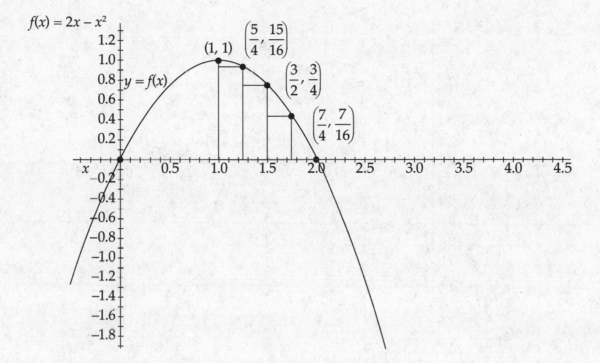

The width of each rectangle is found by taking the difference between the endpoints and dividing by n. Here the width of each rectangle is $\dfrac{2-1}{4} = \dfrac{1}{4}$.

We find the heights of the rectangles by evaluating $y = 2x - x^2$ at the appropriate endpoints.

$$y\left(\dfrac{5}{4}\right) = 2\left(\dfrac{5}{4}\right) - \left(\dfrac{5}{4}\right)^2 = \dfrac{15}{16}; \quad y\left(\dfrac{3}{2}\right) = 2\left(\dfrac{3}{2}\right) - \left(\dfrac{3}{2}\right)^2 = \dfrac{3}{4}; \quad y\left(\dfrac{7}{4}\right) = 2\left(\dfrac{7}{4}\right) - \left(\dfrac{7}{4}\right)^2 = \dfrac{7}{16};$$

and $y(2) = 2(2) - (2)^2 = 0$.

Therefore, the area is $\left(\dfrac{1}{4}\right)\left(\dfrac{15}{16} + \dfrac{3}{4} + \dfrac{7}{16} + 0\right) = \dfrac{17}{32}$.

3. $\dfrac{21}{32}$

First, let's draw a picture.

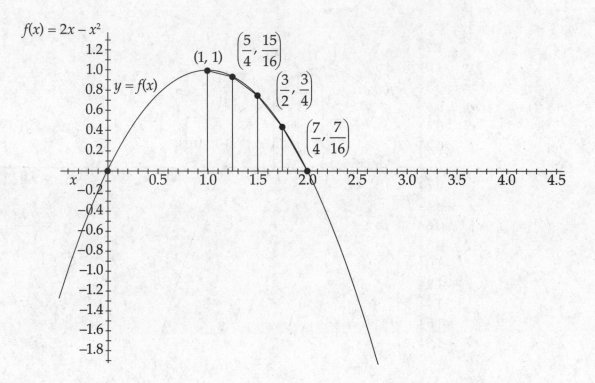

The height of each trapezoid is found by taking the difference between the endpoints and dividing by n. Here the width of each trapezoid is $\dfrac{2-1}{4} = \dfrac{1}{4}$.

We find the bases of the trapezoids by evaluating $y = 2x - x^2$ at the appropriate endpoints.

$y(1) = 2(1) - (1)^2 = 1$; $y\left(\dfrac{5}{4}\right) = 2\left(\dfrac{5}{4}\right) - \left(\dfrac{5}{4}\right)^2 = \dfrac{15}{16}$; $y\left(\dfrac{3}{2}\right) = 2\left(\dfrac{3}{2}\right) - \left(\dfrac{3}{2}\right)^2 = \dfrac{3}{4}$;

$y\left(\dfrac{7}{4}\right) = 2\left(\dfrac{7}{4}\right) - \left(\dfrac{7}{4}\right)^2 = \dfrac{7}{16}$; and $y(2) = 2(2) - (2)^2 = 0$.

Therefore, the area is $\left(\dfrac{1}{2}\right)\left(\dfrac{1}{4}\right)\left[1 + (2)\left(\dfrac{15}{16}\right) + (2)\left(\dfrac{3}{4}\right) + (2)\left(\dfrac{7}{16}\right) + 0\right] = \dfrac{21}{32}$.

4. $\dfrac{43}{64}$

First, let's draw a picture.

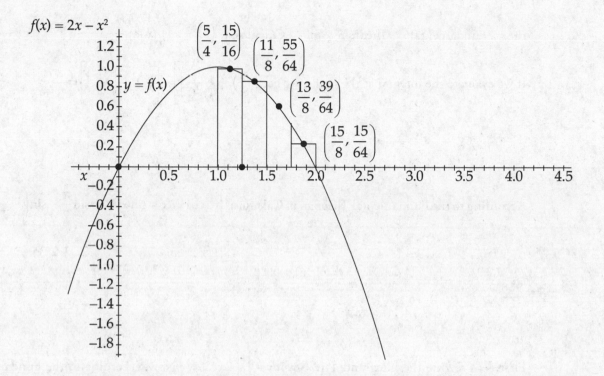

The height of each rectangle is found by taking the difference between the endpoints and dividing by n. Here the width of each rectangle is $\dfrac{2-1}{4} = \dfrac{1}{4}$.

We find the bases of the rectangles by evaluating $y = 2x - x^2$ at the appropriate endpoints.

$y\left(\dfrac{9}{8}\right) = 2\left(\dfrac{9}{8}\right) - \left(\dfrac{9}{8}\right)^2 = \dfrac{63}{64}$; $y\left(\dfrac{11}{8}\right) = 2\left(\dfrac{11}{8}\right) - \left(\dfrac{11}{8}\right)^2 = \dfrac{55}{64}$; $y\left(\dfrac{13}{8}\right) = 2\left(\dfrac{13}{8}\right) - \left(\dfrac{13}{8}\right)^2 = \dfrac{39}{64}$;

and $y\left(\dfrac{15}{8}\right) = 2\left(\dfrac{15}{8}\right) - \left(\dfrac{15}{8}\right)^2 = \dfrac{15}{64}$.

Therefore, the area is $\left(\dfrac{1}{4}\right)\left(\dfrac{63}{64} + \dfrac{55}{64} + \dfrac{39}{64} + \dfrac{15}{64}\right) = \dfrac{43}{64}$.

5. $\dfrac{2}{3}$

We will find the exact area by evaluating the integral $\int_1^2 \left(2x - x^2\right) dx$. According to the Fundamental Theorem of Calculus: $\int_1^2 \left(2x - x^2\right) dx = \left(\dfrac{2x^2}{2} - \dfrac{x^3}{3}\right)\Big|_1^2 = \left(x^2 - \dfrac{x^3}{3}\right)\Big|_1^2$.

If we evaluate the integral at the limits, we get $\left[(2)^2 - \left(\dfrac{2^3}{3}\right)\right] - \left[(1)^2 - \left(\dfrac{1^3}{3}\right)\right] = \dfrac{2}{3}$.

6.   2

According to the Fundamental Theorem of Calculus: $\int_{-\frac{\pi}{2}}^{\frac{\pi}{2}} \cos x \, dx = \sin x \Big|_{-\frac{\pi}{2}}^{\frac{\pi}{2}} = \sin \dfrac{\pi}{2} - \sin\left(-\dfrac{\pi}{2}\right) = 2$.

7. $\dfrac{968}{5}$

First, let's rewrite the integrand: $\int_1^9 \left(2x\sqrt{x}\right) dx = \int_1^9 \left(2x^{\frac{3}{2}}\right) dx$. Now, according to the Fundamental

Theorem of Calculus: $\int_1^9 \left(2x^{\frac{3}{2}}\right) dx = \dfrac{2x^{\frac{5}{2}}}{\frac{5}{2}}\Big|_1^9 = \dfrac{4x^{\frac{5}{2}}}{5}\Big|_1^9 = \left(\dfrac{4(9)^{\frac{5}{2}}}{5} - \dfrac{4(1)^{\frac{5}{2}}}{5}\right) = \dfrac{968}{5}$.

8. $-\dfrac{161}{20}$

According to the Fundamental Theorem of Calculus:

$$\int_0^1 \left(x^4 - 5x^3 + 3x^2 - 4x - 6\right) dx = \left(\dfrac{x^5}{5} - 5\dfrac{x^4}{4} + x^3 - 2x^2 - 6x\right)\Big|_0^1 =$$

$$\left(\dfrac{(1)^5}{5} - 5\dfrac{(1)^4}{4} + (1)^3 - 2(1)^2 - 6(1)\right) - 0 = -\dfrac{161}{20}$$

9. 0

According to the Fundamental Theorem of Calculus:

$$\int_{-\frac{\pi}{2}}^{\frac{\pi}{2}} \sin x \, dx = -\cos x \Big|_{-\frac{\pi}{2}}^{\frac{\pi}{2}} = -\cos\frac{\pi}{2} + \cos\left(-\frac{\pi}{2}\right) = 0$$

10. 216

Recall that the formula for finding the area under the curve using the left end-points is $\left(\frac{b-a}{n}\right)[y_0 + y_1 + y_2 + \ldots + y_{n-1}]$. This formula assumes that the x-values are evenly spaced, but they aren't here, so we will replace the values of $\left(\frac{b-a}{n}\right)$ with the appropriate widths of each rectangle. The width of the first rectangle is $1 - 0 = 1$; the second width is $3 - 1 = 2$; the third is $5 - 3 = 2$; the fourth is $9 - 5 = 4$; and the fifth is $14 - 9 = 5$. We find the height of each rectangle by evaluating $g(x)$ at the appropriate value of x, the left endpoint of each interval on the x-axis. Here, $y_0 = 10$, $y_1 = 8$, $y_2 = 11$, $y_3 = 17$, and $y_4 = 20$. Therefore, we can approximate the integral with:

$$\int_0^{14} g(x) \, dx = (1)(10) + (2)(8) + (2)(11) + (4)(17) + (5)(20) = 216.$$

PRACTICE PROBLEM SET 23

1. $2\sqrt{\dfrac{2}{\pi}}$

We find the average value of the function, $f(x)$, on the interval $[a, b]$ using the formula $f(c) = \dfrac{1}{b-a}\int_a^b f(x)\,dx$. Here we are looking for the average value of $f(x) = 4x \cos x^2$ on the interval $\left[0, \sqrt{\dfrac{\pi}{2}}\right]$. Using the formula, we need to find

$$\frac{1}{\sqrt{\frac{\pi}{2}} - 0}\int_0^{\sqrt{\frac{\pi}{2}}}\left(4x\cos x^2\right)dx = \sqrt{\frac{2}{\pi}}\int_0^{\sqrt{\frac{\pi}{2}}}\left(4x\cos x^2\right)dx.$$ We will need to use u-substitution

to evaluate the integral. Let $u = x^2$ and $du = 2x\ dx$. We need to substitute for $4x\ dx$, so we multiply by 2 to get $2\ du = 4x\ dx$. Now we can substitute into the integral:

$\int (4x \cos x^2)dx = 2 \int \cos u\ du = 2 \sin u$. If we substitute back, we get $2 \sin x^2$. Now we can evaluate the integral: $\sqrt{\dfrac{2}{\pi}} \int_0^{\sqrt{\frac{\pi}{2}}} \left(4x \cos x^2\right) dx = \sqrt{\dfrac{2}{\pi}} \left(2 \sin x^2\right)\Big|_0^{\sqrt{\frac{\pi}{2}}} = \sqrt{\dfrac{2}{\pi}}\left(2 \sin \dfrac{\pi}{2} - 2\sin 0\right) = 2\sqrt{\dfrac{2}{\pi}}$.

2. $\dfrac{8}{3}$

We find the average value of the function, $f(x)$, on the interval $[a, b]$ using the formula $f(c) = \dfrac{1}{b-a}\int_a^b f(x)\ dx$. Here we are looking for the average value of $f(x) = \sqrt{x}$ on the interval $[0, 16]$. Using the formula, we need to find $\dfrac{1}{16-0}\int_0^{16} \left(\sqrt{x}\right) dx$. We get

$$\dfrac{1}{16}\int_0^{16} \left(\sqrt{x}\right) dx = \dfrac{1}{16}\left(\dfrac{x^{\frac{3}{2}}}{\dfrac{3}{2}}\right)\Bigg|_0^{16} = \dfrac{1}{24}\left(x^{\frac{3}{2}}\right)\Bigg|_0^{16} = \dfrac{1}{24}\left((16)^{\frac{3}{2}} - 0\right) = \dfrac{8}{3}.$$

3. 1

We find the average value of the function, $f(x)$, on the interval $[a, b]$ using the formula $f(c) = \dfrac{1}{b-a}\int_a^b f(x)\ dx$. Here we are looking for the average value of $f(x) = 2|x|$ on the interval $[-1, 1]$. Using the formula, we need to find $\dfrac{1}{1-(-1)}\int_{-1}^1 (2|x|)\ dx$. Recall that the absolute value function must be rewritten as a piecewise function: $|x| = \begin{cases} x; x \geq 0 \\ -x; x < 0 \end{cases}$. Thus, we need to split the integral into two separate integrals in order to evaluate it: $\dfrac{1}{1-(-1)}\int_{-1}^1 (2|x|)\ dx = \int_{-1}^0 (-x)\ dx + \int_0^1 x\ dx$.

We get $\int_{-1}^0 (-x)\ dx + \int_0^1 x\ dx = \left(-\dfrac{x^2}{2}\right)\Bigg|_{-1}^0 + \left(\dfrac{x^2}{2}\right)\Bigg|_{-1}^0 = \left(0 - \left(-\dfrac{1}{2}\right)\right) + \left(\dfrac{1}{2} - 0\right) = 1$.

4. $\sin^2 x$

We find the derivative of an integral using the Second Fundamental Theorem of Calculus: $\dfrac{d}{dx}\displaystyle\int_a^x f(t)\, dt = f(x)$. We get $\dfrac{d}{dx}\displaystyle\int_1^x \sin^2 t\, dt = \sin^2 x$.

5. $27x^2 - 9x$

We find the derivative of an integral using the Second Fundamental Theorem of Calculus: $\dfrac{d}{dx}\displaystyle\int_a^x f(t)\, dt = f(x)$. We get $\dfrac{d}{dx}\displaystyle\int_1^{3x} \left(t^2 - t\right) dt = 3\left[(3x)^2 - (3x)\right] = 27x^2 - 9x$. Don't

forget that because the upper limit is a function, we need to multiply the answer by the

derivative of that function.

6. $-2\cos x$

We find the derivative of an integral using the Second Fundamental Theorem of Calculus: $\dfrac{d}{dx}\displaystyle\int_a^x f(t)\, dt = f(x)$. We get $\dfrac{d}{dx}\displaystyle\int_1^x -2\cos t\, dt = -2\cos x$.

PRACTICE PROBLEM SET 24

1. $\ln |\tan x| + C$

Whenever we have an integral in the form of a quotient, we check to see if the solution

is a logarithm. A clue is whether the numerator is the derivative of the denominator, as

it is here. Let's use u-substitution. If we let $u = \tan x$, then $du = \sec^2 x\, dx$. If we substitute

into the integrand, we get $\displaystyle\int \frac{\sec^2 x}{\tan x}\, dx = \int \frac{du}{u}$. Recall that $\displaystyle\int \frac{du}{u} = \ln |u| + C$. Substituting back,

we get $\displaystyle\int \frac{\sec^2 x}{\tan x}\, dx = \ln |\tan x| + C$.

2. $-\ln|1-\sin x| + C$

Whenever we have an integral in the form of a quotient, we check to see if the solution is a logarithm. A clue is whether the numerator is the derivative of the denominator, as it is here. Let's use u-substitution. If we let $u = 1 - \sin x$, then $du = -\cos x\ dx$. If we substitute into the integrand, we get $\int \frac{\cos x}{1-\sin x}\,dx = -\int \frac{du}{u}$. Recall that $\int \frac{du}{u} = \ln|u| + C$. Substituting back, we get $\int \frac{\cos x}{1-\sin x}\,dx = -\ln|1-\sin x| + C$.

3. $\ln|\ln x| + C$

Whenever we have an integral in the form of a quotient, we check to see if the solution is a logarithm. A clue is whether the numerator is the derivative of the denominator, as it is here. Let's use u-substitution. If we let $u = \ln x$, then $du = \frac{1}{x}dx$. If we substitute into the integrand, we get $\int \frac{1}{x\ln x}\,dx = \int \frac{du}{u}$. Recall that $\int \frac{du}{u} = \ln|u| + C$. Substituting back, we get $\int \frac{1}{x\ln x}\,dx = \ln|\ln x| + C$.

4. $-\ln|\cos x| - x + C$

First, let's rewrite the integrand.

$$\int \frac{\sin x - \cos x}{\cos x}\,dx = \int \left(\frac{\sin x}{\cos x} - \frac{\cos x}{\cos x}\right)dx = \int \left(\frac{\sin x}{\cos x} - 1\right)dx = \int \left(\frac{\sin x}{\cos x}\right)dx - \int dx$$

Whenever we have an integral in the form of a quotient, we check to see if the solution is a logarithm.

A clue is whether the numerator is the derivative of the denominator, as it is in the first integral.

Let's use u-substitution. If we let $u = \cos x$, then $du = -\sin x\ dx$. If we substitute into the integrand, we get $\int \left(\frac{\sin x}{\cos x}\right)dx = -\int \frac{du}{u}$. Recall that $\int \frac{du}{u} = \ln|u| + C$. Substituting back, we get $\int \left(\frac{\sin x}{\cos x}\right)dx = -\ln|\cos x| + C$. The second integral is simply $\int dx = x + C$. Therefore, the integral is

$$\int \left(\frac{\sin x}{\cos x}\right)dx - \int dx = -\ln|\cos x| - x + C.$$

5. $\ln\left(1+2\sqrt{x}\right)+C$

Whenever we have an integral in the form of a quotient, we check to see if the solution is a logarithm. A clue is whether the numerator is the derivative of the denominator, as it is here. Let's use u-substitution. If we let $u=1+2\sqrt{x}$, then $du=\dfrac{1}{\sqrt{x}}dx$. If we substitute into the integrand,

we get $\displaystyle\int\dfrac{1}{\sqrt{x}\left(1+2\sqrt{x}\right)}dx=\int\dfrac{du}{u}$.   Recall   that $\displaystyle\int\dfrac{du}{u}=\ln|u|+C$. Substituting back, we get

$\displaystyle\int\dfrac{1}{\sqrt{x}\left(1+2\sqrt{x}\right)}dx=\ln\left(1+2\sqrt{x}\right)+C$. Notice that we don't need the absolute value bars because $1+2\sqrt{x}$ is never negative.

6. $\ln(1+e^x)+C$

Whenever we have an integral in the form of a quotient, we check to see if the solution is a logarithm. A clue is whether the numerator is the derivative of the denominator, as it is here.

Let's use u-substitution. If we let $u=1+e^x$, then $du=e^x\,dx$. If we substitute into the integrand, we get $\displaystyle\int\dfrac{e^x}{1+e^x}dx=\int\dfrac{du}{u}$. Recall that $\displaystyle\int\dfrac{du}{u}=\ln|u|+C$. Substituting back, we get

$\displaystyle\int\dfrac{e^x}{1+e^x}dx=\ln\left(1+e^x\right)+C$. Notice that we don't need the absolute value bars because $1+e^x$ is never negative.

7. $\dfrac{1}{10}e^{5x^2-1}+C$

Recall that $\displaystyle\int e^u\,du=e^u+C$. Let's use u-substitution. If we let $u=5x^2-1$, then $du=10x\,dx$. But we need to substitute for $x\,dx$, so if we divide du by 10, we get $\dfrac{1}{10}du=x\,dx$.

Now, if we substitute into the integrand, we get $\displaystyle\int xe^{5x^2-1}dx=\dfrac{1}{10}\int e^u\,du=\dfrac{1}{10}e^u+C$. Substituting back, we get $\displaystyle\int xe^{5x^2-1}dx=\dfrac{1}{10}e^{5x^2-1}+C$.

8. $\ln \left|e^x - e^{-x}\right| + C$

Recall that $\int e^u \, du = e^u + C$. Let's use u-substitution. If we let $u = e^x - e^{-x}$, then

$du = e^x + e^{-x} \, dx$. If we substitute into the integrand, we get $\int \dfrac{e^x + e^{-x}}{e^x - e^{-x}} \, dx = \int \dfrac{du}{u}$. Recall that

$\int \dfrac{du}{u} = \ln|u| + C$. Substituting back, we get $\int \dfrac{e^x + e^{-x}}{e^x - e^{-x}} \, dx = \ln\left|e^x - e^{-x}\right| + C$.

9. $-\dfrac{4^{-x^2}}{\ln 16} + C$

Recall that $\int a^u \, du = \dfrac{1}{\ln a} a^u + C$. Let's use u-substitution. If we let $u = -x^2$, then

$du = -2x \, dx$. But we need to substitute for $x \, dx$, so if we divide du by -2, we get $-\dfrac{1}{2} du = x \, dx$.

Now, if we substitute into the integrand, we get $\int x 4^{-x^2} \, dx = -\dfrac{1}{2} \int 4^u \, du = -\dfrac{1}{2}\left(\dfrac{1}{\ln 4}\right) 4^u + C$. Sub-

stituting back, we get $\int x 4^{-x^2} \, dx = -\dfrac{1}{2}\left(\dfrac{1}{\ln 4}\right) 4^{-x^2} + C = -\dfrac{1}{\ln 16} 4^{-x^2} + C$.

10. $\dfrac{7^{\sin x}}{\ln 7} + C$

Recall that $\int a^u \, du = \dfrac{1}{\ln a} a^u + C$. Let's use u-substitution. If we let $u = \sin x$, then

$du = \cos x \, dx$. If we substitute into the integrand, we get $\int 7^{\sin x} \cos x \, dx = \int 7^u \, du = \dfrac{1}{\ln 7} 7^u + C$.

Substituting back, we get $\int 7^{\sin x} \cos x \, dx = \dfrac{7^{\sin x}}{\ln 7} + C$.

PRACTICE PROBLEM SET 25

1. $-x \cot x + \ln\left|\sin x\right| + C$

Recall that *integration by parts* is a way of evaluating integrals of the form $\int u \, dv$, where both u

and v are functions of x. The formula for evaluating the integral is: $\int u \, dv = uv - \int v \, du$. Here,

if we let $u = x$ and $dv = \csc^2 x \, dx$, then we can find du by taking the derivative of u, and we

can find v by taking the integral of dv. We get $du = dx$ and $v = \int \csc^2 x \, dx = -\cot x$. Plugging

into the formula, we get $\int x \csc^2 x \, dx = -x \cot x - \int (-\cot x) \, dx = -x \cot x + \int \cot x \, dx$. Now we just have to find $\int \cot x \; dx$. We do this by rewriting the integrand in terms of sin x and cos x:

$\int \cot x \; dx = \int \dfrac{\cos x}{\sin x} dx = \ln|\sin x| + C$. Therefore, $\int x \csc^2 x \; dx = -x \cot x + \ln|\sin x| + C$.

2. $\dfrac{x}{2} e^{2x} - \dfrac{1}{4} e^{2x} + C$

Recall that *integration by parts* is a way of evaluating integrals of the form $\int u \; dv$, where both u

and v are functions of x. The formula for evaluating the integral is $\int u \; dv = uv - \int v \; du$. Here,

if we let $u = x$ and $dv = e^{2x}dx$, then we can find du by taking the derivative of u, and we can find v

by taking the integral of dv. We get $du = dx$ and $v = \int e^{2x}dx = \dfrac{1}{2} e^{2x}$. Plugging into the formula,

we get $\int xe^{2x}dx = x\left(\dfrac{1}{2} e^{2x}\right) - \dfrac{1}{2}\int e^{2x}dx$. Now we just have to find $\dfrac{1}{2}\int e^{2x}dx : \dfrac{1}{2}\int e^{2x}dx = \dfrac{1}{4} e^{2x} + C$.

Therefore, $\int xe^{2x}dx = \dfrac{x}{2} e^{2x} - \dfrac{1}{4} e^{2x} + C$.

3. $-\dfrac{1}{x} \ln x - \dfrac{1}{x} + C$

Recall that *integration by parts* is a way of evaluating integrals of the form $\int u \; dv$, where both u

and v are functions of x. The formula for evaluating the integral is $\int u \; dv = uv - \int v \; du$. Here,

if we let $u = \ln x$ and $dv = \dfrac{1}{x^2} dx$, then we can find du by taking the derivative of u, and we can

find v by taking the integral of dv. We get $du = \dfrac{1}{x} dx$ and $v = \int \dfrac{1}{x^2} dx = -\dfrac{1}{x}$. Plugging into the

formula, we get $\int \dfrac{\ln x}{x^2} dx = -\dfrac{\ln x}{x} - \int \left(-\dfrac{1}{x}\right)\left(\dfrac{1}{x}\right) dx = -\dfrac{\ln x}{x} + \int x^{-2}dx$. Now we just have to find

$\int x^{-2}dx : \int x^{-2}dx = -\dfrac{1}{x} + C$. Therefore, $\int \dfrac{\ln x}{x^2} dx = -\dfrac{\ln x}{x} - \dfrac{1}{x} + C$.

4. $x^2\sin x + 2x\cos x - 2\sin x + C$

Recall that *integration by parts* is a way of evaluating integrals of the form $\int u\,dv$, where both u and v are functions of x. The formula for evaluating the integral is $\int u\,dv = uv - \int v\,du$. Here, if we let $u = x^2$ and $dv = \cos x\,dx$, then we can find du by taking the derivative of u, and we can find v by taking the integral of dv. We get $du = 2x\,dx$ and $v = \int \cos x\,dx = \sin x$. Plugging into the formula, we get $\int x^2 \cos x\,dx = x^2 \sin x - 2\int x\sin x\,dx$. Now we have to find $\int x\sin x\,dx$. Once again we will have to use integration by parts. This time, if we let $u = x$ and $dv = \sin x\,dx$, then $du = dx$ and $v = \int \sin x\,dx = -\cos x$. Plugging into the formula, we get $\int x\sin x\,dx = -x\cos x + \int \cos x\,dx$. Finally, we have to find $\int \cos x\,dx = \sin x + C$. Putting everything together, we get $\int x^2 \cos x\,dx = x^2 \sin x - 2\left(-x\cos x + \sin x\right) + C = x^2 \sin x + 2x\cos x - 2\sin x + C$.

5. $-\dfrac{1}{2}x\cos 2x + \dfrac{1}{4}\sin 2x + C$

Recall that *integration by parts* is a way of evaluating integrals of the form $\int u\,dv$, where both u and v are functions of x. The formula for evaluating the integral is $\int u\,dv = uv - \int v\,du$. Here, if we let $u = x$, and $du = \sin 2x\,dx$, then we can find du by taking the derivative of u, and we can find v by taking the integral of dv. We get $du = dx$ and $v = \int \sin 2x\,dx = -\dfrac{1}{2}\cos 2x$. Plugging into the formula, we get $\int x\sin 2x\,dx = -\dfrac{1}{2}x\cos 2x + \dfrac{1}{2}\int \cos 2x\,dx$. Now we just have to find $\dfrac{1}{2}\int \cos 2x\,dx = \dfrac{1}{4}\sin 2x + C$. Therefore, $\int x\sin 2x\,dx = -\dfrac{1}{2}x\cos 2x + \dfrac{1}{4}\sin 2x + C$.

6. $x\ln^2 x - 2x\ln x + 2x + C$

Recall that *integration by parts* is a way of evaluating integrals of the form $\int u\,dv$, where both u and v are functions of x. The formula for evaluating the integral is $\int u\,dv = uv - \int v\,du$.

Here, if we let $u = \ln^2 x$ and $dv = dx$, then we can find du by taking the derivative of u, and we can find v by taking the integral of dv. We get $du = 2\ln x\left(\dfrac{1}{x}\right)dx$ and $v = \int dx = x$. Plugging into the formula, we get $\int \ln^2 x\, dx = x\ln^2 x - 2\int x\ln x\left(\dfrac{1}{x}\right)dx = x\ln^2 x - 2\int \ln x\, dx$. Now we have to find $\int \ln x\, dx$. Once again we will have to use integration by parts. This time, if we let $u = \ln x$ and $dv = dx$, then $du = \dfrac{1}{x}dx$ and $v = \int dx = x$. Plugging into the formula, we get $\int \ln x\, dx = x\ln x - \int x\left(\dfrac{1}{x}\right)dx = x\ln x - \int dx$. Finally, we have to find $\int x = x + C$. Putting everything together, we get $\int \ln^2 x\, dx = x\ln^2 x - 2\left(x\ln x - x\right) = x\ln^2 x - 2x\ln x + 2x + C$.

7. $\quad x\tan x + \ln\left|\cos x\right| + C$

Recall that *integration by parts* is a way of evaluating integrals of the form $\int u\, dv$, where both u and v are functions of x. The formula for evaluating the integral is $\int u\, dv = uv - \int v\, du$. Here, if we let $u = x$ and $dv = \sec^2 x\, dx$, then we can find du by taking the derivative of u, and we can find v by taking the integral of dv. We get $du = dx$ and $v = \int \sec^2 x\, dx = \tan x$. Plugging into the formula, we get $\int x\sec^2 x\, dx = x\tan x - \int \tan x\, dx$. Now we just have to find $\int \tan x\, dx$. We do this by rewriting the integrand in terms of $\sin x$ and $\cos x$: $\int \tan x\, dx = \int \dfrac{\sin x}{\cos x}\, dx = -\ln\left|\cos x\right| + C$. Therefore, $\int x\sec^2 x\, dx = x\tan x + \ln\left|\cos x\right| + C$.

PRACTICE PROBLEM SET 26

1. $\quad \dfrac{1}{16 + x^2}$

We find the derivative of the inverse tangent using the formula $\dfrac{d}{dx}\left(\tan^{-1} u\right) = \dfrac{1}{1 + u^2}\dfrac{du}{dx}$. Here we have $u = \dfrac{x}{4}$, so $\dfrac{du}{dx} = \dfrac{1}{4}$. Therefore, $\dfrac{d}{dx}\left(\dfrac{1}{4}\tan^{-1}\dfrac{x}{4}\right) = \dfrac{1}{4}\dfrac{1}{1 + \left(\dfrac{x}{4}\right)^2}\left(\dfrac{1}{4}\right) = \dfrac{1}{16}\dfrac{1}{1 + \dfrac{x^2}{16}} = \dfrac{1}{16 + x^2}$.

2. $\dfrac{-1}{|x|\sqrt{x^2-1}}$

We find the derivative of the inverse sine using the formula

$\dfrac{d}{dx}\left(\sin^{-1}u\right)=\dfrac{1}{\sqrt{1-u^2}}\dfrac{du}{dx}$. Here we have $u=\dfrac{1}{x}$, so $\dfrac{du}{dx}=-\dfrac{1}{x^2}$. Therefore,

$\dfrac{d}{dx}\sin^{-1}\left(\dfrac{1}{x}\right)=\dfrac{1}{\sqrt{1-\left(\dfrac{1}{x}\right)^2}}\left(-\dfrac{1}{x^2}\right)=\dfrac{-1}{\sqrt{1-\dfrac{1}{x^2}}}\left(\dfrac{1}{x^2}\right)=\dfrac{-1}{|x|\sqrt{x^2-1}}$.

3. $\dfrac{e^x}{1+e^{2x}}$

We find the derivative of the inverse tangent using the formula $\dfrac{d}{dx}\left(\tan^{-1}u\right)=\dfrac{1}{1+u^2}\dfrac{du}{dx}$. Here we

have $u=e^x$, so $\dfrac{du}{dx}=e^x$. Therefore, $\dfrac{d}{dx}\left(\tan^{-1}e^x\right)=\dfrac{1}{1+\left(e^x\right)^2}\left(e^x\right)=\dfrac{e^x}{1+e^{2x}}$.

4. $\dfrac{1}{\sqrt{\pi}}\sec^{-1}\dfrac{x}{\sqrt{\pi}}+C$

Recall that $\displaystyle\int\dfrac{du}{u\sqrt{u^2-1}}=\sec^{-1}u+C$. Here we have $\displaystyle\int\dfrac{dx}{x\sqrt{x^2-\pi}}$, and we just need to rearrange

the integrand so that it is in the proper form to use the integral formula. If we factor π out of

the radicand, we get $\displaystyle\int\dfrac{dx}{x\sqrt{\pi}\sqrt{\dfrac{x^2}{\pi}-1}}$. Next, we do u-substitution. Let $u=\dfrac{x}{\sqrt{\pi}}$ and $du=\dfrac{1}{\sqrt{\pi}}dx$.

Multiply both by $\sqrt{\pi}$ so that $u\sqrt{\pi}=x$ and $du\sqrt{\pi}=dx$. Substituting into the integrand, we get

$\displaystyle\int\dfrac{du\sqrt{\pi}}{\pi u\sqrt{u^2-1}}=\dfrac{1}{\sqrt{\pi}}\int\dfrac{du}{u\sqrt{u^2-1}}$. Now we get $\dfrac{1}{\sqrt{\pi}}\displaystyle\int\dfrac{du}{u\sqrt{u^2-1}}=\dfrac{1}{\sqrt{\pi}}\sec^{-1}u+C$. Substituting back

we get $\dfrac{1}{\sqrt{\pi}}\sec^{-1}\dfrac{x}{\sqrt{\pi}}+C$.

5. $\dfrac{1}{\sqrt{7}}\tan^{-1}\dfrac{x}{\sqrt{7}}+C$

Recall that $\displaystyle\int\dfrac{du}{1+u^{2}}=\tan^{-1}u+C$. Here we have $\displaystyle\int\dfrac{dx}{7+x^{2}}$, and we just need to rearrange the

integrand so that it is in the proper form to use the integral formula. If we factor 7 out of the

denominator, we get $\displaystyle\int\dfrac{dx}{7\left(1+\dfrac{x^{2}}{7}\right)}=\dfrac{1}{7}\int\dfrac{dx}{\left(1+\left(\dfrac{x}{\sqrt{7}}\right)^{2}\right)}$. Next, we do u-substitution. Let $u=\dfrac{x}{\sqrt{7}}$

and $du=\dfrac{1}{\sqrt{7}}dx$. Multiply du by $\sqrt{7}$ so that $du\sqrt{7}=dx$. Substituting into the integrand we get

$\dfrac{1}{7}\displaystyle\int\dfrac{du\sqrt{7}}{1+u^{2}}=\dfrac{1}{\sqrt{7}}\int\dfrac{du}{1+u^{2}}$. Now we get $\dfrac{1}{\sqrt{7}}\displaystyle\int\dfrac{du}{1+u^{2}}=\dfrac{1}{\sqrt{7}}\tan^{-1}u+C$. Substituting back, we get

$\dfrac{1}{\sqrt{7}}\tan^{-1}\dfrac{x}{\sqrt{7}}+C$.

6. $\tan^{-1}(\ln x)+C$

Recall that $\displaystyle\int\dfrac{du}{1+u^{2}}=\tan^{-1}u+C$. Here we have $\displaystyle\int\dfrac{dx}{x\left(1+\ln^{2}x\right)}$, and we will need to do u-substi-

tution. Let $u=\ln x$ and $du=\dfrac{1}{x}dx$. Substituting into the integrand, we get $\displaystyle\int\dfrac{du}{1+u^{2}}=\tan^{-1}u+C$.

Substituting back, we get $\tan^{-1}\left(\ln x\right)+C$.

7. $\sin^{-1}(\tan x)+C$

Recall that $\displaystyle\int\dfrac{du}{\sqrt{1-u^{2}}}=\sin^{-1}u+C$. Here we have $\displaystyle\int\dfrac{\sec^{2}x\,dx}{\sqrt{1-\tan^{2}x}}$, and we will need to do

u-substitution. Let $u=\tan x$ and $du=\sec^{2}x\,dx$. Substituting into the integrand, we get

$\displaystyle\int\dfrac{du}{\sqrt{1-u^{2}}}=\sin^{-1}u+C$. Substituting back, we get $\sin^{-1}\left(\tan x\right)+C$.

8. $\frac{1}{3}\tan^{-1}\left(e^{3x}\right)+C$

Recall that $\int \frac{du}{1+u^2} = \tan^{-1} u + C$. Here we have $\int \frac{e^{3x}\,dx}{1+e^{6x}}$, and we will need to do u-substitution.

Let $u = e^{3x}$ and $du = 3e^{3x}\,dx$. Divide du by 3 so that $\frac{1}{3}du = e^{3x}\,dx$. Substituting into the integrand,

we get $\int \frac{e^{3x}\,dx}{1+e^{6x}} = \frac{1}{3}\int \frac{du}{1+u^2} = \frac{1}{3}\tan^{-1} u + C$. Substituting back, we get $\frac{1}{3}\tan^{-1}\left(e^{3x}\right)+C$.

PRACTICE PROBLEM SET 27

1. $\frac{3x}{8} - \frac{\sin 2x}{4} + \frac{\sin 4x}{32} + C$

Here we use the substitution $\sin^2 x = \dfrac{1-\cos 2x}{2}$ into the integrand. We get

$\int \sin^4 x \, dx = \int \left(\dfrac{1-\cos 2x}{2}\right)^2 dx = \int \left(\dfrac{1-2\cos 2x+\cos^2 2x}{4}\right)dx$. We can break this into three

integrals: $\int \left(\dfrac{1-2\cos 2x+\cos^2 2x}{4}\right)dx = \dfrac{1}{4}\int dx - \dfrac{1}{2}\int \cos 2x \, dx + \dfrac{1}{4}\int \cos^2 2x \, dx$. The first two

integrals are easy: $\dfrac{1}{4}\int dx - \dfrac{1}{2}\int \cos 2x \, dx = \dfrac{x}{4} - \dfrac{\sin 2x}{4}$. For the third integral, we use the substitu-

tion $\cos^2 x = \dfrac{1+\cos 2x}{2}$. We get $\int \cos^2 2x \, dx = \int \dfrac{1+\cos 4x}{2} dx = \dfrac{1}{2}\int dx + \dfrac{1}{2}\int \cos 4x \, dx$. Now we

integrate these: $\dfrac{1}{2}\int dx + \dfrac{1}{2}\int \cos 4x \, dx = \dfrac{1}{2}x + \dfrac{1}{8}\sin 4x + C$. Now we put everything together to

get $\dfrac{x}{4} - \dfrac{\sin 2x}{4} + \dfrac{1}{4}\left(\dfrac{1}{2}x + \dfrac{1}{8}\sin 4x\right) + C = \dfrac{3x}{8} - \dfrac{\sin 2x}{4} + \dfrac{\sin 4x}{32} + C$.

2. $\frac{3x}{8} + \frac{\sin 2x}{4} + \frac{\sin 4x}{32} + C$

Here we use the substitution $\cos^2 x = \dfrac{1+\cos 2x}{2}$ into the integrand. We get

$\int \cos^4 x \, dx = \int \left(\dfrac{1+\cos 2x}{2}\right)^2 dx = \int \left(\dfrac{1+2\cos 2x+\cos^2 2x}{4}\right)dx$. We can break this into three

integrals: $\int \left(\dfrac{1+2\cos 2x+\cos^2 2x}{4}\right)dx = \dfrac{1}{4}\int dx + \dfrac{1}{2}\int \cos 2x \, dx + \dfrac{1}{4}\int \cos^2 2x \, dx$. The first two integrals are easy: $\dfrac{1}{4}\displaystyle\int dx + \dfrac{1}{2}\displaystyle\int \cos 2x\; dx = \dfrac{x}{4} + \dfrac{\sin 2x}{4}$. For the third integral, we again use the

substitution $\cos^2 x = \dfrac{1+\cos 2x}{2}$. We get $\displaystyle\int \cos^2 2x\; dx = \int \dfrac{1+\cos 4x}{2}\; dx = \dfrac{1}{2}\int dx + \dfrac{1}{2}\int \cos 4x\; dx$.

Now we integrate these: $\dfrac{1}{2}\displaystyle\int dx + \dfrac{1}{2}\displaystyle\int \cos 4x\; dx = \dfrac{1}{2}x + \dfrac{1}{8}\sin 4x + C$. Now we put everything

together to get $\dfrac{x}{4} + \dfrac{\sin 2x}{4} + \dfrac{1}{4}\left(\dfrac{1}{2}x + \dfrac{1}{8}\sin 4x\right) + C = \dfrac{3x}{8} + \dfrac{\sin 2x}{4} + \dfrac{\sin 4x}{32} + C.$

3. $\qquad -\dfrac{\cos^5 x}{5} + C$

Here we can use u-substitution. If we let $u = \cos x$, then $du = -\sin x\; dx$. Substituting

into the integrand, we get $\displaystyle\int \cos^4 x \sin x\; dx = -\int u^4 du = -\dfrac{u^5}{5} + C$. Now we substitute

back: $\displaystyle\int \cos^4 x \sin x\; dx = -\dfrac{\cos^5 x}{5} + C$.

4. $\qquad \dfrac{x}{8} - \dfrac{\sin 4x}{32} + C$

First, we use the trig identity $\sin^2 x = 1 - \cos^2 x$, and substitute into the integrand:

$\displaystyle\int \sin^2 x \cos^2 x\; dx = \int \left(1 - \cos^2 x\right)\cos^2 x\; dx = \int \left(\cos^2 x - \cos^4 x\right)dx$. Next, we use the substitu-

tion $\cos^2 x = \dfrac{1+\cos 2x}{2}$ into the integrand: $\displaystyle\int \left(\dfrac{1+\cos 2x}{2}\right) - \left(\dfrac{1+\cos 2x}{2}\right)^2 dx$. We use a little

algebra:

$$\int \left(\dfrac{1+\cos 2x}{2}\right) - \left(\dfrac{1+\cos 2x}{2}\right)^2 dx =$$

$$\int \left(\dfrac{1+\cos 2x}{2}\right) - \left(\dfrac{1+2\cos 2x+\cos^2 2x}{4}\right)dx =$$

$$\int \left(\dfrac{1}{4} - \dfrac{\cos^2 2x}{4}\right)dx$$

Now we break this into two integrals: $\dfrac{1}{4}\displaystyle\int dx - \int \dfrac{\cos^2 2x}{4}\,dx$. The first integral is easy:

$\dfrac{1}{4}\displaystyle\int dx = \dfrac{x}{4}$. For the second integral, we again use the substitution $\cos^2 x = \dfrac{1+\cos 2x}{2}$

into the integrand: $\dfrac{1}{4}\displaystyle\int \dfrac{1+\cos 2x}{2}\,dx = \dfrac{1}{8}\int dx + \dfrac{1}{8}\int \cos 4x\,dx$. These are both easy

to integrate: $\dfrac{1}{8}\displaystyle\int dx + \dfrac{1}{8}\int \cos 4x\,dx = \dfrac{x}{8} + \dfrac{\sin 4x}{32}$. Putting it all together, we get

$\dfrac{x}{4} - \left(\dfrac{x}{8} + \dfrac{\sin 4x}{32} \right) = \dfrac{x}{8} - \dfrac{\sin 4x}{32} + C$.

There is another way to do this integral that is easier, but only if you spot it. Recall that

$\sin 2x = 2\sin x \cos x$. This means that we could rewrite the integrand as

$\displaystyle\int \sin^2 x \cos^2 x\,dx = \int \left(\dfrac{\sin 2x}{2} \right)^2 dx = \dfrac{1}{4}\int \sin^2 2x\,dx$. Substitute in $\sin^2 x = \dfrac{1-\cos 2x}{2}$ to

rewrite the integrand: $\dfrac{1}{4}\displaystyle\int \sin^2 2x\,dx = \dfrac{1}{4}\int \dfrac{1-\cos 4x}{2}\,dx = \dfrac{1}{8}\int (1-\cos 4x)\,dx$. This is easy to inte-

grate: $\dfrac{1}{8}\displaystyle\int (1-\cos 4x)\,dx = \dfrac{1}{8}\left(x - \dfrac{\sin 4x}{4} \right) = \dfrac{x}{8} - \dfrac{\sin 4x}{32} + C$. This is why you should know your

trigonometric identities really well before you take calculus!

5. $\dfrac{\tan^4 x}{4} + C$

Here we can use u-substitution. If we let $u = \tan x$, then $du = \sec^2 x\,dx$. Substituting

into the integrand, we get $\displaystyle\int \tan^3 x \sec^2 x\,dx = \int u^3\,du = \dfrac{u^4}{4} + C$. Now we substitute back:

$\displaystyle\int \tan^3 x \sec^2 x\,dx = \dfrac{\tan^4 x}{4} + C$.

6. $\dfrac{\tan^4 x}{4} - \dfrac{\tan^2 x}{2} - \ln|\cos x| + C$

First, break up the $\tan^5 x$ in the integrand into $\int \tan^5 x\, dx = \int \tan^3 x \left(\tan^2 x \right) dx$.

Next, we use the trigonometric identity $\tan^2 x = \sec^2 x - 1$, and rewrite the integrand:

$\int \left(\tan^3 x \right) \left(\sec^2 x - 1 \right) dx = \int \left(\tan^3 x \sec^2 x - \tan^3 x \right) dx$. Now, we break this into two inte-

grals: $\int \left(\tan^3 x \sec^2 x \right) dx - \int \tan^3 x\, dx$. Next, break up the integrand of the second integral into

$\int \tan^3 x\, dx = \int \tan x \left(\tan^2 x \right) dx$. Once again, use the trigonometric identity $\tan^2 x = \sec^2 x - 1$ to

rewrite the integrand: $\int \tan x \left(\sec^2 x - 1 \right) dx = \int \tan x \sec^2 x\, dx - \int \tan x\, dx$. Putting the integrals

together, we get $\int \tan^3 x \sec^2 x\, dx - \int \tan x \sec^2 x\, dx + \int \tan x\, dx$. We can use u-substitution to find

the first two integrals. If we let $u = \tan x$, then $du = \sec^2 x\, dx$. Substituting into the integrand,

we get $\int \tan^3 x \sec^2 x\, dx - \int \tan x \sec^2 x\, dx = \int u^3 du - \int u\, du = \dfrac{u^4}{4} - \dfrac{u^2}{2}$. Substituting back, we get

$\int \tan^3 x \sec^2 x\, dx - \int \tan x \sec^2 x\, dx = \dfrac{\tan^4 x}{4} - \dfrac{\tan^2 x}{2}$. The last integral we have done before (see

page 335): $\int \tan x\, dx = -\ln|\cos x| + C$. Therefore, $\int \tan^5 x\, dx = \dfrac{\tan^4 x}{4} - \dfrac{\tan^2 x}{2} - \ln|\cos x| + C$.

7. $-\csc x + C$

First, rewrite the integrand in terms of $\sin x$ and $\cos x$:

$\int \cot^2 x \sec x\, dx = \int \dfrac{\cos^2 x}{\sin^2 x} \left(\dfrac{1}{\cos x} \right) dx = \int \dfrac{\cos x}{\sin^2 x}\, dx$. Next, we do u-substitution. If we let

$u = \sin x$, then $du = \cos x\, dx$. Substituting into the integrand, we get

$\int \dfrac{\cos x}{\sin^2 x}\, dx = \int u^{-2} du = -\dfrac{1}{u} + C$. Substituting back, we get $\int \dfrac{\cos x}{\sin^2 x}\, dx = -\dfrac{1}{\sin x} + C = -\csc x + C$.

PRACTICE PROBLEM SET 28

1. $\dfrac{32}{3}$

We find the area of a region bounded by $f(x)$ above and $g(x)$ below at all points of the interval $[a, b]$ using the formula $\int_a^b [f(x) - g(x)]\,dx$. Here, $f(x) = 2$ and $g(x) = x^2 - 2$. First, let's make a sketch of the region.

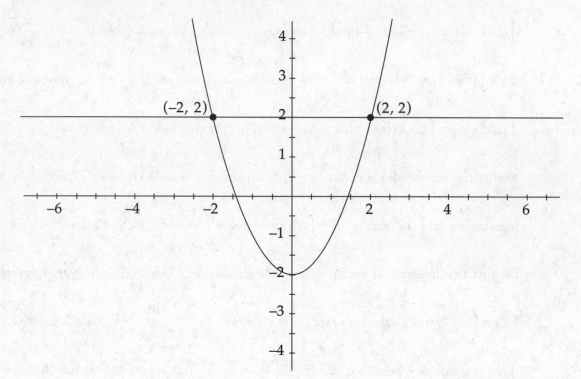

Next, we need to find where the two curves intersect, which will be the endpoints of the region.

We do this by setting the two curves equal to each other. We get $x^2 - 2 = 2$. The solutions are

$(-2, 0)$ and $(2, 0)$. Therefore, in order to find the area of the region, we need to evaluate the integral $\int_{-2}^2 \left(2 - \left(x^2 - 2\right)\right)dx = \int_{-2}^2 \left(4 - x^2\right)dx$. We get

$$\int_{-2}^2 \left(4 - x^2\right)dx = \left(4x - \frac{x^3}{3}\right)\Bigg|_{-2}^2 = \left(4(2) - \frac{(2)^3}{3}\right) - \left(4(-2) - \frac{(-2)^3}{3}\right) = \frac{32}{3}.$$

2. $\dfrac{8}{3}$

We find the area of a region bounded by $f(x)$ above and $g(x)$ below at all points of the interval $[a, b]$

using the formula $\displaystyle\int_{a}^{b}\left[f(x) - g(x)\right]dx$. Here, $f(x) = 4x - x^2$ and $g(x) = x^2$.

First, let's make a sketch of the region.

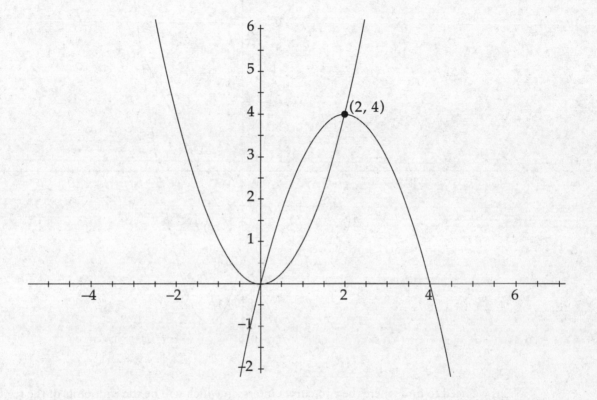

Next, we need to find where the two curves intersect, which will be the end-

points of the region. We do this by setting the two curves equal to each other. We get

$4x - x^2 = x^2$. The solutions are $(0, 0)$ and $(2, 4)$. Therefore, in order to find the area of the

region, we need to evaluate the integral $\displaystyle\int_{0}^{2}\left(\left(4x - x^2\right) - x^2\right)dx = \int_{0}^{2}\left(4x - 2x^2\right)dx$. We get

$$\int_{0}^{2}\left(4x - 2x^2\right)dx = \left(2x^2 - \dfrac{2x^3}{3}\right)\Bigg|_{0}^{2} = \left(2(2)^2 - \dfrac{2(2)^3}{3}\right) - 0 = \dfrac{8}{3}.$$

3.   36

We find the area of a region bounded by $f(x)$ above and $g(x)$ below at all points of the interval $[a, b]$ using the formula $\int_a^b \left[f(x) - g(x) \right] dx$. Here, $f(x) = 2x - 5$ and $g(x) = x^2 - 4x - 5$.

First, let's make a sketch of the region.

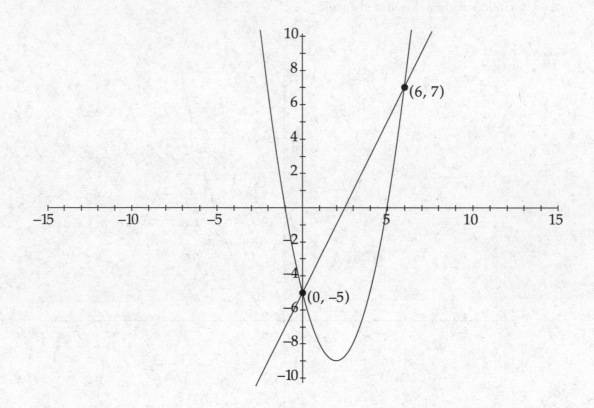

Next, we need to find where the two curves intersect, which will be the endpoints of the region. We

do this by setting the two curves equal to each other. We get $x^2 - 4x - 5 = 2x - 5$. The solutions are

$(0, -5)$ and $(6, 7)$. Therefore, in order to find the area of the region, we need to evaluate the integral

$\int_0^6 \left[(2x - 5) - \left(x^2 - 4x - 5 \right) \right] dx = \int_0^6 \left(-x^2 + 6x \right) dx$. We get

$\int_0^6 \left(-x^2 + 6x \right) dx = \left(-\dfrac{x^3}{3} + 3x^2 \right) \Bigg|_0^6 = \left(-\dfrac{(6)^3}{3} + 3(6)^2 \right) - 0 = 36$.

4. $\dfrac{17}{4}$

We find the area of a region bounded by $f(x)$ above and $g(x)$ below at all points of the interval $[a,\ b]$ using the formula $\int_a^b \left[f(x) - g(x) \right] dx$.

First, let's make a sketch of the region.

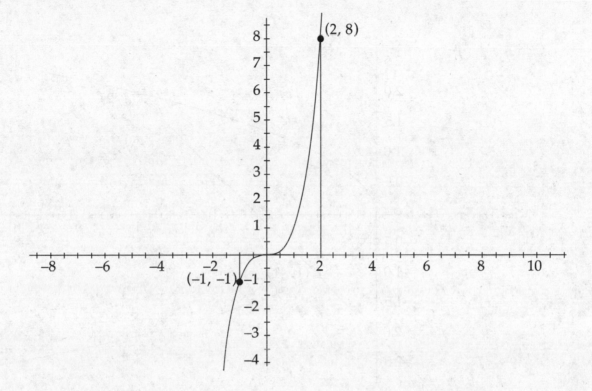

Notice that in the region from $x = -1$ to $x = 0$ the top curve is $f(x) = 0$ (the x-axis), and the

bottom curve is $g(x) = x^3$, but from $x = 0$ to $x = 2$ the situation is reversed, so the top curve is

$f(x) = x^3$, and the bottom curve is $g(x) = 0$. Thus, we split the region into two pieces and find the

area by evaluating two integrals and adding the answers: $\int_{-1}^0 \left(0 - x^3 \right) dx$ and $\int_0^2 \left(x^3 - 0 \right) dx$. We get

$\int_{-1}^0 \left(0 - x^3 \right) dx = \left(-\dfrac{x^4}{4} \right) \Big|_{-1}^0 = 0 - \left(-\dfrac{(-1)^4}{4} \right) = \dfrac{1}{4}$ and $\int_0^2 \left(x^3 - 0 \right) dx = \left(\dfrac{x^4}{4} \right) \Big|_0^2 = \dfrac{(2)^4}{4} - 0 = 4$. There-

fore, the area of the region is $\dfrac{17}{4}$.

5. $\dfrac{9}{2}$

We find the area of a region bounded by $f(y)$ on the right and $g(y)$ on the left at all points of the interval $[c,\ d]$ using the formula $\int_c^d \left[f(y) - g(y)\right]dy$. Here $f(y) = y + 2$ and $g(y) = y^2$.

First, let's make a sketch of the region.

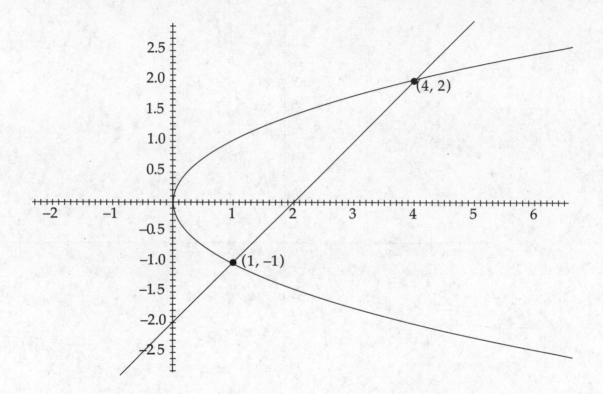

Next, we need to find where the two curves intersect, which will be the endpoints of the region.

We do this by setting the two curves equal to each other. We get $y^2 = y + 2$. The solutions are $(4,\ 2)$

and $(1,\ -1)$. Therefore, in order to find the area of the region, we need to evaluate the integral

$\int_{-1}^{2}\left(y + 2 - y^2\right)dy$. We get

$$\int_{-1}^{2}\left(y + 2 - y^2\right)dy = \left(\dfrac{y^2}{2} + 2y - \dfrac{y^3}{3}\right)\Bigg|_{-1}^{2} = \left(\dfrac{(2)^2}{2} + 2(2) - \dfrac{(2)^3}{3}\right) - \left(\dfrac{(-1)^2}{2} + 2(-1) - \dfrac{(-1)^3}{3}\right) = \dfrac{9}{2}.$$

6.   4

We find the area of a region bounded by $f(y)$ on the right and $g(y)$ on the left at all points of the interval $[c,\ d]$ using the formula $\int_c^d \left[f(y) - g(y) \right] dy$. Here $f(y) = 3 - 2y^2$ and $g(y) = y^2$.

First, let's make a sketch of the region.

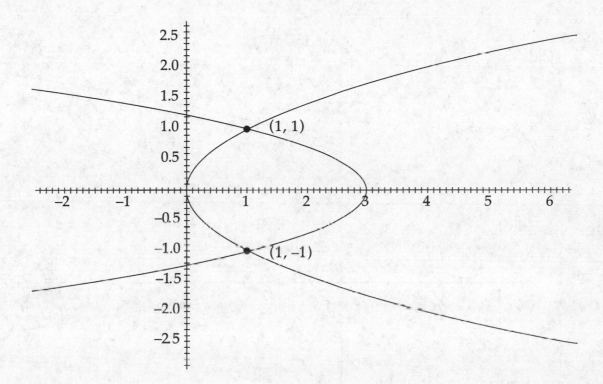

Next, we need to find where the two curves intersect, which will be the endpoints of the region. We do this by setting the two curves equal to each other. We get $y^2 = 3 - 2y^2$. The solutions are $(1, 1)$ and $(1, -1)$. Therefore, in order to find the area of the region, we need to evaluate the integral $\int_{-1}^{1} \left(3 - 2y^2 - y^2 \right) dy = \int_{-1}^{1} \left(3 - 3y^2 \right) dy$. We get

$$\int_{-1}^{1} \left(3 - 3y^2 \right) dy = \left(3y - y^3 \right) \Big|_{-1}^{1} = \left(3(1) - (1)^3 \right) - \left(3(-1) - (-1)^3 \right) = 4.$$

7. $\dfrac{9}{2}$

We find the area of a region bounded by $f(y)$ on the right and $g(y)$ on the left at all points of the

interval $[c,\ d]$ using the formula $\int_c^d \left[f(y) - g(y) \right] dy$. Here $f(y) = y - 2$ and $g(y) = y^2 - 4y + 2$. First,

let's make a sketch of the region.

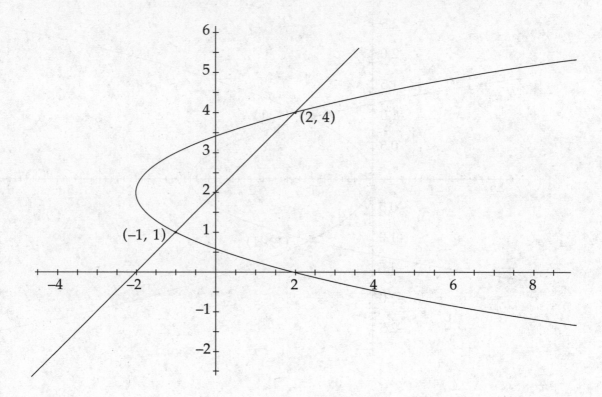

Next, we need to find where the two curves intersect, which will be the endpoints of the region.

We do this by setting the two curves equal to each other. We get $y^2 - 4y + 2 = y - 2$. The solutions

are $(2,\ 4)$ and $(-1,\ 1)$. Therefore, in order to find the area of the region, we need to evaluate the

integral $\int_1^4 \left[(y-2) - \left(y^2 - 4y + 2 \right) \right] dy = \int_1^4 \left[\left(-y^2 + 5y - 4 \right) \right] dy$. We get $\int_1^4 \left[\left(-y^2 + 5y - 4 \right) \right] dy =$

$\left(-\dfrac{y^3}{3} + \dfrac{5y^2}{2} - 4y \right)\Bigg|_1^4 = \left(-\dfrac{(4)^3}{3} + \dfrac{5(4)^2}{2} - 4(4) \right) - \left(-\dfrac{(1)^3}{3} + \dfrac{5(1)^2}{2} - 4(1) \right) = \dfrac{9}{2}$.

8. $\dfrac{12}{5}$

We find the area of a region bounded by $f(y)$ on the right and $g(y)$ on the left at all points of the interval $[c,\ d]$ using the formula $\int_{c}^{d}\left[f(y)-g(y)\right]dy$. Here $f(y) = 2 - y^4$ and $g(y) = y^{\frac{2}{3}}$. First, let's make a sketch of the region.

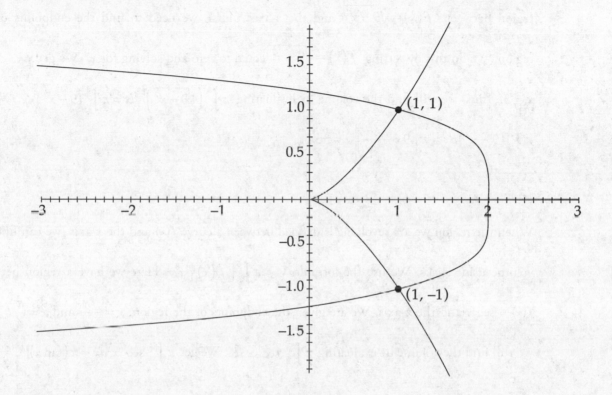

Next, we need to find where the two curves intersect, which will be the end-points of the region. We do this by setting the two curves equal to each other. We get $2 - y^4 = y^{\frac{2}{3}}$. The solutions are $(1,\ 1)$ and $(1,\ -1)$. Therefore, in order to find the area of the region, we need to evaluate the integral $\int_{-1}^{1}\left(2-y^4-y^{\frac{2}{3}}\right)dy$. We get

$$\int_{-1}^{1}\left(2-y^4-y^{\frac{2}{3}}\right)dy = \left(2y-\frac{y^5}{5}-\frac{3}{5}y^{\frac{5}{3}}\right)\Bigg|_{-1}^{1} = \left(2(1)-\frac{(1)^5}{5}-\frac{3}{5}(1)^{\frac{5}{3}}\right)-\left(2(-1)-\frac{(-1)^5}{5}-\frac{3}{5}(-1)^{\frac{5}{3}}\right) = \frac{12}{5}.$$

PRACTICE PROBLEM SET 29

1. 36π

When the region we are revolving is defined between a curve $f(x)$ and the x-axis, we can find the volume using disks. We use the formula $V = \pi \int_a^b \left[f(x) \right]^2 dx$. Here we have a region between $f(x) = \sqrt{9 - x^2}$ and the x-axis. First, we need to find the endpoints of the region. We do this by setting $f(x) = \sqrt{9 - x^2}$ equal to zero and solving for x. We get $x = -3$ and $x = 3$. Thus, we will find the volume by evaluating $\pi \int_{-3}^3 \left(\sqrt{9 - x^2} \right)^2 dx = \pi \int_{-3}^3 \left(9 - x^2 \right) dx$. We get $\pi \int_{-3}^3 \left(9 - x^2 \right) dx = \pi \left(9x - \frac{x^3}{3} \right) \Big|_{-3}^3 = 36\pi$.

2. 2π

When the region we are revolving is defined between a curve $f(x)$ and the x-axis, we can find the volume using disks. We use the formula $V = \pi \int_a^b \left[f(x) \right]^2 dx$. Here we have a region between $f(x) = \sec x$ and the x-axis. We are given the endpoints of the region: $x = -\frac{\pi}{4}$ and $x = \frac{\pi}{4}$. Thus, we will find the volume by evaluating $\pi \int_{-\frac{\pi}{4}}^{\frac{\pi}{4}} \sec^2 x \, dx$. We get $\pi \int_{-\frac{\pi}{4}}^{\frac{\pi}{4}} \sec^2 x \, dx = \pi \left(\tan x \right) \Big|_{-\frac{\pi}{4}}^{\frac{\pi}{4}} = 2\pi$.

3. $\dfrac{16\pi}{15}$

When the region we are revolving is defined between a curve $f(y)$ and the y-axis, we can find the volume using disks. We use the formula $V = \pi \int_a^b \left[f(y) \right]^2 dy$ (see page 359 and note that when $g(y) = 0$ we get disks instead of washers). Here we have a region between $f(y) = 1 - y^2$ and the y-axis. First, we need to find the endpoints of the region. We do this by setting $f(y) = 1 - y^2$ equal to zero and solving for y. We get $y = -1$ and $y = 1$.

Thus, we will find the volume by evaluating $\pi \int_{-1}^{1} \left(1 - y^2\right)^2 dy = \pi \int_{-1}^{1} \left(1 - 2y^2 + y^4\right) dy$. We get

$$\pi \int_{-1}^{1} \left(1 - 2y^2 + y^4\right) dy = \pi \left(y - \frac{2y^3}{3} + \frac{y^5}{5}\right)\Big|_{-1}^{1} = \frac{16\pi}{15}.$$

4. 2π

When the region we are revolving is defined between a curve $f(y)$ and the y-axis, we can find the

volume using disks. We use the formula $V = \pi \int_{a}^{b} \left[f(y)\right]^2 dy$ (see page 359 and note that when

$g(y) = 0$ we get disks instead of washers). Here we have a region between $f(y) = \sqrt{5}y^2$ and the

y-axis. We are given the endpoints of the region: $y = -1$ and $y = 1$. Thus, we will find the volume

by evaluating $\pi \int_{-1}^{1} \left(\sqrt{5}y^2\right)^2 dy = \pi \int_{-1}^{1} \left(5y^4\right) dy$. We get $\pi \int_{-1}^{1} \left(5y^4\right) dy = \pi \left(y^5\right)\Big|_{-1}^{1} = 2\pi$.

5. 8π

When the region we are revolving is defined between a curve $f(x)$ and $g(x)$, we can find the vol-

ume using cylindrical shells. We use the formula $V = 2\pi \int_{a}^{b} x\left[f(x) - g(x)\right] dx$. Here we have

$f(x) = x$ and $g(x) = -\dfrac{x}{2}$. Thus, the height of each shell is $f(x) - g(x) = x - \left(-\dfrac{x}{2}\right) = \dfrac{3x}{2}$,

and the radius is simply x. The endpoints of our region are $x = 0$ and $x = 2$. Therefore, we

will find the volume by evaluating: $2\pi \int_{0}^{2} \left(x\left(\dfrac{3x}{2}\right)\right) dx = 2\pi \int_{0}^{2} \left(\dfrac{3x^2}{2}\right) dx = 3\pi \int_{0}^{2} x^2 \, dx$. We get

$3\pi \int_{0}^{2} x^2 \, dx = 3\pi \left(\dfrac{x^3}{3}\right)\Big|_{0}^{2} = 8\pi.$

6. $\dfrac{7\pi}{15}$

When the region we are revolving is defined between a curve $f(x)$ and $g(x)$, we can find the volume

using cylindrical shells. We use the formula $V = 2\pi \int_{a}^{b} x\left[f(x) - g(x)\right] dx$. Here we have

$f(x) = \sqrt{x}$ and $g(x) = 2x - 1$. Thus, the height of each shell is $f(x) - g(x) = \sqrt{x} - (2x - 1)$, and

the radius is simply x. The left endpoint of our region is $x = 0$, and we find the right endpoint by finding where $f(x) = \sqrt{x}$ intersects $g(x) = 2x - 1$. We get $x = 1$. Therefore, we will find the

volume by evaluating: $2\pi\int_0^1 \left(x\left(\sqrt{x} - 2x + 1\right)\right)dx = 2\pi\int_0^1 \left(x^{\frac{3}{2}} - 2x^2 + x\right)dx$. We get

$$2\pi\int_0^1 \left(x^{\frac{3}{2}} - 2x^2 + x\right)dx = 2\pi\left(\frac{2x^{\frac{5}{2}}}{5} - \frac{2x^3}{3} + \frac{x^2}{2}\right)\Bigg|_0^1 = \frac{7\pi}{15}.$$

7. $\dfrac{128\pi}{5}$

When the region we are revolving is defined between a curve $f(y)$ and $g(y)$, we can find the volume using cylindrical shells. We use the formula $V = 2\pi\int_a^b y\left[f(y) - g(y)\right]dy$. Here we have

$f(y) = \sqrt{y}$ (which we get by solving $y = x^2$ for x) and $g(y) = 0$ (the y-axis). We can easily see that

the top endpoint is $y = 4$ and the bottom one is $y = 0$. Therefore, we will find the volume by evaluat-

ing: $2\pi\int_0^4 \left(y\sqrt{y}\right)dy = 2\pi\int_0^4 \left(y^{\frac{3}{2}}\right)dy$. We get $2\pi\int_0^4 \left(y^{\frac{3}{2}}\right)dy = 2\pi\left(\frac{2y^{\frac{5}{2}}}{5}\right)\Bigg|_0^4 = \frac{128\pi}{5}$.

8. $\dfrac{256\pi}{5}$

When the region we are revolving is defined between a curve $f(x)$ and $g(x)$, we can find the volume using cylindrical shells. We use the formula $V = 2\pi\int_a^b x\left[f(x) - g(x)\right]dx$. Here we have $f(x) = 2\sqrt{x}$ and $g(x) = 0$. Thus, the height of each shell is $f(x) - g(x) = 2\sqrt{x}$, and the

radius is simply x. We can easily see that the left endpoint is $x = 0$ and that the right endpoint is

$x = 4$. Therefore, we will find the volume by evaluating: $2\pi\int_0^4 \left(x\left(2\sqrt{x}\right)\right)dx = 2\pi\int_0^4 \left(2x^{\frac{3}{2}}\right)dx$. We

get $2\pi\int_0^4 \left(2x^{\frac{3}{2}}\right)dx = 4\pi\left(\frac{2x^{\frac{5}{2}}}{5}\right)\Bigg|_0^4 = \frac{256\pi}{5}.$

9. $\dfrac{256}{3}$

To find the volume of a solid with a cross-section of an isosceles right triangle, we integrate the area of the square ($side^2$) over the endpoints of the interval. Here, the sides of the squares are found by $f(x) - g(x) = \sqrt{16 - x^2} - 0$, and the intervals are found by setting $y = \sqrt{16 - x^2}$ equal to zero. We get $x = -4$ and $x = 4$. Thus, we find the volume by evaluating the integral $\int_{-4}^{4} \left(\sqrt{16 - x^2} \right)^2 dx = \int_{-4}^{4} \left(16 - x^2 \right) dx$. We get $\int_{-4}^{4} \left(16 - x^2 \right) dx = \left(16x - \dfrac{x^3}{3} \right)\Bigg|_{-4}^{4} = \dfrac{256}{3}$.

10. $\dfrac{128}{15}$

To find the volume of a solid with a cross-section of a square, we integrate the area of the isosceles right triangle $\dfrac{hypotenuse^2}{4}$ over the endpoints of the interval. Here, the hypotenuses of the triangles are found by $f(x) - g(x) = 4 - x^2$, and the intervals are found by setting $y = x^2$ equal to $y = 4$. We get $x = -2$ and $x = 2$. Thus, we find the volume by evaluating the integral $\int_{-2}^{2} \dfrac{\left(4 - x^2 \right)^2}{4} dx = \int_{-2}^{2} \dfrac{\left(16 - 8x^2 + x^4 \right)}{4} dx = \int_{-2}^{2} \left(4 - 2x^2 + \dfrac{x^4}{4} \right) dx$.

We get $\int_{-2}^{2} \left(4 - 2x^2 + \dfrac{x^4}{4} \right) dx = \left(4x - \dfrac{2x^3}{3} + \dfrac{x^5}{20} \right)\Bigg|_{-2}^{2} = \dfrac{128}{15}$.

PRACTICE PROBLEM SET 30

1. $\dfrac{13}{12}$

Recall that the length of the curve $y = f(x)$ from $x = a$ to $x = b$ is $L = \int_a^b \sqrt{1 + \left(\dfrac{dy}{dx}\right)^2}\, dx$. Here

we have $y = \dfrac{x^3}{12} + \dfrac{1}{x}$ from $x = 1$ to $x = 2$. First, we need to find the derivative: $\dfrac{dy}{dx} = \dfrac{x^2}{4} - \dfrac{1}{x^2}$.

Now, we plug this into the formula for the length: $\int_1^2 \sqrt{1 + \left(\dfrac{x^2}{4} - \dfrac{1}{x^2}\right)^2}\, dx$. Now, we need to use

some algebra to simplify the integrand.

Get a common denominator: $\int_1^2 \sqrt{1 + \left(\dfrac{x^2}{4} - \dfrac{1}{x^2}\right)^2}\, dx = \int_1^2 \sqrt{1 + \left(\dfrac{x^4 - 4}{4x^2}\right)^2}\, dx$.

Square the right-hand term: $\int_1^2 \sqrt{1 + \left(\dfrac{x^4 - 4}{4x^2}\right)^2}\, dx = \int_1^2 \sqrt{1 + \left(\dfrac{x^8 - 8x^4 + 16}{16x^4}\right)}\, dx$.

Get a common denominator: $\int_1^2 \sqrt{1 + \left(\dfrac{x^8 - 8x^4 + 16}{16x^4}\right)}\, dx = \int_1^2 \sqrt{\left(\dfrac{x^8 + 8x^4 + 16}{16x^4}\right)}\, dx$.

Factor the numerator: $\int_1^2 \sqrt{\left(\dfrac{x^8 + 8x^4 + 16}{16x^4}\right)}\, dx = \int_1^2 \sqrt{\dfrac{\left(x^4 + 4\right)^2}{16x^4}}\, dx$.

Take the square root: $\int_1^2 \sqrt{\dfrac{\left(x^4 + 4\right)^2}{16x^4}}\, dx = \int_1^2 \dfrac{\left(x^4 + 4\right)}{4x^2}\, dx$.

And break up the fraction: $\int_1^2 \dfrac{\left(x^4 + 4\right)}{4x^2} dx = \int_1^2 \left(\dfrac{x^2}{4} + \dfrac{1}{x^2} \right) dx$.

Now we can integrate easily: $\int_1^2 \left(\dfrac{x^2}{4} + \dfrac{1}{x^2} \right) dx = \left(\dfrac{x^3}{12} - \dfrac{1}{x} \right) \Big|_1^2 = \dfrac{13}{12}$.

2. $\int_{-\frac{\pi}{6}}^0 \sqrt{1 + \sec^4 x}\, dx$

Recall that the length of the curve $y = f(x)$ from $x = a$ to $x = b$ is $L = \int_a^b \sqrt{1 + \left(\dfrac{dy}{dx} \right)^2}\, dx$. Here we

have $y = \tan x$ from $x = -\dfrac{\pi}{6}$ to $x = 0$. First, we need to find the derivative: $\dfrac{dy}{dx} = \sec^2 x$. Now, we

plug this into the formula for the length: $L = \int_{-\frac{\pi}{6}}^0 \sqrt{1 + \left(\sec^2 x \right)^2}\, dx = \int_{-\frac{\pi}{6}}^0 \sqrt{1 + \sec^4 x}\, dx$.

3. $\int_0^{\frac{1}{4}} \sqrt{\dfrac{1}{1 - x^2}}\, dx$

Recall that the length of the curve $y = f(x)$ from $x = a$ to $x = b$ is $L = \int_a^b \sqrt{1 + \left(\dfrac{dy}{dx} \right)^2}\, dx$.

Here we have $y = \sqrt{1 - x^2}$ from $x = 0$ to $x = \dfrac{1}{4}$. First, we need to find the derivative:

$\dfrac{dy}{dx} = \dfrac{1}{2}\left(1 - x^2\right)^{-\frac{1}{2}}(-2x) = -\dfrac{x}{\sqrt{1 - x^2}}$. Now, we plug this into the formula for the length:

$L = \int_0^{\frac{1}{4}} \sqrt{1 + \left(-\dfrac{x}{\sqrt{1 - x^2}} \right)^2}\, dx = \int_0^{\frac{1}{4}} \sqrt{1 + \dfrac{x^2}{1 - x^2}}\, dx = \int_0^{\frac{1}{4}} \sqrt{\dfrac{1}{1 - x^2}}\, dx$.

4. $\dfrac{47}{36}$

Recall that the length of the curve $x = f(y)$ from $y = c$ to $y = d$ is $L = \int_c^d \sqrt{1 + \left(\dfrac{dx}{dy} \right)^2}\, dy$. Here we

have $x = \dfrac{y^3}{18} + \dfrac{3}{2y}$ from $y = 2$ to $y = 3$. First, we need to find the derivative: $\dfrac{dx}{dy} = \dfrac{y^2}{6} - \dfrac{3}{2y^2}$. Now,

we plug this into the formula for the length: $\int_2^3 \sqrt{1 + \left(\dfrac{y^2}{6} - \dfrac{3}{2y^2} \right)^2}\, dy$. Now, we need to use some

algebra to simplify the integrand.

Get a common denominator: $\int_2^3 \sqrt{1+\left(\dfrac{y^2}{6}-\dfrac{3}{2y^2}\right)^2}\,dy = \int_2^3 \sqrt{1+\left(\dfrac{y^4-9}{6y^2}\right)^2}\,dy$.

Square the right-hand term: $\int_2^3 \sqrt{1+\left(\dfrac{y^4-9}{6y^2}\right)^2}\,dy = \int_2^3 \sqrt{1+\left(\dfrac{y^8-18y^4+81}{36y^4}\right)}\,dy$.

Get a common denominator: $\int_2^3 \sqrt{1+\left(\dfrac{y^8-18y^4+81}{36y^4}\right)}\,dy = \int_2^3 \sqrt{\left(\dfrac{y^8+18y^4+81}{36y^4}\right)}\,dy$.

Factor the numerator: $\int_2^3 \sqrt{\left(\dfrac{y^8+18y^4+81}{36y^4}\right)}\,dy = \int_2^3 \sqrt{\dfrac{\left(y^4+9\right)^2}{36y^4}}\,dy$.

Take the square root: $\int_2^3 \sqrt{\dfrac{\left(y^4+9\right)^2}{36y^4}}\,dy = \int_2^3 \left(\dfrac{y^4+9}{6y^2}\right)\,dy$.

And break up the fraction: $\int_2^3 \left(\dfrac{y^4+9}{6y^2}\right)\,dy = \int_2^3 \left(\dfrac{y^2}{6}+\dfrac{3}{2}y^{-2}\right)\,dy$.

Now we can integrate easily: $\int_2^3 \left(\dfrac{y^2}{6}+\dfrac{3}{2}y^{-2}\right)\,dy = \left(\dfrac{y^3}{18}-\dfrac{3}{2y}\right)\Bigg|_2^3 = \dfrac{47}{36}$.

5. $\displaystyle\int_{-\frac{1}{2}}^{\frac{1}{2}} \sqrt{\dfrac{1}{1-y^2}}\,dy$

Recall that the length of the curve $x = f(y)$ from $y = c$ to $y = d$ is $L = \int_c^d \sqrt{1+\left(\dfrac{dx}{dy}\right)^2}\,dy$.

Here we have $x = \sqrt{1-y^2}$ from $y = -\dfrac{1}{2}$ to $y = \dfrac{1}{2}$. First, we need to find the

derivative: $\dfrac{dx}{dy} = \dfrac{1}{2}\left(1-y^2\right)^{-\frac{1}{2}}(-2y) = -\dfrac{y}{\sqrt{1-y^2}}$. Now, we plug this into the formula for the

length: $L = \int_{-\frac{1}{2}}^{\frac{1}{2}} \sqrt{1+\left(-\dfrac{y}{\sqrt{1-y^2}}\right)^2}\,dy = \int_{-\frac{1}{2}}^{\frac{1}{2}} \sqrt{1+\dfrac{y^2}{1-y^2}}\,dy = \int_{-\frac{1}{2}}^{\frac{1}{2}} \sqrt{\dfrac{1}{1-y^2}}\,dy$.

6. $\int_0^\pi \sqrt{1 + y^2 \sin^2 y}\; dy$

Recall that the length of the curve $x = f(y)$ from $y = c$ to $y = d$ is $L = \int_c^d \sqrt{1 + \left(\dfrac{dx}{dy}\right)^2}\; dy$.

Here we have $x = \sin y - y \cos y$ from $y = 0$ to $y = \pi$. First, we need to find the derivative:

$\dfrac{dx}{dy} = \cos y - (-y \sin y + \cos y) = y \sin y$. Now, we plug this into the formula for the length:

$L = \int_0^\pi \sqrt{1 + y^2 \sin^2 y}\; dy$.

7. $\dfrac{1}{2} \int_1^4 \dfrac{\sqrt{t^7 + 36}}{t^4}\; dt$

Recall that if a curve is defined parametrically, in terms of t, from $t = a$ to $t = b$, its length is

found by evaluating: $\int_a^b \sqrt{\left(\dfrac{dx}{dt}\right)^2 + \left(\dfrac{dy}{dt}\right)^2}\; dt$. Here we have $x = \sqrt{t}$ and $y = \dfrac{1}{t^3}$ from $t = 1$ to

$t = 4$. First, we need to find the derivatives: $\dfrac{dx}{dt} = \dfrac{1}{2\sqrt{t}}$ and $\dfrac{dy}{dt} = -\dfrac{3}{t^4}$. Now, we plug this into

the formula for the length: $\int_1^4 \sqrt{\left(\dfrac{1}{2\sqrt{t}}\right)^2 + \left(\dfrac{-3}{t^4}\right)^2}\; dt = \int_1^4 \sqrt{\dfrac{1}{4t} + \dfrac{9}{t^8}}\; dt$. With a little algebra, we get

$\int_1^4 \sqrt{\dfrac{1}{4t} + \dfrac{9}{t^8}}\; dt = \int_1^4 \sqrt{\dfrac{t^7 + 36}{4t^8}}\; dt = \dfrac{1}{2} \int_1^4 \dfrac{\sqrt{t^7 + 36}}{t^4}\; dt$.

8. $2 \int_1^2 \sqrt{9t^2 + 1}\; dt$

Recall that if a curve is defined parametrically, in terms of t, from $t = a$ to $t = b$, its length

is found by evaluating: $\int_a^b \sqrt{\left(\dfrac{dx}{dt}\right)^2 + \left(\dfrac{dy}{dt}\right)^2}\; dt$. Here we have $x = 3t^2$ and $y = 2t$ from

$t = 1$ to $t = 2$. First, we need to find the derivatives: $\dfrac{dx}{dt} = 6t$ and $\dfrac{dy}{dt} = 2$. Now, we plug this into the

formula for the length: $\int_1^2 \sqrt{(6t)^2 + 4}\; dt = \int_1^2 \sqrt{36t^2 + 4}\; dt = 2 \int_1^2 \sqrt{9t^2 + 1}\; dt$.

PRACTICE PROBLEM SET 31

1. $\dfrac{5}{7}\ln|x-1|+\dfrac{2}{7}\ln|x+6|+C$

We need to find A and B such that $\dfrac{A}{x-1}+\dfrac{B}{x+6}=\dfrac{x+4}{(x-1)(x+6)}$. First, we multiply through by

$(x-1)(x+6)$: $A(x+6)+B(x-1)=x+4$.

Distribute: $Ax+6A+Bx-B=x+4$.

Group the terms on the left side: $(A+B)x+(6A-B)=x+4$.

Now we have two equations: $A+B=1$ and $6A-B=4$.

If we solve for A and B, we get $A=\dfrac{5}{7}$ and $B=\dfrac{2}{7}$. This means that we can rewrite the

integral as $\displaystyle\int\dfrac{x+4}{(x-1)(x+6)}\,dx=\dfrac{5}{7}\int\dfrac{dx}{(x-1)}+\dfrac{2}{7}\int\dfrac{dx}{(x+6)}$. We can easily integrate these:

$\dfrac{5}{7}\ln|x-1|+\dfrac{2}{7}\ln|x+6|+C$.

2. $\dfrac{3}{4}\ln|x-3|+\dfrac{1}{4}\ln|x+1|+C$

We need to find A and B such that $\dfrac{A}{x-3}+\dfrac{B}{x+1}=\dfrac{x}{(x-3)(x+1)}$. First, we multiply through by

$(x-3)(x+1)$: $A(x+1)+B(x-3)=x$.

Distribute: $Ax+A+Bx-3B=x$.

Group the terms on the left side: $(A+B)x+(A-3B)=x$.

Now we have two equations: $A+B=1$ and $A-3B=0$.

If we solve for A and B, we get $A = \dfrac{3}{4}$ and $B = \dfrac{1}{4}$. This means that we can rewrite the

integral as $\displaystyle\int \frac{x}{(x-3)(x+1)}\,dx = \frac{3}{4}\int \frac{dx}{(x-3)} + \frac{1}{4}\int \frac{dx}{(x+1)}$. We can easily integrate these:

$\dfrac{3}{4}\ln|x-3| + \dfrac{1}{4}\ln|x+1| + C$.

3. $-\dfrac{1}{2}\ln|x| + \dfrac{1}{6}\ln|x+2| + \dfrac{1}{3}\ln|x-1| + C$

First, we need to factor the denominator of the integrand:

$$\int \frac{1}{x^3 + x^2 - 2x}\,dx = \int \frac{1}{x\left(x^2 + x - 2\right)}\,dx = \int \frac{1}{x(x+2)(x-1)}\,dx .$$

Now, we need to find A, B, and C such that $\dfrac{A}{x} + \dfrac{B}{x+2} + \dfrac{C}{x-1} = \dfrac{1}{x(x+2)(x-1)}$.

We multiply through by $x(x + 2)(x - 1)$: $A(x + 2)(x - 1) + Bx(x - 1) + Cx(x + 2) = 1$.

Expand all of the terms: $A(x^2 + x - 2) + B(x^2 - x) + C(x^2 + 2x) = 1$.

Distribute: $Ax^2 + Ax - 2A + Bx^2 - Bx + Cx^2 + 2Cx = 1$.

Group the terms on the left side: $(A + B + C)x^2 + (A - B + 2C)x - 2A = 1$.

Now we have three equations: $A + B + C = 0$, $A - B + 2C = 0$, and $-2A = 1$.

If we solve for A, B, and C, we get $A = -\dfrac{1}{2}$, $B = \dfrac{1}{6}$, and $C = \dfrac{1}{3}$. This means that we can rewrite

the integral as $\displaystyle\int \frac{1}{x(x+2)(x-1)}\,dx = -\frac{1}{2}\int \frac{dx}{x} + \frac{1}{6}\int \frac{dx}{x+2} + \frac{1}{3}\int \frac{dx}{x-1}$. We can easily integrate

these: $-\dfrac{1}{2}\ln|x| + \dfrac{1}{6}\ln|x+2| + \dfrac{1}{3}\ln|x-1| + C$.

4. $-7\ln|x-3| + 9\ln|x-4| + C$

First, we need to factor the denominator of the integrand: $\displaystyle\int \frac{2x+1}{x^2 - 7x + 12}\,dx = \int \frac{2x+1}{(x-3)(x-4)}\,dx$.

Now, we need to find A and B such that $\dfrac{A}{x-3} + \dfrac{B}{x-4} = \dfrac{2x+1}{(x-3)(x-4)}$.

We multiply through by $(x - 3)(x - 4)$: $A(x - 4) + B(x - 3) = 2x + 1$.

Distribute: $Ax - 4A + Bx - 3B = 2x + 1$.

Group the terms on the left side: $(A + B)x + (-4A - 3B) = 2x + 1$.

Now we have two equations: $A + B = 2$ and $-4A - 3B = 1$.

If we solve for A and B, we get $A = -7$ and $B = 9$.

This means that we can rewrite the integral as $\int \dfrac{2x+1}{x^2 - 7x + 12}\, dx = -7\int \dfrac{dx}{x-3} + 9\int \dfrac{dx}{x-4}$. We can

easily integrate these: $-7\ln|x - 3| + 9\ln|x - 4| + C$.

5. $2\ln|x - 1| - \dfrac{1}{x-1} + C$

We need to find A and B such that $\dfrac{A}{x-1} + \dfrac{B}{(x-1)^2} = \dfrac{2x-1}{(x-1)^2}$.

We multiply through by $(x - 1)^2$: $A(x - 1) + B = 2x - 1$.

Distribute: $Ax - A + B = 2x - 1$.

Now we have two equations: $A = 2$ and $-A + B = -1$.

If we solve for A and B, we get $A = 2$ and $B = 1$.

This means that we can rewrite the integral as $\int \dfrac{2x-1}{(x-1)^2}\, dx = 2\int \dfrac{dx}{x-1} + \int \dfrac{dx}{(x-1)^2}$.

We can easily integrate these: $2\ln|x - 1| - \dfrac{1}{x-1} + C$.

6. $\dfrac{1}{2}\ln|x+1| - \dfrac{1}{4}\ln(x^2+1) + \dfrac{1}{2}\tan^{-1}x + C$

We need to find A and $Bx+C$ such that $\dfrac{A}{x+1} + \dfrac{Bx+C}{x^2+1} = \dfrac{1}{(x+1)(x^2+1)}$.

We multiply through by $(x+1)(x^2+1)$: $A(x^2+1) + (Bx+C)(x+1) = 1$.

Expand all of the terms: $Ax^2 + A + Bx^2 + Bx + Cx + C = 1$.

Group the terms on the left side: $(A+B)x^2 + (B+C)x + (A+C) = 1$.

Now we have three equations: $A+B=0$, $B+C=0$, and $A+C=1$.

If we solve for A, B, and C, we get $A = \dfrac{1}{2}$, $B = -\dfrac{1}{2}$, and $C = \dfrac{1}{2}$.

This means that we can rewrite the integral as

$$\int \frac{1}{(x+1)(x^2+1)}\,dx = \frac{1}{2}\int \frac{dx}{x+1} - \frac{1}{2}\int \frac{x}{x^2+1}\,dx + \frac{1}{2}\int \frac{dx}{x^2+1}.$$

If we integrate these, we get $\dfrac{1}{2}\ln|x+1| - \dfrac{1}{4}\ln(x^2+1) + \dfrac{1}{2}\tan^{-1}x + C$. If you had trouble with

these integrals, you should review the sections on logarithmic and trigonometric integrals.

7. $\ln|x-1| + \dfrac{4}{\sqrt{3}}\tan^{-1}\dfrac{2x+1}{\sqrt{3}} + C$

First, we need to factor the denominator of the integrand: $\displaystyle\int \frac{x^2+3x-1}{x^3-1}\,dx = \int \frac{x^2+3x-1}{(x-1)(x^2+x+1)}\,dx.$

We need to find A and $Bx+C$ such that $\dfrac{A}{x-1} + \dfrac{Bx+C}{x^2+x+1} = \dfrac{x^2+3x-1}{(x-1)(x^2+x+1)}$.

We multiply through by $(x-1)(x^2+x+1)$: $A(x^2+x+1) + (Bx+C)(x-1) = x^2+3x-1$.

Expand all of the terms: $Ax^2 + Ax + A + Bx^2 - Bx + Cx - C = x^2 + 3x - 1$.

Group the terms on the left side: $(A + B)x^2 + (A - B + C)x + (A - C) = x^2 + 3x - 1$.

Now we have three equations: $A + B = 1$, $A - B + C = 3$, and $A - C = -1$.

If we solve for A, B, and C, we get $A = 1$, $B = 0$, and $C = 2$.

This means that we can rewrite the integral as $\int \dfrac{x^2 + 3x - 1}{x^3 - 1}\, dx = \int \dfrac{dx}{x - 1} + 2\int \dfrac{dx}{x^2 + x + 1}$. The first

integral is easy: $\int \dfrac{dx}{x - 1} = \ln|x - 1|$. The second takes a little work and will be a good test of your

algebra. First, complete the square to rewrite the denominator: $2\int \dfrac{dx}{x^2 + x + 1} = 2\int \dfrac{dx}{\left(x + \dfrac{1}{2}\right)^2 + \dfrac{3}{4}}$.

Next, multiply the top and bottom of the integrand by $\dfrac{4}{3}$: $2\int \dfrac{\dfrac{4}{3}dx}{\dfrac{4}{3}\left(x + \dfrac{1}{2}\right)^2 + 1} = \dfrac{8}{3}\int \dfrac{dx}{\dfrac{4}{3}\left(x + \dfrac{1}{2}\right)^2 + 1}$.

This can be rewritten as $\dfrac{8}{3}\int \dfrac{dx}{\left(\dfrac{2x + 1}{\sqrt{3}}\right)^2 + 1}$. Now we do u-substitution. If we let $u = \dfrac{2x + 1}{\sqrt{3}}$,

then $du = \dfrac{2}{\sqrt{3}}dx$. Multiply du by $\dfrac{\sqrt{3}}{2}$: $\dfrac{\sqrt{3}}{2}du = dx$. Now we can substitute into the integrand:

$\dfrac{8}{3}\left(\dfrac{\sqrt{3}}{2}\right)\int \dfrac{du}{u^2 + 1} = \dfrac{4}{\sqrt{3}}\tan^{-1} u + C$. Substituting back, we get $\dfrac{4}{\sqrt{3}}\tan^{-1}\left(\dfrac{2x + 1}{\sqrt{3}}\right) + C$. Now we

can put both integrals together: $\ln|x - 1| + \dfrac{4}{\sqrt{3}}\tan^{-1} \dfrac{2x + 1}{\sqrt{3}} + C$.

PRACTICE PROBLEM SET 32

1.   2

Notice that the integrand is undefined at the lower limit of integration $x = 0$. We remedy

this by replacing the lower limit with a and taking the limit of the integral as a approaches 0:

$\int_0^1 \frac{dx}{\sqrt{x}} = \lim_{a \to 0} \int_a^1 \frac{dx}{\sqrt{x}}$. Now we evaluate the integral: $\int_a^1 \frac{dx}{\sqrt{x}} = \left(\frac{x^{\frac{1}{2}}}{\frac{1}{2}} \right)\Bigg|_a^1 = \left(2\sqrt{x} \right)\Big|_a^1 = 2 - 2\sqrt{a}$. Finally,

we take the limit: $\lim_{a \to 0} \left(2 - 2\sqrt{a} \right) = 2$.

2.   6

Notice that although the integrand is defined at both limits of integration, it is

undefined between them at $x = 0$. We remedy this by first splitting the integral into

two parts at $x = 0$, $\int_{-1}^1 \frac{dx}{x^{\frac{2}{3}}} = \int_{-1}^0 \frac{dx}{x^{\frac{2}{3}}} + \int_0^1 \frac{dx}{x^{\frac{2}{3}}}$ and second, replacing the limit with a and tak-

ing the limit of the integral as a approaches 0, $\int_{-1}^0 \frac{dx}{x^{\frac{2}{3}}} + \int_0^1 \frac{dx}{x^{\frac{2}{3}}} = \lim_{a \to 0} \int_{-1}^a \frac{dx}{x^{\frac{2}{3}}} + \lim_{a \to 0} \int_a^1 \frac{dx}{x^{\frac{2}{3}}}$.

Now we evaluate the integrals: $\int_{-1}^a \frac{dx}{x^{\frac{2}{3}}} = \left(3x^{\frac{1}{3}} \right)\Bigg|_{-1}^a = 3\sqrt[3]{a} + 3$ and $\int_a^1 \frac{dx}{x^{\frac{2}{3}}} = \left(3x^{\frac{1}{3}} \right)\Bigg|_a^1 = 3 - 3\sqrt[3]{a}$.

Finally, we take the limits: $\lim_{a \to 0} \left(3\sqrt[3]{a} + 3 \right) = 3$ and $\lim_{a \to 0} \left(3 - 3\sqrt[3]{a} \right) = 3$. Therefore, the integral is

$3 + 3 = 6$.

3.   1

We can't evaluate an integral where a limit of integration is infinity. We remedy this by replacing the

lower limit with a and taking the limit of the integral as a approaches $-\infty$: $\int_{-\infty}^{0} e^x\, dx = \lim_{a \to -\infty} \int_{a}^{0} e^x\, dx$.

Now we evaluate the integral: $\int_{a}^{0} e^x\, dx = \left(e^x\right)\Big|_{a}^{0} = 1 - e^a$. Finally, we take the limit: $\lim_{a \to -\infty} \left(1 - e^a\right) = 1$.

4.   Diverges

Notice that the integrand is undefined at the lower limit of integration $x = 1$. We remedy

this by replacing the lower limit with a and taking the limit of the integral as a approaches 1:

$\int_{1}^{4} \dfrac{dx}{1-x} = \lim_{a \to 1} \int_{a}^{4} \dfrac{dx}{1-x}$. Now we evaluate the integral: $\int_{a}^{4} \dfrac{dx}{1-x} = \left(-\ln|1-x|\right)\Big|_{a}^{4} = -\ln 3 + \ln|1-a|$.

Finally, we take the limit: $\lim_{a \to 1}(-\ln 3 + \ln|1-a|) = D.N.E.$ Therefore, the integral diverges.

5. $\sqrt{3}$

Notice that the integrand is undefined at the lower limit of integration $x = 0$. We remedy

this by replacing the lower limit with a and taking the limit of the integral as a approaches

0: $\int_{0}^{1} \dfrac{x+1}{\sqrt{x^2+2x}}\, dx = \lim_{a \to 0} \int_{a}^{1} \dfrac{x+1}{\sqrt{x^2+2x}}\, dx$. Now we evaluate the integral. If we let $u = x^2 + 2x$,

then $du = (2x + 2)dx$. If we divide du by 2, we get $\dfrac{1}{2}du = (x+1)dx$. Now we can substi-

tute into the integrand. Let's ignore the limits of integration until we substitute back. We get

$\int \dfrac{x+1}{\sqrt{x^2+2x}}\, dx = \dfrac{1}{2}\int u^{-\frac{1}{2}}\, du = u^{\frac{1}{2}}$. Now we substitute back and evaluate at the limits of integra-

tion: $\left(\sqrt{x^2+2x}\right)\Big|_{a}^{1} = \sqrt{3} - \sqrt{a^2+2a}$. Finally, we take the limit: $\lim_{a \to 0}\left[\sqrt{3} - \sqrt{a^2+2a}\right] = \sqrt{3}$.

6. $-\dfrac{1}{4}$

We can't evaluate an integral where a limit of integration is infinity. We remedy this by replacing the lower limit with a and taking the limit of the integral as a approaches $-\infty$: $\displaystyle\int_{-\infty}^{0}\dfrac{dx}{\left(2x-1\right)^{3}}=\lim_{a\to-\infty}\int_{a}^{0}\dfrac{dx}{\left(2x-1\right)^{3}}$. Now we evaluate the integral:

$$\int_{a}^{0}\dfrac{dx}{\left(2x-1\right)^{3}}=\left(-\dfrac{1}{4\left(2x-1\right)^{2}}\right)\Bigg|_{a}^{0}=-\dfrac{1}{4}+\dfrac{1}{4\left(2a-1\right)^{2}}\ .\qquad\text{Finally,}\quad\text{we}\quad\text{take}\quad\text{the}\quad\text{limit:}$$

$$\lim_{a\to-\infty}\left(-\dfrac{1}{4}+\dfrac{1}{4\left(2a-1\right)^{2}}\right)=-\dfrac{1}{4}\ .$$

7. Diverges

Notice that, although the integrand is defined at both limits of integration, it is undefined between them at $x = 2$. We remedy this by first splitting the integral into two parts at $x = 2$, $\displaystyle\int_{0}^{3}\dfrac{dx}{x-2}=\int_{0}^{2}\dfrac{dx}{x-2}+\int_{2}^{3}\dfrac{dx}{x-2}$ and, second, replacing the limit with a and taking the limit of the integral as a approaches 2, $\displaystyle\int_{0}^{2}\dfrac{dx}{x-2}+\int_{2}^{3}\dfrac{dx}{x-2}=\lim_{a\to2}\int_{0}^{a}\dfrac{dx}{x-2}+\lim_{a\to2}\int_{2}^{3}\dfrac{dx}{x-2}$. Now, we evaluate the integrals:

$$\int_{0}^{a}\dfrac{dx}{x-2}=\left(\ln\left|x-2\right|\right)\Big|_{0}^{a}=\ln\left|a-2\right|-\ln 2\quad\text{and}\quad\int_{a}^{3}\dfrac{dx}{x-2}=\left(\ln\left|x-2\right|\right)\Big|_{a}^{3}=-\ln\left|a-2\right|\ .$$

Finally, we take the limits: $\displaystyle\lim_{a\to2}\left(\ln\left|a-2\right|-\ln 2\right)=D.N.E.$ and $\displaystyle\lim_{a\to2}\left(-\ln\left|a-2\right|\right)=D.N.E.$ Therefore, the integral diverges.

8. $\dfrac{9}{2}$

Notice that although the integrand is defined at both limits of integration, it is undefined between them at $x = 0$. We remedy this by first splitting the integral into two parts at $x = 0$, $\int_{-1}^{8} \frac{dx}{\sqrt[3]{x}} = \int_{-1}^{0} \frac{dx}{\sqrt[3]{x}} + \int_{0}^{8} \frac{dx}{\sqrt[3]{x}}$ and second, replacing the limit with a and tak-

ing the limit of the integral as a approaches 0, $\int_{-1}^{0} \frac{dx}{\sqrt[3]{x}} + \int_{0}^{8} \frac{dx}{\sqrt[3]{x}} = \lim_{a \to 0} \int_{-1}^{a} \frac{dx}{\sqrt[3]{x}} + \lim_{a \to 0} \int_{a}^{8} \frac{dx}{\sqrt[3]{x}}$.

Now, we evaluate the integrals: $\int_{-1}^{a} \frac{dx}{\sqrt[3]{x}} = \left(\frac{3x^{\frac{2}{3}}}{2} \right)\Bigg|_{-1}^{a} = \frac{3a^{\frac{2}{3}}}{2} - \frac{3}{2}$ and $\int_{a}^{8} \frac{dx}{\sqrt[3]{x}} = \left(\frac{3x^{\frac{2}{3}}}{2} \right)\Bigg|_{a}^{8} = 6 - \frac{3a^{\frac{2}{3}}}{2}$.

Finally, we take the limits: $\lim_{a \to 0} \left(\frac{3a^{\frac{2}{3}}}{2} - \frac{3}{2} \right) = -\frac{3}{2}$ and $\lim_{a \to 0} \left(6 - \frac{3a^{\frac{2}{3}}}{2} \right) = 6$. Therefore, the integral is

$-\frac{3}{2} + 6 = \frac{9}{2}$.

PRACTICE PROBLEM SET 33

1. $\dfrac{4 \sin 4\theta \sin \theta - \cos 4\theta \cos \theta}{4 \sin 4\theta \cos \theta + \cos 4\theta \sin \theta}$

Recall that the formula for the slope of a polar curve is $\dfrac{dy}{dx} = \dfrac{f'(\theta)\sin\theta + f(\theta)\cos\theta}{f'(\theta)\cos\theta - f(\theta)\sin\theta}$.

Here we have $f(\theta) = 2\cos4\theta$, so $f'(\theta) = -8\sin4\theta$. Plugging into the formula, we get

$\dfrac{dy}{dx} = \dfrac{-8\sin 4\theta \sin\theta + 2\cos 4\theta \cos\theta}{-8\sin 4\theta \cos\theta - 2\cos 4\theta \sin\theta}$, which can be reduced to $\dfrac{4\sin 4\theta \sin\theta - \cos 4\theta \cos\theta}{4\sin 4\theta \cos\theta + \cos 4\theta \sin\theta}$.

2. $\dfrac{2}{3}$

Recall that the formula for the slope of a polar curve is $\dfrac{dy}{dx} = \dfrac{f'(\theta)\sin\theta + f(\theta)\cos\theta}{f'(\theta)\cos\theta - f(\theta)\sin\theta}$.

Here we have $f(\theta) = 2 - 3\sin\theta$, so $f'(\theta) = -3\cos\theta$. Plugging into the formula, we get

$\dfrac{dy}{dx} = \dfrac{-3\cos\theta \sin\theta + (2 - 3\sin\theta)\cos\theta}{-3\cos\theta \cos\theta - (2 - 3\sin\theta)\sin\theta}$. Now we evaluate the slope at $(r, \theta) = (2, \pi)$:

$\dfrac{dy}{dx} = \dfrac{-3\cos\pi \sin\pi + (2 - 3\sin\pi)\cos\pi}{-3\cos\pi \cos\pi - (2 - 3\sin\pi)\sin\pi} = \dfrac{2}{3}$.

3. 18π

Recall that the formula for the area between the origin and the curve $r = f(\theta)$ from $\theta = a$ to $\theta = b$ is $A = \int_a^b \frac{1}{2} r^2 \, d\theta$. Here we have $r = 4 + 2\cos\theta$. First, let's graph the curve.

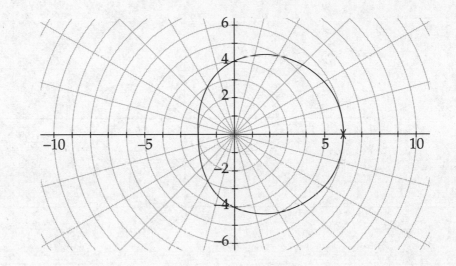

We can see that the loop runs from $\theta = 0$ to $\theta = 2\pi$. Plugging into the formula, we get

$A = \frac{1}{2}\int_0^{2\pi} (4 + 2\cos\theta)^2 \, d\theta = \frac{1}{2}\int_0^{2\pi} (16 + 16\cos\theta + 4\cos^2\theta) \, d\theta = \int_0^{2\pi} (8 + 8\cos\theta + 2\cos^2\theta) \, d\theta$.   We can

break this into three integrals: $\int_0^{2\pi} (8 + 8\cos\theta + 2\cos^2\theta) \, d\theta = 8\int_0^{2\pi} d\theta + 8\int_0^{2\pi} \cos\theta \, d\theta + 2\int_0^{2\pi} \cos^2\theta \, d\theta$.

We can easily evaluate the first two integrals: $8\int_0^{2\pi} d\theta = 8(\theta)\Big|_0^{2\pi} = 16\pi$ and

$8\int_0^{2\pi} \cos\theta \, d\theta = 8(\sin\theta)\Big|_0^{2\pi} = 0$. The third integral requires us to use the trigonometric substitution

$\cos^2\theta = \dfrac{1 + \cos 2\theta}{2}$:

$2\int_0^{2\pi} \cos^2\theta \, d\theta = 2\int_0^{2\pi} \dfrac{1 + \cos\theta}{2} \, d\theta = \int_0^{2\pi} (1 + \cos 2\theta) \, d\theta = \left(\theta + \dfrac{1}{2}\sin 2\theta\right)\Big|_0^{2\pi} = 2\pi$. Therefore, the area

is 18π.

4.   2

Recall that the formula for the area between the origin and the curve $r = f(\theta)$ from $\theta = a$ to $\theta = b$ is $A = \int_a^b \frac{1}{2} r^2 d\theta$. Here we have $r^2 = 4\cos 2\theta$. First, let's graph the curve.

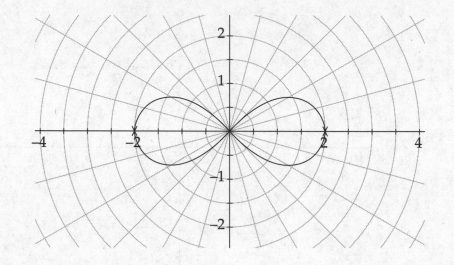

If we plot a few points, we can see that the loop runs from $\theta = -\frac{\pi}{4}$ to $\theta = \frac{\pi}{4}$. Plugging into the formula, we get

$$A = \frac{1}{2} \int_{-\frac{\pi}{4}}^{\frac{\pi}{4}} (4\cos 2\theta) d\theta = 2 \int_{-\frac{\pi}{4}}^{\frac{\pi}{4}} \cos 2\theta \; d\theta = (\sin 2\theta) \Big|_{-\frac{\pi}{4}}^{\frac{\pi}{4}} = 2$$

5. $\dfrac{3\sqrt{3}}{2} - \dfrac{\pi}{2}$

Recall that the formula for the area between the two curves $r_1 = f_1(\theta)$ and $r_2 = f_2(\theta)$, with $0 \le r_1(\theta) \le r_2(\theta)$ from $\theta = a$ to $\theta = b$ is $A = \int_a^b \frac{1}{2}(r_2^2 - r_1^2) d\theta$. Here we have $r_2^2 = 6\cos 2\theta$ and $r_1 = \sqrt{3}$. First, let's graph the curves.

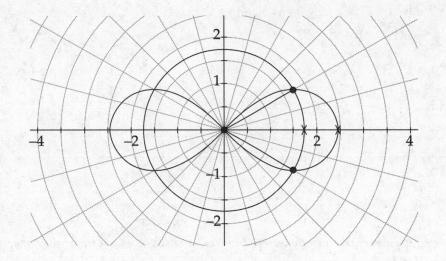

If we solve for the intersection of the two curves, we find that the region runs from $\theta = -\dfrac{\pi}{6}$ to

$\theta = \dfrac{\pi}{6}$. Plugging into the formula, we get $A = \dfrac{1}{2}\displaystyle\int_{-\frac{\pi}{6}}^{\frac{\pi}{6}}\left(6\cos 2\theta - 3\right)d\theta = 3\displaystyle\int_{-\frac{\pi}{6}}^{\frac{\pi}{6}}\left(\cos 2\theta\right)d\theta - \dfrac{3}{2}\displaystyle\int_{-\frac{\pi}{6}}^{\frac{\pi}{6}}d\theta.$

Evaluating the integrals, we get $3\displaystyle\int_{-\frac{\pi}{6}}^{\frac{\pi}{6}}\left(\cos 2\theta\right)d\theta = \left(\dfrac{3\sin 2\theta}{2}\right)\Bigg|_{-\frac{\pi}{6}}^{\frac{\pi}{6}} = \dfrac{3\sqrt{3}}{2}$ and

$\dfrac{3}{2}\displaystyle\int_{-\frac{\pi}{6}}^{\frac{\pi}{6}}d\theta = \dfrac{3}{2}(\theta)\Big|_{-\frac{\pi}{6}}^{\frac{\pi}{6}} = \dfrac{\pi}{2}.$ Therefore, the area is $\dfrac{3\sqrt{3}}{2} - \dfrac{\pi}{2}.$

PRACTICE PROBLEM SET 34

1. $y = \sqrt[4]{\dfrac{28x^3}{3} - 236}$

   We solve this differential equation by separation of variables. We want to get all of the y variables on one side of the equals sign and all of the x variables on the other side. We can do this easily by cross multiplying. We get $y^3\,dy = 7x^2\,dx$. Next, we integrate both sides.

$$\int y^3\,dy = \int 7x^2\,dx$$

$$\frac{y^4}{4} = 7\frac{x^3}{3} + C$$

$$y^4 = \frac{28x^3}{3} + C$$

Now we solve for C. We plug in $x = 3$ and $y = 2$.

$$16 = 252 + C$$

$$C = -236$$

Therefore, $y^4 = \dfrac{28x^3}{3} - 236$.

Now we isolate y. We get the equation $y = \sqrt[4]{\dfrac{28x^3}{3} - 236}$.

2. $\quad y = 6e^{\frac{5x^3}{3}}$

We solve this differential equation by separation of variables. We want to get all of the y variables on one side of the equals sign and all of the x variables on the other side. We can do this easily by dividing both sides by y and multiplying both sides by dx. We get $\dfrac{dy}{y} = 5x^2\,dx$. Next, we integrate both sides.

$$\int \frac{dy}{y} = \int 5x^2\,dx$$

$$\ln y = \frac{5x^3}{3} + C_0$$

Now we isolate y: $y = e^{\frac{5x^3}{3} + C}$. We can rewrite this as $y = e^{\frac{5x^3}{3}}\left(e^{C_0}\right) = Ce^{\frac{5x^3}{3}}$.

Note that we are using the letter C in the last equation. This is to distinguish it from the C_0 in the first equation. Now we solve for C. We plug in $x = 0$ and $y = 6$: $6 = Ce^0 = C$. Therefore, the equation is $y = 6e^{\frac{5x^3}{3}}$.

3. $\quad y = \sqrt[3]{3e^x - 2}$

We solve this differential equation by separation of variables. We want to get all of the y variables on one side of the equals sign and all of the x variables on the other side. We can do this easily by cross multiplying. We get $y^2\,dy = e^x\,dx$. Next, we integrate both sides.

$$\int y^2\,dy = \int e^x\,dx$$

$$\frac{y^3}{3} = e^x + C$$

Now we isolate y.

$$y^3 = 3e^x + C$$

$$y = \sqrt[3]{3e^x + C}$$

Now we solve for C. We plug in $x = 0$ and $y = 1$.

$$1 = \sqrt[3]{3e^0 + C}$$

$$C = -2$$

Therefore, the equation is $y = \sqrt[3]{3e^x - 2}$.

4. $y = 2x^2$

We solve this differential equation by separation of variables. We want to get all of the y variables on one side of the equals sign and all of the x variables on the other side. We can do this easily by dividing both sides by y^2 and multiplying both sides by dx. We get $\dfrac{dy}{y^2} = \dfrac{dx}{x^3}$.

Next, we integrate both sides.

$$\int \frac{dy}{y^2} = \int \frac{dx}{x^3}$$

$$-\frac{1}{y} = -\frac{1}{2x^2} + C$$

Now we isolate y.

$$\frac{1}{y} = \frac{1}{2x^2} + C$$

Now we solve for C. We plug in $x = 1$ and $y = 2$.

$$\frac{1}{2} = \frac{1}{2(1)^2} + C$$

$$C = 0$$

Therefore, the equation is $\dfrac{1}{y} = \dfrac{1}{2x^2} + C$, which can be rewritten as $y = 2x^2$.

5. $y = \sin^{-1}(-\cos x)$

We solve this differential equation by separation of variables. We want to get all of the y variables on one side of the equals sign and all of the x variables on the other side. We can do this easily by cross multiplying. We get $\cos y \, dy = \sin x \, dx$. Next, we integrate both sides.

$$\int \cos y \, dy = \int \sin x \, dx$$

$$\sin y = -\cos x + C$$

Now we solve for C. We plug in $x = 0$ and $y = \dfrac{3\pi}{2}$.

$$\sin\frac{3\pi}{2} = -\cos 0 + C$$

$$-1 = -1 + C$$

$$C = 0$$

Now we isolate y to get the equation $y = \sin^{-1}(-\cos x)$.

6.   20,000 (approximately)

The phrase "exponential growth" means that we can represent the situation with the differential equation $\dfrac{dy}{dt} = ky$, where k is a constant and y is the population at a time t. Here we are also told that $y = 4{,}000$ at $t = 0$ and $y = 6{,}500$ at $t = 3$. We solve this differential equation by separation of variables. We want to get all of the y variables on one side of the equals sign and all of the t variables on the other side. We can do this easily by dividing both sides by y and multiplying both sides by dt. We get $\dfrac{dy}{y} = k\,dt$. Next, we integrate both sides.

$$\int \frac{dy}{y} = \int k\,dt$$

$$\ln y = kt + C_0$$

$$y = Ce^{kt}$$

Next, we plug in $y = 4{,}000$ and $t = 0$ to solve for C.

$$4{,}000 = Ce^0$$

$$C = 4{,}000$$

Now, we have $y = 4{,}000e^{kt}$. Next, we plug in $y = 6{,}500$ and $t = 3$ to solve for k.

$$6{,}500 = 4{,}000e^{3k}$$

$$k = \frac{1}{3}\ln\frac{13}{8} \approx 0.162$$

Therefore, our equation for the population of bacteria, y, at time, t, is $y \approx 4,000e^{0.162t}$, or if we want an exact solution, it is $y = 4,000\left(\dfrac{13}{8}\right)^{\frac{t}{3}}$. Finally, we can solve for the population at time $t = 10$: $y \approx 4,000e^{0.162(10)} \approx 20,212$ or $y = 4,000\left(\dfrac{13}{8}\right)^{\frac{10}{3}} \approx 20,179$. (Notice how even with an "exact" solution, we still have to round the answer. And, if we are concerned with significant figures, the population can be written as 20,000.)

7.   45 m

Because acceleration is the derivative of velocity, we can write $\dfrac{dv}{dt} = -9$. We are also told that $v = 18$ and $h = 45$ when $t = 0$. We solve this differential equation by separation of variables. We want to get all of the v variables on one side of the equals sign and all of the t variables on the other side. We can do this easily by multiplying both sides by dt: $dv = -9\,dt$. Next, we integrate both sides:

$$\int dv = \int -9\,dt$$
$$v = -9 + C_0$$

Now we can solve for C_0 by plugging in $v = 18$ and $t = 0$: $18 = -9(0) + C_0$ so $C_0 = 18$. Thus, our equation for the velocity of the rock is $v = -9t + 18$. Next, because the height of the rock is the derivative of the velocity, we can write $\dfrac{dh}{dt} = -9t + 18$. We again separate the variables by multiplying both sides by dt: $dh = (-9t + 18)\,dt$. We integrate both sides:

$$\int dh = \int (-9t + 18)\,dt$$
$$h = -\frac{9t^2}{2} + 18t + C_1$$

Next, we plug in $h = 45$ and $t = 0$ to solve for C_1.

$45 = -\dfrac{9(0)^2}{2} + 18(0) + C_1$, so $C_1 = 45$

Therefore, the equation for the height of the rock, h, at time, t, is $h = -\dfrac{9t^2}{2} + 18t + 45$.

Finally, we can solve for the height of the rock at time $t = 4$: $h = -\dfrac{9t^2}{2} + 18t + 45$, so

$h = -\dfrac{9(4)^2}{2} + 18(4) + 45 = 45$ m.

8.   8,900 grams (approximately)

We can express this situation with the differential equation $\frac{dm}{dt} = -km$, where m is the mass at time t. We are also given that $m = 10,000$ when $t = 0$ and $m = 5,000$ when $t = 5,750$. We solve this differential equation by separation of variables. We want to get all of the m variables on one side of the equals sign and all of the t variables on the other side. We can do this easily by dividing both sides by m and multiplying both sides by dt. We get $\frac{dm}{m} = -k\,dt$. Next, we integrate both sides.

$$\int \frac{dm}{m} = \int -k\,dt$$

$$\ln m = -kt + C_0$$

$$m = Ce^{-kt}$$

Now we can solve for C by plugging in $m = 10,000$ and $t = 0$: $10,000 = Ce^0$, so $C = 10,000$. This gives us the equation $m = 10,000e^{-kt}$. Next, we can solve for k by plugging in $m = 5,000$ and $t = 5,750$.

$$5,000 = 10,000e^{-5,750k}$$

$$\frac{1}{2} = e^{-5,750k}$$

$$-\frac{1}{5,750} \ln \frac{1}{2} = k \approx 0.000121$$

Therefore, the equation for the mass of the element, m, at time, t, is $m \approx 10,000e^{-.000121t}$, or if we want an exact solution, it is $m = 10,000\left(\frac{1}{2}\right)^{\frac{t}{5,750}}$.

Finally, we can solve for the mass of the element at time $t = 1,000$: $m \approx 10,000e^{-.000121(1,000)} \approx 8,860$ gms or $m = 10,000\left(\frac{1}{2}\right)^{\frac{1,000}{5,750}} \approx 8,864$ gms. (And, if we are concerned with significant figures, the mass can be written as 8,900.)

9.   4.441

We start with $x_0 = 0$ and $y_0 = 2$ and $h = 0.25$. The slope is found by plugging $x_0 = 0$ and $y_0 = 2$ into $y' = y - x$, so we have $y_0' = 2 - 0 = 2$.

Step 1: Increase x_0 by h to get x_1: $x_1 = 0 + 0.25 = 0.25$.

Step 2: Multiply h by y_0' and add y_0 to get y_1: $y_1 = 2 + (0.25)(2) = 2.5$.

Step 3: Find y_1' by plugging y_1 and x_1 into the equation $y' = y - x$: $y_1' = 2.5 - 0.25 = 2.25$.

Now we repeat until we get to $x = 1$.

Step 1: Increase x_1 by h to get x_2: $x_2 = 0.25 + 0.25 = 0.5$.

Step 2: Multiply h by y_1' and add y_1 to get y_2: $y_2 = 2.5 + (0.25)(2.25) = 3.0625$.

Step 3: Find y_2' by plugging y_2 and x_2 into the equation $y' = y - x$: $y_2' = 3.0625 - 0.5 = 2.5625$.

Repeat.

Step 1: Increase x_2 by h to get x_3: $x_3 = 0.25 + 0.5 = 0.75$.

Step 2: Multiply h by y_2' and add y_2 to get y_3: $y_3 = 3.0625 + (0.25)(2.5625) = 3.703125$.

Step 3: Find y_3' by plugging y_3 and x_3 into the equation $y' = y - x$: $y_3' = 3.703125 - 0.75 = 2.953125$.

Repeat one last time.

Step 1: Increase x_3 by h to get x_4: $x_4 = 0.25 + 0.75 = 1$.

Step 2: Multiply h by y_3' and add y_3 to get y_4: $y_4 = 3.703125 + (0.25)(2.953125) = 4.441$.

10.   0.328

We start with $x_0 = 0$ and $h = 0.2$. The slope is found by plugging $x_0 = 0$ and $y_0 = 1$ into $y' = -y$, so we have $y_0' = -1$.

Step 1: Increase by h to get x_1: $x_1 = 0 + 0.2 = 0.2$.

Step 2: Multiply h by y_0' and add y_0 to get y_1: $y_1 = 1 + (0.2)(-1) = 0.8$.

Step 3: Find y_1' by plugging y_1 and x_1 into the equation $y' = -y$: $y_1' = -0.8$.

Now we repeat until we get to $x = 1$.

Step 1: Increase x_1 by h to get x_2: $x_2 = 0.2 + 0.2 = 0.4$.

Step 2: Multiply h by y_1' and add y_1 to get y_2: $y_2 = 0.8 + (0.2)(-0.8) = 0.64$.

Step 3: Find y_2' by plugging y_2 and x_2 into the equation $y' = -y$: $y_2' = -0.64$.

Repeat.

Step 1: Increase x_2 by h to get x_3: $x_3 = 0.2 + 0.4 = 0.6$.

Step 2: Multiply h by y_2' and add y_2 to get y_3: $y_3 = 0.64 + (0.2)(-0.64) = 0.512$.

Step 3: Find y_3' by plugging y_3 and x_3 into the equation $y' = -y$: $y_3' = -0.512$.

Repeat.

Step 1: Increase x_3 by h to get x_4: $x_4 = 0.2 + 0.6 = 0.8$.

Step 2: Multiply h by y_3' and add y_3 to get y_4: $y_4 = 0.512 + (0.2)(-0.512) = 0.4096$.

Step 3: Find y_4' by plugging y_4 and x_4 into the equation $y' = -y$: $y_4' = -0.4096$.

Repeat one last time.

Step 1: Increase x_4 by h to get x_5: $x_5 = 0.2 + 0.8 = 1$.

Step 2: Multiply h by y_4' and add y_4 to get $y_0 = 1$ $y_5 = 0.4096 + (0.2)(-0.4096) = 0.328$.

11.   0.04

We start with $x_0 = 0$ and $y_0 = 0$ and $h = 0.1$. The slope is found by plugging $x_0 = 0$ and $y_0 = 0$ into $y' = 4x^3$, so we have $y_0' = 0$.

Step 1: Increase x_0 by h to get x_1: $x_1 = 0 + 0.1 = 0.1$.

Step 2: Multiply h by y_0' and add y_0 to get y_1: $y_1 = 0 + (0.1)(0) = 0$.

Step 3: Find y_1' by plugging y_1 and x_1 into the equation $y' = 4x^3$: $y_1' = 0.004$.

Now we repeat until we get to $x = 0.5$.

Step 1: Increase x_1 by h to get x_2: $x_2 = 0.1 + 0.1 = 0.2$.

Step 2: Multiply h by y_1' and add y_1 to get y_2: $y_2 = 0 + (0.1)(0.004) = 0.0004$.

Step 3: Find y_2' by plugging y_2 and x_2 into the equation $y' = 4x^3$: $y_2' = 0.032$.

Repeat.

Step 1: Increase x_2 by h to get x_3: $x_3 = 0.1 + 0.2 = 0.3$.

Step 2: Multiply h by y_2' and add y_2 to get y_3: $y_3 = 0.0004 + (0.1)(0.032) = 0.0036$.

Step 3: Find y_3' by plugging y_3 and x_3 into the equation $y' = 4x^3$: $y_3' = 0.108$.

Repeat.

Step 1: Increase x_3 by h to get x_4: $x_4 = 0.1 + 0.3 = 0.4$.

Step 2: Multiply h by y_3' and add y_3 to get y_4: $y_4 = 0.0036 + (0.1)(0.108) = 0.0144$.

Step 3: Find y_4' by plugging y_4 and x_3 into the equation $y' = 4x^3$: $y_4' = 0.256$.

Repeat one last time.

Step 1: Increase x_4 by h to get x_5: $x_5 = 0.1 + 0.4 = 0.5$.

Step 2: Multiply h by y_4' and add y_4 to get y_5: $y_5 = 0.0144 + (0.1)(0.256) = 0.04$.

12.

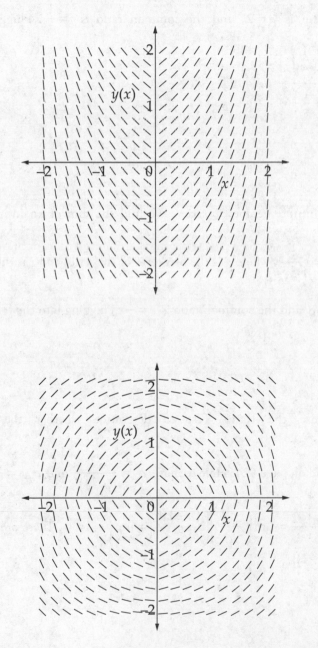

13.

PRACTICE PROBLEM SET 35

1. 2.499

This is a geometric series. We find the sum of the first n terms of a geometric series using the

formula $S_n = \dfrac{a(1-r^n)}{(1-r)}$, where a is the first term of the series and r is the common ratio. Here,

the first term is $a = 2$, and the common ratio is $r = \dfrac{1}{5}$. Plugging into the formula, we get

$$S_n = \frac{2\left(1 - \left(\dfrac{1}{5}\right)^5\right)}{\left(1 - \dfrac{1}{5}\right)} = 2.499.$$

2. $\dfrac{28}{3}$

This is an infinite geometric series. We find the sum of an infinite geometric series using the

formula $S = \dfrac{a}{1-r}$, where a is the first term of the series and r is the common ratio. Here, the first

term is $a = 8$ and the common ratio is $r = \dfrac{1}{7}$. Plugging into the formula, we get $S = \dfrac{8}{1 - \dfrac{1}{7}} = \dfrac{28}{3}$.

3. Converges

We can use the Ratio Test to determine whether the series converges. The test

can be found on page 424. Here we have $a_n = \dfrac{5^n}{(n-1)!}$ and $a_{n+1} = \dfrac{5^{n+1}}{n!}$. Plugging into the test, we

get $\displaystyle\lim_{n\to\infty} \frac{\dfrac{5^{n+1}}{n!}}{\dfrac{5^n}{(n-1)!}} = \lim_{n\to\infty}\left(\frac{5^{n+1}}{n!}\right)\left(\frac{(n-1)!}{5^n}\right) = \lim_{n\to\infty}\left(\frac{5}{n}\right) = 0$. Therefore, the series converges.

4.   Diverges

We can use the Ratio Test to determine whether the series converges. The test can be found

on page 424. Here we have $a_n = \dfrac{5^n}{n^2}$ and $a_{n+1} = \dfrac{5^{n+1}}{(n+1)^2}$. Plugging into the test, we get

$$\lim_{n \to \infty} \dfrac{\dfrac{5^{n+1}}{(n+1)^2}}{\dfrac{5^n}{n^2}} = \lim_{n \to \infty} \left(\dfrac{5^{n+1}}{(n+1)^2} \right)\left(\dfrac{n^2}{5^n} \right) = \lim_{n \to \infty} (5)\left(\dfrac{n}{n+1} \right)^2 = 5. \text{ Therefore, the series diverges.}$$

5.   Yes, converges

If you see a series that is a ratio of polynomials, you can often use the Limit Comparison Test.

For large n, the series will behave like $\dfrac{n}{n^3} = \dfrac{1}{n^2}$, so let's make $b_n = \dfrac{1}{n^2}$. We know that the series $\displaystyle\sum_{n=1}^{\infty} \dfrac{1}{n^2}$,

converges. Let's use the test: $\displaystyle\lim_{n \to \infty} \dfrac{a_n}{b_n} = \lim_{n \to \infty} \dfrac{\dfrac{n-3}{n^3 - n^2 + 3}}{\dfrac{1}{n^2}} = \lim_{n \to \infty} \dfrac{n-3}{n^3 - n^2 + 3} \dfrac{n^2}{1} = \lim_{n \to \infty} \dfrac{n^3 - 3n^2}{n^3 - n^2 + 3} = 1.$

This result corresponds to part 1 of the test. Because the ratio is 1 and the series $\displaystyle\sum_{n=1}^{\infty} \dfrac{1}{n^2}$ converges,

the series $\displaystyle\sum_{n=1}^{\infty} \dfrac{n-3}{n^3 - n^2 + 3}$ also converges.

6.   No, diverges

If you see a series that is a ratio of polynomials, you can often use the Limit Comparison Test. For large

n, the series will behave like $\dfrac{n^2}{n^3} = \dfrac{1}{n}$, so let's make $b_n = \dfrac{1}{n}$. We know that the series $\displaystyle\sum_{n=1}^{\infty} \dfrac{1}{n}$ diverges. Let's

use the test: $\displaystyle\lim_{n \to \infty} \dfrac{a_n}{b_n} = \lim_{n \to \infty} \dfrac{\dfrac{n^2+n}{n^3 - n^2 + n - 1}}{\dfrac{1}{n}} = \lim_{n \to \infty} \dfrac{n^2 + n}{n^3 - n^2 + n - 1} \dfrac{n}{1} = \lim_{n \to \infty} \dfrac{n^3 + n^2}{n^3 - n^2 + n - 1} = 1.$

This result corresponds to part 1 of the test. Because the ratio is 1 and the series $\sum_{n=1}^{\infty} = \frac{1}{n}$ diverges, the series $\sum_{n=1}^{\infty} \frac{n^2 + n}{n^3 - n^2 + n - 1}$ also diverges.

7.   Yes, converges

Here is another place where we can use the Limit Comparison Test. For large n, the series will behave like $\frac{2^n}{4^n} = \frac{1}{2^n}$, so let's make $b_n = \frac{1}{2^n}$. We know that the series $\sum_{n=1}^{\infty} = \frac{1}{2^n}$ converges because it is a geometric series with a ratio less than 1 (and greater than -1). Let's use the test:

$$\lim_{n \to \infty} \frac{a_n}{b_n} = \lim_{n \to \infty} \frac{\frac{2^n}{1 + 4^n}}{\frac{1}{2^n}} = \lim_{n \to \infty} \frac{2^n}{1 + 4^n} \cdot \frac{2^n}{1} = \lim_{n \to \infty} \frac{4^n}{1 + 4^n} = 1.$$

This result corresponds to part 1 of the test. Because the ratio is 1 and the series $\sum_{n=1}^{\infty} = \frac{1}{2^n}$ converges, the series $\sum_{n=1}^{\infty} \frac{2^n}{1 + 4^n}$ also converges.

8.   Converges absolutely

Let's apply the Alternating Series Test:

(a) $b_n > 0$. That is, the terms $\frac{n}{n^3 + 1}$ are all positive.

(b) $b_n > b_{n+1}$ for all n. That is, the terms are decreasing.

(c) $b_n \to 0$ as $n \to \infty$.

Therefore, the series converges conditionally. Now we need to test whether the series converges absolutely. We will test the series $\sum_{n=1}^{\infty} \frac{n}{n^3 + 1}$. Let's apply the Limit Comparison Test. For large n, the series will behave like $\frac{n}{n^3} = \frac{1}{n^2}$, so let's make $b_n = \frac{1}{n^2}$. We know that the series $\sum_{n=1}^{\infty} \frac{1}{n^2}$ converges.

Let's use the test: $\lim_{n \to \infty} \frac{a_n}{b_n} = \lim_{n \to \infty} \frac{\frac{n}{n^3 + 1}}{\frac{1}{n^2}} = \lim_{n \to \infty} \frac{n}{n^3 + 1} \cdot \frac{n^2}{1} = \lim_{n \to \infty} \frac{n^3}{n^3 + 1} = 1.$

This result corresponds to part 1 of the test. Because the ratio is 1 and the series $\displaystyle\sum_{n=1}^{\infty}\frac{1}{n^2}$ converges, the series $\displaystyle\sum_{n=1}^{\infty}\frac{n}{n^3+1}$ also converges. Therefore, the series converges absolutely.

9.   Converges conditionally

(a) $b_n > 0$. That is, the terms $\dfrac{1+n}{n^2}$ are all positive.

(b) $b_n > b_{n+1}$ for all n. That is, the terms are decreasing.

(c) $b_n \to 0$ as $n \to \infty$.

Therefore, the series converges conditionally. Now we need to test whether the series converges absolutely. We will test the series $\displaystyle\sum_{n=1}^{\infty}\frac{1+n}{n^2}$. Let's apply the Limit Comparison Test. For large n, the series will behave like $\dfrac{n}{n^2} = \dfrac{1}{n}$, so let's make $b_n = \dfrac{1}{n}$. We know that the series $\displaystyle\sum_{n=1}^{\infty}\frac{1}{n}$ diverges.

Let's use the test: $\displaystyle\lim_{n\to\infty}\frac{a_n}{b_n} = \lim_{n\to\infty}\frac{\dfrac{1+n}{n^2}}{\dfrac{1}{n}} = \lim_{n\to\infty}\frac{1+n}{n^2}\,\frac{n}{1} = \lim_{n\to\infty}\frac{1+n}{n} = 1.$

This result corresponds to part 1 of the test. Because the ratio is 1 and the series $\displaystyle\sum_{n=1}^{\infty}\frac{1}{n}$ diverges, the series $\displaystyle\sum_{n=1}^{\infty}(-1)^n\frac{1+n}{n^2}$ also diverges. Therefore, the series converges conditionally but not absolutely.

10.   Diverges

Let's apply the Alternating Series Test:

(a) $b_n > 0$. That is, the terms $\dfrac{3+n}{4+n}$ are all positive.

(b) $b_n > b_{n+1}$ for all n. That is, the terms are decreasing.

Therefore, the series diverges. Additionally, $b_n \to 1$ as $n \to \infty$.

11. $\cos x = 1 - \dfrac{x^2}{2!} + \dfrac{x^4}{4!} - \dfrac{x^6}{6!} + \ldots = \displaystyle\sum_{k=0}^{\infty}(-1)^k\frac{x^{2k}}{(2k)!}$

We find a Taylor series by the formula:

$$\sum_{k=0}^{\infty}\frac{f^{(k)}}{k!}(x-a)^k = f(a) + f'(a)(x-a) + \frac{f''(a)}{2!}(x-a)^2 + \ldots + \frac{f^{(n)}(a)}{n!}(x-a)^n + \ldots$$

First, let's find a few derivatives of cos x and evaluate them at $a = 0$.

$f(x) = \cos x;$ $f(0) = \cos 0 = 1$

$f'(x) = -\sin x;$ $f'(0) = -\sin 0 = 0$

$f''(x) = -\cos x;$ $f''(0) = -\cos 0 = -1$

$f'''(x) = \sin x;$ $f'''(0) = \sin 0 = 0$

$f^{(4)}(x) = \cos x;$ $f^{(4)}(0) = \cos 0 = 1$

$f^{(5)}(x) = -\sin x;$ $f^{(5)}(x) = -\sin 0 = 0$

$f^{(6)}(x) = -\cos x;$ $f^{(6)}(0) = -\cos 0 = -1$

Let's see if we find a pattern. Plugging into the formula, we get

$$\cos x = 1 + 0 + \frac{(-1)}{2!}x^2 + 0 + \frac{1}{4!}x^4 + 0 - \frac{1}{6!}x^6$$

We can see that the general formula is: $\cos x = 1 - \dfrac{x^2}{2!} + \dfrac{x^4}{4!} - \dfrac{x^6}{6!} + \ldots = \displaystyle\sum_{k=0}^{\infty}(-1)^k \frac{x^{2k}}{(2k)!}$.

You should memorize the Taylor series that are highlighted on page 434. The AP Exam does not require you to derive them when you use them. You can simply state the formulas.

12. $\ln(1+x) = x - \dfrac{x^2}{2} + \dfrac{x^3}{3} - \dfrac{x^4}{4} + \ldots = \displaystyle\sum_{k=0}^{\infty}(-1)^k \frac{x^{k+1}}{k+1}$

We find a Taylor series by the formula:

$$\sum_{k=0}^{\infty}\frac{f^{(k)}}{k!}(x-a)^k = f(a) + f'(a)(x-a) + \frac{f''(a)}{2!}(x-a)^2 + \ldots + \frac{f^{(n)}(a)}{n!}(x-a)^n + \ldots$$

First, let's find a few derivatives of ln $(1 + x)$ and evaluate them at $a = 0$.

$f(x) = \ln(1+x);$ $f(0) = \ln(1+0) = 0$

$f'(x) = \dfrac{1}{(1+x)};$ $f'(0) = \dfrac{1}{1+0} = 1$

$f''(x) = \dfrac{-1}{(1+x)^2};$ $f''(0) = \dfrac{-1}{(1+0)^2} = -1$

$$f'''(x) = \frac{2}{(1+x)^2}; \quad f'''(0) = \frac{2}{(1+0)^3} = 2$$

$$f^{(4)}(x) = \frac{-6}{(1+x)^3}; \quad f^{(4)}(0) = \frac{-6}{(1+0)^3} = -6$$

Let's see if we find a pattern. Plugging into the formula, we get

$$\ln(1+x) = 0 + (1)(x) + \left(\frac{-1}{2!}\right)(x^2) + \left(\frac{2}{3!}\right)(x^3) + \left(\frac{-6}{4!}\right)(x^4).$$ This simplifies to

$$\ln(1+x) = x - \frac{x^2}{2} + \frac{x^3}{3} - \frac{x^4}{4}.$$

We can see that the general formula is $\ln(1+x) = x - \dfrac{x^2}{2} + \dfrac{x^3}{3} - \dfrac{x^4}{4} + \ldots = \displaystyle\sum_{k=0}^{\infty} (-1)^k \frac{x^{k+1}}{k+1}$.

You should memorize the Taylor series that are highlighted on page 434. The AP Exam does not require you to derive them when you use them. You can simply state the formulas.

13. $$e^{-x} = 1 - x + \frac{x^2}{2!} - \frac{x^3}{3!} + \frac{x^4}{4!} + \ldots = \sum_{k=0}^{\infty} (-1)^k \frac{x^k}{k!}$$

We find a Taylor series by the formula:

$$\sum_{k=0}^{\infty} \frac{f^{(k)}}{k!}(x-a)^k = f(a) + f'(a)(x-a) + \frac{f''(a)}{2!}(x-a)^2 + \ldots + \frac{f^{(n)}(a)}{n!}(x-a)^n + \ldots$$

First, let's find a few derivatives of e^{-x} and evaluate them at $a = 0$.

$f(x) = e^{-x}$; $f(0) = e^0 = 1$

$f'(x) = -e^{-x}$; $f'(0) = -e^0 = -1$

$f''(x) = e^{-x}$; $f''(0) = e^0 = 1$

$f'''(x) = -e^{-x}$; $f'''(0) = -e^0 = -1$

$f^{(4)}(x) = e^{-x}$; $f^{(4)}(0) = e^0 = 1$

Let's see if we find a pattern. Plugging into the formula, we get

$$e^{-x} = 1 - x + \frac{x^2}{2!} - \frac{x^3}{3!} + \frac{x^4}{4!}$$

We can see that the general formula is $e^{-x} = 1 - x + \frac{x^2}{2!} - \frac{x^3}{3!} + \frac{x^4}{4!} + \ldots = \sum_{k=0}^{\infty} (-1)^k \frac{x^k}{k!}$.

You should memorize the Taylor series that are highlighted on page 434. The AP Exam does not require you to derive them when you use them. You can simply state the formulas.

14.　　$\dfrac{\sqrt{3}}{2} + \dfrac{1}{2}\left(x - \dfrac{\pi}{3}\right) - \dfrac{\sqrt{3}}{4}\left(x - \dfrac{\pi}{3}\right)^2$

We find a Taylor series by the formula:

$$\sum_{k=0}^{\infty} \frac{f^{(k)}}{k!}(x-a)^k = f(a) + f'(a)(x-a) + \frac{f''(a)}{2!}(x-a)^2 + \ldots + \frac{f^{(n)}(a)}{n!}(x-a)^n + \ldots$$

First, let's find a few derivatives of $\sin x$ and evaluate them at $a = \dfrac{\pi}{3}$.

$f(x) = \sin x;$ 　　　　　　$f\left(\dfrac{\pi}{3}\right) = \sin\dfrac{\pi}{3} = \dfrac{\sqrt{3}}{2}$

$f'(x) = \cos x;$ 　　　　　　$f'\left(\dfrac{\pi}{3}\right) = \cos\dfrac{\pi}{3} = \dfrac{1}{2}$

$f''(x) = -\sin x;$ 　　　　　　$f''\left(\dfrac{\pi}{3}\right) = -\sin\dfrac{\pi}{3} = -\dfrac{\sqrt{3}}{2}$

These should give us three non-zero terms. Let's plug into the formula. We get

$\sin x = \dfrac{\sqrt{3}}{2} + \dfrac{1}{2}\left(x - \dfrac{\pi}{3}\right) - \dfrac{\sqrt{3}}{2} \dfrac{\left(x - \dfrac{\pi}{3}\right)^2}{2!}$, which simplifies to $\dfrac{\sqrt{3}}{2} + \dfrac{1}{2}\left(x - \dfrac{\pi}{3}\right) - \dfrac{\sqrt{3}}{4}\left(x - \dfrac{\pi}{3}\right)^2$.

You should memorize the Taylor series that are highlighted on page 434. The AP Exam does not require you to derive them when you use them. You can simply state the formulas.

15.　　The radius of convergence is $\dfrac{1}{3}$; the interval of convergence is $\left(-\dfrac{1}{3}, \dfrac{1}{3}\right)$.

The interval of convergence of a series refers to those values of x where the series converges. A

geometric series converges when the common ratio of its terms is $-1 < r < 1$. Here we have the series $\sum_{k=0}^{\infty} 3^n x^n = \sum_{k=0}^{\infty} (3x)^n$, so the common ratio is $3x$. To find the interval of convergence, all

we have to do is set $-1 < 3x < 1$ and solve for x. We get $-\dfrac{1}{3} < x < \dfrac{1}{3}$. Thus, the interval of con-

vergence is $\left(-\dfrac{1}{3}, \dfrac{1}{3}\right)$. The radius of convergence is simply the distance from the center of the

interval to either endpoint, or half the width of the interval. In this case, the radius of convergence

is $\dfrac{1}{3}$.

16.   The radius of convergence is ∞; the interval of convergence is $(-\infty, \infty)$.

The interval of convergence of a series refers to those values of x where the

series converges. Here we will apply the Ratio Test for absolute convergence:

$$ p = \lim_{n \to \infty} \left| \frac{(-1)^{n+1}\left(\dfrac{x^{2n+2}}{(2n+2)!}\right)}{(-1)^n\left(\dfrac{x^{2n}}{(2n)!}\right)} \right| = \lim_{n \to \infty} \left| \left(\frac{x^{2n+2}}{(2n+2)!}\right)\left(\frac{(2n)!}{x^{2n}}\right) \right| = \lim_{n \to \infty} \frac{x^2}{(2n+2)(2n+1)} = 0. $$

This will converge if $p < 1$, which is true here for all values of x. Therefore, the interval of conver-

gence is $(-\infty, \infty)$. The radius of convergence is simply the distance from the center of the interval

to either endpoint, or half the width of the interval. In this case, the radius of convergence is ∞.

17. 0.980067; $8.\overline{8} \times 10^{-8}$

We find a Taylor series by the formula:

$$ \sum_{k=0}^{\infty} \frac{f^{(k)}}{k!}(x-a)^k = f(a) + f'(a)(x-a) + \frac{f''(a)}{2!}(x-a)^2 + \ldots + \frac{f^{(n)}(a)}{n!}(x-a)^n + \ldots $$

First, let's find a few derivatives of $\cos x$ and evaluate them at $a = 0$.

$f(x) = \cos x;$ $\qquad$ $f(0) = \cos 0 = 1$

$f'(x) = -\sin x;$ $\qquad$ $f'(0) = -\sin 0 = 0$

$f''(x) = -\cos x;$ $\qquad$ $f''(0) = -\cos 0 = -1$

$f'''(x) = \sin x;$ $\qquad$ $f'''(0) = -\sin 0 = 0$

$f^{(4)}(x) = \cos x;$ $\qquad$ $f^{(4)}(0) = \cos 0 = 1$

Plugging into the formula, we get $\cos x \approx 1 + 0 + \dfrac{(-1)}{2!}x^2 + 0 + \dfrac{1}{4!}x^4$. This is the fourth degree Taylor polynomial. You should memorize the Taylor series that are highlighted on page 434. The AP Exam does not require you to derive them when you use them. You can simply state the formulas. Now we simply plug $x = 0.2$ into the polynomial. We get $\cos x \approx 1 + 0 + \dfrac{(-1)}{2!}(0.2)^2 + 0 + \dfrac{1}{4!}(0.2)^4 \approx 0.980067$. To find the error bound, we can simply use the next term of the polynomial. Here, the next term is $\dfrac{x^6}{6!}$. If we plug in $x = 0.2$, we get $8.\overline{8} \times 10^{-8}$.

18. $\quad$ 0.264; 0.002025

We find a Taylor series by the formula:

$$\sum_{k=0}^{\infty} \frac{f^{(k)}}{k!}(x-a)^k = f(a) + f'(a)(x-a) + \frac{f''(a)}{2!}(x-a)^2 + \ldots + \frac{f^{(n)}(a)}{n!}(x-a)^n + \ldots$$

First, let's find a few derivatives of $\ln(1+x)$ and evaluate them at $a = 0$.

$f(x) = \ln(1+x);$ $\quad$ $f(0) = \ln(1+0) = 0$

$f'(x) = \dfrac{1}{(1+x)};$ $\qquad$ $f'(0) = \dfrac{1}{1+0} = 1$

$$f''(x) = \frac{-1}{(1+x)^2}; \quad f''(0) = \frac{-1}{(1+0)^2} = -1$$

$$f'''(x) = \frac{2}{(1+x)^3}; \quad f'''(0) = \frac{2}{(1+0)^3} = 2$$

Plugging into the formula, we get

$$\ln(1+x) \approx 0 + (1)(x) + \left(\frac{-1}{2!}\right)(x^2) + \left(\frac{2}{3!}\right)(x^3).$$ This simplifies to $\ln(1+x) \approx x - \frac{x^2}{2} + \frac{x^3}{3}$.

This is the third degree Taylor polynomial.

You should memorize the Taylor series that are highlighted on page 434. The AP Exam does not require you to derive them when you use them. You can simply state the formulas. Now we simply plug $x = 0.3$ into the polynomial. We get $\ln(1+0.3) \approx (0.3) - \frac{(0.3)^2}{2} + \frac{(0.3)^3}{3} \approx 0.264$. To find the error bound, we can simply use the next term of the polynomial. Here, the next term is $\frac{x^4}{4}$. If we plug in $x = 0.3$, we get 0.002025.

19.   Diverges by the Integral Test

There are a variety of tests for convergence of a series. If a series is not alternating and is not geometric, we usually look to see if we can use a Comparison Test, a Ratio Test, or an Integral Test. We can easily integrate this integral using u-substitution, so we will use the Integral Test. Here, the series is $\sum_{n=2}^{\infty} \frac{1}{n \ln n}$, so we need to see if $\int_{2}^{\infty}\left(\frac{1}{x \ln x}\right)dx$ converges. First, we replace the upper limit with a and take the limit as a goes to infinity. We get $\int_{2}^{\infty}\left(\frac{1}{x \ln x}\right)dx = \lim_{a \to \infty} \int_{2}^{a}\left(\frac{1}{x \ln x}\right)dx$. Next, if we let $u = \ln x$, then $du = \frac{1}{x}dx$. Substituting into the integrand, we get $\int\left(\frac{1}{x \ln x}\right)dx = \int \frac{du}{u} = \ln|u|$.

Substituting back, we get $\left(\ln|\ln x|\right)\Big|_{2}^{a} = \ln(\ln a) - \ln(\ln 2)$. Finally, we take the limit:

$\lim\limits_{a\to\infty}\left[\ln(\ln a) - \ln(\ln 2)\right] = D.N.E.$ Therefore, the series diverges by the Integral Test.

20.   Converges by the Integral Test

There are a variety of tests for convergence of a series. If a series is not alternating and is not geo-

metric, we usually look to see if we can use a Comparison Test, a Ratio Test, or an Integral Test.

We can easily integrate this integral, so we will use the Integral Test. Here, the series is $\sum\limits_{n=1}^{\infty}\dfrac{1}{n\sqrt[4]{n}}$,

so we need to see if $\int_{1}^{\infty}\left(\dfrac{1}{x\sqrt[4]{x}}\right)dx = \int_{1}^{\infty}x^{-\frac{5}{4}}\,dx$ converges. First, we replace the upper limit with

a and take the limit as a goes to infinity. We get $\int_{1}^{\infty}x^{-\frac{5}{4}}\,dx = \lim\limits_{a\to\infty}\int_{1}^{a}x^{-\frac{5}{4}}\,dx$. Next, we integrate:

$\int_{1}^{a}x^{-\frac{5}{4}}\,dx = \left(-4x^{-\frac{1}{4}}\right)\Big|_{1}^{a} = -4a^{-\frac{1}{4}} + 4$. Now, we take the limit: $\lim\limits_{a\to\infty}\left(-4a^{-\frac{1}{4}} + 4\right) = 4$. Therefore, the

series converges by the Integral Test.

21.   Diverges by the Comparison Test

There are a variety of tests for convergence of a series. If a series is not alternating and is not

geometric, we usually look to see if we can use a Comparison Test, a Ratio Test, or an Integral

Test. Here, the series is $\sum\limits_{n=2}^{\infty}\dfrac{\ln n}{n+1}$, which we can easily compare to the series $\sum\limits_{n=2}^{\infty}\dfrac{1}{n+1}$. The series

$\sum\limits_{n=2}^{\infty}\dfrac{1}{n+1}$ is essentially the harmonic series (it's just missing the first two terms), which diverges. The

series $\sum\limits_{n=2}^{\infty}\dfrac{\ln n}{n+1} > \sum\limits_{n=2}^{\infty}\dfrac{1}{n+1}$ for all terms. Therefore, the series diverges by the Comparison Test.

22.   Converges by the Comparison Test

There are a variety of tests for convergence of a series. If a series is not alternating and is not geo-

metric, we usually look to see if we can use a Comparison Test, a Ratio Test, or an Integral Test.

Here, the series is $\sum\limits_{n=4}^{\infty}\dfrac{1}{n!}$, which we can easily compare to the series $\sum\limits_{n=4}^{\infty}\dfrac{1}{n^2}$. The series $\sum\limits_{n=4}^{\infty}\dfrac{1}{n^2}$

converges and the series $\sum\limits_{n=4}^{\infty}\dfrac{1}{n!} < \sum\limits_{n=4}^{\infty}\dfrac{1}{n^2}$ for all terms. Therefore, the series converges by the

Comparison Test.

Chapter 24
Answers to End of
Chapter Drills

CHAPTER 3

1. **C**   Simply plug in $x = 2$: $\lim\limits_{x \to 2}\left(x^4 - 3x^2 + 5\right) = (2)^4 - 3(2)^2 + 5 = 16 - 12 + 5 = 9$.

   The answer is (C).

2. **A**   Plug in $x = 2$: $\lim\limits_{x \to 2}\dfrac{x^2 - x - 2}{x^2 - 8x + 15} = \dfrac{(2)^2 - (2) - 2}{(2)^2 - 8(2) + 15} = \dfrac{0}{3} = 0$.

   The answer is (A)

3. **B**   Plug in $x = 7$: $\lim\limits_{x \to 7}\dfrac{(7)^2 - 3(7) - 28}{(7)^2 - 5(7) - 14} = \dfrac{0}{0}$. When we get a limit of $\dfrac{0}{0}$, this does not necessarily

   mean that the limit does not exist. First, let's factor the numerator and denominator of the limit:

   $\lim\limits_{x \to 7}\dfrac{x^2 - 3x - 28}{x^2 - 5x - 14} = \lim\limits_{x \to 7}\dfrac{(x - 7)(x + 4)}{(x - 7)(x + 2)}$. Note that when we plug in $x = 7$, the factor $(x - 7)$ is what

   makes the function zero in the numerator and denominator. Cancel the factors to reduce the limit

   to $\lim\limits_{x \to 7}\dfrac{(x + 4)}{(x + 2)}$. Now plug in $x = 7$: $\lim\limits_{x \to 7}\dfrac{(x + 4)}{(x + 2)} = \dfrac{7 + 4}{7 + 2} = \dfrac{11}{9}$.

   The answer is (B).

4. **C**   Remember that when we are finding the limit at infinity of a rational function, we compare the

   degree of the highest term in the numerator to the degree of the highest term in the denominator.

   The degrees are both 3, so the limit will be the ratio of the coefficients of the terms, namely $\dfrac{4}{5}$.

   The answer is (C).

5. **A**   When we have a limit of the form $\lim\limits_{x \to 0}\dfrac{a\sin bx}{c\tan dx}$, the limit is $\dfrac{ab}{cd}$. Here we get

   $\lim\limits_{x \to 0}\dfrac{4\sin 3x}{2\tan 5x} = \dfrac{(4)(3)}{(2)(5)} = \dfrac{6}{5}$.

   The answer is (A).

6. **D**   Plug in $x = 5$: $\lim\limits_{x \to 5}\dfrac{11}{x^2 - 25} = \dfrac{11}{(5)^2 - 25} = \dfrac{11}{0}$. This is going to give us an infinite limit or the limit

   doesn't exist. We need to take the limit from both sides of 5. First, find $\lim\limits_{x \to 5^-}\dfrac{11}{x^2 - 25}$. The

   denominator will be just less than zero, so $\lim\limits_{x \to 5^-}\dfrac{11}{x^2 - 25} = -\infty$. Next, find $\lim\limits_{x \to 5^+}\dfrac{11}{x^2 - 25}$. Now the

   denominator will be just greater than zero, so $\lim\limits_{x \to 5^+}\dfrac{11}{x^2 - 25} = \infty$. The two limits do not match, so

   the limit does not exist.

   The answer is (D).

CHAPTER 4

1. **C** First, let's factor the numerator and denominator. We get $f(x)=\dfrac{x^2-8x+15}{x^2-9}=\dfrac{(x-3)(x-5)}{(x-3)(x+3)}$.

   At a value of x where the denominator is zero and the numerator is not also zero, we get a vertical asymptote, which is an essential discontinuity. At a value of x where the denominator is zero and the numerator is also zero, we get a hole, which is a removable discontinuity. Thus we have a removable discontinuity at $x=3$ and an essential discontinuity at $x=-3$.

   The answer is (C).

2. **D** The only value where we might have a discontinuity is at $x=4$. Plug $x=4$ into both pieces of the function: $f(x)=\begin{cases}5(4)-11=9\\2x+1=9\end{cases}$. The function is 9 for both pieces, so the function is continuous for all real numbers.

   The answer is (D).

3. **C** The only value where we might have a discontinuity is at $x=2$. Plug $x=2$ into both pieces of the function: $f(x)=\begin{cases}6(2)+1=13\\5(2)+2=12\end{cases}$. The function "jumps" from 12 to 13 at $x=2$, so it has a jump discontinuity at $x-2$.

   The answer is (C).

4. **A** We need to find a value of k where the value of f in the top piece and the bottom piece is the same at $x=1$. Plug $x=1$ into the top piece and we get $(1)^2+k(1)-4=k-3$. Now plug $x=1$ into the bottom piece and we get $6k(1)+7=6k+7$. Set the two pieces equal to each other and solve for k.

$$k-3=6k+7$$
$$5k=-10$$
$$k=-2$$

   The answer is (A).

5. **B** Here, in order for f to be continuous for all real numbers, the value of f in the top piece at $x=2$ has to equal the value of f in the middle piece at $x=2$, and the value of f in the middle piece at $x=-2$ has to equal the value of f in the bottom piece at $x=-2$. So, first let's plug $x=2$ into the top piece: $a(2)^2+2b(2)-3=4a+4b-3$. Next plug $x=2$ into the middle piece: $4b(2)^2-3a(2)+3=16b-6a+3$.

   Now set the two pieces equal to each other:

   $4a+4b-3=16b-6a+3$

   $10a-12b=6$, which simplifies to $5a-6b=3$.

Now, let's plug $x = -2$ into the middle piece: $4b(-2)^2 - 3a(-2) + 3 = 16b + 6a + 3$. Next plug $x = -2$ into the bottom piece: $4a(-2)^2 + 5b(-2) + 25 = 16a - 10b + 25$. Now set the two pieces equal to each other:

$16a - 10b + 25 = 16b + 6a + 3$

$10a - 26b = -22$

Finally, we solve the pair of equations for a and b.

$5a - 6b = 3$

$10a - 26b = -22$

Multiply the top equation by 2:

$10a - 12b = 6$

$10a - 26b = -22$

Subtract the bottom equation from the top equation: $14b = 28$, so $b = 2$. Plug in for b to get $5a - 6(2) = 3$, so $a = 3$.

The answer is (B).

CHAPTER 5

1. **C** We can find the derivative using the formula $f'(x) = \lim\limits_{h \to 0} \dfrac{f(x+h) - f(x)}{h}$. We get $f'(x) = \lim\limits_{h \to 0} \dfrac{8(x+h) - 11 - (8x - 11)}{h}$.

Simplify the numerator: $\lim\limits_{h \to 0} \dfrac{8(x+h) - 11 - (8x - 11)}{h} = \lim\limits_{h \to 0} \dfrac{8x + 8h - 11 - 8x + 11}{h} = \lim\limits_{h \to 0} \dfrac{8h}{h}$.

Cancel h out of the numerator and denominator: $\lim\limits_{h \to 0} \dfrac{8h}{h} = \lim\limits_{h \to 0} 8$.

And take the limit: $\lim\limits_{h \to 0} 8 = 8$.

Therefore, the derivative is 8.

The answer is (C).

2. **C** We can find the derivative using the formula $f'(x) = \lim_{h \to 0} \dfrac{f(x+h) - f(x)}{h}$. We get

$$f'(x) = \lim_{h \to 0} \frac{3(x+h)^2 + 7(x+h) - 2 - (3x^2 + 7x - 2)}{h}.$$

Simplify the numerator: $\lim_{h \to 0} \dfrac{3(x^2 + 2xh + h^2) + 7x + 7h - 2 - 3x^2 - 7x + 2}{h}$

$$\lim_{h \to 0} \frac{3x^2 + 6xh + 3h^2 + 7x + 7h - 2 - 3x^2 - 7x + 2}{h}$$

$$\lim_{h \to 0} \frac{6xh + 3h^2 + 7h}{h}$$

Factor h out of the numerator: $\lim_{h \to 0} \dfrac{h(6x + 3h + 7)}{h}$.

Cancel the h in the numerator and the denominator: $\lim_{h \to 0} 6x + 3h + 7$.

And take the limit: $\lim_{h \to 0} 6x + 3h + 7 = 6x + 7$.

Therefore, the derivative is $6x + 7$.

The answer is (C).

3. **D** We can find the derivative using the formula $f'(x) = \lim_{h \to 0} \dfrac{f(x+h) - f(x)}{h}$. We get

$$f'(x) = \lim_{h \to 0} \frac{(x+h)^3 - 4(x+h)^2 + 11 - (x^3 - 4x^2 + 11)}{h}.$$

Simplify the numerator: $\lim_{h \to 0} \dfrac{x^3 + 3x^2h + 3xh^2 + h^3 - 4(x^2 + 2xh + h^2) + 11 - x^3 + 4x^2 - 11}{h}$

$$\lim_{h \to 0} \frac{x^3 + 3x^2h + 3xh^2 + h^3 - 4x^2 - 8xh - 4h^2 + 11 - x^3 + 4x^2 - 11}{h}$$

$$\lim_{h \to 0} \frac{3x^2h + 3xh^2 + h^3 - 8xh - 4h^2}{h}$$

Factor h out of the numerator: $\lim_{h \to 0} \dfrac{h(3x^2 + 3xh + h^2 - 8x - 4h)}{h}$.

Cancel the h in the numerator and the denominator: $\lim_{h \to 0} 3x^2 + 3xh + h^2 - 8x - 4h$.

And take the limit: $\lim_{h \to 0} 3x^2 + 3xh + h^2 - 8x - 4h = 3x^2 - 8x$.

Therefore, the derivative is $3x^2 - 8x$.

The answer is (D).

4. **A** We can find the derivative using the formula $f'(x) = \lim_{h \to 0} \dfrac{f(x+h) - f(x)}{h}$, and we can plug in

$x = 25$ either after we find the derivative or before. Let's do it afterward.

We get $f'(x) = \lim_{h \to 0} \dfrac{6\sqrt{(x+h)} - 6\sqrt{x}}{h}$.

Multiply the numerator and the denominator by the conjugate of the numerator:

$$\lim_{h \to 0} \frac{\left(6\sqrt{(x+h)} - 6\sqrt{x}\right)}{h} \cdot \frac{\left(6\sqrt{(x+h)} + 6\sqrt{x}\right)}{\left(6\sqrt{(x+h)} + 6\sqrt{x}\right)}.$$

Multiply out the numerator: $\lim_{h \to 0} \dfrac{36(x+h) - 36x}{h\left(6\sqrt{(x+h)} + 6\sqrt{x}\right)}$.

Simplify the numerator and denominator: $\lim_{h \to 0} \dfrac{36x + 36h - 36x}{h\left(6\sqrt{(x+h)} + 6\sqrt{x}\right)} = \lim_{h \to 0} \dfrac{36h}{6h\left(\sqrt{(x+h)} + \sqrt{x}\right)}$.

Cancel the $6h$ in the numerator and the denominator: $\lim_{h \to 0} \dfrac{6}{\sqrt{(x+h)} + \sqrt{x}}$.

And take the limit: $\dfrac{6}{\sqrt{x+h} + \sqrt{x}} = \dfrac{6}{2\sqrt{x}} = \dfrac{3}{\sqrt{x}}$.

Finally, plug in $x = 25$: $\dfrac{3}{\sqrt{25}} = \dfrac{3}{5}$.

The answer is (A).

5. **A** We can find the derivative using the formula $f'(x) = \lim_{h \to 0} \dfrac{f(x+h) - f(x)}{h}$, and we can plug in

$x = 2$ either after we find the derivative or before. Let's do it at the beginning.

We get $f'(x) = \lim_{h \to 0} \dfrac{\dfrac{1}{4(x+h)^2} - \dfrac{1}{4x^2}}{h}$. Plug in $x = 2$: $f'(x) = \lim_{h \to 0} \dfrac{\dfrac{1}{4(2+h)^2} - \dfrac{1}{16}}{h}$.

Combine the two rational expressions in the numerator:

$$\lim_{h \to 0} \frac{\dfrac{1}{4(2+h)^2} - \dfrac{1}{16}}{h} = \lim_{h \to 0} \frac{\dfrac{16}{4(2+h)^2(16)} - \dfrac{4(2+h)^2}{4(2+h)^2 16}}{h} = \lim_{h \to 0} \frac{\dfrac{16 - 4(2+h)^2}{4(2+h)^2(16)}}{h}$$

Multiply out the numerator: $\lim_{h \to 0} \dfrac{\dfrac{16 - 4(4 + 4h + h^2)}{64(2+h)^2}}{h} = \lim_{h \to 0} \dfrac{\dfrac{16 - 16 - 16h - 4h^2}{64(2+h)^2}}{h}$.

Simplify the numerator and denominator: $\lim_{h \to 0} \dfrac{\dfrac{16 - 16 - 16h - 4h^2}{64(2+h)^2}}{h} = \lim_{h \to 0} \dfrac{-16h - 4h^2}{64h(2+h)^2}$.

Factor $-4h$ out of the numerator $\lim\limits_{h\to0}\dfrac{-4h(4+h)}{64h(2+h)^2}$.

Cancel the $4h$ in the numerator and the denominator: $\lim\limits_{h\to0}\dfrac{-(4+h)}{16(2+h)^2}$.

And take the limit: $\dfrac{-4}{64}=-\dfrac{1}{16}$.

The answer is (A).

CHAPTER 6

1. **D** Use the Power Rule:

$$y=5x^3-7x^2+11x+2$$

$$\frac{dy}{dx}=15x^2-14x+11$$

The answer is (D).

2. **B** Use the Chain Rule:

$$\frac{dy}{dx}=4\sec^2(4x)$$

The answer is (B).

3. **C** Use the Chain Rule:

$$\frac{dy}{dx}=2\cos(x^2)(-\sin(x^2))(2x)$$, which simplifies to $\dfrac{dy}{dx}=-4x\cos(x^2)\sin(x^2)$.

The answer is (C).

4. **A** It might help to rewrite the function as $y=4(x-1)^{-\frac{1}{2}}$. Now we can use the Chain Rule:

$$\frac{dy}{dx}=4\left(-\frac{1}{2}\right)(x-1)^{-\frac{3}{2}}(1)$$, which simplifies to $\dfrac{dy}{dx}=-\dfrac{2}{(x-1)^{\frac{3}{2}}}$.

The answer is (A).

5. **D** It might help to rewrite the function as $y=\left(\dfrac{x^2}{\sec x}\right)^{\frac{1}{2}}$. Now we can use the Chain Rule and the

Quotient Rule: $\dfrac{dy}{dx}=\dfrac{1}{2}\left(\dfrac{x^2}{\sec x}\right)^{-\frac{1}{2}}\left(\dfrac{\sec x(2x)-(x^2)\sec x\tan x}{\sec^2 x}\right)$.

The answer is (D).

CHAPTER 7

1. **B** Take the derivative of each term with respect to x. Remember that when you are taking the derivative of a function of y, we have to multiply that derivative by $\dfrac{dy}{dx}$. We get $2x + 6y^2\dfrac{dy}{dx} = 3x^2 - 8y\dfrac{dy}{dx}$.

   Now we have to isolate $\dfrac{dy}{dx}$. Group the terms that contain $\dfrac{dy}{dx}$ on one side of the equals sign and the terms that do not contain $\dfrac{dy}{dx}$ on the other side: $6y^2\dfrac{dy}{dx} + 8y\dfrac{dy}{dx} = 3x^2 - 2x$.

   Factor out $\dfrac{dy}{dx}$: $\left(6y^2 + 8y\right)\dfrac{dy}{dx} = 3x^2 - 2x$.

   And divide: $\dfrac{dy}{dx} = \dfrac{3x^2 - 2x}{6y^2 + 8y}$.

   The answer is (B).

2. **A** Take the derivative of each term with respect to x. Remember that when you are taking the derivative of a function of y, we have to multiply that derivative by $\dfrac{dy}{dx}$. We get $2x + 3x\left(2y\dfrac{dy}{dx}\right) + 3y^2 - 3y^2\dfrac{dy}{dx} = 0$.

   Now we have to isolate $\dfrac{dy}{dx}$. Group the terms that contain $\dfrac{dy}{dx}$ on one side of the equals sign and the terms that do not contain $\dfrac{dy}{dx}$ on the other side: $6xy\dfrac{dy}{dx} - 3y^2\dfrac{dy}{dx} = -2x - 3y^2$.

   Factor out $\dfrac{dy}{dx}$: $\left(6xy - 3y^2\right)\dfrac{dy}{dx} = -2x - 3y^2$.

   And divide: $\dfrac{dy}{dx} = \dfrac{-2x - 3y^2}{6xy - 3y^2}$.

   The answer is (A).

3. **C** Take the derivative of each term with respect to x. Remember that when you are taking the derivative of a function of y, we have to multiply that derivative by $\dfrac{dy}{dx}$. We get $4\cos\left(x^2\right)(2x) = 3\dfrac{dy}{dx} - 2y\dfrac{dy}{dx}$.

   Now, simply plug in $\left(\sqrt{\dfrac{\pi}{6}}, 2\right)$: $4\cos\left(\dfrac{\pi}{6}\right)\left(2\sqrt{\dfrac{\pi}{6}}\right) = 3\dfrac{dy}{dx} - 4\dfrac{dy}{dx}$.

   This simplifies to $4\dfrac{\sqrt{3}}{2}\left(2\sqrt{\dfrac{\pi}{6}}\right) = -\dfrac{dy}{dx}$

   $$4\sqrt{3}\left(\sqrt{\dfrac{\pi}{6}}\right) = -\dfrac{dy}{dx}$$

   $$-4\sqrt{\dfrac{\pi}{2}} = \dfrac{dy}{dx}$$

   The answer is (C).

4. **A** Take the derivative of each term with respect to x. Remember that when you are taking the derivative of a function of y, we have to multiply that derivative by $\dfrac{dy}{dx}$. We get $x\left(2y\dfrac{dy}{dx}\right)+y^2-4x^3\left(2y\dfrac{dy}{dx}\right)-12x^2y^2=0$. Now, simply plug in (1,1): $2\dfrac{dy}{dx}+1-8\dfrac{dy}{dx}-12=0$.

This simplifies to $-6\dfrac{dy}{dx}=11$

$$\dfrac{dy}{dx}=-\dfrac{11}{6}$$

The answer is (A).

5. **A** Take the derivative of each term with respect to x. Remember that when you are taking the derivative of a function of y, we have to multiply that derivative by $\dfrac{dy}{dx}$. We get $\dfrac{dy}{dx}=3y^2\dfrac{dy}{dx}-2x$.

Now we have to isolate $\dfrac{dy}{dx}$. Group the terms that contain $\dfrac{dy}{dx}$ on one side of the equals sign and the terms that do not contain $\dfrac{dy}{dx}$ on the other side: $2x=3y^2\dfrac{dy}{dx}-\dfrac{dy}{dx}$.

Factor out $\dfrac{dy}{dx}$: $2x=\left(3y^2-1\right)\dfrac{dy}{dx}$.

And divide: $\dfrac{dy}{dx}=\dfrac{2x}{3y^2-1}$.

Now we need to find the second derivative so do implicit differentiation again:

$$\dfrac{d^2y}{dx^2}=\dfrac{\left(3y^2-1\right)(2)-(2x)\left(6y\dfrac{dy}{dx}\right)}{\left(3y^2-1\right)^2}$$

Next, we substitute $\dfrac{dy}{dx}=\dfrac{2x}{3y^2-1}$: $\dfrac{d^2y}{dx^2}=\dfrac{\left(3y^2-1\right)(2)-(12xy)\left(\dfrac{2x}{3y^2-1}\right)}{\left(3y^2-1\right)^2}$. Which simplifies to

$\dfrac{d^2y}{dx^2}=\dfrac{\left(3y^2-1\right)^2(2)-\left(24x^2y\right)}{\left(3y^2-1\right)^3}$.

The answer is (A).

CHAPTER 8

1. **C** First, find the y-coordinate: $y(2) = 5(2)^3 - 20(2) + 10 = 10$.

    Second, take the derivative: $\dfrac{dy}{dx} = 15x^2 - 20$. Plug in $x = 2$ to get the slope of the tangent line:

    $\dfrac{dy}{dx} = 15(2)^2 - 20 = 40$.

    Now we have a slope and a point, so the equation is $y - 10 = 40(x - 2)$.

    The answer is (C).

2. **D** First, find the y-coordinate: $y\left(\dfrac{\pi}{8}\right) = 4\sec\left(\dfrac{\pi}{4}\right) = 4\sqrt{2}$.

    Second, take the derivative: $\dfrac{dy}{dx} = 8\sec(2x)\tan(2x)$. Plug in $x = \dfrac{\pi}{8}$ to get the slope of the tangent

    line: $\dfrac{dy}{dx} = 8\sec\left(\dfrac{\pi}{4}\right)\tan\left(\dfrac{\pi}{4}\right) = 8\sqrt{2}$. Thus, the slope of the normal line is the negative reciprocal of

    the slope of the tangent line: $-\dfrac{1}{8\sqrt{2}}$.

    Now we have a slope and a point, so the equation is $y - 4\sqrt{2} = -\dfrac{1}{8\sqrt{2}}\left(x - \dfrac{\pi}{8}\right)$.

    The answer is (D).

3. **B** First, find $f(3)$ and $f(8)$: $f(3) = 3^2 - 6(3) + 5 = -4$ and $f(8) = 8^2 - 6(8) + 5 = 21$. Then

    $\dfrac{21 - (-4)}{8 - 3} = 5 = f'(c)$. Next, find $f'(x)$: $f'(x) = 2x - 6$. Then, according to the MVTD:

    $2c - 6 = 5$, so $c = \dfrac{11}{2}$.

    The answer is (B).

4. **D** First, because the function is not continuous on the interval, there may not be a solution. Let's

    show that this is true. First, find $f(2)$ and $f(6)$: $f(2) = \dfrac{2}{2 - 4} = -1$ and $f(6) = \dfrac{2}{6 - 4} = 1$.

    Then $\dfrac{1 - (-1)}{6 - 2} = \dfrac{1}{2} = f'(c)$. Next, find $f'(x)$: $f'(x) = \dfrac{-2}{(x - 4)^2}$. Then, according to the MVTD:

    $\dfrac{-2}{(c - 4)^2} = \dfrac{1}{2}$. This gives us $(c - 4)^2 = -4$, which has no solution.

    The answer is (D).

5. **D**   First, find $f(-3)$ and $f(3)$: $f(-3) = (-3)^4 - 5(-3)^2 + 4 = 40$ and $f(3) = (3)^4 - 5(3)^2 + 4 = 40$.

Next, find $f'(x)$: $f'(x) = 4x^3 - 10x$. By setting $f'(c) = 4c^3 - 10c = 0$, we get

$c(4c^2 - 10) = 0$, and $c = 0$ or $c = \pm\sqrt{\dfrac{5}{2}}$, all of which are in the interval.

The answer is (D).

CHAPTER 9

1. **C**   First, let's draw a picture:

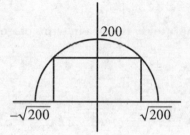

Note that the base of the rectangle is $2x$ and the height of the rectangle is y. Thus, the area of the rectangle is $A = 2xy$. Because $y = \sqrt{200 - x^2}$, we can substitute and the area is $A = 2x\sqrt{200 - x^2}$.

Take the derivative: $\dfrac{dA}{dx} = 2\sqrt{200 - x^2} + 2x\left(\dfrac{1}{2}\left(200 - x^2\right)^{-\frac{1}{2}}(-2x)\right) = 2\sqrt{200 - x^2} + \dfrac{-2x^2}{\sqrt{200 - x^2}}$.

Set the derivative equal to zero and solve for x: $2\sqrt{200 - x^2} + \dfrac{-2x^2}{\sqrt{200 - x^2}} = 0$

$$2\sqrt{200 - x^2} = \dfrac{2x^2}{\sqrt{200 - x^2}}$$

$$2(200 - x^2) = 2x^2$$

$$200 - x^2 = x^2$$

$$200 = 2x^2$$

$$x = \pm 10$$

We can throw out the negative answer so the maximum is at $x = 10$.

The answer is (C).

2.   **D**   First, let's draw a picture:

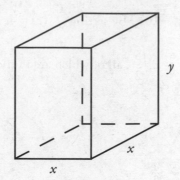

Note that the area of the base is x^2 and the height is y, so the volume of the box is $V = x^2 y = 108$

and the surface area is $A = x^2 + 4xy$. If we solve the volume for y, we get $y = \dfrac{108}{x^2}$. We can then

plug this into the area formula to get the area in terms of x only: $A = x^2 + 4x\left(\dfrac{108}{x^2}\right) = x^2 + \dfrac{432}{x}$.

Take the derivative: $\dfrac{dA}{dx} = 2x - \dfrac{432}{x^2}$.

Set it equal to zero and solve for x: $2x - \dfrac{432}{x^2} = 0$.

$$2x^3 = 432$$

$$x^3 = 216$$

$x = 6$ in. and thus $y = \dfrac{108}{6^2} = 3$ in.

The answer is (D).

3.   **B**   First, take the derivative: $\dfrac{dy}{dx} = 3x^2 - 6x - 24$. Set the derivative equal to zero and solve for x:

$$3x^2 - 6x - 24 = 0$$

$$x^2 - 2x - 8 = 0$$
$$(x - 4)(x + 2) = 0$$
$$x = 4, -2$$

Now that we have the critical values, we can find the y-coordinates by plugging the critical values back into the original equation. We get $y = 4^3 - 3(4)^2 - 24(4) + 6 = -74$.

$$y = (-2)^3 - 3(-2)^2 - 24(-2) + 6 = 34$$

Finally, we can determine which gives us the maximum and which gives us the minimum by using

the second derivative test. The second derivative is $\dfrac{d^2 y}{dx^2} = 6x - 6$.

At $x = 4$, the second derivative is $6(4) - 6 > 0$, so $(4, -74)$ is at a minimum.

At $x = -2$, the second derivative is $6(-2) - 6 < 0$, so $(-2, 34)$ is at a maximum.

The answer is (B).

4. **A** First, find the second derivative.

$$\frac{dy}{dx} = 3x^2 - 24x$$

$$\frac{d^2y}{dx^2} = 6x - 24$$

Now, set the second derivative equal to zero and solve for x: $6x - 24 = 0$, and $x = 4$.

Now that we have the x-coordinate of the point of inflection, we can find the y-coordinate by plugging x into the original equation. We get $y = (4)^3 - 12(4)^2 + 120 = -8$.

Therefore, the point of inflection is at $(4, -8)$.

The answer is (A).

5. **B** First, take the derivative: $\dfrac{dy}{dx} = \cos x - \sqrt{3} \sin x$. Next, set the derivative equal to zero and solve for x: $\cos x - \sqrt{3} \sin x = 0$.

$$\cos x = \sqrt{3} \sin x$$

$$\frac{1}{\sqrt{3}} = \frac{\sin x}{\cos x} = \tan x$$

$$x - \frac{\pi}{6}$$

Now we need to figure out the interval where the function is increasing or decreasing. Put the critical value on a number line:

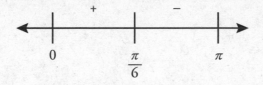

We can test values in each interval to find the sign of the derivative. Let's try a value between $x = 0$

and $x = \dfrac{\pi}{6}$: $\dfrac{dy}{dx} = \cos \dfrac{\pi}{12} - \sqrt{3} \sin \dfrac{\pi}{12} > 0$.

And a value between $x = \dfrac{\pi}{6}$ and $x = \pi$: $\dfrac{dy}{dx} = \cos \dfrac{\pi}{2} - \sqrt{3} \sin \dfrac{\pi}{2} < 0$.

Thus the function is increasing on the interval $\left(0, \dfrac{\pi}{6}\right)$ and decreasing on $\left(\dfrac{\pi}{6}, \pi\right)$.

The answer is (B).

CHAPTER 10

1. **C** First, let's draw a picture:

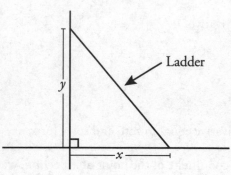

Let's call the distance from the top of the ladder to the ground y and the distance from the wall to the bottom of the ladder x. The relationship between x and y is $x^2 + y^2 = 25^2$ (Pythagorean Theorem). We are given $\dfrac{dy}{dt} = -6$ and we are looking for $\dfrac{dx}{dt}$ when $y = 20$.

Let's take the derivative with respect to t: $2x\dfrac{dx}{dt} + 2y\dfrac{dy}{dt} = 0$, which simplifies to $x\dfrac{dx}{dt} + y\dfrac{dy}{dt} = 0$.

We have $y = 20$ and $\dfrac{dy}{dt}$, so in order to find $\dfrac{dx}{dt}$, we will need to find x. Plugging into the original equation, we get $x^2 + (20)^2 = 25^2$, so $x = 15$. Now we can plug this into the derivative and solve for

$\dfrac{dx}{dt}$: $15\dfrac{dx}{dt} + (20)(-6) = 0$.

Therefore, $\dfrac{dx}{dt} = \dfrac{120}{15} = 8 \text{ ft/s}$.

The answer is (C).

2. **A**   First, let's draw a picture:

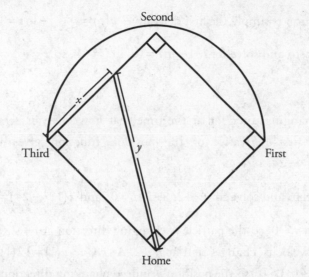

Call the distance from the runner to third base x and the distance from the runner to home plate y.

Then $\dfrac{dx}{dt} = -30$ (it's negative because the distance is shrinking), and we are looking for $\dfrac{dy}{dt}$ when

$x = 45$ (halfway from second base to third base). The relationship between x and y is $x^2 + 90^2 = y^2$

(Pythagorean Theorem). Take the derivative with respect to t: $2x\dfrac{dx}{dt} = 2y\dfrac{dy}{dt}$, which simplifies to

$x\dfrac{dx}{dt} = y\dfrac{dy}{dt}$. We will need to find y next. We have $45^2 + 90^2 = y^2$, so $y = 45\sqrt{5}$. Plugging every-

thing in, we get $(45)(-30) = \left(45\sqrt{5}\right)\dfrac{dy}{dt}$. Thus $\dfrac{dy}{dt} = -\dfrac{30}{\sqrt{5}}\,\text{ft/s}$.

The answer is (A).

3. **B**   We are given $\dfrac{dV}{dt} = -24\pi$ and we are looking for $\dfrac{dS}{dt}$ when $r = 6$. We know that the volume

of a sphere is $V = \dfrac{4}{3}\pi r^3$ and the surface area is $S = 4\pi r^2$. We could solve the volume for r

and plug it into the surface area equation to get S in terms of V, but this is messy. It's easier

to work with the equations separately. First, let's take the derivative of V with respect to t:

$\dfrac{dV}{dt} = \dfrac{4}{3}\pi\left(3r^2\dfrac{dr}{dt}\right) = 4\pi r^2\dfrac{dr}{dt}$. Next we plug in to solve for $\dfrac{dr}{dt}$:

$-24\pi = 4\pi(6)^2\dfrac{dr}{dt}$

$\dfrac{dr}{dt} = -\dfrac{1}{6}$

Next, let's take the derivative of S with respect to t: $\dfrac{dS}{dt} = 8\pi r\dfrac{dr}{dt}$. Now we can plug in $r = 6$ and

$\dfrac{dr}{dt} = -\dfrac{1}{6}$: $\dfrac{dS}{dt} = 8\pi(6)\dfrac{-1}{6} = -8\pi\,\text{in.}^2/\text{s}$.

The answer is (B).

4.   **C**   The velocity function is simply the first derivative: $v(t) = t\left(\dfrac{1}{t}\right) + \ln t = 1 + \ln t$.

Set this equal to zero and solve for t: $1 + \ln t = 0$. $\ln t = -1$, so $t = e^{-1} = \dfrac{1}{e}$.

The answer is (C).

5.   **D**   The particle is changing direction at the time that its velocity is zero and the velocity changes sign there. The first derivative of the position function gives us the velocity function: $x'(t) = v(t) = 12t^2 - 24t$.

Set this equal to zero and solve for t: $12t^2 - 24t = 0$, and $t(12t - 24) = 0$, so $t = 0$ or $t = 2$.

We can throw out $t = 0$, so the particle is changing direction at $t = 2$. We can verify this by making sure that the velocity changes sign at $t = 2$. At $t = 1$, $v(1) = 12(1)^2 - 24(1) < 0$ and at $t = 3$, $v(3) = 12(3)^2 - 24(3) > 0$, so the position is indeed changing direction at $t = 2$.

The answer is (D).

CHAPTER 11

1.   **C** $\dfrac{dy}{dx} = \dfrac{3x^2 - 7}{x^3 - 7x + 5}$

The answer is (C).

2.   **A**   First, use the log laws to rewrite the function:

$$y = \ln\left(\frac{\sqrt{x^2 - 11}}{x^2 + 9}\right) = \ln\sqrt{x^2 - 11} - \ln\left(x^2 + 9\right) = \frac{1}{2}\ln\left(x^2 - 11\right) - \ln\left(x^2 + 9\right)$$

Now take the derivative: $\dfrac{dy}{dx} = \dfrac{1}{2}\dfrac{2x}{x^2 - 11} - \dfrac{2x}{x^2 + 9} = \dfrac{x}{x^2 - 11} - \dfrac{2x}{x^2 + 9}$.

The answer is (A).

3.   **D** $\dfrac{dy}{dx} = 2\cos x e^{2\sin x}$

The answer is (D).

4.   **B** $\dfrac{dy}{dx} = \dfrac{4x^3 + \sec^2 x}{x^4 + \tan x}\dfrac{1}{\ln 8}$

The answer is (B).

5.   **C** $\dfrac{dy}{dx} = 7^{x^2 + 1}\left(2x\ln 7\right)$

The answer is (C).

CHAPTER 12

1. **C** First, by inspection we can see that when $y = -7$, $x = 0$. Now let's use the formula for the derivative

of the inverse. The derivative is $\dfrac{dy}{dx} = 15x^2 + 1$. Now we use the formula to find the derivative of

the inverse: $\dfrac{1}{\left.\dfrac{dy}{dx}\right|_{x=0}} = \dfrac{1}{15x^2 + 1\big|_{x=0}} = 1$.

The answer is (C).

2. **D** First, let's find the slope of the tangent line using $\dfrac{dy}{dx} = \dfrac{\dfrac{dy}{dt}}{\dfrac{dx}{dt}}$: $\dfrac{dy}{dt} = -2\sin t$ and $\dfrac{dx}{dt} = 8\sin t \cos t$.

Therefore, $\dfrac{dy}{dx} = \dfrac{-2\sin t}{8\sin t \cos t} = \dfrac{-1}{4\cos t}$. Next, plug in $t = \dfrac{\pi}{3}$ to get the slope of the tan-

gent line: $\dfrac{-1}{4\cos\left(\dfrac{\pi}{3}\right)} = -\dfrac{1}{2}$. We will also need to find a pair of coordinates. At $t = \dfrac{\pi}{3}$,

$x = 4\sin^2\dfrac{\pi}{3} = 4\left(\dfrac{\sqrt{3}}{2}\right)^2 = 4\left(\dfrac{3}{4}\right) = 3$ and $y = 2\cos\dfrac{\pi}{3} = 1$. Therefore, the equation of the tangent

line is $y - 1 = -\dfrac{1}{2}(x - 3)$.

The answer is (D).

3. **B** Because $\lim\limits_{x \to \infty} x^{-\frac{4}{3}} = 0$ and $\lim\limits_{x \to \infty} \sin\left(\dfrac{1}{x}\right) = 0$, the limit is an indeterminate form $\dfrac{0}{0}$. We can try to

find the limit using L'Hôpital's Rule. Take the derivative of the numerator and the denominator:

$\lim\limits_{x \to \infty} \dfrac{x^{-\frac{4}{3}}}{\sin\left(\dfrac{1}{x}\right)} = \lim\limits_{x \to \infty} \dfrac{-\dfrac{4}{3}x^{-\frac{7}{3}}}{\left(-\dfrac{1}{x^2}\right)\cos\left(\dfrac{1}{x}\right)}$.

This simplifies to $\lim\limits_{x\to\infty}\dfrac{-\dfrac{4}{3}x^{-\frac{7}{3}}}{\left(-\dfrac{1}{x^2}\right)\cos\left(\dfrac{1}{x}\right)}=\lim\limits_{x\to\infty}\dfrac{\dfrac{4}{3}x^{-\frac{1}{3}}}{\cos\left(\dfrac{1}{x}\right)}$. Now when we take the limit, we get

$\lim\limits_{x\to\infty}\dfrac{\dfrac{4}{3}x^{-\frac{1}{3}}}{\cos\left(\dfrac{1}{x}\right)}=\dfrac{0}{1}=0$.

The answer is (B).

4. **C** Let $x=2$, $\Delta x=+0.03$, and $f(x)=x^5$. Using differentials, we get $f(2+.03)\approx f(2)+f'(2)(0.03)$. Take the derivative: $f'(x)=5x^4$. At $x=2$, we get $f(2)=2^5=32$ and $f'(2)=5(2)^4=80$, so $(2.03)^5\approx32+(80)(0.03)\approx34.4$.

The answer is (C).

5. **A** Take the log of both sides: $\ln y=\ln\sqrt[4]{\dfrac{(8x+1)^3}{(2x-11)^5}}$. Simplify using the log rules:

$\ln y=\dfrac{1}{4}\ln\dfrac{(8x+1)^3}{(2x-11)^5}=\dfrac{1}{4}\ln(8x+1)^3-\dfrac{1}{4}\ln(2x-11)^5=\dfrac{3}{4}\ln(8x+1)-\dfrac{5}{4}\ln(2x-11)$.

Now take the derivative of both sides: $\dfrac{1}{y}\dfrac{dy}{dx}=\dfrac{3}{4}\dfrac{8}{8x+1}-\dfrac{5}{4}\dfrac{2}{2x-11}$.

Multiply by y: $\dfrac{dy}{dx}=y\left(\dfrac{3}{4}\dfrac{8}{8x+1}-\dfrac{5}{4}\dfrac{2}{2x-11}\right)$.

Note that we could now replace y with $\sqrt[4]{\dfrac{(8x+1)^3}{(2x-11)^5}}$, but shouldn't do so unless asked to.

The answer is (A).

CHAPTER 13

1. **C** Using the Power Rule, we get $\int\left(3x^4-6x^3+2x+7\right)dx=\dfrac{3x^5}{5}-\dfrac{6x^4}{4}+x^2+7x+C$.

The answer is (C).

2. **B** $\int 3\cos x-\dfrac{4}{\sqrt{x}}\,dx=3\sin x-8\sqrt{x}+C$

The answer is (B).

3. **A** Use u-substitution. Let $u=1-x^3$ and $du=-3x^2dx$, so $-3du=9x^2dx$.

Substitute: $\int\dfrac{9x^2}{\left(1-x^3\right)^2}\,dx=-3\int\dfrac{du}{u^2}=-3\int u^{-2}\,du=3u^{-1}+C=\dfrac{3}{1-x^3}+C$.

The answer is (A).

4. **D** Use u-substitution. Let $u = \dfrac{1}{x}$ and $du = -\dfrac{1}{x^2}dx$.

Substitute: $\displaystyle\int \dfrac{\cos\left(\dfrac{1}{x}\right)}{x^2}dx = -\int \cos u\, du = -\sin u = -\sin\left(\dfrac{1}{x}\right) + C$.

The answer is (D).

5. **B** Use u-substitution. Let $u = \cos(3x)$ and $du = -3\sin(3x)dx$, so $-\dfrac{1}{3}du = \sin(3x)dx$. Substitute:

$$\int \cos^3(3x)\sin(3x)\,dx = -\dfrac{1}{3}\int u^3\,du = -\dfrac{u^4}{12} = -\dfrac{\cos^4(3x)}{12} + C.$$

The answer is (B).

CHAPTER 14

1. **C** The width of each rectangle is $\dfrac{5-1}{4} = 1$. Using the formula, the area is approximately:

$$(1)\big[y(1) + y(2) + y(3) + y(4)\big] =$$

$$\left[\left(3+(1)^2\right) + \left(3+(2)^2\right) + \left(3+(3)^2\right) + \left(3+(4)^2\right)\right] =$$

$$[4 + 7 + 12 + 19] = 42$$

The answer is (C).

2. **B** The width of each trapezoid is $\dfrac{8-0}{4} = 2$. Using the formula, the area is approximately:

$$\dfrac{1}{2}(2)\big[y(0) + 2y(2) + 2y(4) + 2y(6) + y(8)\big] =$$

$$\left[\left(0^3+4\right) + 2\left(2^3+4\right) + 2\left(4\left(4^3+4\right)\right) + 2\left(6^3+4\right) + \left(8^3+4\right)\right] =$$

$$\big[4 + 2(12) + 2(68) + 2(220) + (516)\big] = 1{,}120$$

The answer is (B).

3. **D** The width of each rectangle is $\dfrac{16-0}{4} = 4$. Using the formula, the area is approximately:

$$(4)\big[y(2) + y(6) + y(10) + y(14)\big] =$$

$$(4)\left[\dfrac{3}{2} + \dfrac{3}{6} + \dfrac{3}{10} + \dfrac{3}{14}\right] = (4)\left(\dfrac{88}{35}\right) = \dfrac{352}{35}$$

The answer is (D).

4. **B** We need to evaluate $\dfrac{1}{4-2}\displaystyle\int_2^4 \dfrac{x^2}{\left(1-x^3\right)^2}\,dx = \dfrac{1}{2}\displaystyle\int_2^4 \dfrac{x^2}{\left(1-x^3\right)^2}\,dx$. We can evaluate $\displaystyle\int \dfrac{x^2}{\left(1-x^3\right)^2}\,dx$ using

u-substitution. Let $u = 1-x^3$ and $du = -3x^2\,dx$, so $-\dfrac{1}{3}\,du = x^2\,dx$.

Substitute: $\displaystyle\int \dfrac{x^2}{\left(1-x^3\right)^2}\,dx = -\dfrac{1}{3}\int \dfrac{du}{u^2} = -\dfrac{1}{3}\int u^{-2}\,du = \dfrac{1}{3}u^{-1} + C = \dfrac{1}{3\left(1-x^3\right)} + C$. This is a definite in-

tegral, so we evaluate: $\dfrac{1}{2}\left(\dfrac{1}{3\left(1-x^3\right)}\right)\Bigg|_2^4 = \dfrac{1}{2}\left(\dfrac{1}{3(-63)} - \dfrac{1}{3(-7)}\right) = \dfrac{4}{189}$.

The answer is (B).

5. **D** $\dfrac{d}{dx}\displaystyle\int_2^{x^2} \sin^3 t\,dt = \left(\sin^3 x\right)(2x) = 2x\sin^3 x$

The answer is (D).

CHAPTER 15

1. **A** Use u-substitution. Let $u = 4x^3 + 1$ and $du = 12x^2\,dx$, so $\dfrac{1}{12}\,du = x^2\,dx$.

Substitute: $\displaystyle\int \dfrac{x^2}{\left(4x^3 + 1\right)}\,dx = \dfrac{1}{12}\int \dfrac{du}{u}$. And integrate: $\dfrac{1}{12}\int \dfrac{du}{u} = \dfrac{1}{12}\ln|u| + C$. Finally, we substitute

back and get $\dfrac{1}{12}\ln\left|4x^3 + 1\right| + C$.

The answer is (A).

2. **C** Use u-substitution. Let $u = x^2$ and $du = 2x\,dx$, so $\dfrac{1}{2}\,du = x\,dx$.

Substitute: $\displaystyle\int x\tan x^2\,dx = \dfrac{1}{2}\int \tan u\,du$. And integrate: $\dfrac{1}{2}\int \tan u\,du = -\dfrac{1}{2}\ln|\cos u| + C$. Finally, we

substitute back and get $-\dfrac{1}{2}\ln\left|\cos x^2\right| + C$.

The answer is (C).

3. **D** Use u-substitution. Let $u = 1 + x^4$ and $du = 4x^3\,dx$, so $\dfrac{1}{4}\,du = x^3\,dx$.

Substitute: $\displaystyle\int x^3 e^{1+x^4}\,dx = \dfrac{1}{4}\int e^u\,du$. And integrate: $\dfrac{1}{4}\int e^u\,du = \dfrac{1}{4}e^u + C$. Finally, we substitute back

and get $\dfrac{1}{4}e^{1+x^4} + C$.

The answer is (D).

4. **A**   Use u-substitution. Let $u = 2 + x^3$ and $du = 3x^2 dx$, so $\dfrac{1}{3} du = x^2 dx$.

Substitute: $\displaystyle\int x^2 5^{2+x^3} dx = \dfrac{1}{3}\int 5^u \, du$. And integrate: $\dfrac{1}{3}\int 5^u \, du = \dfrac{1}{3\ln 5} 5^u + C$. Finally, we substitute

back and get $\dfrac{1}{3\ln 5} 5^{2+x^3} + C$.

The answer is (A).

5. **B**   Use u-substitution. Let $u = x^2$ and $du = 2x dx$, so $2 du = 4x dx$.

Substitute: $\displaystyle\int 4x \sec\left(x^2\right) dx = 2\int \sec(u)\, du$. And integrate: $2\int \sec(u)\, du = 2\ln\left|\sec(u) + \tan(u)\right| + C$.

Finally, we substitute back and get $2\ln\left|\sec\left(x^2\right) + \tan\left(x^2\right)\right| + C$.

The answer is (B).

CHAPTER 16

1. **D**   Use Integration by Parts. Let $u = \ln x$ and $dv = x^2 dx$. Then $du = \dfrac{1}{x} dx$ and $v = \dfrac{x^3}{3}$. We can

now rewrite the integral: $\displaystyle\int x^2 \ln x\, dx = \dfrac{x^3}{3}\ln x - \int \dfrac{1}{x}\left(\dfrac{x^3}{3}\right) dx = \dfrac{x^3}{2}\ln x - \dfrac{1}{3}\int x^2 dx$. Now evaluate

the right hand integral: $\displaystyle\int x^2 \ln x\, dx = \dfrac{x^3}{2}\ln x - \dfrac{1}{3}\dfrac{x^3}{3} + C = \dfrac{x^3}{2}\ln x - \dfrac{x^3}{9} + C$.

The answer is (D).

2. **B**   Use Integration by Parts. Let $u = x$ and $dv = \sec^2 x\, dx$. Then $du = dx$ and $v = \tan x$. We can

now rewrite the integral: $\displaystyle\int x \sec^2 x\, dx = x\tan x - \int \tan x\, dx = x\tan x - \ln|\cos x| + C$.

The answer is (B).

3. **A**   Use Integration by Parts. Let $u = x$ and $dv = e^{4x} dx$. Then $du = dx$ and $v = \dfrac{e^{4x}}{4}$. We can now

rewrite the integral: $\displaystyle\int xe^{4x}\, dx = \dfrac{xe^{4x}}{4} - \dfrac{1}{4}\int e^{4x} dx$. Integrate the right hand term and we get

$\dfrac{xe^{4x}}{4} - \dfrac{1}{4}\int e^{4x} dx = \dfrac{xe^{4x}}{4} - \dfrac{1}{4}\dfrac{e^{4x}}{4} + C = \dfrac{xe^{4x}}{4} - \dfrac{e^{4x}}{16} + C$.

The answer is (A).

4. **C** Use Integration by Parts (IBP). Let $u = x^2$ and $dv = \sin(3x)\,dx$. Then $du = 2x\,dx$ and $v = -\dfrac{\cos(3x)}{3}$. We can now rewrite the integral: $\displaystyle\int x^2 \sin(3x)\,dx = -\dfrac{x^2\cos(3x)}{3} + \dfrac{2}{3}\int x\cos(3x)\,dx$.

Use IBP again. Let $u = x$ and $dv = \cos(3x)\,dx$. Then $du = dx$ and $v = \dfrac{\sin(3x)}{3}$. We can now rewrite the integral: $\displaystyle\int x^2 \sin(3x)\,dx = -\dfrac{x^2\cos(3x)}{3} + \dfrac{2}{3}\left(\dfrac{x\sin(3x)}{3} - \int \dfrac{\sin(3x)}{3}\,dx\right)$, which simplifies to $-\dfrac{x^2\cos(3x)}{3} + \dfrac{2x\sin(3x)}{9} - \dfrac{2}{9}\int \sin(3x)\,dx$.

Now integrate the final term: $-\dfrac{x^2\cos(3x)}{3} + \dfrac{2x\sin(3x)}{9} + \dfrac{2\cos(3x)}{27} + C$.

The answer is (C).

5. **D** Use Integration by Parts. Let $u = e^{\frac{x}{2}}$ and $dv = \cos(2x)\,dx$. Then $du = \dfrac{1}{2}e^{\frac{x}{2}}\,dx$ and $v = \dfrac{\sin(2x)}{2}$.

We can now rewrite the integral: $\displaystyle\int e^{\frac{x}{2}}\cos(2x)\,dx = \dfrac{e^{\frac{x}{2}}\sin(2x)}{2} - \dfrac{1}{4}\int e^{\frac{x}{2}}\sin(2x)\,dx$.

Do IBP again. Let $u = e^{\frac{x}{2}}$ and $dv = \sin(2x)\,dx$. Then $du = \dfrac{1}{2}e^{\frac{x}{2}}\,dx$ and $v = -\dfrac{\cos(2x)}{2}$. We can now rewrite the integral again: $\displaystyle\int e^{\frac{x}{2}}\cos(2x)\,dx = \dfrac{e^{\frac{x}{2}}\sin(2x)}{2} - \dfrac{1}{4}\left[-\dfrac{e^{\frac{x}{2}}\cos(2x)}{2} - \dfrac{1}{4}\int e^{\frac{x}{2}}(-\cos(2x))\,dx\right]$,

which simplifies to $\displaystyle\int e^{\frac{x}{2}}\cos(2x)\,dx = \dfrac{e^{\frac{x}{2}}\sin(2x)}{2} + \dfrac{e^{\frac{x}{2}}\cos(2x)}{8} - \dfrac{1}{16}\int e^{\frac{x}{2}}\cos(2x)\,dx$. (Be careful

with your minus signs and constants!)

Now we can add $\dfrac{1}{16}\displaystyle\int e^{\frac{x}{2}}\cos(2x)\,dx$ to both sides to get

$\dfrac{17}{16}\displaystyle\int e^{\frac{x}{2}}\cos(2x)\,dx = \dfrac{e^{\frac{x}{2}}\sin(2x)}{2} + \dfrac{e^{\frac{x}{2}}\cos(2x)}{8}$.

Multiply through by $\dfrac{16}{17}$: $\displaystyle\int e^{\frac{x}{2}}\cos(2x)\,dx = \dfrac{16}{17}\left[\dfrac{e^{\frac{x}{2}}\sin(2x)}{2} + \dfrac{e^{\frac{x}{2}}\cos(2x)}{8}\right] + C$.

The answer is (D).

CHAPTER 17

1. **B** First, break up the integrand into $\int \sin^{19} x \cos^3 x \, dx = \int \sin^{19} x \cos^2 x \cos x \, dx$. Next, use the trig identity $\cos^2 x = 1 - \sin^2 x$.

   Rewrite the integrand as $\int \sin^{19} x \left(1 - \sin^2 x\right) \cos x \, dx$.

   Now we can use u-substitution. Let $u = \sin x$ and $du = \cos x \, dx$.

   Substitute: $\int \sin^{19} x \left(1 - \sin^2 x\right) \cos x \, dx = \int u^{19} \left(1 - u^2\right) du = \int u^{19} - u^{21} \, du$.

   Integrate: $\int u^{19} - u^{21} \, du = \dfrac{u^{20}}{20} - \dfrac{u^{22}}{22} = \dfrac{\sin^{20} x}{20} - \dfrac{\sin^{22} x}{22} + C$.

   The answer is (B).

2. **D** Use u-substitution. Let $u = \cos(3x)$ and $du = -3\sin(3x)\,dx$, so $-\dfrac{1}{3} du = \sin(3x)\,dx$. Substitute:

   $\int \cos^3(3x) \sin(3x)\, dx = -\dfrac{1}{3} \int u^3\, du = -\dfrac{u^4}{12} = -\dfrac{\cos^4(3x)}{12} + C$.

   The answer is (D).

3. **A** Use the substitutions $\sin^2 2x = \dfrac{1}{2}(1 - \cos 4x)$ and $\cos^2 2x = \dfrac{1}{2}(1 + \cos 4x)$. Substitute into the integrand: $\int 16 \sin^2(2x) \cos^2(2x)\, dx = 16 \int \dfrac{1}{2}(1 - \cos 4x) \dfrac{1}{2}(1 + \cos 4x)\, dx$, which simplifies to $4 \int (1 - \cos 4x)(1 + \cos 4x)\, dx = 4 \int \left(1 - \cos^2 4x\right) dx$.

   Break this into two integrals: $4 \int \left(1 - \cos^2(4x)\right) dx = 4 \int dx - 4 \int \cos^2 4x \, dx$.

   Now use the substitution $\cos^2(4x) = \dfrac{1}{2}(1 + \cos 8x)$:

   $4 \int dx - 4 \int \cos^2 4x = 4 \int dx - 4 \int \dfrac{1}{2}(1 + \cos(8x))\, dx$.

   This simplifies to $4 \int dx - 2 \int (1 + \cos(8x))\, dx = 4 \int dx - 2 \int dx - 2 \int \cos(8x)\, dx$.

   And integrate: $4x - 2x - 2\dfrac{\sin 8x}{8} = 2x - \dfrac{\sin 8x}{4} + C$.

   The answer is (A).

4. **C** Break up the integrand into $\int \sec^4 x \, dx = \int \sec^2 x \sec^2 x \, dx$. Use the trig identity $1 + \tan^2 x = \sec^2 x$ for one of the $\sec^2 x$ terms: $\int \left(1 + \tan^2 x\right) \sec^2 x \, dx$.

   Now use u-substitution. Let $u = \tan x$ and $du = \sec^2 x \, dx$.

   Substitute: $\int 1 + u^2 \, du = u + \dfrac{u^3}{3} = \tan x + \dfrac{\tan^3 x}{3} + C$.

   The answer is (C).

5. **D** Use Integration by Parts. Let $u = \arctan x$ and $dv = dx$. Then $du = \dfrac{dx}{1+x^2}$ and $v = x$. We can

now rewrite the integral: $\displaystyle\int \arctan x\, dx = x \arctan x - \int \dfrac{x}{1+x^2}\, dx$. Use u-substitution on the right

hand integral. Let $u = 1 + x^2$ and $du = 2x\, dx$, so $\dfrac{1}{2}\, du = x\, dx$.

Substitute: $\displaystyle\int \dfrac{x}{1+x^2}\, dx = \dfrac{1}{2}\int \dfrac{du}{u} = \dfrac{1}{2}\ln u = \dfrac{1}{2}\ln\left(1+x^2\right)$.

Therefore, $\displaystyle\int \arctan x\, dx = x \arctan x - \dfrac{1}{2}\ln\left(1+x^2\right) + C$. (BTW, we don't need the absolute value

bars for the logarithm because $1 + x^2$ is always positive.)

The answer is (D).

CHAPTER 18

1. **B** First, let's draw a picture.

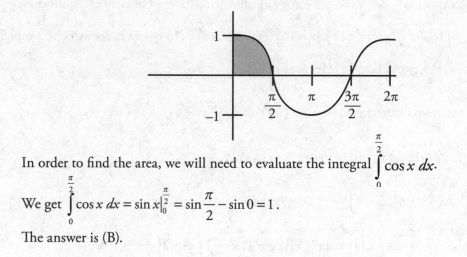

In order to find the area, we will need to evaluate the integral $\displaystyle\int_0^{\frac{\pi}{2}} \cos x\, dx$.

We get $\displaystyle\int_0^{\frac{\pi}{2}} \cos x\, dx = \sin x\Big|_0^{\frac{\pi}{2}} = \sin\dfrac{\pi}{2} - \sin 0 = 1$.

The answer is (B).

2. **B** First, let's draw a picture.

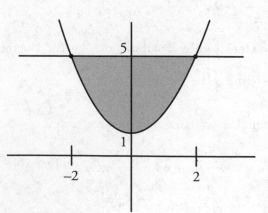

Note that the curve $y = 5$ is on top and the curve $y = x^2 + 1$ is on the bottom. Next, we need to find where the curves intersect. Set them equal to each other and solve for x:

$$x^2 + 1 = 5$$

$$x^2 = 4$$

$$x = \pm 2$$

Thus, in order to find the area, we will need to evaluate the integral $\int_{-2}^{2} 5 - \left(x^2 + 1\right) dx = \int_{-2}^{2} 4 - x^2 \, dx$.

We get $\int_{-2}^{2} 4 - x^2 \, dx = \left(4x - \dfrac{x^3}{3}\right)\Big|_{-2}^{2} = \left(8 - \dfrac{8}{3}\right) - \left(-8 + \dfrac{8}{3}\right) = \dfrac{32}{3}$.

The answer is (B).

3. **B**   First, let's draw a picture.

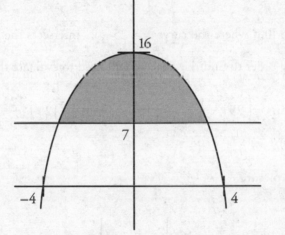

Note that the curve $y = 16 - x^2$ is on top and the curve $y = 7$ is on the bottom. Next, we need to find where the curves intersect. Set them equal to each other and solve for x:

$$16 - x^2 = 7$$
$$x^2 = 9$$
$$x = \pm 3$$

Thus, in order to find the area, we will need to evaluate the integral $\int_{-3}^{3} \left(16 - x^2\right) - 7 \, dx = \int_{-3}^{3} 9 - x^2 \, dx$.

We get $\int_{-3}^{3} 9 - x^2 \, dx = \left(9x - \dfrac{x^3}{3}\right)\Big|_{-3}^{3} = (27 - 9) - (-27 + 9) = 36$.

The answer is (B).

4.   **D**   First, let's draw a picture.

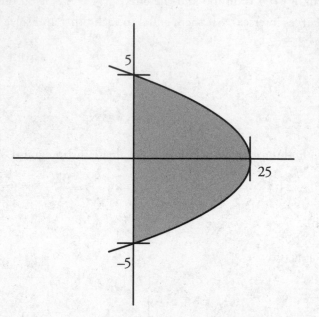

Next, we need to find where the curve $x = 25 - y^2$ intersects the y-axis. Set $25 - y^2 = 0$. We get

$y = \pm 5$. Thus, in order to find the area, we will need to evaluate the integral $\int\limits_{-5}^{5} 25 - y^2 \, dy$.

We get $\int\limits_{-5}^{5} 25 - y^2 \, dy = \left(25y - \dfrac{y^3}{3} \right)\Bigg|_{-5}^{5} = \left(125 - \dfrac{125}{3} \right) - \left(-125 + \dfrac{125}{3} \right) = \dfrac{500}{3}$.

The answer is (D).

5.   **C**   First, let's draw a picture.

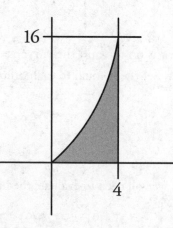

Note that the curve $x = 4$ is farther away from the y-axis than the curve $x = \sqrt{y}$. Next, we need

to find where the two curves intersect. Set them equal to each other and solve for y: $\sqrt{y} = 4$, so

$y = 16$. Thus, in order to find the area, we will need to evaluate the integral $\int\limits_0^{16} 4 - \sqrt{y} \, dy$. We get

$$\int\limits_0^{16} 4 - \sqrt{y} \, dy = \left(4y - \frac{2}{3} y^{\frac{3}{2}} \right)\Bigg|_0^{16} = \left(4y - \frac{2}{3} \sqrt{y^3} \right)\Bigg|_0^{16} = \left(64 - \frac{128}{3} \right) - 0 = \frac{64}{3}.$$

The answer is (C).

CHAPTER 19

1. **B**   First, let's draw a picture.

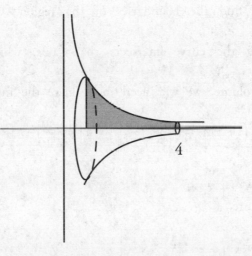

Next, we need to find the boundaries of the region. We are given that they are $x = 1$ and $x = 4$.

Thus, in order to find the volume, we will need to evaluate the integral $\pi \int\limits_1^4 \left(\frac{1}{x} \right)^2 dx$. We get

$$\pi \int\limits_1^4 \left(\frac{1}{x} \right)^2 dx = \pi \int\limits_1^4 x^{-2} \, dx = \pi \left(\frac{x^{-1}}{-1} \right)\Bigg|_1^4 = \pi \left(-\frac{1}{x} \right)\Bigg|_1^4 = \frac{3\pi}{4}.$$

The answer is (B).

2.   **C**   First, let's draw a picture.

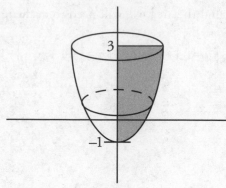

Next, we need to find the boundaries of the region. One of them is $y = 3$ and the other is where the curve intersects the y-axis, which is at $y = -1$. Thus, in order to find the volume, we will need to evaluate the integral $\pi \int_{-1}^{3} \left(\sqrt{1+y} \right)^2 dy$. We get

$$\pi \int_{-1}^{3} \left(\sqrt{1+y} \right)^2 dy = \pi \int_{-1}^{3} \left(1+y \right) dy = \pi \left(y + \frac{y^2}{2} \right) \Bigg|_{-1}^{3} = \pi \left(3 + \frac{9}{2} \right) - \pi \left(-1 + \frac{1}{2} \right) = 8\pi \ .$$

The answer is (C).

3.   **D**   First, let's draw a picture.

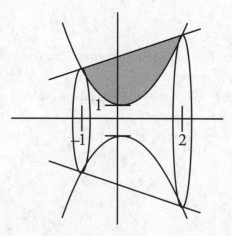

Next, we need to find the boundaries of the region. Set the curves equal to each other and solve for x: $x^2 + 1 = x + 3$

$$x^2 - x - 2 = 0$$

$$(x - 2)(x + 1) = 0$$

$$x = -1, 2$$

Note that the curve $y = x + 3$ is always on top and the curve $y = x^2 + 1$ is always on the bottom.

Thus, in order to find the volume, we will need to evaluate the integral $\pi \int_{-1}^{2} (x+3)^2 - (x^2+1)^2 \, dx$.

We get $\pi \int_{-1}^{2} (x+3)^2 - (x^2+1)^2 \, dx = \pi \int_{-1}^{2} -x^4 - x^2 + 6x + 8 \, dx = \pi \left(-\dfrac{x^5}{5} - \dfrac{x^3}{3} + 3x^2 + 8x \right) \Bigg|_{-1}^{2} = \dfrac{117\pi}{5}$.

The answer is (D).

4. **C** First, let's draw a picture.

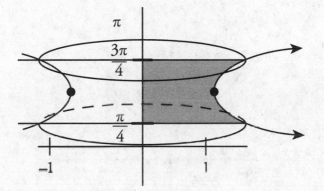

Next, we need to find the boundaries of the region. We are given that they are $y = \dfrac{\pi}{4}$ and $y = \dfrac{3\pi}{4}$.

Thus, in order to find the volume, we will need to evaluate the integral $\pi \int_{\frac{\pi}{4}}^{\frac{3\pi}{4}} (\csc y)^2 \, dy$. We get

$$\pi \int_{\frac{\pi}{4}}^{\frac{3\pi}{4}} (\csc y)^2 \, dy = -\pi \cot y \Big|_{\frac{\pi}{4}}^{\frac{3\pi}{4}} = 2\pi.$$

The answer is (C).

5. **A** First, let's draw a picture.

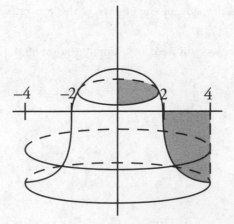

We need to rewrite the curve in terms of x. We get $y = \left(4 - x^2\right)^{\frac{1}{3}}$ Next, we need to find the boundaries of the region. We are given that they are $x = 0$ and $x = 4$, but note that from $x = 0$ to $x = 2$ the curve is above the x-axis and that from $x = 2$ to $x = 4$ the curve is below the x-axis. Thus, in order to find the volume, we will need to evaluate the integrals $2\pi \int_0^2 x\left(4 - x^2\right)^{\frac{1}{3}} dx - 2\pi \int_2^4 x\left(4 - x^2\right)^{\frac{1}{3}} dx$.

The answer is (A).

6. **C** First, let's draw a picture.

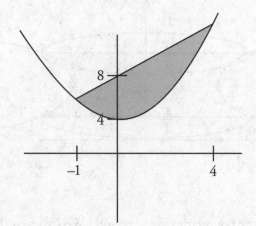

Note that the curve $y = 3x + 8$ is always on top and $y = x^2 + 4$ is always on the bottom of the region. Next, we need to find the boundaries of the region. Set the curves equal to each other and solve for x:

$x^2 + 4 = 3x + 8$

$x^2 - 3x - 4 = 0$

$(x - 4)(x + 1) = 0$

$x = -1, 4$

The "slice" that will be the side of the square is the top curve minus the bottom curve, so in order to find the volume, we will need to evaluate the integral $\int_{-1}^4 \left[(3x + 8) - \left(x^2 + 4\right)\right]^2 dx$. We get

$\int_{-1}^4 \left[(3x + 8) - \left(x^2 + 4\right)\right]^2 dx = \int_{-1}^4 \left[-x^2 + 3x + 4\right]^2 dx = \int_{-1}^4 x^4 - 6x^3 + x^2 + 24x + 16 \, dx = \dfrac{625}{6}$.

The answer is (C).

CHAPTER 20

1. **D** $\dfrac{A}{x-2}+\dfrac{B}{x+1}=\dfrac{8x-7}{(x-2)(x+1)}$

   $A(x+1)+B(x-2)=8x-7$

   $Ax+A+Bx-2B=8x-7$

   $(A+B)x+(A-2B)=8x-7$

   So $\begin{matrix}A+B=8\\A-2B=-7\end{matrix}$ and $A=3$, $B=5$

   Thus, we can rewrite $\displaystyle\int\dfrac{8x-7}{x^2-x-2}\,dx=\int\dfrac{3}{x-2}\,dx+\int\dfrac{5}{x+1}\,dx=3\ln|x-2|+5\ln|x+1|+C$

   The answer is (D).

2. **B** First, rewrite the integral as a limit: $\displaystyle\int_{10}^{\infty}\dfrac{e^{-x}\,dx}{1-e^{-x}}=\lim_{c\to\infty}\int_{10}^{c}\dfrac{e^{-x}\,dx}{1-e^{-x}}$.

   Let's evaluate the indefinite integral. Use u-substitution: let $u=1-e^{-x}$ and $du=e^{-x}dx$.

   And integrate: $\displaystyle\int\dfrac{e^{-x}\,dx}{1-e^{-x}}=\int\dfrac{du}{u}=\ln|u|=\ln\left|1-e^{-x}\right|$. Now evaluate the integral at the limits:

   $\displaystyle\lim_{c\to\infty}\ln\left|1-e^{-x}\right|\Big|_{10}^{c}=\lim_{c\to\infty}\ln\left|1-e^{-c}\right|-\lim_{c\to\infty}\ln\left|1-e^{-10}\right|$.

   And take the limit: $=\ln 1-\ln\left|1-e^{-10}\right|=-\ln\left|1-e^{-10}\right|$.

   The answer is (B).

3. **D** First, rewrite the integral as a limit: $\displaystyle\int_{0}^{1}\dfrac{dx}{2x-1}=\lim_{d\to\frac{1}{2}^{-}}\int_{0}^{d}\dfrac{dx}{2x-1}+\lim_{d\to\frac{1}{2}^{+}}\int_{d}^{1}\dfrac{dx}{2x-1}$.

   Integrate: $\displaystyle\lim_{d\to\frac{1}{2}^{-}}\dfrac{\ln|2x-1|}{2}\Big|_{0}^{d}+\lim_{d\to\frac{1}{2}^{+}}\dfrac{\ln|2x-1|}{2}\Big|_{d}^{1}=\lim_{d\to\frac{1}{2}^{-}}\dfrac{\ln|2d-1|}{2}-\dfrac{\ln 1}{2}+\lim_{d\to\frac{1}{2}^{+}}\dfrac{\ln 1}{2}-\dfrac{\ln|2d-1|}{2}$.

   The limit of the logarithm is infinite, so the integral diverges.

   The answer is (D).

4. **A** First, let's find $\dfrac{dr}{d\theta}$: $\dfrac{dr}{d\theta}=-3\sin(3\theta)$. At $\theta=\dfrac{3\pi}{4}$, $\dfrac{dr}{d\theta}=-3\sin\left(\dfrac{9\pi}{4}\right)=-\dfrac{3\sqrt{2}}{2}$. Next, at $\theta=\dfrac{3\pi}{4}$,

   $\tan\left(\dfrac{3\pi}{4}\right)=-1$. Thus, the slope of the tangent line is $m=\dfrac{\dfrac{\sqrt{2}}{2}+(-1)\left(\dfrac{-3\sqrt{2}}{2}\right)}{-\left(\dfrac{\sqrt{2}}{2}\right)(-1)+\left(\dfrac{-3\sqrt{2}}{2}\right)}=-2$.

   The answer is (A).

5. **C** First, let's find the intersection. Set the equations equal to each other and solve for θ:

$$4 + 4\cos\theta = 6$$

$$4\cos\theta = 2$$

$$\cos\theta = \frac{1}{2}, \text{ so } \theta = \frac{\pi}{3} \text{ and } \theta = -\frac{\pi}{3}$$

Thus, in order to find the area, we need to evaluate:

$$\frac{1}{2}\int_{-\frac{\pi}{3}}^{\frac{\pi}{3}}(4+4\cos\theta)^2\,d\theta - \frac{1}{2}\int_{-\frac{\pi}{3}}^{\frac{\pi}{3}}(6)^2\,d\theta$$

$$\frac{1}{2}\int_{-\frac{\pi}{3}}^{\frac{\pi}{3}}16\cos\theta + 8\cos^2\theta - 10\theta\,d\theta =$$

$$\frac{1}{2}\int_{-\frac{\pi}{3}}^{\frac{\pi}{3}}16\cos\theta + 4(1+\cos2\theta) - 10\theta\,d\theta =$$

$$16\cos\theta + (4\theta + 2\sin2\theta) - 10\theta\Big|_{-\frac{\pi}{3}}^{\frac{\pi}{3}} = 18\sqrt{3} - 4\pi$$

The answer is (C).

CHAPTER 21

1. **C** First, separate the variables: $\frac{dy}{y^2} = -4x\,dx$. Integrate both sides: $\int\frac{dy}{y^2} = \int -4x\,dx$, which can be rewritten as $\int y^{-2}\,dy = -4\int x\,dx$.

$\frac{y^{-1}}{-1} = -4\frac{x^2}{2} + C$, which can be rewritten as $-\frac{1}{y} = -2x^2 + C$.

Isolate y: $\frac{1}{2x^2 + C} = y$. Substitute using the initial condition: $\frac{1}{2(0)^2 + C} = 1$, so $C = 1$. Therefore, the equation is $y = \frac{1}{2x^2 + 1}$.

The answer is (C).

2. **D** First, separate the variables: $\frac{1}{\tan y}\,dy = \frac{4}{\sec x}\,dx$. This can be rewritten as $\cot y\,dy = 4\cos x\,dx$. Integrate both sides: $\int\cot y\,dy = 4\int\cos x\,dx$.

$\ln|\sin y| = 4\sin x + C$. Exponentiate both sides: $\sin y = e^{4\sin x + C} = Ae^{4\sin x}$.

Substitute using the initial condition: $\sin\left(\dfrac{\pi}{6}\right) = Ae^0 = A$, so $A = \dfrac{1}{2}$. Therefore, the equation is $\sin y = \dfrac{1}{2}e^{4\sin x}$.

The answer is (D).

3. **B** The exponential decay means that $\dfrac{dy}{dt} = -ky$, where k is a constant. Separate the variables: $\dfrac{dy}{y} = -kdt$. Integrate both sides: $\displaystyle\int \dfrac{dy}{y} = -k\int dt$.

$\ln|y| = -kt + C$. Exponentiate both sides: $y = e^{-kt+C} = Ae^{-kt}$. Next, we are given that the half-life is 140 days, so $50 = 100e^{-k(140)}$. Now we can solve for k: $50 = 100e^{-k(140)}$.

$\dfrac{1}{2} = e^{-k(140)}$, and $\ln\dfrac{1}{2} = -140k$, so $k = -\dfrac{1}{140}\ln\dfrac{1}{2}$. If we substitute this for k into the equation, we get $y = 100e^{\left(\frac{1}{140}\ln\frac{1}{2}\right)t}$. Use the rules of logs to rewrite the exponent: $y = 100e^{\left(\ln\frac{1}{2}^{\frac{1}{140}}\right)t}$, and

$y = 100\dfrac{1}{2}^{\frac{t}{140}}$.

Finally, plug in $t = 70$ (10 weeks): $y = 100\dfrac{1}{2}^{\frac{1}{2}} \approx 71\,\text{g}$.

The answer is (B).

4. **C** We start with $x_0 = 0$ and $y_0 = 3$. The slope is $\dfrac{dy}{dx} = 0 + 3 = 3$, $y_0' = 3$.

Increase x_0 by $h = 0.25$: $x_1 = 0.25$.

Multiply $h = 0.25$ by $y_0' = 3$ and add to y_0: $y_1 = 3 + (3)(0.25) = 3.75$.

Plug into the differential equation to get $y_1' = 0.25 + 3.75 = 4$.

Repeat:

$x_2 = 0.25 + 0.25 = 0.5$

$y_2 = 4 + (4)(0.25) = 5$

$y_2' = 0.5 + 5 = 5.5$

Repeat:

$x_3 = 0.5 + 0.25 = 0.75$

$y_2 = 5 + (5)(0.25) = 6.25$

$y_2' = 0.75 + 6.25 = 7$

One more time!

$$x_4 = 0.75 + 0.25 = 1$$

$$y_3 = 6.25 + (6.25)(0.25) = 7.8125$$

$$y_3' = 1 + 7.8125 = 8.8125$$

The answer is (C).

5. **B** The exponential growth means that $\frac{dy}{dt} = ky$, where k is a constant. Separate the variables $\frac{dy}{y} = kdt$. Integrate both sides: $\int \frac{dy}{y} = k \int dt$

$\ln|y| = kt + C$. Exponentiate both sides: $y = e^{kt+C} = Ae^{kt}$. Next, we are given that initially there are 16 cm², so $16 = Ae^{k(0)}$ and thus $A = 16$. Now the equation is $y = 16e^{kt}$. Next, we can solve for k:

$$60 = 16e^{k(30)}$$

$$\frac{60}{16} = \frac{15}{4} = e^{k(30)}$$

$\ln\frac{15}{4} = 30k$, so $k = \frac{1}{30}\ln\frac{15}{4}$. If we substitute this for k into the equation, to get $y = 16e^{\left(\frac{1}{30}\ln\frac{15}{4}\right)t}$. Use the rules of logs to rewrite the exponent: $y = 16e^{\left(\frac{1}{30}\ln\frac{15}{4}\right)t} = 16e^{\left(\ln\frac{15}{4}\right)^{\frac{t}{30}}} = 16\left(\frac{15}{4}\right)^{\frac{t}{30}}$.

Finally, plug in $t = 92$: $y = 16\left(\frac{15}{4}\right)^{\frac{92}{30}} \approx 921\,\mathrm{cm}^2$.

The answer is (B).

CHAPTER 22

1. **C** This is a geometric series with a ratio of $r = \frac{1}{3}$ and an initial term of 6, so the sum is $\frac{6}{1-\frac{1}{3}} = \frac{6}{\frac{2}{3}} = 9$.

The answer is (C).

2. Use the Ratio Test:

$$\rho = \lim_{n\to\infty} \frac{\dfrac{9^{n+1}}{n+1!}}{\dfrac{9^n}{n!}} = \lim_{n\to\infty} \frac{9^{n+1}}{n+1!} \frac{n!}{9^n} = \lim_{n\to\infty} \frac{9}{n+1} = 0$$

Therefore, the series converges.

3. Use the Integral Test. We will need to evaluate the integral $\int_1^\infty \frac{x^3}{1+x^4}dx$. First, use limits to rewrite the

    integral: $\lim_{c\to\infty}\int_1^c \frac{x^3}{1+x^4}dx$. Next, evaluate the indefinite integral using u-substitution. Let $u = 1+x^4$

    and $du = 4x^3dx$, so $\frac{1}{4}du = x^3dx$. Substitute: $\int \frac{x^3}{1+x^4}dx = \frac{1}{4}\int\frac{du}{u} = \frac{1}{4}\ln|u|+C$. Substitute back:

    $\frac{1}{4}\ln|1+x^4|+C$. And now evaluate the definite integral: $\frac{1}{4}\ln|1+x^4|\Big|_1^c = \frac{1}{4}\ln(1+c^4)-\frac{1}{4}\ln 2$.

    Take the limit: $\lim_{c\to\infty}\left|\frac{1}{4}\ln(1+c^4)-\frac{1}{4}\ln 2\right| = \infty$ so the series diverges.

4. **B** First, find the derivatives of $f(x) = \sqrt{1+x}$ and compute their values at $x = 0$.

    $f(x) = \sqrt{1+x} = (1+x)^{\frac{1}{2}}$ $\qquad\qquad$ $f(0) = (1+0)^{\frac{1}{2}} = 1$

    $f'(x) = \frac{1}{2}(1+x)^{-\frac{1}{2}}$ $\qquad\qquad$ $f'(0) = \frac{1}{2}(1+0)^{-\frac{1}{2}} = \frac{1}{2}$

    $f''(x) = -\frac{1}{4}(1+x)^{-\frac{3}{2}}$ $\qquad\qquad$ $f''(0) = -\frac{1}{4}(1+0)^{-\frac{3}{2}} = -\frac{1}{4}$

    $f'''(x) = \frac{3}{8}(1+x)^{-\frac{5}{2}}$ $\qquad\qquad$ $f'''(0) = \frac{3}{8}(1+0)^{-\frac{5}{2}} = \frac{3}{8}$

    Next, plug into the formula to generate the Taylor series:

    $$\sqrt{1+x} \approx 1 + \frac{x}{2} + \left(\frac{-1}{4}\right)\frac{x^2}{2!} + \left(\frac{3}{8}\right)\left(\frac{x^3}{3!}\right) = 1 + \frac{x}{2} - \frac{x^2}{8} + \frac{x^3}{16}$$

    The answer is (B).

5. **A** This is a geometric series with $r = 5x$; thus it converges if $-1 < 5x < 1$, which simplifies to

    $\frac{-1}{5} < x < \frac{1}{5}$. Therefore, the radius of convergence is $R = \frac{1}{5}$ and the interval of convergence is

    $\left(-\frac{1}{5}, \frac{1}{5}\right)$.

    The answer is (A).

Part VI
Practice Tests

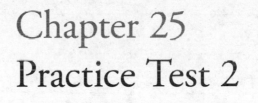

Chapter 25
Practice Test 2

AP® Calculus BC Exam

DO NOT OPEN THIS BOOKLET UNTIL YOU ARE TOLD TO DO SO.

At a Glance

Total Time
1 hour and 45 minutes
Number of Questions
45
Percent of Total Grade
50%
Writing Instrument
Pencil required

Instructions

Section I of this examination contains 45 multiple-choice questions. Fill in only the ovals for numbers 1 through 45 on your answer sheet.

CALCULATORS MAY NOT BE USED IN THIS PART OF THE EXAMINATION.

Indicate all of your answers to the multiple-choice questions on the answer sheet. No credit will be given for anything written in this exam booklet, but you may use the booklet for notes or scratch work. After you have decided which of the suggested answers is best, completely fill in the corresponding oval on the answer sheet. Give only one answer to each question. If you change an answer, be sure that the previous mark is erased completely. Here is a sample question and answer.

Sample Question   Sample Answer

Chicago is a
(A) state
(B) city
(C) country
(D) continent

Use your time effectively, working as quickly as you can without losing accuracy. Do not spend too much time on any one question. Go on to other questions and come back to the ones you have not answered if you have time. It is not expected that everyone will know the answers to all the multiple-choice questions.

About Guessing

Many candidates wonder whether or not to guess the answers to questions about which they are not certain. Multiple choice scores are based on the number of questions answered correctly. Points are not deducted for incorrect answers, and no points are awarded for unanswered questions. Because points are not deducted for incorrect answers, you are encouraged to answer all multiple-choice questions. On any questions you do not know the answer to, you should eliminate as many choices as you can, and then select the best answer among the remaining choices.

CALCULUS BC

SECTION I, Part A

Time—60 Minutes

Number of questions—30

A CALCULATOR MAY NOT BE USED ON THIS PART OF THE EXAMINATION

Directions: Solve each of the following problems, using the available space for scratchwork. After examining the form of the choices, decide which is the best of the choices given and fill in the corresponding oval on the answer sheet. No credit will be given for anything written in the test book. Do not spend too much time on any one problem.

In this test: Unless otherwise specified, the domain of a function f is assumed to be the set of all real numbers x for which $f(x)$ is a real number.

1. What is the slope of the line tangent to the curve $x^2 + 2xy + 3y^2 = 2$ when $y = 1$?

 (A) $-\dfrac{1}{8}$

 (B) -1

 (C) 0

 (D) $\dfrac{1}{8}$

---

2. $\displaystyle\int_{-1}^{1} xe^{x^2}\, dx =$

 (A) $-\dfrac{e}{2}$

 (B) 0

 (C) $\dfrac{e}{2}$

 (D) e

GO ON TO THE NEXT PAGE.

3. If, for $t > 0$, $x = t^2$ and $y = \cos(t^2)$, then $\dfrac{dy}{dx} =$

    (A) $\cos(t^2)$
    (B) $-\sin(t^2)$
    (C) $-\sin(2t)$
    (D) $\sin(t^2)$

---

4. The function $f(x) = 4x^3 - 8x^2 + 1$ on the interval $[-1, 1]$ has an absolute minimum at $x =$

    (A) -11

    (B) -1

    (C) 0

    (D) $\dfrac{4}{3}$

GO ON TO THE NEXT PAGE.

5. $\int \dfrac{x\,dx}{x^2+5x+6} =$

(A) $\ln\left|\dfrac{(x+3)^3}{(x+2)^2}\right| + C$

(B) $\ln\left|(x+3)^3(x+2)^2\right| + C$

(C) $\ln\left|\dfrac{(x+2)^2}{(x+3)^3}\right| + C$

(D) $\ln\left|(x+3)^2(x+2)^3\right| + C$

6. $\dfrac{d}{dx}(x^2\sin^2 x) =$

(A) $2x\sin 2x$
(B) $2x\cos^2 x$
(C) $2x\sin^2 x + x^2\cos^2 x$
(D) $2x\sin^2 x + x^2\sin 2x$

GO ON TO THE NEXT PAGE.

7. The line normal to the curve $y = \dfrac{x^2 - 1}{x^2 + 1}$ at $x = 2$ has slope

(A) $-\dfrac{8}{25}$

(B) $-\dfrac{25}{8}$

(C) 1

(D) $\dfrac{8}{25}$

8. $\displaystyle\int \dfrac{1 + 2x}{\sqrt{1 - x^2}}\, dx =$

(A) $\sin^{-1} x - 2\sqrt{1 - x^2} + C$

(B) $\sin^{-1} x + 2\sqrt{1 - x^2} + C$

(C) $-2\sqrt{1 - x^2} + C$

(D) $2\sqrt{1 - x^2} + C$

GO ON TO THE NEXT PAGE.

9.

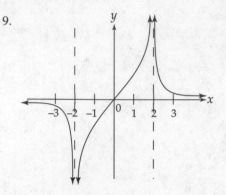

The graph of $y = f(x)$ is shown above. Which of the following could be the graph of $y = f'(x)$?

(A)

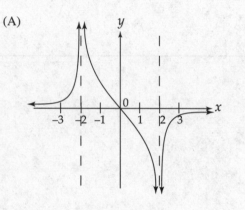

(C)

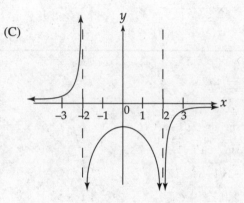

(B)

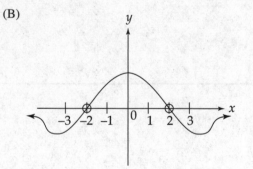

(D)

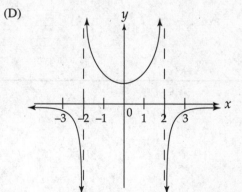

GO ON TO THE NEXT PAGE.

10. $\displaystyle\int_1^{e^2} x\ln x \, dx =$

(A) $\dfrac{e^2}{4} - \dfrac{1}{4}$

(B) $\dfrac{5e^4}{4} + \dfrac{1}{4}$

(C) $\dfrac{3e^2}{4} + \dfrac{1}{4}$

(D) $\dfrac{3e^4}{4} + \dfrac{1}{4}$

11. $\displaystyle\int_4^{\infty} \dfrac{dx}{x^2 + 16} =$

(A) $\dfrac{\pi}{16}$

(B) $\dfrac{\pi}{4}$

(C) $\dfrac{\pi}{2}$

(D) Divergent

GO ON TO THE NEXT PAGE.

12. What is the equation of the line tangent to the graph of $y = \sin^2 x$ at $x = \dfrac{\pi}{4}$?

(A) $\quad y - \dfrac{1}{2} = -\left(x - \dfrac{\pi}{4}\right)$

(B) $\quad y - \dfrac{1}{2} = \left(x - \dfrac{\pi}{4}\right)$

(C) $\quad y - \dfrac{1}{\sqrt{2}} = \left(x - \dfrac{\pi}{4}\right)$

(D) $\quad y - \dfrac{1}{2} = \dfrac{1}{2}\left(x - \dfrac{\pi}{4}\right)$

13. $\displaystyle\int \sin^5 x \cos^3 x \, dx =$

(A) $\quad \dfrac{\sin^9 x}{9} + C$

(B) $\quad \dfrac{\sin^6 x}{6} - \dfrac{\sin^8 x}{8} + C$

(C) $\quad \dfrac{\sin^6 x}{6} + C$

(D) $\quad \dfrac{\sin^6 x}{6} - \dfrac{\cos^4 x}{4} + C$

GO ON TO THE NEXT PAGE.

14. Evaluate $\int_2^\infty \dfrac{dt}{t^3}$.

(A) 0

(B) ∞

(C) $\dfrac{1}{8}$

(D) $-\dfrac{1}{2}$

15. If $f(x) = \dfrac{x^2 + 5x - 24}{x^2 + 10x + 16}$, then $\lim\limits_{x \to -8} f(x)$ is

(A) 0

(B) $-\dfrac{3}{2}$

(C) $\dfrac{11}{6}$

(D) Nonexistent

GO ON TO THE NEXT PAGE.

16. What is the approximation of the value of e^3 obtained by using the fourth-degree Taylor polynomial about $x = 0$ for e^x?

    (A) $1 + 3 + \dfrac{3^2}{2} + 3^2 + \dfrac{3^4}{4}$

    (B) $1 + 3 + \dfrac{3^2}{2!} + \dfrac{3^3}{3!} + \dfrac{3^4}{4!}$

    (C) $1 - 3 + \dfrac{3^2}{2!} - \dfrac{3^3}{3!} + \dfrac{3^4}{4!}$

    (D) $1 - \dfrac{3^2}{2!} + \dfrac{3^4}{4!}$

---

17. A rock is thrown straight upward with an initial velocity of 50 m/s from a point 100 m above the ground. If the acceleration of the rock at any time t is $a = -10$ m/s^2, what is the maximum height of the rock (in meters)?

    (A) 150
    (B) 175
    (C) 200
    (D) 225

---

18. The sum of the infinite geometric series $2 - \dfrac{2}{3} + \dfrac{2}{9} - \dfrac{2}{27} + \dots$ is

    (A) -6

    (B) -3

    (C) $\dfrac{3}{7}$

    (D) $\dfrac{3}{2}$

GO ON TO THE NEXT PAGE.

19. What are all values of x for which the series $\displaystyle\sum_{n=1}^{\infty} \frac{(x-3)^n}{n^2(5^n)}$ converges?

    (A) $-2 \leq x \leq 8$

    (B) $-2 < x \leq 8$

    (C) $-5 \leq x \leq 5$

    (D) $-5 \leq x < 5$

---

20. Find the area inside one loop of the curve $r = \sin 2\theta$.

    (A) $\dfrac{\pi}{16}$

    (B) $\dfrac{\pi}{8}$

    (C) $\dfrac{\pi}{4}$

    (D) π

---

21. The average value of $\sec^2 x$ on the interval $\left[\dfrac{\pi}{6},\ \dfrac{\pi}{4}\right]$ is

    (A) $\dfrac{12\sqrt{3} - 12}{\pi}$

    (B) $\dfrac{12 - 4\sqrt{3}}{\pi}$

    (C) $\dfrac{6\sqrt{2} - 6}{\pi}$

    (D) $\dfrac{6 - 6\sqrt{2}}{\pi}$

GO ON TO THE NEXT PAGE.

22. Find the length of the arc of the curve defined by $x = \frac{1}{2}t^2$ and $y = \frac{1}{9}(6t+9)^{\frac{3}{2}}$, from $t = 0$ to $t = 2$.

(A) 8
(B) 10
(C) 12
(D) 16

23. The function f is given by $f(x) = x^4 + 4x^3$. On which of the following intervals is f decreasing?

(A) $(0, \infty)$
(B) $(-3, \infty)$
(C) $(-\infty, -3)$
(D) $(-\infty, 0)$

24. Which of the following series converge(s)?

I. $\sum_{n=1}^{\infty} \frac{(-1)^n}{n}$

II. $\sum_{n=1}^{\infty} \frac{1}{\sqrt{n^3}}$

III. $\sum_{n=1}^{\infty} \frac{1}{\sqrt[3]{n^2}}$

(A) I only
(B) II only
(C) I and II
(D) I and III

GO ON TO THE NEXT PAGE.

25. Given the differential equation $\dfrac{dz}{dt} = z\left(4 - \dfrac{z}{100}\right)$, where $z(0) = 50$, what is $\lim_{t \to \infty} z(t)$?

(A) 400

(B) 200

(C) 100

(D) 4

26.

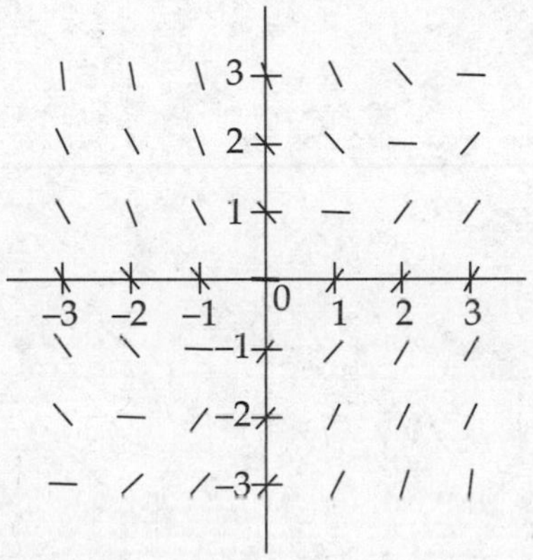

The slope field shown above corresponds to which of the following differential equations?

(A) $\dfrac{dy}{dx} = \dfrac{x}{y}$

(B) $\dfrac{dy}{dx} = \dfrac{y}{x}$

(C) $\dfrac{dy}{dx} = x - y$

(D) $\dfrac{dy}{dx} = x + y$

GO ON TO THE NEXT PAGE.

27. The value of c that satisfies the Mean Value Theorem for derivatives on the interval $[0, 5]$ for the function $f(x) = x^3 - 6x$ is

(A) $-\dfrac{5}{\sqrt{3}}$

(B) 0

(C) 1

(D) $\dfrac{5}{\sqrt{3}}$

28.

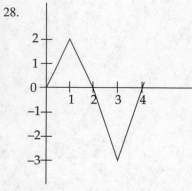

The graph of f is shown in the figure above. If $g(x) = \displaystyle\int_0^x f(t)\, dt$, for what positive value of x does $g(x)$ have a minimum?

(A) 1
(B) 2
(C) 3
(D) 4

GO ON TO THE NEXT PAGE.

29. $\int \sec^4 x \, dx =$

   (A) $\dfrac{\sec^5 x}{5} + C$

   (B) $\tan^2 x + C$

   (C) $\sec^2 x - \dfrac{\sec^5 x}{5} + C$

   (D) $\tan x + \dfrac{\tan^3 x}{3} + C$

---

30. If f is a vector-valued function defined by $f(t) = (\sin 2t, \sin^2 t)$, then $f''(t) =$

   (A) $(-4\sin 2t, 2\cos 2t)$
   (B) $(-\sin 2t, -\cos^2 t)$
   (C) $(4\sin 2t, -2\cos 2t)$
   (D) $(2\cos 2t, -4\sin 2t)$

END OF PART A, SECTION I

IF YOU FINISH BEFORE TIME IS CALLED, YOU MAY CHECK YOUR WORK ON PART A ONLY.

DO NOT GO ON TO PART B UNTIL YOU ARE TOLD TO DO SO.

CALCULUS BC

SECTION I, Part B

Time—45 Minutes

Number of questions—15

A GRAPHING CALCULATOR IS REQUIRED FOR SOME QUESTIONS ON THIS PART OF THE EXAMINATION

Directions: Solve each of the following problems, using the available space for scratchwork. After examining the form of the choices, decide which is the best of the choices given and fill in the corresponding oval on the answer sheet. No credit will be given for anything written in the test book. Do not spend too much time on any one problem.

In this test:

(1) The **exact** numerical value of the correct answer does not always appear among the choices given. When this happens, select from among the choices the number that best approximates the exact numerical value.

(2) Unless otherwise specified, the domain of a function f is assumed to be the set of all real numbers x for which $f(x)$ is a real number.

31. If $f(x)$ is the function given by $f(x) = e^{3x} + 1$, at what value of x is the slope of the tangent line to $f(x)$ equal to 2 ?

   (A) −0.135
   (B)   0.231
   (C) −0.366
   (D)   0.693

32. The side of a square is increasing at a constant rate of 0.4 cm/sec. In terms of the perimeter, P, what is the rate of change of the area of the square, in cm²/sec?

   (A) 0.05P
   (B) 0.2P
   (C) 0.4P
   (D) 6.4P

GO ON TO THE NEXT PAGE.

33. The height of a mass hanging from a spring at time t seconds, where $t > 0$, is given by $h(t) = 12 - 4 \cos(2t)$. In the first two seconds, how many times is the velocity of the mass equal to 0 ?

    (A) 1
    (B) 2
    (C) 3
    (D) 4

34. $\displaystyle \lim_{h \to 0} \frac{\tan^{-1}(1+h) - \dfrac{\pi}{4}}{h} =$

    (A) 2

    (B) $\dfrac{4}{4 + \pi^2}$

    (C) $\dfrac{1}{2}$

    (D) Nonexistent

35. What is the trapezoidal approximation of $\displaystyle \int_0^3 e^x \, dx$ using $n = 4$ subintervals?

    (A)   6.407
    (B) 19.972
    (C) 27.879
    (D) 34.944

GO ON TO THE NEXT PAGE.

36. Given $x^2y + x^2 = y^2 + 1$, find $\dfrac{d^2y}{dx^2}$ at $(1, 1)$.

    (A) 36
    (B) 12
    (C) −12
    (D) −36

37. If $\displaystyle\int_{-2}^{4} f(x)\,dx = a$ and $\displaystyle\int_{3}^{4} f(x)\,dx = b$, then $\displaystyle\int_{3}^{-2} f(x)\,dx =$

    (A) $a - 2b$
    (B) $a - b$
    (C) $b - a$
    (D) $2b - a$

GO ON TO THE NEXT PAGE.

38. If $\int_{1}^{11} f(x)\, dx = 24$ and $\int_{4}^{11} f(x)\, dx = 20$ find $\int_{1}^{4} f(x)\, dx$.

(A) 44

(B) 4

(C) −4

(D) −44

---

39. Using the Taylor series about $x = 0$ for $\sin x$, approximate $\sin(0.4)$ to four decimal places.

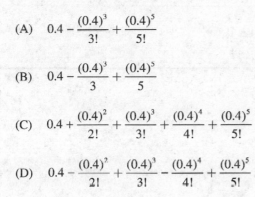

(A) $\quad 0.4 - \dfrac{(0.4)^3}{3!} + \dfrac{(0.4)^5}{5!}$

(B) $\quad 0.4 - \dfrac{(0.4)^3}{3} + \dfrac{(0.4)^5}{5}$

(C) $\quad 0.4 + \dfrac{(0.4)^2}{2!} + \dfrac{(0.4)^3}{3!} + \dfrac{(0.4)^4}{4!} + \dfrac{(0.4)^5}{5!}$

(D) $\quad 0.4 - \dfrac{(0.4)^2}{2!} + \dfrac{(0.4)^3}{3!} - \dfrac{(0.4)^4}{4!} + \dfrac{(0.4)^5}{5!}$

GO ON TO THE NEXT PAGE.

40. Let R be the region in the first quadrant between the graphs of $y = e^{-x}$, $y = \sin x$, and the y-axis. The volume of the solid that results when R is revolved about the x-axis is

    (A) −0.888
    (B) −0.869
    (C)   0.869
    (D)   0.888

41. Use Euler's Method, with $h = 0.2$ to estimate $y(3)$, if $\dfrac{dy}{dx} = 2y - 4x$ and $y(2) = 6$.

    (A)   9.684
    (B) 10.442
    (C) 12.378
    (D) 18.426

GO ON TO THE NEXT PAGE.

42. $\int \tan^4 x \, dx =$

(A) $\dfrac{\sec^8 x}{8} + C$

(B) $\dfrac{\tan^5 x}{5} + C$

(C) $\dfrac{\tan^3 x}{3} - \tan x + x + C$

(D) $\dfrac{\tan^3 x}{3} - \tan x - x + C$

43. Let $f(x) = \int \cot x \, dx$; $0 < x < \pi$. If $f\left(\dfrac{\pi}{6}\right) = 1$, then $f(1) =$

(A) -1.861
(B) -0.480
(C) 0.524
(D) 1.521

GO ON TO THE NEXT PAGE.

44. $\int \sqrt{4 - x^2} \, dx =$

   (A) $\frac{2}{3}(4 - x^2)^{\frac{3}{2}} + C$

   (B) $2\sin^{-1}\left(\frac{x}{2}\right) + x\sqrt{4 - x^2} + C$

   (C) $2\sin^{-1}\left(\frac{x}{2}\right) + \frac{x}{2}\sqrt{4 - x^2} + C$

   (D) $\frac{2}{3}(4 - x^3)^{\frac{3}{2}} + C$

---

45. A force of 250 N is required to stretch a spring 5 m from rest. Using Hooke's Law, $F = kx$, how much work, in Joules, is required to stretch the spring 7 m from rest?

   (A)   14.286
   (B)   71.429
   (C)   490
   (D)   1,225

STOP
END OF PART B, SECTION I
IF YOU FINISH BEFORE TIME IS CALLED, YOU MAY CHECK YOUR WORK ON PART B ONLY.
DO NOT GO ON TO SECTION II UNTIL YOU ARE TOLD TO DO SO.

SECTION II
GENERAL INSTRUCTIONS

You may wish to look over the problems before starting to work on them, since it is not expected that everyone will be able to complete all parts of all problems. All problems are given equal weight, but the parts of a particular problem are not necessarily given equal weight.

A GRAPHING CALCULATOR IS REQUIRED FOR SOME PROBLEMS OR PARTS OF PROBLEMS ON THIS SECTION OF THE EXAMINATION.

- You should write all work for each part of each problem in the space provided for that part in the booklet. Be sure to write clearly and legibly. If you make an error, you may save time by crossing it out rather than trying to erase it. Erased or crossed-out work will not be graded.

- Show all your work. You will be graded on the correctness and completeness of your methods as well as your answers. Correct answers without supporting work may not receive credit.

- Justifications require that you give mathematical (noncalculator) reasons and that you clearly identify functions, graphs, tables, or other objects you use.

- You are permitted to use your calculator to solve an equation, find the derivative of a function at a point, or calculate the value of a definite integral. However, you must clearly indicate the setup of your problem, namely the equation, function, or integral you are using. If you use other built-in features or programs, you must show the mathematical steps necessary to produce your results.

- Your work must be expressed in standard mathematical notation rather than calculator syntax. For example, $\int_1^5 x^2\,dx$ may not be written as fnInt (X^2, X, 1, 5).

- Unless otherwise specified, answers (numeric or algebraic) need not be simplified. If your answer is given as a decimal approximation, it should be correct to three places after the decimal point.

- Unless otherwise specified, the domain of a function f is assumed to be the set of all real numbers x for which $f(x)$ is a real number.

GO ON TO THE NEXT PAGE.

SECTION II, PART A
Time—30 minutes
Number of problems—2

A graphing calculator is required for some problems or parts of problems.

During the timed portion for Part A, you may work only on the problems in Part A.

On Part A, you are permitted to use your calculator to solve an equation, find the derivative of a function at a point, or calculate the value of a definite integral. However, you must clearly indicate the setup of your problem, namely the equation, function, or integral you are using. If you use other built-in features or programs, you must show the mathematical steps necessary to produce your results.

1. An object moving along a curve in the xy-plane has its position given by $(x(t), y(t))$ at time t seconds, $0 \leq t \leq 1$, with
$\dfrac{dx}{dt} = 8t\cos t$ units/sec and $\dfrac{dy}{dt} = 8t\sin t$ units/sec.

At time $t = 0$, the object is located at $(5, 11)$.

(a) Find the speed of the object at time $t = 1$.

(b) Find the length of the arc described by the curve's position from time $t = 0$ to time $t = 1$.

(c) Find the location of the object at time $t = \dfrac{\pi}{2}$.

GO ON TO THE NEXT PAGE.

2.

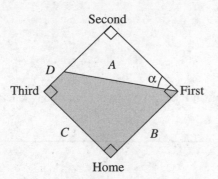

Second

D   A

α

Third   First

C   B

Home

A baseball diamond is a square with each side 90 feet in length. A player runs from second base to third base at a rate of 18 ft/sec.

(a) At what rate is the player's distance from first base, A, changing when his distance from third base, D, is 22.5 feet?

(b) At what rate is angle α increasing when D is 22.5 feet?

(c) At what rate is the area of the trapezoidal region, formed by line segments A, B, C, and D, changing when D is 22.5 feet?

GO ON TO THE NEXT PAGE.

SECTION II, PART B
Time—1 hour
Number of problems—4

No calculator is allowed for these problems.

During the timed portion for Part B, you may continue to work on the problems in Part A without the use of any calculator.

---

3. Consider the equation $x^2 - 2xy + 4y^2 = 64$.

   (a) Write an expression for the slope of the curve at any point (x, y).

   (b) Find the equations of the tangent lines to the curve at the point $x = 2$.

   (c) Find $\dfrac{d^2y}{dx^2}$ at $(0, 4)$.

(a) $x^2 - 2xy + 4y^2 = 64$

$2x - (2xy' + 2y) + 8yy' = 0$

$2x - 2xy' - 2y + 8yy' = 0$

$2x - 2y = 2xy' - 8yy'$

$\dfrac{2x - 2y}{2x - 8y} = y'$

$\dfrac{2(x-y)}{2(x-4y)} = y'$

$\boxed{\dfrac{x-y}{x-4y} = y'}$

(b) $x = 2$

$4 - 4y + 4y^2 = 64$

$-60 - 4y + 4y^2 = 0$

$4y(-1 + y) = 60$

GO ON TO THE NEXT PAGE.

4.

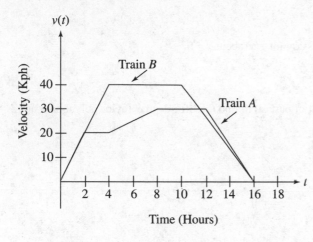

Time (Hours)

Three trains, *A, B,* and *C* each travel on a straight track for $0 \le t \le 16$ hours. The graphs above, which consist of line segments, show the velocities, in kilometers per hour, of trains *A* and *B*. The velocity of *C* is given by $v(t) = 8t - 0.25t^2$.

(Indicate units of measure for all answers.)

(a) Find the velocities of *A* and *C* at time $t = 6$ hours.

look at graph   *plug and chug*

(b) Find the accelerations of *B* and *C* at time $t = 6$ hours.

look at graph (slope)   *du derivative, pling & chug*

(c) Find the positive difference between the total distance that *A* traveled and the total distance that *B* traveled in 16 hours.

area under each curve

(d) Find the total distance that *C* traveled in 16 hours.

integral of v(t) from 0 to 16 and solve

5. Let *y* be the function satisfying $f'(x) = x(1 - f(x))$; $f(0) = 10$.   *(0,10)*

(a) Use Euler's Method, starting at $x = 0$, with step size of 0.5 to approximate $f(x)$ at $x = 1$.

.5(1-10) = -4.5 · .5 = -2.25   (1, 7.75)

(b) Find an exact solution for $f(x)$ when $x = 1$.

(c) Evaluate $\int_0^{\infty} x(1 - f(x))\, dx$.

GO ON TO THE NEXT PAGE.

6. Given $f(x) = \tan^{-1}(x)$ and $g(x) = \dfrac{1}{1+x}$, for $|x| \le 1$

   (a) Find the fifth-degree Taylor polynomial and general expression for $g(x)$ about $x = 0$.

   (b) Given that $\dfrac{d}{dx} \tan^{-1} x = \dfrac{1}{1+x^2}$, for $|x| \le 1$, use the result of part (a) to find the fifth-degree Taylor polynomial and general expression for $f(x)$ about $x = 0$.

   (c) Use the fifth-degree Taylor polynomial to estimate $f\left(\dfrac{1}{10}\right)$.

(a) $g(x) = 1$

$g'(x) = -1(1+x)^{-2} = -1$

$g''(x) = 2(1+x)^{-3} = 2$

$g'''(x) = -6(1+x)^{-4} = -6$

$g''''(x) = 24(1+x)^{-5} = 24$

$g^5(x) = -120(1+x)^{-6} = -120$

$g(x) = 1 - \dfrac{1}{1!}x + \dfrac{2}{2!}x^2 - \dfrac{6}{3!}x^3 + \dfrac{24}{4!}x^4 - \dfrac{120}{5!}x^5$

$1 - x + x^2 - x^3 + x^4 - x^5 \cdots \boxed{(-1)^n x^n}$

(b) $\dfrac{d}{dx} \tan^{-1} = \dfrac{1}{1+x^2}$

$\dfrac{d}{dx} \tan^{-1} = 1 - x^2 + x^4 - x^6 + x^8$

$\tan^{-1}(x) = \int 1 - x^2 + x^4 \, dx$

$\boxed{\tan^{-1}(x) = x - \dfrac{x^3}{3} + \dfrac{x^5}{5} + C}$

$\cdots \dfrac{x^{2n+1}}{2n+1}$

(c) $\boxed{f\left(\tfrac{1}{10}\right) = \dfrac{1}{10} - \dfrac{(\tfrac{1}{10})^3}{3} + \dfrac{(\tfrac{1}{10})^5}{5}}$

STOP

END OF EXAM

Chapter 26
Practice Test 2
Answers and
Explanations

ANSWER KEY TO SECTION I

1.	C	11.	A	21.	B	31.	A	41.	C
2.	B	12.	B	22.	A	32.	B	42.	C
3.	B	13.	B	23.	C	33.	A	43.	D
4.	B	14.	C	24.	C	34.	C	44.	C
5.	A	15.	C	25.	A	35.	B	45.	D
6.	D	16.	B	26.	C	36.	C		
7.	B	17.	D	27.	D	37.	C		
8.	A	18.	D	28.	D	38.	B		
9.	D	19.	A	29.	D	39.	A		
10.	D	20.	B	30.	A	40.	D		

ANSWERS AND EXPLANATIONS TO SECTION I

1. **C** You need to use implicit differentiation to find $\dfrac{dy}{dx}$.

$$2x + 2\left(x\frac{dy}{dx} + y \right) + 6y\frac{dy}{dx} = 0$$

$$2x + 2x\frac{dy}{dx} + 2y + 6y\frac{dy}{dx} = 0$$

Now, if you wanted to solve for $\dfrac{dy}{dx}$ in terms of x and y, you would have to do some algebra to isolate $\dfrac{dy}{dx}$. But, because you are asked to solve for $\dfrac{dy}{dx}$ at a specific value of x, you don't need to simplify.

You need to find the x-coordinate that corresponds to the y-coordinate $y = 1$. Plug $y = 1$ into the equation and solve for x.

$$x^2 + 2x(1) + 3(1)^2 = 2$$

$$x^2 + 2x + 3 = 2$$

$$x^2 + 2x + 1 = 0$$

$$(x + 1)^2 = 0$$

$$x - -1$$

Finally, plug $x = -1$ and $y = 1$ into the derivative to get

$$2(-1) + 2(-1)\frac{dy}{dx} + 2(1) + 6(1)\frac{dy}{dx} = 0$$

$$-2 - 2\frac{dy}{dx} + 2 + 6\frac{dy}{dx} = 0$$

$$\frac{dy}{dx} = 0$$

2. **B** You can use u-substitution to evaluate the integral.

Let $u = x^2$ and $du = 2x\,dx$. Then $\dfrac{1}{2}\,du = x\,dx$.

Now substitute into the integral $\dfrac{1}{2}\displaystyle\int e^u\,du$, leaving out the limits of integration for the moment.

Evaluate the integral to get $\dfrac{1}{2}\displaystyle\int e^u \, du = \dfrac{1}{2}e^u$.

Now substitute back to get $\dfrac{1}{2}e^{x^2}$.

Finally, evaluate at the limits of integration to get $\dfrac{1}{2}e^{x^2}\Big|_{-1}^{1} = \dfrac{1}{2}e - \dfrac{1}{2}e = 0$.

3. **B** If you have a pair of parametric equations, $x(t)$ and $y(t)$, then $\dfrac{dy}{dx} = \dfrac{\frac{dy}{dt}}{\frac{dx}{dt}}$.

Here you get $\dfrac{dy}{dt} = -2t\sin(t^2)$ and $\dfrac{dx}{dt} = 2t$.

Then $\dfrac{dy}{dx} = \dfrac{-2t\sin(t^2)}{2t} = -\sin(t^2)$.

4. **B** If you want to find the minimum, take the derivative and find where the derivative is zero.

$$f'(x) = 12x^2 - 16x$$

Next, set the derivative equal to zero and solve for x, in order to find the critical values.

$$12x^2 - 16x = 0$$

$$4x(3x - 4) = 0$$

$$x = 0 \text{ or } x = \frac{4}{3}$$

Next, you can use the second derivative test to determine which critical value is a minimum and which is a maximum.

Remember the second derivative test: if the sign of the second derivative at a critical value is positive, then the curve has a local minimum there. If the sign of the second derivative is negative, then the curve has a local maximum there.

Take the second derivative: $f'(x) = 24x - 16$.

This is negative at $x = 0$ and positive at $x = \dfrac{4}{3}$. This means that the curve has a relative minimum at $x = \dfrac{4}{3}$, but this value is outside of the interval $[-1, 1]$. So, in order to find where it has an absolute *minimum*, plug the endpoints of the interval into the original equation, and the smaller value will be the answer.

At $x = -1$, the value is $f(-1) = -11$. At $x = 1$, the value is $f(1) = -3$.

5.   **A**   Whenever you have an integrand that is a rational expression, you can often use the method of partial fractions to rewrite the integral in a form in which it's easy to evaluate.

First, separate the denominator into its two components and place the constants A and B in the numerators of the fractions and the components into the denominators. Set their sum equal to the original rational expression.

$$\frac{A}{x+3} + \frac{B}{x+2} = \frac{x}{(x+3)(x+2)}$$

Now, you want to solve for the constants A and B. First, multiply through by $(x + 3)(x + 2)$ to clear the denominators.

$$(x+3)(x+2)\left[\frac{A}{x+3} + \frac{B}{x+2}\right] = x$$

$$A(x + 2) + B(x + 3) = x$$

Now, distribute, then group, the terms on the left side.

$$Ax + 2A + Bx + 3B = x$$

$$Ax + Bx + 2A + 3B = x$$

$$(A + B)x + (2A + 3B) = x$$

In order for this last equation to be true, you need $A + B = 1$ and $2A + 3B = 0$.

If you solve these simultaneous equations, you get $A = 3$ and $B = -2$.

Now that you have done the partial fraction decomposition, you can rewrite the original integral as $\int\left[\frac{3}{x+3} - \frac{2}{x+2}\right]dx$. This is now simple to evaluate.

$$\int\left[\frac{3}{x+3} - \frac{2}{x+2}\right]dx = 3\int\frac{dx}{x+3} - 2\int\frac{dx}{x+2} = 3\ln|x+3| - 2\ln|x+2| + C$$

Using the rules of logarithms, the answer can be rewritten as $\ln\left|\frac{(x+3)^3}{(x+2)^2}\right| + C$.

6. **D** Here you need to use the Product Rule, which is: If $f(x) = uv$, where u and v are both functions of x, then $f'(x) = u\dfrac{dv}{dx} + v\dfrac{du}{dx}$.

You get $\dfrac{d}{dx}(x^2\sin^2 x) = 2x\sin^2 x + x^2(2\sin x \cos x)$.

This can be simplified to $2x\sin^2 x + x^2\sin 2x$.

7. **B** The normal line to a curve at a point is perpendicular to the tangent line at the same point. Thus, the slope of the normal line is the negative reciprocal of the slope of the tangent line. You find the slope of the tangent line by finding the derivative and evaluating it at the point.

You need to use the Quotient Rule, which is

Given $y = \dfrac{g(x)}{h(x)}$, then $\dfrac{dy}{dx} = \dfrac{h(x)g'(x) - g(x)h'(x)}{[h(x)]^2}$.

Here you have $\dfrac{dy}{dx} = \dfrac{(x^2+1)(2x) - (x^2-1)(2x)}{(x^2+1)^2}$.

Next, plug in $x = 2$ and solve $\dfrac{dy}{dx}\bigg|_{x=2} = \dfrac{(4+1)(4) - (4-1)(4)}{(4+1)^2} = \dfrac{8}{25}$.

Therefore, the slope of the normal line is $-\dfrac{25}{8}$.

8. **A** First, break up the integrand: $\displaystyle\int \dfrac{1+2x}{\sqrt{1-x^2}}\,dx = \int \dfrac{1}{\sqrt{1-x^2}}\,dx + \int \dfrac{2x}{\sqrt{1-x^2}}\,dx$.

The first integral is one that you should have memorized: $\displaystyle\int \dfrac{1}{\sqrt{1-x^2}}\,dx = \sin^{-1} x$.

You can do the second integral using u-substitution. Let $u = 1 - x^2$ and $du = -2x\,dx$.

Now substitute to get $\displaystyle\int \dfrac{2x}{\sqrt{1-x^2}}\,dx = -\int u^{-\frac{1}{2}}\,du$.

Integrate: $-\displaystyle\int u^{-\frac{1}{2}}\,du = -\dfrac{u^{\frac{1}{2}}}{\frac{1}{2}} = -2\sqrt{u}$. Now you just have to substitute back: $-2\sqrt{1-x^2}$. Then

combine the two results: $\sin^{-1} x - 2\sqrt{1-x^2} + C$.

9. **D** Here you want to examine the slopes of various pieces of the graph of $f(x)$. Notice that the graph starts with a slope of approximately zero and has a negative slope from $x = -\infty$ to $x = -2$, where the slope is $-\infty$. Thus, you are looking for a graph of $f'(x)$ that is negative from $x = -\infty$ to $x = -2$ and undefined at $x = -2$. Next, notice that the graph of $f(x)$ has a positive slope from

$x = -2$ to $x = 2$ and that the slope shrinks from ∞ to approximately one and then grows to ∞. Thus, you are looking for a graph of $f'(x)$ that is positive from $x = -2$ to $x = 2$ and approximately equal to one at $x = 0$. Finally, notice that the graph of $f(x)$ has a negative slope from $x = 2$ to $x = \infty$, where the slope starts at $-\infty$ and grows to approximately zero. Thus, you are looking for a graph of $f'(x)$ that is negative from $x = 2$ to $x = \infty$, where it is approximately zero. Graph (D) satisfies all of these requirements.

10. **D** You can integrate this using Integration by Parts. Let $u = \ln x$ and $dv = x\,dx$. Then $du = \frac{1}{x}dx$ and $v = \frac{x^2}{2}$. Plug them into the formula to get $\int x \ln x\,dx = \frac{x^2}{2}\ln x - \int \frac{x^2}{2} \cdot \frac{1}{x}\,dx$ (ignore the limits of integration until you are done integrating). This simplifies to $\frac{x^2}{2}\ln x - \int \frac{x^2}{2} \cdot \frac{1}{x}\,dx = \frac{x^2}{2}\ln x - \int \frac{x}{2}\,dx$.

Integrate again: $\frac{x^2}{2}\ln x - \frac{x^2}{4}$. Now evaluate this using the limits of integration. You get

$$\left(\frac{x^2}{2}\ln x - \frac{x^2}{4}\right)\Bigg|_1^e = \left(\frac{e^4}{2}\ln e^2 - \frac{e^4}{4}\right) - \left(\frac{1}{2}\ln 1 - \frac{1}{4}\right) = \frac{3e^4}{4} + \frac{1}{4}$$

11. **A** Whenever you have an integral of the form $\int \frac{dx}{a^2 + x^2}$, where a is a constant, the integral is going to be an inverse tangent. So put the integrand into the desired form. Also, you're going to ignore the limits of integration until after you have done the antidifferentiation.

First, divide the numerator and the denominator by 16: $\int \frac{dx}{x^2 + 16} = \frac{1}{16}\int \frac{dx}{\left(1 + \frac{x^2}{16}\right)}$.

Next, do u-substitution. Let $u = \frac{x}{4}$ and $du = \frac{dx}{4}$ or $4\,du = dx$.

Substitute into the integrand: $\frac{1}{16}\int \frac{dx}{\left(1 + \frac{x^2}{16}\right)} = \frac{1}{16}\int \frac{4\,du}{\left(1 + u^2\right)} = \frac{1}{4}\int \frac{du}{\left(1 + u^2\right)}$.

Evaluate the integral: $\frac{1}{4}\int \frac{du}{\left(1 + u^2\right)} = \frac{1}{4}\tan^{-1} u$.

Substitute back to get $\frac{1}{4}\tan^{-1}\left(\frac{x}{4}\right)$.

Now, evaluate the function at the limits of integration:

$$\left[\frac{1}{4}\tan^{-1}\left(\frac{x}{4}\right)\right]_4^\infty = \frac{1}{4}\left[\tan^{-1}\infty - \tan^{-1}1\right] = \frac{1}{4}\left[\frac{\pi}{2} - \frac{\pi}{4}\right] = \frac{\pi}{16}$$

12. **B** If you want to find the equation of the tangent line, you first need to find the y-coordinate that corresponds to $x = \dfrac{\pi}{4}$. It is $y = \sin^2\left(\dfrac{\pi}{4}\right) = \left(\dfrac{1}{\sqrt{2}}\right)^2 = \dfrac{1}{2}$.

Next, find the derivative of the curve at $x = \dfrac{\pi}{4}$, using the Chain Rule.

You get $\dfrac{dy}{dx} = 2\sin x \cos x$. At $x = \dfrac{\pi}{4}$, $\dfrac{dy}{dx}\Big|_{x=\frac{\pi}{4}} = 2\sin\left(\dfrac{\pi}{4}\right)\cos\left(\dfrac{\pi}{4}\right) = 2\left(\dfrac{1}{\sqrt{2}}\right)\left(\dfrac{1}{\sqrt{2}}\right) = 1$.

Now, you have the slope of the tangent line and a point that it goes through. You can use the point-slope formula for the equation of a line, $(y - y_1) = m(x - x_1)$, and plug in what you have just found. You get $\left(y - \dfrac{1}{2}\right) = (1)\left(x - \dfrac{\pi}{4}\right)$.

13. **B** First, factor $\cos^2 x$ out of the integrand to get $\int \sin^5 x \, \cos^2 x \cos x \, dx$. Next, replace $\cos^2 x$ with $1 - \sin^2 x$ to get $\int u^5 (1 - u^2) \, du = \int u^5 - u^7 \, du$.

Integrate: $\int u^5 - u^7 \, du = \dfrac{u^6}{6} - \dfrac{u^8}{8} + C$. And substitute back: $\dfrac{\sin^6 x}{6} - \dfrac{\sin^8 x}{8} + C$.

14. **C** This is an improper integral. You first rewrite the integral as $\displaystyle\lim_{n\to\infty} \int_2^n t^{-3} dt$. Now integrate:

$\displaystyle\lim_{n\to\infty} \int_2^n t^{-3} dt = \lim_{n\to\infty} \left(\dfrac{t^{-2}}{-2}\right)\Big|_2^n = \lim_{n\to\infty} \left(-\dfrac{1}{2t^2}\right)\Big|_2^n$. Now evaluate the integral at the limits:

$\displaystyle\lim_{n\to\infty} \left(-\dfrac{1}{2t^2}\right)\Big|_2^n = \lim_{n\to\infty} \left(-\dfrac{1}{2n^2} + \dfrac{1}{8}\right)$. Then take the limit: $\displaystyle\lim_{n\to\infty} \left(-\dfrac{1}{2n^2} + \dfrac{1}{8}\right) = \dfrac{1}{8}$.

15. **C** First, try plugging $x = -8$ into $f(x) = \dfrac{x^2 + 5x - 24}{x^2 + 10x + 16}$.

You get $f(x) = \dfrac{(-8)^2 + 5(-8) - 24}{(-8)^2 + 10(-8) + 16} = \dfrac{0}{0}$. This does NOT necessarily mean that the limit does not exist. When you get a limit of the form $\dfrac{0}{0}$, you first try to simplify the function by factoring and canceling like terms. Here you get

$$f(x) = \dfrac{x^2 + 5x - 24}{x^2 + 10x + 16} = \dfrac{(x+8)(x-3)}{(x+8)(x+2)} = \dfrac{(x-3)}{(x+2)}$$

Now, if you plug in $x = -8$, you get $f(x) = \dfrac{(-8-3)}{(-8+2)} = \dfrac{-11}{-6} = \dfrac{11}{6}$.

16. **B** The Taylor series for e^x about $x = 0$ is $e^x = 1 + x + \dfrac{x^2}{2!} + \dfrac{x^3}{3!} + \dfrac{x^4}{4!} + \dots$

Here, simply substitute 3 for x in the series to get

$$e^3 = 1 + 3 + \frac{3^2}{2!} + \frac{3^3}{3!} + \frac{3^4}{4!} + \dots$$

17. **D** Because the derivative of velocity with respect to time is acceleration, you have

$$v(t) = \int -10 \, dt = -10t + C$$

Now plug in the initial condition to solve for the constant.

$$50 = -10(0) + C$$

$$C = 50$$

Therefore, the velocity function is $v(t) = -10t + 50$.

Note that the velocity is zero at $t = 5$.

Next, because the derivative of position with respect to time is velocity, you have

$$s(t) = \int \left(-10t + 50\right) dt = -5t^2 + 50t + C$$

Now, plug in the initial condition to solve for the constant.

$$100 = -5(0)^2 + 50(0) + C$$
$$C = 100$$

Therefore, the position function is $s(t) = -5t^2 + 50t + 100$.

The maximum height occurs when the velocity is zero, so we plug $t = 5$ into the position function to get $s(5) = -5(5)^2 + 50(5) + 100 = 225$.

18. **D** This is the geometric series $\displaystyle\sum_{n=0}^{\infty} 2\left(-\frac{1}{3}\right)^n$. The sum of an infinite series of the form $\displaystyle\sum_{n=0}^{\infty} ar^n$ is

$S = \dfrac{a}{1-r}$. Here, the sum is $S = \dfrac{2}{1-\left(-\dfrac{1}{3}\right)} = \dfrac{2}{\dfrac{4}{3}} = \dfrac{3}{2}$.

19. **A** Use the Ratio Test to determine the interval of convergence.

You get $\displaystyle\lim_{n\to\infty} \dfrac{\dfrac{(x-3)^{n+1}}{(n+1)^2(5^{n+1})}}{\dfrac{(x-3)^n}{(n)^2(5^n)}} = \lim_{n\to\infty} \dfrac{(x-3)^{n+1}}{(n+1)^2(5^{n+1})} \dfrac{(n)^2(5^n)}{(x-3)^n} = \lim_{n\to\infty} \dfrac{x-3}{5} \dfrac{(n)^2}{(n+1)^2} = \dfrac{x-3}{5}$.

This converges if $\left|\dfrac{x-3}{5}\right| < 1$ or $-1 < \dfrac{x-3}{5} < 1$ and diverges if $\left|\dfrac{x-3}{5}\right| > 1$.

Thus, the series converges when $-2 < x < 8$.

Now you need to test whether the series converges at the endpoints of this interval.

When $x = 8$, you get the series $\displaystyle\sum_{n=1}^{\infty} \dfrac{1}{n^2}$, which converges, and when $x = -2$, you get the alternating series $\displaystyle\sum_{n=1}^{\infty} \dfrac{(-1)^n}{n^2}$, which also converges. Thus, the series converges when $-2 \le x \le 8$.

20. **B** First, graph the curve.

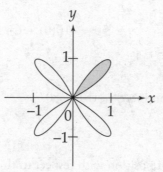

Find the area of the loop in the first quadrant, which is the interval from $\theta = 0$ to $\theta = \dfrac{\pi}{2}$. You find the area of a polar graph by evaluating $A = \dfrac{1}{2}\displaystyle\int_0^{\frac{\pi}{2}} r^2\, d\theta$.

Thus, evaluate: $A = \dfrac{1}{2}\displaystyle\int_0^{\frac{\pi}{2}} \sin^2(2\theta)\, d\theta$.

Next, you need to do a trigonometric substitution to evaluate this integral. Recall that $\cos 2\theta = 1 - 2\sin^2\theta$. You can rearrange this to obtain $\sin^2\theta = \dfrac{1 - \cos(2\theta)}{2}$, or in this case, $\sin^2 2\theta = \dfrac{1 - \cos(4\theta)}{2}$.

Thus, rewrite the integral: $A = \dfrac{1}{2}\displaystyle\int_0^{\frac{\pi}{2}} \dfrac{1 - \cos(4\theta)}{2}\, d\theta$.

Now evaluate this integral.

$$\int \dfrac{1 - \cos(4\theta)}{2}\, d\theta = \int \dfrac{d\theta}{2} - \int \dfrac{\cos(4\theta)}{2}\, d\theta = \dfrac{\theta}{2} - \dfrac{\sin(4\theta)}{8}$$

Now evaluate the integral at the limits of integration.

$$\frac{1}{2}\left(\frac{\theta}{2} - \frac{\sin(4\theta)}{8}\right)\Bigg|_0^{\frac{\pi}{2}} = \frac{1}{2}\left(\frac{\pi}{4} - \frac{\sin(2\pi)}{8}\right) - \frac{1}{2}\left(0 - \frac{\sin(0)}{8}\right) = \frac{\pi}{8}$$

21. **B** In order to find the average value, you use the Mean Value Theorem for integrals, which says that the average value of $f(x)$ on the interval $[a, b]$ is $\dfrac{1}{b-a}\displaystyle\int_a^b f(x)\,dx$.

Here you have $\dfrac{1}{\dfrac{\pi}{4} - \dfrac{\pi}{6}}\displaystyle\int_{\frac{\pi}{6}}^{\frac{\pi}{4}} \sec^2 x\,dx$.

Next, recall that $\dfrac{d}{dx} = \tan x = \sec^2 x$.

You evaluate the integral: $\dfrac{1}{\dfrac{\pi}{4} - \dfrac{\pi}{6}}(\tan x)\Big|_{\frac{\pi}{6}}^{\frac{\pi}{4}} = \dfrac{1}{\dfrac{\pi}{4} - \dfrac{\pi}{6}}\left[\tan\dfrac{\pi}{4} - \tan\dfrac{\pi}{6}\right] = \dfrac{1}{\dfrac{\pi}{4} - \dfrac{\pi}{6}}\left(1 - \dfrac{\sqrt{3}}{3}\right)$.

Next, you need to do a little algebra. Get a common denominator for each of the two expressions.

$$\frac{1}{\dfrac{\pi}{4} - \dfrac{\pi}{6}}\left(1 - \frac{\sqrt{3}}{3}\right) - \frac{1}{\dfrac{6\pi}{24} - \dfrac{4\pi}{24}}\left(\frac{3}{3} - \frac{\sqrt{3}}{3}\right)$$

You can simplify this to $\dfrac{1}{\dfrac{2\pi}{24}}\left(\dfrac{3 - \sqrt{3}}{3}\right) = \dfrac{12}{\pi}\left(\dfrac{3 - \sqrt{3}}{3}\right) = \dfrac{12 - 4\sqrt{3}}{\pi}$.

22. **A** You can find the length of a parametric curve on the interval $[a, b]$ by evaluating the integral.

$$\int_a^b \sqrt{\left(\frac{dx}{dt}\right)^2 + \left(\frac{dy}{dt}\right)^2}\,dt$$

First, take the derivatives: $\dfrac{dx}{dt} = t$ and $\dfrac{dy}{dt} = \sqrt{(6t + 9)}$.

Now, square the derivatives: $\left(\dfrac{dx}{dt}\right)^2 = t^2$ and $\left(\dfrac{dy}{dt}\right)^2 = 6t + 9$.

Now, plug this into the formula to get

$$\int_0^2 \sqrt{(t^2 + 6t + 9)}\,dt = \int_0^2 \sqrt{(t + 3)^2}\,dt = \int_0^2 (t + 3)\,dt = \left(\frac{t^2}{2} + 3t\right)\Bigg|_0^2 = 8$$

23. **C** A function is decreasing on an interval where the derivative is negative.

The derivative is $f'(x) = 4x^3 + 12x^2$.

Next, you want to determine on which intervals the derivative of the function is positive and on which it is negative. You do this by finding where the derivative is zero.

$$4x^3 + 12x^2 = 0$$
$$4x^2(x + 3) = 0$$
$$x = -3 \text{ or } x = 0$$

You can test where the derivative is positive and negative by picking a point in each of the three regions $-\infty < x < -3$, $-3 < x < 0$, and $0 < x < \infty$, plugging the point into the derivative, and seeing what the sign of the answer is. You should find that the derivative is negative on the interval $-\infty < x < -3$.

24. **C** Take a look at each series (the series are duplicated below).

I. $\displaystyle\sum_{n=1}^{\infty} \frac{(-1)^n}{n}$ II. $\displaystyle\sum_{n=1}^{\infty} \frac{1}{\sqrt{n^3}}$ III. $\displaystyle\sum_{n=1}^{\infty} \frac{1}{\sqrt[3]{n^2}}$

<u>Series I:</u> You might recognize this as the alternating harmonic series, which converges. If you don't, use the alternating series test, which says, in order for an alternating series $\displaystyle\sum_{n=1}^{\infty} (-1)^n a_n$ to converge: (1) $a_{n+1} < a_n$; (2) $\displaystyle\lim_{n \to \infty} a_n = 0$; and (3) $a_n > 0$. This series satisfies these conditions, so it converges.

<u>Series II:</u> You can rewrite this series as $\displaystyle\sum_{n=1}^{\infty} \frac{1}{n^{\frac{3}{2}}}$, which is a p-series. A p-series is a series of the form $\displaystyle\sum_{n=1}^{\infty} \frac{1}{n^p}$, which converges if $p > 1$ and diverges if $p < 1$. Thus, this series converges.

<u>Series III:</u> You can rewrite this series as $\displaystyle\sum_{n=1}^{\infty} \frac{1}{n^{\frac{2}{3}}}$, which is also a p-series. Because $p < 1$, this series diverges.

25. **A** You can solve this differential equation by separation of variables.

$$\int \frac{dz}{z\left(4 - \dfrac{z}{100}\right)} = \int dt$$

The integral on the right is trivial. We get $t + C$.

The one on the left will require the Method of Partial Fractions. First, separate the denominator into its two components and place the constants A and B in the numerators of the fractions and the components into the denominators. Set their sum equal to the original rational expression:

$$\frac{A}{z} + \frac{B}{4 - \frac{z}{100}} = \frac{1}{z\left(4 - \frac{z}{100}\right)}$$

Now, solve for the constants A and B. First, multiply through by $z\left(4 - \frac{z}{100}\right)$ to clear the denominators.

$$z\left(4 - \frac{z}{100}\right)\left[\frac{A}{z} + \frac{B}{\left(4 - \frac{z}{100}\right)}\right] = 1$$

$$A\left(4 - \frac{z}{100}\right) + Bz = 1$$

Now, distribute, then group, the terms on the left side.

$$4A - A\frac{z}{100} + Bz = 1$$

$$z\left(B - \frac{A}{100}\right) + 4A = 1$$

In order for this last equation to be true, you need $4A = 1$ and $B - \frac{A}{100} = 0$.

If you solve these simultaneous equations, you get $A = \frac{1}{4}$ and $B = \frac{1}{400}$.

Now that you have done the partial fraction decomposition, you can rewrite the original integral as

$\int \frac{\frac{1}{4}}{z} + \frac{\frac{1}{400}}{\left(4 - \frac{z}{100}\right)} \, dz$. This is now simple to evaluate.

$$\int \frac{\frac{1}{4}}{z} + \frac{\frac{1}{400}}{\left(4 - \frac{z}{100}\right)} \, dz = \frac{1}{4}\int \frac{dz}{z} + \frac{1}{400} \int \frac{dz}{\left(4 - \frac{z}{100}\right)} = \frac{1}{4}\ln|z| - \frac{1}{4}\ln\left|4 - \frac{z}{100}\right| + C$$

You can rewrite this with the laws of logarithms to get

$$\frac{1}{4}\ln|z| - \frac{1}{4}\ln\left|4 - \frac{z}{100}\right| = \ln\left(\frac{100z}{400 - z}\right)^{\frac{1}{4}}$$

Thus, the solution to the differential equation is $\ln\left(\dfrac{100z}{400-z}\right)^{\frac{1}{4}} = t + C$. You will need to do some

algebra to rearrange the equation. First, exponentiate both sides to base e.

$$\left(\dfrac{100z}{400-z}\right)^{\frac{1}{4}} = e^{t+C}$$

Then, because $e^{t+C} = e^{t}e^{C}$, and because e^{C} is a constant, you get $\left(\dfrac{100z}{400-z}\right)^{\frac{1}{4}} = Ce^{t}$.

Next, raise both sides to the fourth power: $\left(\dfrac{100z}{400-z}\right) = Ce^{4t}$.

Invert both sides: $\left(\dfrac{400-z}{100z}\right) = Ce^{-4t}$.

Break the left side into two fractions: $\dfrac{4}{z} - \dfrac{1}{100} = Ce^{-4t}$.

Add $\dfrac{1}{100}$ to both sides: $\dfrac{4}{z} = \dfrac{1}{100} + Ce^{-4t} = \dfrac{Ce^{-4t}+1}{100}$.

Invert both sides (again): $\dfrac{z}{4} = \dfrac{100}{Ce^{-4t}+1}$.

Finally: $z(t) = \dfrac{400}{Ce^{-4t}+1}$ (Whew!).

Now, take the limit: $\displaystyle\lim_{t\to\infty} \dfrac{400}{Ce^{-4t}+1} = 400$.

26. **C**

Notice that the slope of the differential equation is zero (horizontal tangent) at the origin. This eliminates (A) and (B) because they are undefined at the origin (so they would show a vertical tangent there). Finally, notice that the slope is negative at $(0, 1)$, which eliminates (D), which is positive there.

27. **D** The Mean Value Theorem for derivatives says that, given a function $f(x)$ which is continuous and differentiable on $[a, b]$, then there exists some value c on (a, b) where

$$\frac{f(b) - f(a)}{b - a} = f'(c)$$

Here you have $\dfrac{f(b) - f(a)}{b - a} = \dfrac{f(5) - f(0)}{5 - 0} = \dfrac{95 - 0}{5} = 19$.

Plus, $f'(c) = 3c^2 - 6$, so simply set $3c^2 - 6 = 19$. If you solve for c, you get $c = \pm\dfrac{5}{\sqrt{3}}$. Both of these values satisfy the Mean Value Theorem for derivatives, but only the positive value, $c = \dfrac{5}{\sqrt{3}}$, is in the interval.

28. **D** If you want to find where $g(x)$ is a minimum, you can look at $g'(x)$. The Second Fundamental Theorem of Calculus tells you how to find the derivative of an integral: $\dfrac{d}{dx}\displaystyle\int_c^x f(t)\,dt = f(x)$, where c is a constant. Thus, $g'(x) = f(x)$. The graph of f is zero at $x = 0$, $x = 2$, and $x = 4$. You can eliminate $x = 0$ because you are looking for a *positive* value of x. Next, notice that f is negative to the left of $x = 4$ and positive to the right of $x = 4$. Thus, $g(x)$ has a minimum at $x = 4$.

You also could have found the answer geometrically. The function $g(x) = \displaystyle\int_0^x f(t)\,dt$ is called an *accumulation function* and stands for the area between the curve and the x-axis to the point x. Thus, the value of g grows from $x = 0$ to $x = 2$. Then, because you subtract the area under the x-axis from the area above it, the value of g shrinks from $x = 2$ to $x = 4$. The value begins to grow again after $x = 4$.

29. **D** First, break up the integrand into $\displaystyle\int \sec^4 x\,dx = \int \sec^2 x \sec^2 x\,dx$. Next, use the trig identity $1 + \tan^2 x = \sec^2 x$ to rewrite the integrand: $\displaystyle\int \sec^2 x \sec^2 x\,dx = \int (1 + \tan^2 x)\sec^2 x\,dx$.

Now use u-substitution. Let and $u = \tan x$ and $du = \sec^2 x\,dx$.

You get $\displaystyle\int (1 + \tan^2 x)\sec^2 x\,dx = \int (1 + u^2)\,du$.

Integrate: $\displaystyle\int (1 + u^2)\,du = u + \dfrac{u^3}{3} + C$. And substitute back: $\tan x + \dfrac{\tan^3 x}{3} + C$.

30. **A** The acceleration vector is the second derivative of the position vector (the velocity vector is the first derivative).

The velocity vector of this particle is $(2\cos 2t, 2\sin t \cos t)$.

This can be simplified to $(2\cos 2t, \sin 2t)$.

The acceleration vector is $(-4\sin 2t, 2\cos 2t)$.

31.   **A**   The slope of the tangent line is the derivative of the function. You get $f'(x) = 3e^{3x}$. Now set the derivative equal to 2 and solve for x.

$$3e^{3x} = 2$$

$$e^{3x} = \frac{2}{3}$$

$$3x = \ln\frac{2}{3}$$

$$x = \frac{1}{3}\ln\frac{2}{3} \approx -0.135$$

(Remember to round all answers to three decimal places on the AP Exam.)

32.   **B**   The formula for the perimeter of a square is $P = 4s$, where s is the length of a side of the square.

If you differentiate this with respect to t, you get $\dfrac{dP}{dt} = 4\dfrac{ds}{dt}$. We plug in $\dfrac{ds}{dt} = 0.4$, and you get $\dfrac{dP}{dt} = 4(0.4) = 1.6$.

The formula for the area of a square is $A = s^2$. If we solve the perimeter equation for s in terms of P and substitute it into the area equation, you get

$$s = \frac{P}{4}, \text{ so } A = \left(\frac{P}{4}\right)^2 = \frac{P^2}{16}$$

If you differentiate this with respect to t, you get $\dfrac{dA}{dt} = \dfrac{P}{8}\dfrac{dP}{dt}$.

Now plug in $\dfrac{dP}{dt} = 1.6$ to get $\dfrac{dA}{dt} = \dfrac{P}{8}(1.6) = 0.2P$.

33.   **A**   The velocity of the mass is the first derivative of the height: $v(t) = 8\sin(2t)$.

Now graph the equation to find how many times the graph of $v(t)$ crosses the t-axis between $t = 0$ and $t = 2$. Or you should know that this is a sine graph with an amplitude of 8 and a period of π, which will cross the t-axis once on the interval at $t = \dfrac{\pi}{2}$.

34.   **C**   Notice how this limit takes the form of the definition of the derivative, which is

$$f'(x) = \lim_{h \to 0} \frac{f(x+h) - f(x)}{h}$$

Here, if you think of $f(x)$ as $\tan^{-1} x$, then this expression gives the derivative of $\tan^{-1} x$ at the point $x = 1$.

The derivative of $\tan^{-1} x$ is $f'(x) = \dfrac{1}{1 + x^2}$. At $x = 1$, you get

$$f'(1) = \frac{1}{2}$$

35.   **B**   The Trapezoid Rule enables us to approximate the area under a curve with a fair degree of accuracy. The rule says that the area between the x-axis and the curve $y = f(x)$, on the interval $[a, b]$, with n trapezoids, is

$$\frac{1}{2} \frac{b-a}{n} [y_0 + 2y_1 + 2y_2 + 2y_3 + \ldots + 2y_{n-1} + y_n]$$

Using the rule here, with $n = 4$, $a = 0$, and $b = 3$, you get

$$\frac{1}{2} \cdot \frac{3}{4} \left[e^0 + 2e^{\frac{3}{4}} + 2e^{\frac{6}{4}} + 2e^{\frac{9}{4}} + e^3 \right] \approx 19.972$$

36.   **C**   Use implicit differentiation to find $\dfrac{dy}{dx}$: $x^2 \dfrac{dy}{dx} + 2xy + 2x = 2y \dfrac{dy}{dx}$.

Now isolate $\dfrac{dy}{dx}$, which will take some algebra.

First, put all of the terms containing $\dfrac{dy}{dx}$ on one side of the equals sign and all of the other terms on the other side: $2xy + 2x = 2y \dfrac{dy}{dx} - x^2 \dfrac{dy}{dx}$.

Next, factor $\dfrac{dy}{dx}$ out of the right hand side: $2xy + 2x = \dfrac{dy}{dx} (2y - x^2)$.

Finally, divide both sides by $(2y - x^2)$ to isolate $\dfrac{dy}{dx}$: $\dfrac{dy}{dx} = \dfrac{2xy + 2x}{2y - x^2}$.

At $(1, 1)$, you get $\dfrac{dy}{dx} = \dfrac{2 + 2}{2 - 1} = 4$.

Now, find the second derivative by again performing implicit differentiation.

$$\frac{d^2 y}{dx^2} = \frac{\left(2y - x^2\right)\left(2x \dfrac{dy}{dx} + 2y + 2\right) - \left(2xy + 2x\right)\left(2 \dfrac{dy}{dx} - 2x\right)}{\left(2y - x^2\right)^2}$$

At (1, 1), you get $\dfrac{d^2y}{dx^2} = \dfrac{(2-1)(8+2+2)-(2+2)(8-2)}{(2-1)^2} = \dfrac{12-24}{1} = -12$.

37. **C** Because the integral of a function can be interpreted as the area between the function and the curve, you can say that $\int_a^b f(x)\,dx = \int_a^c f(x)\,dx + \int_c^b f(x)\,dx$, where c is a point in the interval $[a, b]$. Here you have $\int_{-2}^4 f(x)\,dx = \int_{-2}^3 f(x)\,dx + \int_3^4 f(x)\,dx$. You can rearrange this to

$\int_{-2}^4 f(x)\,dx - \int_3^4 f(x)\,dx = \int_{-2}^3 f(x)\,dx$. This means that $a - b = \int_{-2}^3 f(x)\,dx$. Finally, you know

that $\int_a^b f(x)\,dx = -\int_b^a f(x)\,dx$, so $b - a = \int_3^{-2} f(x)\,dx$.

38. **B** Remember that $\int_a^b f(x)\,dx + \int_b^c f(x)\,dx = \int_a^c f(x)\,dx$. You can use that here: $\int_1^4 f(x)\,dx + \int_4^{11} f(x)\,dx = \int_1^{11} f(x)\,dx$. Now simply substitute the values for the respective integrals: $\int_1^4 f(x)\,dx + 20 = 24$. Therefore, $\int_1^4 f(x)\,dx = 4$.

39. **A** You are given that $\sin x = x - \dfrac{x^3}{3!} + \dfrac{x^5}{5!} + \ldots$

Here, simply substitute 0.4 for x in the series to get

$$\sin 0.4 = 0.4 - \dfrac{(0.4)^3}{3!} + \dfrac{(0.4)^5}{5!} + \ldots$$

Find the value on your calculator. Keep adding terms until you get four decimal places of accuracy. You should get $\sin(0.4) = 0.3894$. If you got only 0.3893, you didn't use enough terms (you need to use the fifth power term).

40. **D** First, graph the curves.

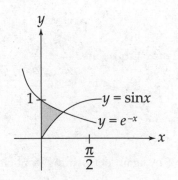

You can find the volume by taking a vertical slice of the region. The formula for the volume of a solid of revolution around the y-axis, using a vertical slice bounded from above by the curve $f(x)$ and from below by $g(x)$, on the interval $[a, b]$, is

$$\pi \int_a^b \left[f(x)^2 - g(x)^2 \right] dx$$

The upper curve is $y = e^{-x}$, and the lower curve is $y = \sin x$.

Next, you need to find the point(s) of intersection of the two curves with a calculator (Good luck doing it by hand!), which you do by setting them equal to each other and solving for x. You should get approximately: $x = 0.589$ (remember to round to three decimal places on the AP Exam).

Thus, the limits of integration are $x = 0$ and $x = 0.589$.

Now, evaluate the integral.

$$\pi \int_0^{0.589} \left[(e^{-x})^2 - (\sin^2 x) \right] dx = \pi \int_0^{0.589} \left[e^{-2x} - \sin^2 x \right] dx$$

Evaluate this integral by hand but, because you will need a calculator to find the answer, you might as well use it to evaluate the integral.

You get $\pi \int_0^{0.589} (e^{-2x} - \sin^2 x)\, dx = 0.888$ (rounded to three decimal places).

41. **C** You can use Euler's Method to find an approximate answer to the differential equation. The method is quite simple. First, you need a starting point, (x_0, y_0) and an initial slope, y_0'. Next, use increments of h to come up with approximations. Each new approximation will use the following rules:

$$x_n = x_{n-1} + h$$

$$y_n = y_{n-1} + h \cdot y_{n-1}'$$

Repeat for $n = 1, 2, 3, \ldots$

You are given that the curve goes through the point $(2, 6)$. You will call the coordinates of this point $x_0 = 2$ and $y_0 = 6$. The slope is found by plugging these coordinates into $y' = 2y - 4x$, so you have an initial slope of $y_0' = 4$.

Now, find the next set of points.

Step 1: Increase x_0 by h to get x_1: $x_1 = 2.2$.

Step 2: Multiply h by y_0' and add to y_0 to get y_1: $y_1 = 6 + 0.2(4) = 6.8$.

Step 3: Find y_1' by plugging x_1 and y_1 into the equation for y'.

$$y_1' = 2(6.8) - 4(2.2) = 4.8$$

Repeat until you get to the desired point (in this case $x = 3$).

Step 1: Increase x_1 by h to get x_2: $x_2 = 2.4$.

Step 2: Multiply h by y_1' and add to y_1 to get y_2: $y_2 = 6.8 + 0.2(4.8) = 7.76$.

Step 3: Find y_2' by plugging x_2 and y_2 into the equation for y'.

$$y_2' = 2(7.76) - 4(2.4) = 5.92$$

Step 1: $x_3 = x_2 + h$: $x_3 = 2.6$

Step 2: $y_3 = y_2 + h(y_2')$: $y_3 = 7.76 + 0.2(5.92) = 8.944$

Step 3: $y_3' = 2(y_3) - 4(x_3)$: $y_3' = 2(8.944) - 4(2.6) = 7.488$

Step 1: $x_4 = x_3 + h$: $x_4 = 2.8$

Step 2: $y_4 = y_3 + h(y_3')$: $y_4 = 8.944 + 0.2(7.488) = 10.4416$

Step 3: $y_4' = 2(y_4) - 4(x_4)$: $y_4' = 2(10.4416) - 4(2.8) = 9.6832$

Step 1: $x_5 = x_4 + h$: $x_5 = 3$

Step 2: $y_5 = y_4 + h(y_4')$: $y_5 = 10.4416 + 0.2(9.6832) = 12.378$

42. **C** First, break up the integrand into: $\int \tan^4 x \, dx = \int \tan^2 x \tan^2 x \, dx$. Next, use the trig

identity $\tan^2 x = \sec^2 x - 1$ to replace one of the $\tan^2 x$ terms:

$\int \tan^2 x \tan^2 x \, dx = \int \tan^2 x (\sec^2 x - 1) \, dx = \int \tan^2 x \sec^2 x \, dx - \int \tan^2 x \, dx$. Then, use the identity

to replace the $\tan^2 x$ in the second integral: $\int \tan^2 x \sec^2 x \, dx - \int \sec^2 x - 1 \, dx$. We can

evaluate the first integral using u substitution. Let $u = \tan x$ and $du = \sec^2 x$. Substitute:

$\int \tan^2 x \sec^2 x \, dx = \int u^2 \, du$. Integrate: $\int u^2 \, du = \dfrac{u^3}{3}$. And substitute back: $\dfrac{u^3}{3} = \dfrac{\tan^3 x}{3}$.

The second integral is simple: $\int \sec^2 x - 1 \, dx = \tan x - x$. Putting them together, we get:

$\int \tan^2 x \sec^2 x \, dx - \int \sec^2 x - 1 \, dx = \dfrac{\tan^3 x}{3} - \tan x + x + C$.

The answer is (C).

43. **D** You find $\int \cot x \, dx$ by rewriting the integral as $\int \dfrac{\cos x}{\sin x} \, dx$. Then use u-substitution. Let $u = \sin x$ and $du = \cos x$. Substituting, you can get

$$\int \frac{\cos x}{\sin x} \, dx = \int \frac{du}{u} = \ln|u| + C$$

Then, substituting back, you get $\ln(\sin x) + C$. (You can get rid of the absolute value bars because sine is always positive on the interval.) Next, use $f\left(\dfrac{\pi}{6}\right) = 1$ to solve for C.

You get

$$1 = \ln\left(\sin\frac{\pi}{6}\right) + C$$

$$1 = \ln\left(\frac{1}{2}\right) + C$$

$$1 - \ln\left(\frac{1}{2}\right) = C = 1.693147$$

Thus, $f(x) = \ln(\sin x) + 1.693147$.

At $x = 1$, you get $f(1) = \ln(\sin 1) + 1.693147 = 1.521$ (rounded to three decimal places).

44. **C** You solve an integral of the form $\sqrt{a^2 - x^2}$ by performing the trig substitution $x = a\sin\theta$. Here use $x = 2\sin\theta$, which means that $dx = 2\cos\theta \, d\theta$. You get $\int\left[\sqrt{4 - 4\sin^2\theta}\,(2\cos\theta)\right] d\theta$. Next, factor out the 4: $\int 4\sqrt{1 - \sin^2\theta}\,\cos\theta \, d\theta$.

Next, simplify the radicand: $\int 4\sqrt{\cos^2\theta}\,\cos\theta \, d\theta$, which gives $4\int\cos^2\theta \, d\theta$. Now, you need to use another trig identity. Recall that $\cos 2\theta = 2\cos^2\theta - 1$, which can be rewritten as $\cos^2\theta = \dfrac{1 + \cos 2\theta}{2}$.

Now, rewrite the integral: $4\int\cos^2\theta \, d\theta = 4\int\dfrac{1 + \cos 2\theta}{2} \, d\theta = 2\int(1 + \cos 2\theta) \, d\theta$.

Integrate: $2\int(1 + \cos 2\theta) \, d\theta = 2\theta + \sin 2\theta + C$.

Now, you have to substitute back.

Because $x = 2\sin\theta$, you know that $\theta = \sin^{-1}\left(\dfrac{x}{2}\right)$ and that $\cos\theta = \dfrac{\sqrt{4 - x^2}}{2}$.

This gives $2\theta + \sin 2\theta + C = 2\sin^{-1}\left(\dfrac{x}{2}\right) + \dfrac{x}{2}\sqrt{4 - x^2} + C$.

45. **D** Hooke's Law says that the force needed to compress or stretch a spring from its natural state is $F = kx$, where k is the spring constant. You can find the value of k from the initial information, namely, $F = 250$ N for a stretch of 5 m. Thus, solve to find the value of k: $k = 250/5 = 50$ N/m.

You find the work done by a variable force along the x-axis from $x = a$ to $x = b$ by evaluating the integral for the Work, $W = \int_a^b F(x)\, dx$.

Using the information you have, $W = \int_a^b F(x)\, dx = \frac{1}{2}kx^2\Big|_0^7 = \frac{1}{2}50(7)^2 = 1{,}225$ N.

ANSWERS AND EXPLANATIONS TO SECTION II

1. An object moving along a curve in the xy-plane has its position given by $(x(t),\, y(t))$ at time t seconds, $0 \le t \le 1$, with $\dfrac{dx}{dt} = 8t\cos t$ units/sec and $\dfrac{dy}{dt} = 8t\sin t$ units/sec.

   At time $t = 0$, the object is located at $(5, 11)$.

   (a) Find the speed of the object at time $t = 1$.

   The equation for the speed of an object is $speed = \sqrt{\left(\dfrac{dx}{dt}\right)^2 + \left(\dfrac{dy}{dt}\right)^2}$.

   Here, you get $speed = \sqrt{\left(8t\cos t\right)^2 + \left(8t\sin t\right)^2}$.

   This can be simplified to $speed = \sqrt{64t^2\cos^2 t + 64t^2\sin^2 t} = 8t\sqrt{\cos^2 t + \sin^2 t} = 8t$.

   Thus, the speed of the object at time $t = 1$ is 8.

   (b) Find the length of the arc described by the curve's position from time $t = 0$ to time $t = 1$.

   You can find the length of a parametric curve $(x(t),\, y(t))$, on the interval $[a, b]$, by evaluating the integral: $\int_a^b \sqrt{\left(\dfrac{dx}{dt}\right)^2 + \left(\dfrac{dy}{dt}\right)^2}\, dt$.

   You found the integrand in part (a): $\sqrt{\left(\dfrac{dx}{dt}\right)^2 + \left(\dfrac{dy}{dt}\right)^2} = 8t$. Thus, all you have to do is integrate this with respect to t from $t = 0$ to $t = 1$.

   You get $\int_0^1 8t\, dt = (4t^2)\Big|_0^1 = 4$.

(c) Find the location of the object at time $t = \dfrac{\pi}{2}$.

First, you will need to find the equations for the coordinates, which you can do by solving the differential equations. Let's find x: $\dfrac{dx}{dt} = 8t\cos t$.

You can solve this by separation of variables. First, move dt to the right side of the equals sign: $dx = 8t\cos t\, dt$.

Integrate both sides (pulling the constant out of the integrand): $\displaystyle\int dx = 8\int t\cos t\, dt$.

The integral on the left is trivial: $\displaystyle\int dx = x$.

You will need to use integration by parts to solve the integral on the right.

The rule for integration by parts says that $\displaystyle\int u\, dv = uv - \int v\, du$.

Here, let $u = t$ and $dv = \cos t\, dt$. Then, $du = dt$ and $v = \sin t$.

Substituting the terms you get $8\displaystyle\int t\cos t\, dt = 8t\sin t - 8\int \sin t\, dt$.

Now, integrate the second term, which gives

$$8\int t\cos t\, dt = 8t\sin t + 8\cos t + C$$

Thus, the x-coordinate is $x(t) = 8t\sin t + 8\cos t + C$.

Now, plug in the initial condition: $5 = (8)(0)\sin(0) + (8)\cos(0) + C$.

This means that $C = -3$, so the equation for the x-coordinate is

$$x(t) = 8t\sin t + 8\cos t - 3$$

Now, find y: $\dfrac{dy}{dt} = 8t\sin t$.

First, move dt to the right side of the equals sign: $dy = 8t\sin t\, dt$.

Integrate both sides (pulling the constant out of the integrand): $\displaystyle\int dy = 8\int t\sin t\, dt$.

The integral on the left is trivial: $\displaystyle\int dy = y$.

Again, you will need to use integration by parts to solve the integral on the right.

Here, let $u = t$ and $dv = \sin t\, dt$. Then, $du = dt$ and $v = -\cos t$.

Substituting the terms, you get $8\displaystyle\int t\sin t\, dt = -8t\cos t + 8\int \cos t\, dt$.

Now, integrate the second term, which gives

$$8 \int t\sin t \, dt = -8t\cos t + 8 \sin t + C$$

Thus, the y-coordinate is $y(t) = -8t\cos t + 8\sin t + C$.

Now, plug in the initial condition: $11 = -(8)(0)\cos(0) + (8)\sin(0) + C$.

This means that $C = 11$, so the equation for the y-coordinate is

$$y(t) = -8t\cos t + 8\sin t + 11$$

Finally, plug $t = \dfrac{\pi}{2}$ into the equations for the coordinates.

$$x\left(\frac{\pi}{2}\right) = (8)\left(\frac{\pi}{2}\right)\sin\left(\frac{\pi}{2}\right) + (8)\cos\left(\frac{\pi}{2}\right) - 3 = 4\pi - 3$$

$$y\left(\frac{\pi}{2}\right) = -(8)\left(\frac{\pi}{2}\right)\cos\left(\frac{\pi}{2}\right) + (8)\sin\left(\frac{\pi}{2}\right) + 11 = 19$$

Therefore, the object's location at time $t = \dfrac{\pi}{2}$ is $(4\pi - 3, 19)$.

2.

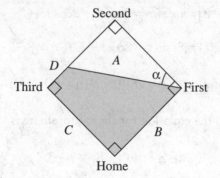

A baseball diamond is a square with each side 90 feet in length. A player runs from second base to third base at a rate of 18 ft/sec.

(a) At what rate is the player's distance from first base, A, changing when his distance from third base, D, is 22.5 feet?

A is related to D by the Pythagorean Theorem: $90^2 + (90 - D)^2 = A^2$. This can be simplified to $16{,}200 - 180D + D^2 = A^2$.

Take the derivative of both sides with respect to t: $\quad -180\dfrac{dD}{dt} + 2D\dfrac{dD}{dt} = 2A\dfrac{dA}{dt}$.

Now, you are given that $\dfrac{dD}{dt} = -18$ (it's negative because D is shrinking) and $D = 22.5$.

Next, you need to solve for A: $90^2 + (90 - 22.5)^2 = A^2$. You should get $A = 112.5$.

Now, plug in and solve for $\dfrac{dA}{dt}$.

$$-180(-18) + 2(22.5)(-18) = 2(112.5)\dfrac{dA}{dt}$$

$$3,240 - 810 = 225\dfrac{dA}{dt}$$

$$\dfrac{dA}{dt} = 10.8 \text{ ft/sec}$$

(b) At what rate is angle α increasing when D is 22.5 feet?

Notice that $\tan \alpha = \dfrac{90 - D}{90} = 1 - \dfrac{D}{90}$. You differentiate both sides with respect to t.

$$\sec^2 \alpha \, \dfrac{d\alpha}{dt} = -\dfrac{1}{90}\dfrac{dD}{dt}$$

Next, you need to solve for $\sec^2 \alpha$ when $D = 22.5$. From part (a), you know that $A = 112.5$, so $\sec^2 \alpha = \dfrac{112.5}{90}$, so $\sec^2 \alpha = \dfrac{25}{16}$.

Now, plug in to solve for $\dfrac{d\alpha}{dt}$: $\left(\dfrac{25}{16}\right)\dfrac{d\alpha}{dt} = -\dfrac{1}{90}(-18)$.

$$\dfrac{d\alpha}{dt} = \dfrac{16}{125} = 0.128 \text{ radians/sec}$$

(c) At what rate is the area of the trapezoidal region, formed by line segments A, B, C, and D, changing when D is 22.5 feet?

The area of the trapezoid is $a = \dfrac{1}{2}C(B + D)$. Notice that B and C are constants. Differentiate both sides with respect to t: $\dfrac{da}{dt} = \dfrac{1}{2}C\dfrac{dD}{dt}$.

Now, plug in and solve for $\dfrac{da}{dt}$: $\dfrac{da}{dt} = \dfrac{1}{2}(90)(-18) = -810 \text{ ft}^2/\text{sec}$.

3. Consider the equation $x^2 - 2xy + 4y^2 = 64$.

   (a) Write an expression for the slope of the curve at any point (x, y).

   Step 1: The slope of the curve is just the derivative. But, here you have to use implicit differentiation to find the derivative. If you take the derivative of each term with respect to x, you get

   $$2x\frac{dx}{dx} - 2\left(x\frac{dy}{dx} + y\frac{dx}{dx}\right) + 8y\frac{dy}{dx} = 0$$

   Remember that $\frac{dx}{dx} = 1$, which gives

   $$2x - 2\left(x\frac{dy}{dx} + y\right) + 8y\frac{dy}{dx} = 0$$

   Step 2: Now just simplify and solve for $\frac{dy}{dx}$.

   $$2x - 2x\frac{dy}{dx} - 2y + 8y\frac{dy}{dx} = 0$$

   $$x - x\frac{dy}{dx} - y + 4y\frac{dy}{dx} = 0$$

   $$-x\frac{dy}{dx} + 4y\frac{dy}{dx} = y - x$$

   $$\left(4y - x\right)\frac{dy}{dx} = y - x$$

   $$\frac{dy}{dx} = \frac{y - x}{4y - x}$$

   (b) Find the equation of the tangent lines to the curve at the point $x = 2$.

   Step 1: You are going to use the point-slope form of a line, $y - y_1 = m(x - x_1)$, where (x_1, y_1) is a point on the curve, and the derivative at that point is the slope m. First, you need to know the value of y when $x = 2$. If you plug 2 for x into the original equation, you get

   $$4 - 4y + 4y^2 = 64$$

   $$4y^2 - 4y - 60 = 0$$

   Using the quadratic formula, you get

   $$y = \frac{1 \pm \sqrt{61}}{2} \approx 4.41, -3.41$$

Notice that there are two values of y when $x = 2$, which is why there are two tangent lines.

Step 2: Now that you have our points, you need the slope of the tangent line at $x = 2$.

$$\frac{dy}{dx} = \frac{y - x}{4y - x}$$

At $y = 4.41$, $\dfrac{dy}{dx} = \dfrac{4.41 - 2}{4(4.41) - 2} = 0.15$

At $y = -3.41$, $\dfrac{dy}{dx} = \dfrac{-3.41 - 2}{4(-3.41) - 2} = 0.35$

Step 3: Plugging into our equation for the tangent line, you get

$$y - 4.41 = 0.15(x - 2)$$

$$y + 3.41 = 0.35(x - 2)$$

It is not necessary to simplify these equations.

(c)   Find $\dfrac{d^2 y}{dx^2}$ at $(0, 4)$.

Step 1: Once you have the first derivative, you have to differentiate again to find $\dfrac{d^2 y}{dx^2}$.

But you have to use implicit differentiation again.

$$\frac{dy}{dx} - \frac{y - x}{4y - x}$$

Using the Quotient Rule,

$$\frac{d^2 y}{dx^2} = \frac{(4y - x)\left(\dfrac{dy}{dx} - \dfrac{dx}{dx}\right) - (y - x)\left(4\dfrac{dy}{dx} - \dfrac{dx}{dx}\right)}{(4y - x)^2}$$

Simplifying, you get

$$\frac{d^2 y}{dx^2} = \frac{(4y - x)\left(\dfrac{dy}{dx} - 1\right) - (y - x)\left(4\dfrac{dy}{dx} - 1\right)}{(4y - x)^2}$$

Now plug in $\dfrac{y-x}{4y-x}$ for $\dfrac{dy}{dx}$, which gives

$$\frac{d^2y}{dx^2} = \frac{(4y-x)\left(\dfrac{y-x}{4y-x}-1\right)-(y-x)\left(4\dfrac{y-x}{4y-x}-1\right)}{(4y-x)^2}$$

Now you would have to use a lot of algebra to simplify this but, fortunately, you can just plug (0, 4) in immediately for x and y, and solve from there.

$$\frac{d^2y}{dx^2} = \frac{(16)\left(\dfrac{4}{16}-1\right)-(4)\left(4\dfrac{4}{16}-1\right)}{(16)^2} = \frac{-3}{64}$$

4.

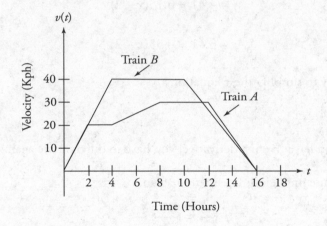

Three trains, A, B, and C, each travel on a straight track for $0 \le t \le 16$ hours. The graphs above, which consist of line segments, show the velocities, in kilometers per hour, of trains A and B. The velocity of C is given by $v(t) = 8t - 0.25t^2$.

(Indicate units of measure for all answers.)

(a) Find the velocities of A and C at time $t = 6$ hours.

You can find the velocity of train A at time $t = 6$ simply by reading the graph. You get $v_A(6) = 25$ km/hr. Find the velocity of train C at time $t = 6$ by plugging $t = 6$ into the formula. You get $v_C(6) = 8(6) - 0.25(6^2) = 39$ kilometers per hour.

(b) Find the accelerations of B and C at time $t = 6$ hours.

Acceleration is the derivative of velocity with respect to time. For train B, you look at the slope of the graph at time $t = 6$. You get $a_B(6) = 0$ km/hr^2. For train C, take the derivative of v. You get $a(t) = 8 - 0.5t$. At time $t = 6$, you get $a_C(6) = 5$ km/hr^2.

(c) Find the positive difference between the total distance that A traveled and the total distance that B traveled in 16 hours.

In order to find the total distance that train A traveled in 16 hours, you need to find the area under the graph. You can find this area by adding up the areas of the different geometric objects that are under the graph. From time $t = 0$ to $t = 2$, you need to find the area of a triangle with a base of 2 and a height of 20. The area is 20. Next, from time $t = 2$ to $t = 4$, you need to find the area of a rectangle with a base of 2 and a height of 20. The area is 40. Next, from time $t = 4$ to $t = 8$, you need to find the area of a trapezoid with bases of 20 and 30 and a height of 4. The area is 100. Next, from time $t = 8$ to $t = 12$, you need to find the area of a rectangle with a base of 4 and a height of 30. The area is 120. Finally, from time $t = 12$ to $t = 16$, you need to find the area of a triangle with a base of 4 and a height of 30. The area is 60. Thus, the total distance that train A traveled is 340 km.

Let's repeat the process for train B. From time $t = 0$ to $t = 4$, you need to find the area of a triangle with a base of 4 and a height of 40. The area is 80. Next, from time $t = 4$ to $t = 10$, you need to find the area of a rectangle with a base of 6 and a height of 40. The area is 240. Finally, from time $t = 10$ to $t = 16$, you need to find the area of a triangle with a base of 6 and a height of 40. The area is 120. Thus, the total distance that train B traveled is 440 km.

Therefore, the positive difference between their distances is 100 km.

(d) Find the total distance that C traveled in 16 hours.

First, note that the graph of train C's velocity, $v(t) = 8t - 0.25t^2$, is above the x-axis on the entire interval. Therefore, in order to find the total distance traveled, you integrate $v(t)$ over the interval.

You get $\int_0^{16} (8t \quad 0.25t^2)\, dt$.

Evaluate the integral: $\int_0^{16} (8t - 0.25t^2)\, dt = \left(4t^2 - \frac{t^3}{12} \right)_0^{16} = \frac{2{,}048}{3}$ km.

5. Let y be the function satisfying $f'(x) = x(1 - f(x)); f(0) = 10$.

(a) Use Euler's Method, starting at $x = 0$, with step size of 0.5 to approximate $f(x)$ at $x = 1$.

You use Euler's Method to find an approximate answer to the differential equation. First, you need a starting point (x_0, y_0) and an initial slope, y_0'. (Note that here $f(x) = y$.) Next, use increments of h to come up with approximations. Each new approximation will use the following rules:

$$x_n = x_{n-1} + h$$

$$y_n = y_{n-1} + h \cdot y_{n-1}'$$

Repeat for $n = 1, 2, 3, \ldots$

You are given that the curve goes through the point (0, 10). You will call the coordinates of this point $x_0 = 0$ and $y_0 = 10$. The slope is found by plugging these coordinates into $y' = x(1 - y)$, so you have an initial slope of $y'_0 = 0$.

Now find the next set of points.

Step 1: Increase x_0 by h to get x_1: $x_1 = 0.5$.

Step 2: Multiply h by y'_0 and add to y_0 to get y_1: $y_1 = 10 + 0.5(0) = 10$.

Step 3: Find y'_1 by plugging x_1 and y_1 into the equation for y'.

$$y'_1 = (0.5)(1 - 10) = -4.5$$

Repeat until you get to the desired point (in this case $x = 1$).

Step 1: Increase x_1 by h to get x_2: $x_2 = 1$.

Step 2: Multiply h by y'_1 and add to y_1 to get y_2.

$$y_2 = 10 + 0.5(-4.5) = 7.75$$

(b)   Find an exact solution for $f(x)$ when $x = 1$.

Here you need to solve the differential equation $\dfrac{dy}{dx} = x(1 - y)$. You can use separation of variables. Move the term containing y to the left side and the dx to the right side.

You get $\dfrac{dy}{(1 - y)} = x\,dx$.

Integrate both sides.

$$\int \frac{dy}{(1 - y)} = \int x\,dx$$

$$-\ln|1 - y| = \frac{x^2}{2} + C$$

It's traditional to isolate y. First, exponentiate both sides to base e: $|1 - y| = e^{-\frac{x^2}{2} + C}$.

Next, use the rules of exponents to rewrite the following: $|1 - y| = e^C e^{-\frac{x^2}{2}}$.

Because e^C is a constant, you can rewrite this as $|1 - y| = Ce^{-\frac{x^2}{2}}$.

Finally, isolate y: $y = 1 - Ce^{-\frac{x^2}{2}}$.

Now, you use the initial condition to solve for C: $10 = 1 - Ce^0$.

Therefore, $C = -9$.

Thus, because $f(x) = y$, the exact solution is $f(x) = 1 + 9e^{-\frac{x^2}{2}}$.

Therefore, $f(1) = 1 + 9e^{-\frac{1}{2}}$.

(c) Evaluate $\int_0^\infty x(1 - f(x))\, dx$.

Using the Fundamental Theorem of Calculus, you know that

$$\int_a^b f'(x)\, dx = f(b) - f(a)$$

In part (b), you found that $f(x) = 1 + 9e^{-\frac{x^2}{2}}$, so here you need to evaluate f at infinity and at zero.

You can find f at infinity using limits: $\lim_{x\to\infty} f(x) = \lim_{x\to\infty} 1 + 9e^{-\frac{x^2}{2}} = 1$.

You can find f at zero by plugging in: $f(0) = 1 + 9e^0 = 10$.

Therefore, $\int_0^\infty x(1 - f(x))\, dx = 1 - 10 = -9$.

6. Given $f(x) = \tan^{-1}(x)$ and $g(x) = \dfrac{1}{1+x}$, for $|x| \le 1$

(a) Find the fifth-degree Taylor polynomial and general expression for $g(x)$ about $x = 0$.

The formula for a Taylor polynomial of order n about $x = a$ is

$$P_n(x) = f(a) + f'(a)(x - a) + \frac{f''(a)}{2!}(x - a)^2 + \ldots + \frac{f^{(k)}(a)}{k!}(x - a)^k + \ldots + \frac{f^{(n)}(a)}{n!}(x - a)^n$$

First, find the derivatives of $g(x) = \dfrac{1}{1+x}$.

$$g'(x) = -(1 + x)^{-2}$$
$$g''(x) = 2(1 + x)^{-3}$$
$$g^{(3)}(x) = -6(1 + x)^{-4}$$
$$g^{(4)}(x) = 24(1 + x)^{-5}$$
$$g^{(5)}(x) = -120(1 + x)^{-6}$$

Next, evaluate the derivatives about $x = 0$.

$$g(0) = 1;\ g'(0) = -1;\ g''(0) = 2;\ g^{(3)}(0) = -6;\ g^{(4)}(0) = 24;\ g^{(5)}(0) = -120$$

Now, if you plug in the formula for the Taylor polynomial, you get

$$g(x) = \frac{1}{1+x} = 1 - x + x^2 - x^3 + x^4 - x^5$$

The general expression is $g(x) = \sum_{n=0}^{\infty} (-1)^n x^n$.

(b) Given that $\frac{d}{dx} \tan^{-1} x = \frac{1}{1+x^2}$, for $|x| \leq 1$, use the result of part (a) to find the fifth-degree Taylor polynomial and general expression for $f(x)$ about $x = 0$.

You can find the Taylor polynomial for $\frac{1}{1+x^2}$ by substituting x^2 for x in the formula above.

You get $\frac{1}{1+x^2} = 1 - x^2 + x^4 - x^6 + \ldots$ (actually, you don't need the last term). Then, because

$\frac{d}{dx} \tan^{-1} x = \frac{1}{1+x^2}$, if you perform term-by-term integration on the series for $\frac{1}{1+x^2}$, you will

get the Taylor polynomial for $\tan^{-1} x$.

You get $\tan^{-1} x = \int (1 - x^2 + x^4)\, dx = x - \frac{x^3}{3} + \frac{x^5}{5} + C$. Because $\tan^{-1}(0) = 0$, $C = 0$, and thus

the general expression for $f(x)$ is $\tan^{-1} x = \sum_{n=0}^{\infty} (-1)^n \frac{x^{2n+1}}{2n+1}$.

(c) Use the fifth-degree Taylor polynomial to estimate $f\left(\frac{1}{10}\right)$.

You can estimate $f\left(\frac{1}{10}\right)$ simply by plugging into the expression that we found in part (b). You get

$$f\left(\frac{1}{10}\right) = \left(\frac{1}{10}\right) - \frac{\left(\frac{1}{10}\right)^3}{3} + \frac{\left(\frac{1}{10}\right)^5}{5} = \frac{1}{10} - \frac{1}{3,000} + \frac{1}{500,000}, \text{ which you don't need to simplify.}$$

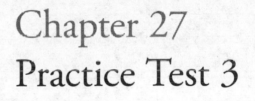

Chapter 27
Practice Test 3

AP® Calculus BC Exam

SECTION I: Multiple-Choice Questions

DO NOT OPEN THIS BOOKLET UNTIL YOU ARE TOLD TO DO SO.

At a Glance

Total Time
1 hour and 45 minutes
Number of Questions
45
Percent of Total Grade
50%
Writing Instrument
Pencil required

Instructions

Section I of this examination contains 45 multiple-choice questions. Fill in only the ovals for numbers 1 through 45 on your answer sheet.

CALCULATORS MAY NOT BE USED IN THIS PART OF THE EXAMINATION.

Indicate all of your answers to the multiple-choice questions on the answer sheet. No credit will be given for anything written in this exam booklet, but you may use the booklet for notes or scratch work. After you have decided which of the suggested answers is best, completely fill in the corresponding oval on the answer sheet. Give only one answer to each question. If you change an answer, be sure that the previous mark is erased completely. Here is a sample question and answer.

Sample Question   Sample Answer

Chicago is a
(A) state
(B) city
(C) country
(D) continent

Use your time effectively, working as quickly as you can without losing accuracy. Do not spend too much time on any one question. Go on to other questions and come back to the ones you have not answered if you have time. It is not expected that everyone will know the answers to all the multiple-choice questions.

About Guessing

Many candidates wonder whether or not to guess the answers to questions about which they are not certain. Multiple choice scores are based on the number of questions answered correctly. Points are not deducted for incorrect answers, and no points are awarded for unanswered questions. Because points are not deducted for incorrect answers, you are encouraged to answer all multiple-choice questions. On any questions you do not know the answer to, you should eliminate as many choices as you can, and then select the best answer among the remaining choices.

CALCULUS BC

SECTION I, Part A

Time—60 Minutes

Number of questions—30

A CALCULATOR MAY NOT BE USED ON THIS PART OF THE EXAMINATION

<u>Directions</u>: Solve each of the following problems, using the available space for scratchwork. After examining the form of the choices, decide which is the best of the choices given and fill in the corresponding oval on the answer sheet. No credit will be given for anything written in the test book. Do not spend too much time on any one problem.

<u>In this test</u>: Unless otherwise specified, the domain of a function f is assumed to be the set of all real numbers x for which $f(x)$ is a real number.

1. If $y = \tan(2\pi x)$, then $\dfrac{dy}{dx} = $?

   (A) $(2\pi)\cot(2\pi x)$

   (B) $(2\pi)\sec^2(2\pi x)$

   (C) $\dfrac{\sec^2(2\pi x)}{2\pi}$

   (D) $\dfrac{\cot(2\pi x)}{2\pi}$

2. $\displaystyle\lim_{x\to 0} \dfrac{\sin x - e^x + 1}{x} =$

   (A)   0
   (B)   1
   (C)   −1
   (D)   The limit does not exist.

GO ON TO THE NEXT PAGE.

3. $\int \tan^2 x \, dx =$

(A) $\tan x - x + C$

(B) $\dfrac{\tan^3 x}{3} + C$

(C) $\tan x + C$

(D) $\tan x + x + C$

---

4. For $0 \le t \le \dfrac{11\pi}{6}$, an object travels along the ellipse given by $x = 2\sin t$, $y = -3\cos t$. At $t = \dfrac{11\pi}{6}$, the object leaves the ellipse and travels along the line tangent to the path at that point. What is the slope of the tangent line at $t = \dfrac{11\pi}{6}$?

(A) $\dfrac{2\sqrt{3}}{3}$

(B) $\dfrac{-3}{2\sqrt{3}}$

(C) $\dfrac{-3\sqrt{3}}{2}$

(D) $\dfrac{-2\sqrt{3}}{3}$

GO ON TO THE NEXT PAGE.

5. If $f(x) = \begin{cases} x^3 - 4x, & x \geq 2 \\ 4x - 2x^2, & x < 2 \end{cases}$, which of the following is true?

(A) $f(x)$ is continuous and differentiable at $x = 2$.
(B) $f(x)$ is differentiable but not continuous at $x = 2$.
(C) $f(x)$ is continuous but not differentiable at $x = 2$.
(D) $f(x)$ is neither continuous nor differentiable at $x = 2$.

6. $\int \tan^2 x \sec^2 x \, dx =$

(A) $\dfrac{\tan^3 x \sec^3 x}{3} + C$

(B) $\tan^2 x + C$

(C) $\dfrac{\sec^3 x}{3} + C$

(D) $\dfrac{\tan^3 x}{3} + C$

GO ON TO THE NEXT PAGE.

7. The position of a particle at time $t \geq 0$, in seconds, is given by $x = 2t^3 + 3t^2 - 12t$ and $y = 2t^3 - 15t^2 + 24t$. At what values of t is the particle at rest?

    (A) $t = 1$
    (B) $t = 1, 2, 4$
    (C) $t = 2$
    (D) $t = 2, -3$

8. $\int 3x \csc^2 x^2 \, dx =$

    (A) $\dfrac{3}{2} \sec^2(x^2) + C$

    (B) $\dfrac{\csc x^3}{x} + C$

    (C) $-\dfrac{3}{2} \cot(x^2) + C$

    (D) $3\sec^2(x^2) + C$

GO ON TO THE NEXT PAGE.

9. If $f(x) = \log_5(x^3 - x - 1)$, then $f'(2) =$

(A) $\dfrac{11}{5\ln 5}$

(B) 11

(C) e^5

(D) $\dfrac{11}{5}$

10. What is the value of $\displaystyle\sum_{n=1}^{\infty} \dfrac{5^{n+1}}{4^n}$?

(A) 1

(B) $\dfrac{2}{3}$

(C) $\dfrac{1}{3}$

(D) Diverges

GO ON TO THE NEXT PAGE.

11. If the MacLaurin series for e^x is $\displaystyle\sum_{n=0}^{\infty} \frac{x^n}{n!}$, what is a power series expansion for e^{x^3} ?

(A) $1 + \dfrac{x^3}{3!} + \dfrac{x^6}{6!} + \dfrac{x^9}{9!} + ...$

(B) $1 + x^3 + \dfrac{x^6}{2!} + \dfrac{x^9}{3!} + ...$

(C) $x^3 + \dfrac{x^6}{3!} + \dfrac{x^9}{6!} + ...$

(D) $1 - \dfrac{x^3}{3!} + \dfrac{x^6}{6!} - \dfrac{x^9}{9!} + ...$

12. The rate of growth of algae, A, in a pond with respect to time, t, is directly proportional to one half the amount of algae in the pond. Which differential equation describes this relationship?

(A) $A(t) = \dfrac{1}{2}kA$

(B) $\dfrac{dA}{dt} = \dfrac{1}{2}kA$

(C) $\dfrac{dA}{dt} = \dfrac{1}{2}k\dfrac{A}{t}$

(D) $\dfrac{dA}{dt} = \dfrac{1}{2}kAt$

GO ON TO THE NEXT PAGE.

13. If $\int_0^{10} f(x)\, dx = 100$ and $\int_{10}^{6} f(x)\, dx = 75$, find $\int_0^{6} f(x)\, dx$.

   (A) 175
   (B) 25
   (C) −25
   (D) −175

GO ON TO THE NEXT PAGE.

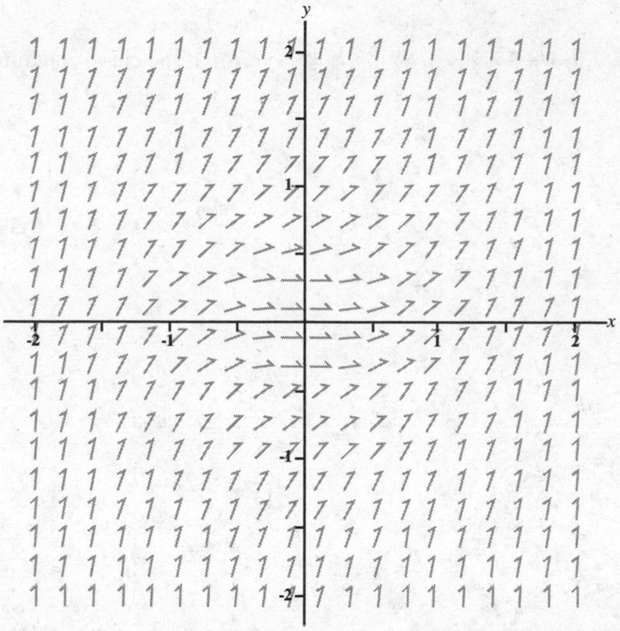

14. Which of the following is the differential equation of the slope field above?

(A) $\dfrac{dy}{dx} = x^2 + y^2$

(B) $\dfrac{dy}{dx} = x^2 - y^2$

(C) $\dfrac{dy}{dx} = x - y$

(D) $\dfrac{dy}{dx} = (x + y)^2$

GO ON TO THE NEXT PAGE.

15. The length of a curve from $x = 2$ to $x = 3$ is given by $\int_2^3 \sqrt{1 + 25x^6}\, dx$. If the curve contains the point $(2, 10)$, which of the following could be the curve?

(A) $y = \dfrac{5x^4}{4}$

(B) $y = \dfrac{5x^4}{4} - 10$

(C) $y = 25x^2 - 90$

(D) $y = 25x^3 + 10$

16. If the line tangent to the graph of the function f at the point $(8, 2)$ passes through the point $(2, 5)$, then $f'(8) =$

(A) 2

(B) −2

(C) $\dfrac{1}{2}$

(D) $-\dfrac{1}{2}$

GO ON TO THE NEXT PAGE.

17. A curve is defined by the parametric equations $x = t^3 + 2t - 1$ and $y = t^2$. Which of the following is an equation of the line tangent to the graph at $(11, 4)$?

   (A) $2y - 7x = 69$
   (B) $7y - 2x = 6$
   (C) $2x - 7y = 6$
   (D) $7x + 2y = 69$

18. The graph of the function f is shown below.

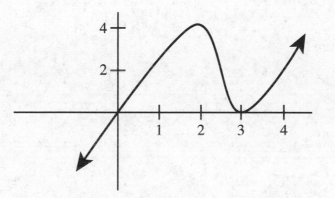

If $g(x) = \int_0^{x^2} f(t) \, dt$, what is $g'(2)$?

   (A) 32
   (B) 16
   (C)   8
   (D)   4

GO ON TO THE NEXT PAGE.

19. If the graph of $y = f(x)$ has slope $3x^2 + 5$ at each point (x, y) on the curve and it passes through (2, 18), which of the following is the equation of $f(x)$?

    (A) $y = 17x - 16$
    (B) $y = x^3 + 5x$
    (C) $y = x^3 + 5x - 1$
    (D) $y = 17x - 52$

---

20. Which of the following are the first four nonzero terms of the MacLaurin series for $f(x) = \sin(x^2)$?

    (A) $x^2 - \dfrac{x^6}{6!} + \dfrac{x^8}{8!} - \dfrac{x^{10}}{10!}$

    (B) $x^2 - \dfrac{x^6}{6!} + \dfrac{x^{10}}{10!} - \dfrac{x^{14}}{14!}$

    (C) $x^2 - \dfrac{x^6}{3!} + \dfrac{x^{10}}{5!} - \dfrac{x^{14}}{7!}$

    (D) $x^2 + \dfrac{x^6}{3!} + \dfrac{x^{10}}{5!} + \dfrac{x^{14}}{7!}$

GO ON TO THE NEXT PAGE.

21. The number of new members of a social media site is modeled by the function y that satisfies the differential equation $\frac{dy}{dt} = 25y(1 - \frac{y}{10000})$, where t is the time in days and $y(0) = 100$. What is $\lim_{t \to \infty} y(t)$?

    (A) 100
    (B) 2500
    (C) 10,000
    (D) 100,000

22. For what values of x does the series $\sum_{n=1}^{\infty} \frac{5^n}{n^2} x^n$ converge?

    (A) $-\frac{1}{5} < x < \frac{1}{5}$

    (B) $x > \frac{1}{5}$

    (C) $-5 < x < 5$

    (D) All real numbers

GO ON TO THE NEXT PAGE.

23. $\int xe^{6x}\,dx =$

    (A) $6xe^{6x} - 36e^{6x} + C$

    (B) $\dfrac{x}{6}e^{6x} - \dfrac{1}{36}e^{6x} + C$

    (C) $\dfrac{x}{6}e^{6x} + C$

    (D) $\dfrac{x}{36}e^{6x} + C$

24. Which of the following series converge(s)?

    (I) $\displaystyle\sum_{n=0}^{\infty}\left(\frac{5}{\pi}\right)^{n}$

    (II) $\displaystyle\sum_{n=0}^{\infty} n^{-\frac{3}{2}}$

    (III) $\displaystyle\sum_{n=0}^{\infty}\frac{n!}{n^{2}}$

    (A) I only
    (B) II only
    (C) I and II
    (D) II and III

GO ON TO THE NEXT PAGE.

25. The function f is continuous on the closed interval $[0, 10]$ and has the values in the following table:

x	0	2	5	7	10
$f(x)$	8	11	20	29	34

Using the subintervals $[0, 2]$, $[2, 5]$, $[5, 7]$, and $[7, 10]$, approximate $\int_0^{10} f(x)\, dx$ using a left Riemann sum.

(A) 102

(B) 176

(C) 204

(D) 242

26. $\int \dfrac{x+8}{(x-2)(x+3)}\, dx =$

(A) $2\ln|x-2| - \ln|x+3| + C$

(B) $2\ln|x-2| + \ln|x+3| + C$

(C) $\ln|x^2 + x - 6| - 8 + C$

(D) $\ln|x^2 + x - 6| + 8 + C$

GO ON TO THE NEXT PAGE.

27. Evaluate $\int_{10}^{\infty} \frac{e^{-x}\,dx}{1-e^{-x}}$.

   (A) 0

   (B) $\ln \left| e^{-10} \right|$

   (C) $\ln \left| 1 - e^{-10} \right|$

   (D) Diverges

28. What is the coefficient of x^8 in the Taylor series for e^{x^2} about $x = 0$?

   (A) $\dfrac{1}{24}$

   (B) $\dfrac{1}{6}$

   (C) 6

   (D) 24

29. If $P(x) = x - 2x^2 + 3x^4 - 4x^6$ is the sixth degree Taylor polynomial for the function f about $x = 0$, what is $f''(0)$?

    (A) −120

    (B) −4

    (C) 1

    (D) 72

30.

x	$f(x)$	$f'(x)$	$g(x)$	$g'(x)$
0	2	−6	−3	−4
1	0	−1	4	6
2	1	−3	3	2

In the table above, if $h(x) = g(f(x))$, then $h'(2) =$

    (A) −18

    (B) −6

    (C) 2

    (D) 4

END OF PART A, SECTION I

IF YOU FINISH BEFORE TIME IS CALLED, YOU MAY CHECK YOUR WORK ON PART A ONLY.

DO NOT GO ON TO PART B UNTIL YOU ARE TOLD TO DO SO.

CALCULUS BC

SECTION I, Part B

Time—45 Minutes

Number of questions—15

A GRAPHING CALCULATOR IS REQUIRED FOR SOME QUESTIONS ON THIS PART OF THE EXAMINATION

Directions: Solve each of the following problems, using the available space for scratchwork. After examining the form of the choices, decide which is the best of the choices given and fill in the corresponding oval on the answer sheet. No credit will be given for anything written in the test book. Do not spend too much time on any one problem.

In this test:

(1) The **exact** numerical value of the correct answer does not always appear among the choices given. When this happens, select from among the choices the number that best approximates the exact numerical value.

(2) Unless otherwise specified, the domain of a function f is assumed to be the set of all real numbers x for which $f(x)$ is a real number.

31. Let $y = f(x)$ be the solution to $\dfrac{dy}{dx} = 2x - y$ with $f(1) = 5$. Use Euler's method to approximate $f(3)$, starting at $x = 1$ with a step size of 0.5.

    (A) 3.625
    (B) 3.75
    (C) 3.84375
    (D) 4.3125

32. The side of a square is decreasing at a constant rate of 0.4 cm/s. What is the rate of decrease of the perimeter of the square when the area of the square is 16 m² ?

    (A) −3.2 m/s
    (B) −1.6 m/s
    (C)   1.6 m/s
    (D)   3.2 m/s

GO ON TO THE NEXT PAGE.

33. $\lim\limits_{x \to 1} \dfrac{\sin 3x}{2x} =$

    (A) 0.071

    (B) 1.500

    (C) 3

    (D) The limit does not exist.

---

34. Profits, in dollars, for a small business can be modeled by $P(t) = \dfrac{1000}{1 + e^{\frac{-3t}{10}}}$, where t is measured in years. What are the total profits in years $5 \le t \le 8$, to the nearest dollar?

    (A) $2,618

    (B) $2,824

    (C) $5,442

    (D) $5,959

---

35. The position of a particle at any time $t \ge 0$ is given by $x(t) = 4\cos 2t$ and $y(t) = 3\sin 4t$. What is the particle's speed at time $t = 5$ seconds?

    (A) 6.551

    (B) 8.393

    (C) 12.848

    (D) 21.313

GO ON TO THE NEXT PAGE.

36. What is the minimum value of $f(x) = 2xe^{-x^2}$ on the closed interval $[-1, 2]$?

 (A) −0.858
 (B) −0.736
 (C)   0.073
 (D)   0.736

37. Which of the following integrals gives the area of the region enclosed by $r = 3 - 2\cos 4\theta$ on the closed interval $\left[0, \dfrac{\pi}{2}\right]$?

 (A) $\displaystyle\int_0^{\frac{\pi}{x}} 3 - 2\cos 4\theta \; d\theta$

 (B) $\displaystyle\int_0^{\frac{\pi}{x}} 8\sin 4\theta \, d\theta$

 (C) $\dfrac{1}{2}\displaystyle\int_0^{\frac{\pi}{x}} (3 - 2\cos 4\theta)^2 \, d\theta$

 (D) $\displaystyle\int_0^{\frac{\pi}{x}} (3 - 2\cos 4\theta)^2 \, d\theta$

38. Use the Trapezoidal Sum to approximate $\displaystyle\int_0^2 (3 + x + x^2)\, dx$ using $n = 4$ subintervals.

 (A) 10.75
 (B) 13.75
 (C) 21.5
 (D) 27.5

GO ON TO THE NEXT PAGE.

39. A particle moves along the x-axis so that its acceleration at any time $t \geq 0$ is given by $a(t) = 8\sin 2t + 4\cos 2t$. If its initial velocity is 8 units/s and its initial position is $x = 4$, what is its position, in units, at time $t = 2$ seconds?

   (A)   9.101
   (B)   15.167
   (C)   26.833
   (D)   31.167

40. $\int_0^1 x\left(\sqrt{x} + x^2\right) dx =$

   (A)   0

   (B) $\dfrac{13}{20}$

   (C) $\dfrac{11}{4}$

   (D) $\dfrac{11}{12}$

GO ON TO THE NEXT PAGE.

41. $g(x) = \dfrac{d}{dx}\displaystyle\int_{-5}^{x^2} \cos t \, dt =$

    (A) $\cos(x^2) \cdot 2x$

    (B) $\cos(x^2)$

    (C) $\sin(x^2) \cdot 2x$

    (D) $\sin(x^2)$

---

42. The region in the first quadrant bounded by $y = 4x - x^2$ and the x-axis is revolved around the x-axis. What is the volume of the resulting solid?

    (A)   33.510
    (B)   34.133
    (C)   53.617
    (D)   107.233

---

43. What is the length of the curve $f(x) = \ln(x + 1)$ from $x = 1$ to $x = 3$?

    (A)   1.614
    (B)   1.870
    (C)   2.121
    (D)   2.320

GO ON TO THE NEXT PAGE.

44. Let $f(x) = 3x - x^3$. What is the value of c on the closed interval $[1, 3]$ that satisfies the conclusion of the Mean Value Theorem?

(A) 1.577
(B) 2
(C) 2.082
(D) 2.613

45. If $f(x) = \arctan x$, what is the average value of $f(x)$ on the closed interval $\left[0, \dfrac{\pi}{4} \right]$?

(A) 0.167
(B) 0.360
(C) 16.195
(D) 20.620

STOP
END OF PART B, SECTION I
IF YOU FINISH BEFORE TIME IS CALLED, YOU MAY CHECK YOUR WORK ON PART B ONLY.
DO NOT GO ON TO SECTION II UNTIL YOU ARE TOLD TO DO SO.

SECTION II
GENERAL INSTRUCTIONS

You may wish to look over the problems before starting to work on them, since it is not expected that everyone will be able to complete all parts of all problems. All problems are given equal weight, but the parts of a particular problem are not necessarily given equal weight.

A GRAPHING CALCULATOR IS REQUIRED FOR SOME PROBLEMS OR PARTS OF PROBLEMS ON THIS SECTION OF THE EXAMINATION.

- You should write all work for each part of each problem in the space provided for that part in the booklet. Be sure to write clearly and legibly. If you make an error, you may save time by crossing it out rather than trying to erase it. Erased or crossed-out work will not be graded.

- Show all your work. You will be graded on the correctness and completeness of your methods as well as your answers. Correct answers without supporting work may not receive credit.

- Justifications require that you give mathematical (noncalculator) reasons and that you clearly identify functions, graphs, tables, or other objects you use.

- You are permitted to use your calculator to solve an equation, find the derivative of a function at a point, or calculate the value of a definite integral. However, you must clearly indicate the setup of your problem, namely the equation, function, or integral you are using. If you use other built-in features or programs, you must show the mathematical steps necessary to produce your results.

- Your work must be expressed in standard mathematical notation rather than calculator syntax. For example, $\int_1^5 x^2 \, dx$ may not be written as fnInt (X^2, X, 1, 5).

- Unless otherwise specified, answers (numeric or algebraic) need not be simplified. If your answer is given as a decimal approximation, it should be correct to three places after the decimal point.

- Unless otherwise specified, the domain of a function f is assumed to be the set of all real numbers x for which $f(x)$ is a real number.

GO ON TO THE NEXT PAGE.

SECTION II, PART A
Time—30 minutes
Number of problems—2

A graphing calculator is required for some problems or parts of problems.

During the timed portion for Part A, you may work only on the problems in Part A.

On Part A, you are permitted to use your calculator to solve an equation, find the derivative of a function at a point, or calculate the value of a definite integral. However, you must clearly indicate the setup of your problem, namely the equation, function, or integral you are using. If you use other built-in features or programs, you must show the mathematical steps necessary to produce your results.

1. Let R be the region bounded by the graphs of $y = 4 \ln x$, $y = \dfrac{12}{x+1}$, and $x = 1$, as shown in the figure below.

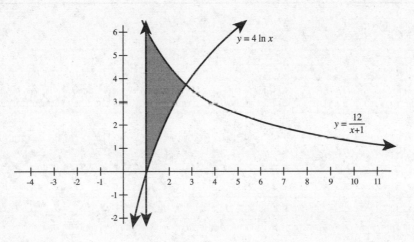

(a) Find the area of R.

(b) Find the volume of the solid that results when R is revolved about the vertical line $x = 1$.

(c) Let R be the base of a solid. Every cross-section of the solid perpendicular to the x-axis is an equilateral triangle with one side across the base. Find the volume of the solid.

GO ON TO THE NEXT PAGE.

2. For $t \geq 0$, a particle is moving along a curve with its position at time t given by $(x(t), y(t))$ and $\dfrac{dx}{dt} = 3t^2 - 4$, $\dfrac{dy}{dt} = 2t$. At time $t = 1$, its position is $(-3, 1)$.

(a) Find the slope of the particle's path at time $t = 3$.

(b) Find the x-coordinate of the particle's position at time $t = 5$.

(c) Find the speed of the particle at time $t = 5$.

(d) Find the acceleration vector of the particle at time $t = 5$.

GO ON TO THE NEXT PAGE.

SECTION II, PART B
Time—1 hour
Number of problems—4

No calculator is allowed for these problems.

During the timed portion for Part B, you may continue to work on the problems in Part A without the use of any calculator.

3. A water storage tank is in the shape of an inverted cone whose height is 16 meters and radius is 4 meters. The tank initially contains 36π m. The water is filling from the tank at a rate of $2\pi h$ m/min.

    (a) Find $\dfrac{dh}{dt}$ in terms of h.

    (b) If $h = 12$ m at time $t = 0$, find an equation for h as a function of t.

    (c) At what time is the storage tank full?

GO ON TO THE NEXT PAGE.

4. The function f is differentiable everywhere and the MacLaurin series is $\displaystyle\sum_{n=0}^{\infty}(-1)^n\frac{x^{2n}}{2n+1}=1-\frac{x^2}{3}+\frac{x^4}{5}-\frac{x^6}{7}+\ldots$

   (a) Use the Ratio Test to find the interval of convergence of the MacLaurin series of f.

   (b) Approximate $f\left(\dfrac{1}{3}\right)$ using the first three nonzero terms.

   (c) Write the first four nonzero terms and the general term for the MacLaurin series for $f'(x)$.

GO ON TO THE NEXT PAGE.

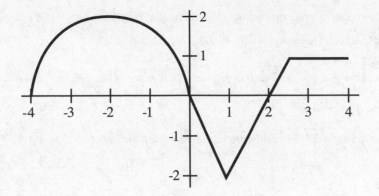

5. The figure above shows the graph of f, which is a continuous function defined on $-4 \le x \le 4$. The graph of f from $x = -4$ to $x = 0$ is a semi-circle, and from $x = 0$ to $x = 4$ consists of three line segments. Let $g(x) = \int_0^x f(t)\, dt$.

(a) Find $g(2)$ and $g(-2)$.

(b) Evaluate $g'(1)$.

(c) Find all values of x where $g(x)$ has a point of inflection.

(d) At what value of x is $g(x)$ a minimum?

GO ON TO THE NEXT PAGE.

6. Bacteria is growing in a Petri dish at a rate that satisfies the differential equation $\dfrac{dA}{dt} = \dfrac{3}{10}(A - 40)$, where A is the number of bacteria and t is measured in days. There are initially 1200 bacteria in the dish.

   (a) Use the line tangent to the graph of A at $t = 0$ to estimate the number of bacteria the end of $t = 5$ days.

   (b) Find $\dfrac{d^2 A}{dt^2}$ in terms of A.

   (c) Find the particular solution to the above differential equation with the initial condition $A(0) = 1200$.

STOP

END OF EXAM

Chapter 28
Practice Test 3
Answers and
Explanations

ANSWER KEY TO SECTION I

1.	B	11.	B	21.	C	31.	D	41.	A
2.	A	12.	B	22.	A	32.	B	42.	D
3.	A	13.	A	23.	B	33.	A	43.	C
4.	B	14.	A	24.	B	34.	A	44.	C
5.	C	15.	B	25.	B	35.	A	45.	B
6.	D	16.	D	26.	A	36.	A		
7.	A	17.	B	27.	C	37.	C		
8.	C	18.	C	28.	A	38.	A		
9.	A	19.	B	29.	B	39.	D		
10.	D	20.	C	30.	A	40.	B		

ANSWERS AND EXPLANATIONS TO SECTION I

1. **B** Remember that the derivative of tan x is sec^2 x. You take the derivative, using the Chain Rule and

   get $\dfrac{dy}{dx} = (2\pi)\sec^2 (2\pi x)$.

2. **A** $\lim\limits_{x \to 0} \dfrac{\sin x - e^x + 1}{x} =$

   Notice that if you plug in 0 for x, you will get $\dfrac{0}{0}$. This is an indeterminate form and you can

   use L'Hôpital's rule to find the limit. You take the derivative of the numerator and the denomi-

   nator separately to get $\lim\limits_{x \to 0} \dfrac{\sin x - e^x + 1}{x} = \lim\limits_{x \to 0} \dfrac{\cos x - e^x}{1}$. Now if you plug in 0 for x, you get

   $\lim\limits_{x \to 0} \dfrac{\cos x - e^x}{1} = \dfrac{1-1}{1} = 0$.

3. **A** Use the trig identity $\tan^2 x = \sec^2 x - 1$ to rewrite the integrand: $\int \tan^2 x \, dx = \int (\sec^2 x - 1) \, dx$.

   Integrate: $\int (\sec^2 x - 1) \, dx = \tan x - x + C$.

4. **B** You can find the slope of the tangent line by $\dfrac{dy}{dx} = \dfrac{\frac{dy}{dt}}{\frac{dx}{dt}}$ and evaluating it at $t = \dfrac{11\pi}{6}$. First, find

   the derivatives separately: $\dfrac{dx}{dt} = 2\cos t$ and $\dfrac{dy}{dt} = 3\sin t$. Thus, $\dfrac{dy}{dx} = \dfrac{\frac{dy}{dt}}{\frac{dx}{dt}} = \dfrac{3\sin t}{2\cos t}$. Now plug

   in $t = \dfrac{11\pi}{6}$ to get $\dfrac{dy}{dx} = \dfrac{3\sin\frac{11\pi}{6}}{2\cos\frac{11\pi}{6}} = \dfrac{3\left(-\frac{1}{2}\right)}{2\left(\frac{\sqrt{3}}{2}\right)} = -\dfrac{3}{2\sqrt{3}}$. Note that you also could have written

   $\dfrac{dy}{dx} = \dfrac{3}{2}\tan\dfrac{11\pi}{6} = \dfrac{3}{2}\left(-\dfrac{1}{\sqrt{3}}\right) = -\dfrac{3}{2\sqrt{3}}$.

5. **C** First, determine if $f(x)$ is continuous at $x = 2$. The $\lim\limits_{x \to 2^-} f(x) = 8 - 8 = 0$ and $\lim\limits_{x \to 2^+} f(x) = 8 - 8 = 0$,

   so $\lim\limits_{x \to 2} f(x) = 0$. Next, $f(2) = 0$. Because $\lim\limits_{x \to 2} f(x) = f(2)$, you know that $f(x)$ is continuous at

   $x = 2$. So eliminate (B) and (D). Now, determine if $f(x)$ is differentiable at $x = 2$. You can take the derivative of both pieces of (x): $f'(x) = \begin{cases} 3x^2 - 4, \ x > 2 \\ 4 - 4x, \ x < 2 \end{cases}$. Now evaluate both pieces of

$f'(x)$ to see if they are the same. For $x < 2$, you get $f'(2^-) = 4 - 4(2) = -4$, and for $x > 2$, you get

$f'(2^+) = 3(2^2) - 4 = 8$. The two pieces of $f'(x)$ are not the same so $f(x)$ is continuous but not differentiable

at $x = 2$.

6. **D**   You can use u-substitution. Let $u = \tan x$ and $du = \sec^2 x \ dx$. We get $\int \tan^2 x \sec^2 x \ dx = \int u^2 \ du$.

Integrate: $\int u^2 \ du = \dfrac{u^3}{3} + C$. Substitute back: $\dfrac{\tan^3 x}{3} + C$.

7. **A**   The particle will be at rest when it is not moving in either the x or y direction. You can find these times

by finding where the respective derivatives are zero. First, find $\dfrac{dx}{dt}$: $\dfrac{dx}{dt} = 6t^2 + 6t - 12$. Now you set it

equal to zero: $6t^2 + 6t - 12 = 0$. Divide through by 6: $t^2 + t - 2 = 0$, which factors to $(t + 2)(t - 1) = 0$.

Thus, $\dfrac{dx}{dt} = 0$ at time $t = 1$ (you can throw out the other answer because t can't be negative). Now repeat

the process for $\dfrac{dy}{dt}$: $\dfrac{dy}{dt} = 6t^2 + 30t + 24$. Set it equal to zero: $6t^2 + 30t + 24 = 0$. Divide through by 6:

$t^2 - 5t + 4 = 0$, which factors into: $(t - 4)(t - 1) = 0$. Thus, $\dfrac{dy}{dt} = 0$ at times $t = 1, 4$. In order for the particle

to be at rest, both $\dfrac{dx}{dt}$ and $\dfrac{dy}{dt}$ must be zero. The only time that this occurs is at $t = 1$.

8. **C**   Here, you can use u-substitution. Let $u = x^2$. Then $du = 2x \ dx$, which you can divide through by 2

to get $\dfrac{1}{2} du = x \ dx$. Now substitute into the integrand: $\int 3x \csc^2 x^2 \ dx = \dfrac{3}{2} \int \csc^2 u \ du$. Now you can

integrate this to get $\dfrac{3}{2} \int \csc^2 u \ du = -\dfrac{3}{2} \cot u + C$. Substituting back, you get $-\dfrac{3}{2} \cot(x^2) + C$.

9. **A**   Remember that $\dfrac{d}{dx}(\log_b x) = \dfrac{1}{x \ln b}$. Here, we use the Chain Rule to get

$f'(x) = \dfrac{1}{(x^3 - x - 1)\ln 5} \cdot (3x^2 - 1)$. Now plug in $x = 2$:

$f'(2) = \dfrac{1}{(2^3 - 2 - 1)\ln 5} \cdot (3(2)^2 - 1) = \dfrac{11}{5\ln 5}$.

10. **D** If you factor a 5 out of the top, you can rewrite the sum as a Geometric series:

$$\sum_{n=1}^{\infty} \frac{5^{n+1}}{4^n} = 5\sum_{n=1}^{\infty} \frac{5^n}{4^n} = 5\sum_{n=1}^{\infty} \left(\frac{5}{4}\right)^n.$$ Remember that a Geometric series converges when the common

ratio $-1 < r < 1$, and diverges otherwise. Here, r is $\frac{5}{4}$, so the series diverges.

11. **B** To find the MacLaurin series for e^{x^3}, simply take the series for e^x, and replace x with x^3. You get

$$e^{x^3} = \sum_{n=0}^{\infty} \frac{x^{3n}}{n!} = 1 + \frac{x^3}{1!} + \frac{x^6}{2!} + \frac{x^9}{3!} + \dots = 1 + x^3 + \frac{x^6}{2!} + \frac{x^9}{3!} + \dots$$

12. **B** You can write the rate of growth of the Algae as $\frac{dA}{dt}$. When you are told that a variable is directly proportional to another one, you can think of it as an equation where the one variable is equal to a constant times the other one. Thus, you get $\frac{dA}{dt} = \frac{1}{2}kA$, where k is the constant of proportionality.

13. **A** Remember that $\int_a^b f(x)\,dx + \int_b^c f(x)\,dx = \int_a^c f(x)\,dx$. Also, remember that $\int_a^b f(x)\,dx = -\int_b^a f(x)\,dx$.

First, you can rewrite $\int_{10}^6 f(x)\,dx = 75$ as $\int_6^{10} f(x)\,dx = -75$. Now you get

$\int_{10}^6 f(x)\,dx + \int_6^{10} f(x)\,dx = \int_0^{10} f(x)\,dx$. Now simply substitute the values for the respective integrals:

$\int_0^6 f(x)\,dx - 75 = 100$. Therefore, $\int_0^6 f(x)\,dx = 175$.

14. **A** You can find the correct differential equation by testing some values of x and y and seeing what the slope is at those values. At the origin, the slope is 0, which unfortunately, does not eliminate any of the choices. At $(1, 1)$, the slope is positive, which eliminates (B) and (C). At $(-1, 1)$, the slope is again positive, which eliminates (D). This leaves (A).

15. **B** The formula for the length of the curve $y = f(x)$ from $x = a$ to $x = b$ is given by $\int_a^b \sqrt{1 + \left(\frac{dy}{dx}\right)^2}\,dx$. So in this case, $\frac{dy}{dx}$ must be $5x^3$. The derivative of (A) and (B) are both $5x^3$, but only (B) goes through the point $(2, 10)$.

16. **D** The tangent line passes through the points $(8, 2)$ and $(2, 5)$, so the slope of the tangent line is $\frac{5-2}{2-8} = -\frac{1}{2}$. Thus, $f'(8) = -\frac{1}{2}$.

17. **B** For a parametric equation, $\dfrac{dy}{dx} = \dfrac{\frac{dy}{dt}}{\frac{dx}{dt}}$. Here, $\dfrac{dy}{dt} = 2t$ and $\dfrac{dx}{dt}$ $3t^2 + 2$. This means that the slope of

the tangent line, $\dfrac{dy}{dx} = \dfrac{2t}{3t^2 + 2}$. Now you have to figure out what value of t to plug in. Note that

the equation for $y = t^2$, and $y = 4$. Thus $t = \pm 2$. If you plug $t = 2$ into our equation for x, you get

$x = 11$. So plug $t = 2$ into $\dfrac{dy}{dx} = \dfrac{2t}{3t^2 + 2}$, to get the slope of the tangent line. You get $\dfrac{dy}{dx} = \dfrac{2}{7}$.

Now plug the slope and the point into the equation for a line to get $y - 4 = \dfrac{2}{7}(x - 11)$,

which you can rearrange to $7y - 2x = 6$.

18. **C** The Second Fundamental Theorem of Calculus says that $\dfrac{d}{dx}\displaystyle\int_c^x f(t)\,dt = f'(x)$. Here, you get

$g'(x) = \dfrac{d}{dx}\displaystyle\int_0^{x^2} f(t)\,dt = f(x^2)\cdot 2x$ (the $2x$ term comes from applying the Chain Rule to the limit in

the integral). Therefore, $g'(2) = f(4)\cdot 4$. From the graph, you can see that $f(4) = 2$, so $g'(2) = 2\cdot 4 = 8$.

19. **B** Simply take the derivative of each of the answer choices. Only (B) and (C) have a derivative of

$3x^2 + 5$, and of those, only (B) goes through the point $(2, 18)$.

20. **C** The first four terms of the MacLaurin series for $f(x) = \sin x$ are $\sin x \approx x - \dfrac{x^3}{3!} + \dfrac{x^5}{5!} - \dfrac{x^7}{7!}$.

To find the series for $f(x) = \sin(x^2)$, you replace x with x^2: $\sin x^2 \approx x^2 - \dfrac{\left(x^2\right)^3}{3!} + \dfrac{\left(x^2\right)^5}{5!} - \dfrac{\left(x^2\right)^7}{7!}$,

which can be simplified to $x^2 - \dfrac{x^6}{3!} + \dfrac{x^{10}}{5!} - \dfrac{x^{14}}{7!}$.

21. **C** Note that this differential equation is a *logistic equation*. Recall that the solution of a differential

equation of the form $\dfrac{dy}{dt} = ky(1 - \dfrac{y}{M})$ is $y = \dfrac{M}{Ae^{-kt} + 1}$, where $A = \dfrac{M - y_0}{y_0}$. Here you get

$y = \dfrac{10,000}{Ae^{-25t} + 1}$ and $A = \dfrac{10,000 - 100}{100} = 99$. Thus, the equation becomes $y = \dfrac{10,000}{99e^{-25t} + 1}$. Now

find $\displaystyle\lim_{t \to \infty} \dfrac{10,000}{99e^{-25t} + 1} = 10,000$.

22. **A** You can apply the Ratio Test: $\rho = \displaystyle\lim_{n \to \infty} \dfrac{\dfrac{5^{n+1}}{(n+1)^2} x^{n+1}}{\dfrac{5^n}{n^2} x^n} = \lim_{n \to \infty} \dfrac{n^2}{(n+1)^2} \dfrac{5^{n+1} x^{n+1}}{5^n x^n} = \lim_{n \to \infty} \dfrac{n^2}{(n+1)^2} 5x$.

Now you need $|\rho| < 1$. If you evaluate the limit, you get $\rho = \displaystyle\lim_{n \to \infty} \dfrac{n^2}{(n+1)^2} 5x = 5x \lim_{n \to \infty} \dfrac{n^2}{(n+1)^2} = 5x(1)$,

so you are only concerned with $|5x| < 1$, which is true when $-\dfrac{1}{5} < x < \dfrac{1}{5}$.

23. **B** You can find the answer using Integration By Parts, which says that you can evaluate an integral of

the form $\displaystyle\int u\, dv$ by the formula $\displaystyle\int u\, dv = uv - \int v\, du$. Let $u = x$ and $dv = e^{6x}\, dx$. Then $du = dx$ and

$v = \dfrac{e^{6x}}{6}$. Plug it into the formula: $\displaystyle\int xe^{6x}\, dx = \dfrac{xe^{6x}}{6} - \int \dfrac{e^{6x}}{6}\, dx$. Now you just have to integrate

$\displaystyle\int \dfrac{e^{6x}}{6}\, dx$. You get $\displaystyle\int \dfrac{e^{6x}}{6}\, dx = \dfrac{e^{6x}}{36} + C$. Therefore, the answer is $\displaystyle\int xe^{6x}\, dx = \dfrac{xe^{6x}}{6} - \dfrac{e^{6x}}{36} + C$.

24. **B** Series I is a geometric series. These series converge if the ratio is $-1 < r < 1$ and diverge

otherwise. Here, the ratio is $r = \dfrac{5}{\pi}$, which is greater than 1, so it diverges. You can

now eliminate (A) and (C). You can rewrite series II as $\displaystyle\sum_{n=0}^{\infty} n^{-\frac{3}{2}} = \sum_{n=0}^{\infty} \dfrac{1}{n^{\frac{3}{2}}}$. This is a

p-series. These series converge if $p > 1$ and diverge if $p \leq 1$. Because $p > 1$, the series converges.

Now you have to test series III. Here, you can determine whether it converges by the Ratio Test.

$\rho = \displaystyle\lim_{n \to \infty} \dfrac{\dfrac{(n+1)!}{(n+1)^2}}{\dfrac{n!}{n^2}} = \lim_{n \to \infty} \dfrac{(n+1)!}{(n+1)^2} \dfrac{n^2}{n!} = \lim_{n \to \infty} \dfrac{n^2}{(n+1)^2} \dfrac{(n+1)!}{n!} = \lim_{n \to \infty} \dfrac{n^2}{(n+1)^2} (n+1) = \infty$. Therefore,

series III diverges. Thus, only series II converges.

25. **B** You need to add up the areas of the four rectangles whose widths are the distances between the x-coordinates and whose heights are determined by evaluating f at the left coordinate of each interval. You get $R = (2) f(0) + (3) f(2) + (2) f(5) + (3) f(7)$. Now you can use the table to find the appropriate values of f. You get $R = (2)(8) + (3)(11) + (2)(20) + (3)(29) = 16 + 33 + 40 + 87 = 176$.

26. **A** You can evaluate the integral using the Method of Partial Fractions. First, rewrite the integrand as

$\dfrac{A}{x-2} + \dfrac{B}{x+3} = \dfrac{x+8}{(x-2)(x+3)}$. Next, multiply both sides by $(x-2)(x+3)$ to clear the denomi-

nators: $(x-2)(x+3)\dfrac{A}{x-2} + (x-2)(x+3)\dfrac{B}{x+3} = (x-2)(x+3)\dfrac{x+8}{(x-2)(x+3)}$.

This simplifies to $A(x+3) + B(x-2) = (x+8)$. Now distribute the terms on the left side:

$Ax + 3A + Bx - 2B = x + 8$. Group the terms: $Ax + Bx + 3A - 2B = x + 8$. This tells you that $A + B = 1$ and

$3A - 2B = 8$. Using a little algebra, you get $A = 2$ and $B = -1$. Now that you have solved for A and

B, you can rewrite the integral as $\displaystyle\int \dfrac{x+8}{(x-2)(x+3)} dx = \int \dfrac{2}{x-2} dx - \int \dfrac{dx}{x+3}$. Evaluating the inte-

grals, you get $\displaystyle\int \dfrac{2}{x-2} dx - \int \dfrac{dx}{x+3} = 2\ln|x-2| - \ln|x+3| + C$.

27. **C** This is an improper integral. You first rewrite the integral as $\displaystyle\lim_{n\to\infty} \int_{10}^{n} \dfrac{e^{-x}}{1-e^{-x}} dx$. Next, use

u-substitution. Let $u = 1 - e^{-x}$ and $du = e^{-x} dx$. Substitute (ignore the limits for now): $-\displaystyle\int \dfrac{du}{u}$.

Integrate: $-\displaystyle\int \dfrac{du}{u} = -\ln|u|$. And substitute back: $-\ln|u| = -\ln|1 - e^{-x}|$. Now you can

evaluate the limits: $\displaystyle\lim_{n\to\infty} \int_{10}^{n} \dfrac{e^{-x} dx}{1-e^{-x}} = \lim_{n\to\infty} \left(-\ln|1-e^{-x}|\right)\Big|_{10}^{n}$. Now you evaluate the integral

at the limits: $\displaystyle\lim_{n\to\infty} \left(-\ln|1-e^{-x}|\right)\Big|_{10}^{n} = \lim_{n\to\infty}\left(-\ln|1-e^{-n}| + \ln|1-e^{-10}|\right)$. Then take the limit:

$\displaystyle\lim_{n\to\infty}\left(-\ln|1-e^{-n}| + \ln|1-e^{-10}|\right) = -\ln|1| + \ln|1-e^{-10}| = \ln|1-e^{-10}|$.

28. **A** The Taylor series for e^x is $e^x \approx 1 + x + \dfrac{x^2}{2!} + \dfrac{x^3}{3!} + \dfrac{x^4}{4!} + \cdots$, so you can find the Taylor series for e^{x^2}

by replacing x with x^2: $e^{x^2} \approx 1 + x^2 + \dfrac{(x^2)^2}{2!} + \dfrac{(x^2)^3}{3!} + \dfrac{(x^2)^4}{4!} + \ldots = 1 + x^2 + \dfrac{x^4}{2!} + \dfrac{x^6}{3!} + \dfrac{x^8}{4!} + \ldots$.

Therefore, the coefficient of x^8 is $\dfrac{1}{4!} = \dfrac{1}{24}$.

29. **B** Remember that the formula for the Taylor polynomial about $x = 0$ is: $P(x) = f(0) + f'(0)\dfrac{x}{1!} + f''(0)$

$\dfrac{x^2}{2!} + f'''(0)\dfrac{x^3}{3!} + \ldots$. If you look at the terms of the polynomial, you can see that $f(0) = 0$ and thus

$\dfrac{f''(0)}{2!} = -2$. Therefore, $f''(0) = -4$.

30. **A** You can find $h'(x)$ using the Chain Rule. You get $h'(x) = g'(f(x)) \cdot f'(x)$. Now, you just have to evalu-

ate $h'(2)$. You get $h'(2) = g'(f(2)) \cdot f'(2) = 6 \cdot -3 = -18$.

31. **D** Euler's Method is quite simple. First, you need a starting point (x_0, y_0) and an initial slope y_0'. Next, use increments of h to come up with approximations. Each new approximation will use the following rules:

$$x_n = x_{n-1} + h$$

$$y_n = y_{n-1} + h \cdot y_{n-1}'$$

Repeat for $n = 1, 2, 3, \ldots$

You are given that the curve goes through the point $(1, 5)$. You will call the coordinates of this point $x_0 = 1$ and $y_0 = 6$. The slope is found by plugging these coordinates into $y' = 2x - y$, so you have an initial slope of $y_0' = 2 - 5 = -3$. Now find the next set of points.

Step 1: Increase x_0 by h to get: $x_1 = 1.5$.

Step 2: Multiply h by y_0' and add y_0 to get y_1: $y_1 = 5 - 3(.5) = 3.5$.

Step 3: Find y_1' by plugging x_1 and y_1 into the equation for y': $y_1' = 2(1.5) - 3.5 = -0.5$.

Repeat these three steps until we get to $x = 3$.

Step 1: $x_2 = 2$

Step 2: $y_2 = 3.5 - 0.5(0.5) = 3.25$

Step 3: $y_2' = 2(2) - 3.25 = 0.75$

Step 1: $x_3 = 2.5$

Step 2: $y_3 = 3.25 + 0.5(0.75) = 3.625$

Step 3: $y_3' = 2(2.5) - 3.625 = 1.375$

Step 1: $x_4 = 3$

Step 2: $y_4 = 3.625 + 0.5(1.375) = 4.3125$

32. **B** You need to find a relationship between the side of a square and its perimeter. If you call the length

of a side of the square x, then you know that $\dfrac{dx}{dt} = -0.4$. You also know that when the area of the

square is 16 m, the side of the square is 4 m. The perimeter of the square is $P = 4x$, so $\dfrac{dP}{dt} = 4\dfrac{dx}{dt}$.

Plugging in, you get $\dfrac{dP}{dt} = 4(-0.4) = -1.6$ m/s.

33. **A** Note that this is NOT the limit as x approaches 0. It is the limit as x approaches 1. If you plug in

1, you get $\displaystyle\lim_{x \to 1} \dfrac{\sin 3x}{2x} = \dfrac{\sin 3}{2} \approx 0.071$. Make sure that your calculator is in radian mode!

34. **A** If you want to find the total profits, you need to evaluate the integral of $P(t)$, with the limits of

integration being from $t = 5$ to $t = 8$. We get $\displaystyle\int_5^8 \dfrac{1000}{1 + e^{\frac{-3t}{10}}} \, dt$. You can evaluate this on a calculator

to get \$2,618.

35. **A** The speed of the particle is given by $v = \sqrt{\left(\dfrac{dx}{dt}\right)^2 + \left(\dfrac{dy}{dt}\right)^2}$, which you then evaluate at

$t = 5$. You get $\dfrac{dx}{dt} = -8\sin 2t$ and $\dfrac{dy}{dt} = 12\cos 4t$. Now plug in $t = 5$ and find v:

$v = \sqrt{(-8\sin 10)^2 + (12\cos 20)^2} \approx 6.551$.

36. **A** First, you need to find $f'(x)$ to get the critical points. You get $f'(x) = 2e^{-x^2} - 4x^2 e^{-x^2} = e^{-x^2}(2 - 4x^2)$.

Next, you set the derivative equal to zero to find the critical points. Note that the exponential term

is always positive, so you only need to find when $2 - 4x^2 = 0$. You get $x = \pm\sqrt{\dfrac{1}{2}}$. Now, you can

just evaluate f at the critical points and the end points, and the smallest value is the minimum. You

get $f(-1) \approx -0.736$, $f\left(-\sqrt{\dfrac{1}{2}}\right) \approx -0.858$, $f\left(\sqrt{\dfrac{1}{2}}\right) \approx -0.858$, and $f(2) \approx 0.073$, so the minimum is

$f\left(-\sqrt{\dfrac{1}{2}}\right) \approx -0.858$.

37. **C** You can find the area of the polar curve by evaluating the integral $\displaystyle\int_a^b \dfrac{1}{2} r^2 \, d\theta$. Here you get

$\dfrac{1}{2}\displaystyle\int_0^{\frac{\pi}{2}} (3 - 2\cos 4\theta)^2 \, d\theta$.

38. **A** First, you need to find the width of each subinterval. You find this by subtracting the endpoints and

dividing by $n = 4$. You get $\dfrac{2-0}{4} = \dfrac{1}{2}$. Now, add up the areas of all of the trapezoids. Remem-

ber that the formula for the area of a trapezoid is $\dfrac{1}{2}(b_1 + b_2)h$. The height of each trapezoid is the

width of the subinterval, and the bases are found by evaluating $f(x)$ at the left and right endpoints of

the subinterval. You get $S = \dfrac{1}{2} \cdot \dfrac{1}{2}\left[f(0) + 2f\left(\dfrac{1}{2}\right) + 2f(1) + 2f\left(\dfrac{3}{2}\right) + f(2) \right]$. Plugging into the

formula, you get $S = \dfrac{1}{4}\left[3 + 2(3.75) + 2(5) + 2(6.75) + 9 \right] = 10.75$

39. **D** You can find the particle's velocity by integrating its acceleration and using the initial condition.

$v(t) = \int 8\sin 2t + 4\cos 2t\, dt = -4\cos 2t + 2\sin 2t + C$. Now you use the fact that its velocity

is 8 units/sec when $t = 0$ to solve for C. You get $8 = -4 + 0 + C$, so $C = 12$. Now that you have

the particle's velocity, you can solve for its position by integrating the velocity and using the ini-

tial condition. $x(t) = \int -4\cos 2t + 2\sin 2t + 12\, dt = 2\sin 2t - \cos 2t + 12t + C$. Now you can

use the fact that the position is 4 units when $t = 0$ to solve for C. You get $4 = 0 - 1 + 0 + C$,

so $C = 5$. Thus, the position function is $x(t) = -2\sin 2t - \cos 2t + 12t + 5$. Now you can find

$x(2) = -2\sin 4 - \cos 4 + 24 + 5 \approx 31.167$.

40. **B** First, distribute the x in the integrand: $\displaystyle\int_{10}^{1} x\left(\sqrt{x} + x^2\right)dx = \int_{0}^{1}\left(x^{\frac{3}{2}} + x^3\right)dx$. Then integrate:

$\displaystyle\int_{0}^{1}\left(x^{\frac{3}{2}} + x^3\right)dx = \left(\dfrac{x^{\frac{5}{2}}}{\frac{5}{2}} + \dfrac{x^4}{4}\right)\Bigg|_{0}^{1} = \left(\dfrac{2x^{\frac{5}{2}}}{5} + \dfrac{x^4}{4}\right)\Bigg|_{0}^{1} = \left(\dfrac{2}{5} + \dfrac{1}{4}\right) - 0 = \dfrac{13}{20}$.

41. **A** Here, you use the Second Fundamental Theorem of Calculus, which says that $\dfrac{d}{dx}\displaystyle\int_{c}^{x} f(t)\, dt = f(x)$.

You get $\dfrac{d}{dx}\displaystyle\int_{-5}^{x^2} \cos t\, dt = \cos(x^2) \cdot 2x$.

Remember that you need to use the Chain Rule when you have a function in the limits of

integration.

42.   **D**   First, make a sketch of the region:

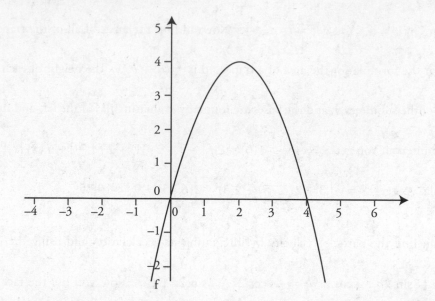

You can see that the bounds of the region are $x = 0$ and $x = 4$. You can find the volume using

Washers. You get $V = \pi \int_0^4 (4x - x^2)^2 \, dx$. You could evaluate this by hand, but you are in the calcu-

lator portion of the AP Exam, so simply plug it into the calculator. You get $V = \pi \int_0^4 (4x - x^2)^2 \, dx$

≈ 107.233.

43.   **C**   The formula for the length of the curve $y = f(x)$ from $x = a$ to $x = b$ is given by $\int_a^b \sqrt{1 + \left(\dfrac{dy}{dx}\right)^2} \, dx$. In

this case, $\dfrac{dy}{dx} = \dfrac{1}{x + 1}$. You get $\int_1^3 \sqrt{1 + \left(\dfrac{1}{x + 1}\right)^2} \, dx \approx 2.121$.

44.   **C**   The Mean Value Theorem states that if $f(x)$ is continuous on the interval $[a, b]$ and differ-

entiable on the interval (a, b), then there exists at least one value c, on the interval (a, b),

such that $f'(c) = \dfrac{f(b) - f(a)}{b - a}$. Find the various terms. You get $f(3) = 9 - 27 = -18$ and

$f(1) = 3 - 1 = 2$, so $\dfrac{f(3) - f(1)}{3 - 1} = \dfrac{-18 - 2}{2} = -10$. Now you need to find $f'(c) = 3 - 3c^2$. If you

set them equal to each other and solve, you get $-10 = 3 - 3c^2$ and $c = \pm\sqrt{\dfrac{13}{3}} \approx 2.082$. You can

throw out the negative answer because it is not in the interval.

45. **B** You can find the average value of a function $f(x)$ with the formula $f_{avg} = \dfrac{1}{b-a} \int_a^b f(x)\,dx$. Here,

you get $f_{avg} = \dfrac{1}{\frac{\pi}{4} - 0} \int_0^{\frac{\pi}{4}} \arctan x\, dx$. You can evaluate this with a calculator to get 0.360.

ANSWERS AND EXPLANATIONS TO SECTION II

1. Let R be the region bounded by the graphs of $y = 4 \ln x$, $y = \dfrac{12}{x+1}$, and $x = 1$, as shown in the figure below.

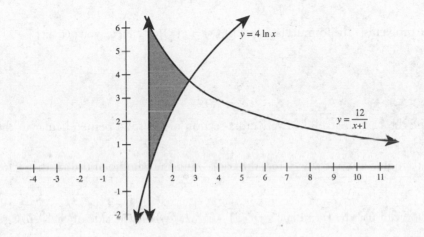

(a) Find the area of R.

Remember that the area of the region can be found by evaluating the integral $\int_a^a f(x) - g(x)\,dx$. Here,

$f(x) = \dfrac{12}{x+1}$, $g(x) = 4\ln x$, and $a = 1$. You now need to find b, which is the intersection of the

two curves. You can find the intersection using the calculator. You get $x = 2.410$. Thus, you need

to evaluate the integral $\int_1^{2.410} \left(\dfrac{12}{x+1} - 4\ln x \right) dx$. You can evaluate this with the calculator and get

$\int_1^{2.410} \left(\dfrac{12}{x+1} - 4\ln x \right) dx \approx 3.563$. Remember that the AP Exam expects three decimal places for

answers unless it states otherwise.

(b)   Find the volume of the solid that results when R is revolved about the vertical line $x = 1$.

You can find the volume of the solid either using washers or cylindrical shells. The difficulty using washers is twofold: First, you would have to convert the equations from being in terms of x to being in terms of y, and second, you would have to use two integrals because the region consists of $f(x) = \dfrac{12}{x+1}$ on the right and $x = 1$ on the left until they intersect, and then switches to $g(x) = 4 \ln x$ on the right and $x = 1$ on the left. However, if you use cylindrical shells, you can avoid both of these problems. The formula is $2\pi \int_a^b x\big(f(x) - g(x)\big) dx$. Here, you get $2\pi \int_1^{2.410} x\left(\dfrac{12}{x+1} - 4\ln x\right) dx$

≈ 32.091.

(c)   Let R be the base of a solid. Every cross-section of the solid perpendicular to the x-axis is an equilateral triangle with one side across the base. Find the volume of the solid.

Here, you can use the formula $V = \dfrac{\sqrt{3}}{4}\int_a^b side^2\, dx$, where the side of each cross-section is found by $f(x) - g(x)$. You get $\dfrac{\sqrt{3}}{4}\int_1^{2.410}\left(\dfrac{12}{x+1} - 4\ln x\right)^2 dx \approx 5.662$.

2.   For $t \geq 0$, a particle is moving along a curve with its position at time t given by $(x(t), y(t))$ and $\dfrac{dx}{dt} = 3t^2 - 4$, $\dfrac{dy}{dt} = 2t$. At time $t = 1$, its position is $(-3, 1)$.

(a)   Find the slope of the particle's path at time $t = 3$.

You can find the slope of the particle's path by $\dfrac{dy}{dx} = \dfrac{\dfrac{dy}{dt}}{\dfrac{dx}{dt}} = \dfrac{2t}{3t^2 - 4}$. At time $t = 3$, $\dfrac{dy}{dx} = \dfrac{6}{23}$.

(b)   Find the x-coordinate of the particle's position at time $t = 5$.

You can find the x-coordinate of the particle's position by evaluating the integral $\int \dfrac{dx}{dt}\, dt$ and using the information about its location at time $t = 1$ to solve for the constant. You get $x(t) = \int 3t^2 - 4\, dt = t^3 - 4t + C$. Now plug in $t = 1$ to solve for C: $-3 = 1 - 4 + C$, so $C = 0$. Thus, $x(t) = t^3 - 4t$. Finally, plug in $t = 5$: $x(5) = 5^3 - 20 = 105$.

(c)   Find the speed of the particle at time $t = 5$.

You can find the speed of the particle by $v = \sqrt{\left(\dfrac{dx}{dt}\right)^2 + \left(\dfrac{dy}{dt}\right)^2}$. At time $t = 5$, $\dfrac{dx}{dt} = 71$ and $\dfrac{dy}{dt} = 10$. Therefore, the speed is $v = \sqrt{71^2 + 10^2} \approx 71.701$.

(d)   Find the acceleration vector of the particle at time $t = 5$.

The acceleration vector is given by $\left(\dfrac{d^2x}{dt^2}, \dfrac{d^2y}{dt^2}\right)$. Here, $\dfrac{d^2x}{dt^2} = 6t$ and $\dfrac{d^2y}{dt^2} = 2$. So, the acceleration vector is $(36t^2, 4)$.

3.   A water storage tank is in the shape of an inverted cone whose height is 16 meters and radius is 4 meters. The tank initially contains 36π m³. The water is filling the tank at a rate of $2\pi h$ m³/min.

(a)   Find $\dfrac{dh}{dt}$ in terms of h.

The volume of a cone is $V = \dfrac{1}{3}\pi r^2 h$. You also know that, because the height of the cone is

16 meters and radius is 4 meters, the ratio of the height to the radius of the water in the cone is

always $\dfrac{h}{r} = 4$, or $r = \dfrac{h}{4}$. You can substitute this ratio into the equation for volume to get it in

terms of h only. You get $V = \dfrac{1}{3}\pi\left(\dfrac{h}{4}\right)^2 h = \dfrac{\pi}{48}h^3$. Now find $\dfrac{dh}{dt}$ by taking the derivative of the

volume with respect to t: $\dfrac{dV}{dt} = \dfrac{\pi h^2}{16}\dfrac{dh}{dt}$. You are given that $\dfrac{dV}{dt} = 2\pi h^3$, so $2\pi h^3 = \dfrac{\pi h^2}{16}\dfrac{dh}{dt}$. With

a little algebra, you get $\dfrac{dh}{dt} = 32h$.

(b)   If $h = 12$ m at time $t = 0$, find an equation for h as a function of t.

You know that $\dfrac{dh}{dt} = 32h$. If you separate the variables and integrate, you get $\displaystyle\int \dfrac{dh}{h} = \int 32\,dt$,

which gives $\ln h = 32t + C$ (h can't be negative so you don't need the absolute value). If you expo-

nentiate both sides, you get $h = Ce^{32t}$. Now solve for C: $12 = C$. Therefore, you get $h = 12e^{32t}$.

(c) At what time is the storage tank full?

Here, you simply need to find when $h = 16$. You get $16 = 12e^{32t}$. You can solve this for t: $\dfrac{4}{3} = e^{32t}$,

$\ln\dfrac{4}{3} = 32t$, so $t = \dfrac{1}{32}\ln\dfrac{4}{3}$ min.

4. The function f is differentiable everywhere and the MacLaurin series is

$$\sum_{n=0}^{\infty}(-1)^n\frac{x^{2n}}{2n+1} = 1 - \frac{x^2}{3} + \frac{x^4}{5} - \frac{x^6}{7} + \dots$$

(a) Use the Ratio Test to find the interval of convergence of the MacLaurin series of f.

The Ratio Test gives

$$\rho = \lim_{n\to\infty}\frac{\dfrac{x^{2(n+1)}}{2(n+1)+1}}{\dfrac{x^{2n}}{2n+1}} = \lim_{n\to\infty}\frac{\dfrac{x^{2n+2}}{2n+3}}{\dfrac{x^{2n}}{2n+1}} = \lim_{n\to\infty}\frac{x^{2n+2}}{2n+3}\frac{2n+1}{x^{2n}} = \lim_{n\to\infty}\frac{2n+1}{2n+3}x^2 = x^2.$$ In order for the

series to converge, $-1 < \rho < 1$. Here, you get $x^2 < 1$. If you take the square root, we get $-1 < x < 1$.

(b) Approximate $f\left(\dfrac{1}{3}\right)$ using the first three nonzero terms.

All you have to do is plug $\dfrac{1}{3}$ in for x. You get $1 - \dfrac{\left(\dfrac{1}{3}\right)^2}{3} + \dfrac{\left(\dfrac{1}{3}\right)^4}{5} = 1 - \dfrac{1}{27} + \dfrac{1}{405} = \dfrac{391}{405}$.

(c) Write the first four nonzero terms and the general term for the MacLaurin series for $f'(x)$.

You can find $f'(x)$ by differentiating each of the terms. You get $-\dfrac{2x}{3} + \dfrac{4x^3}{5} - \dfrac{6x^5}{7} + \dfrac{8x^7}{9}$.

5.

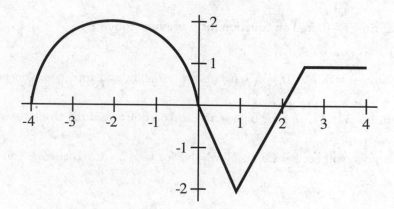

The figure above shows the graph of f, which is a continuous function defined on $-4 \leq x \leq 4$. The graph of f from $x = -4$ to $x = 0$ is a semi-circle, and from $x = 0$ to $x = 4$ consists of three line segments.

Let $g(x) = \int_0^x f(t)\, dt$.

(a) Find $g(2)$ and $g(-2)$.

From the definition of g, you get $g(2) = \int_0^2 f(t)\,dt$. This is the area under $f(t)$ from $x = 0$ to $x = 2$. This is simply the area of the triangle: $\frac{1}{2}(2)(-2) = -2$. Similarly, $g(-2) = \int_0^{-2} f(t)\,dt$.

Note that here, the greater limit of integration is on the bottom of the integral sign and the smaller one is on top. You can reverse the order of the limits and change the sign of the integral:

$g(-2) = \int_0^{-2} f(t)\,dt = -\int_{-2}^0 f(t)\,dt$. This time, you are finding the area under $f(t)$ from $x = -2$ to $x = 0$. This is simply the area of the quarter circle: $\frac{1}{4}\pi(2)^2 = \pi$, so $g(-2) = -\pi$.

(b) Evaluate $g'(1)$.

You can use the Second Fundamental Theorem of Calculus to find the derivative of g:

$g'(x) = \frac{d}{dx}\int_0^x f(t)\,dt = f(x)$. So, $g'(1) = f(1) = -2$.

(c) Find all values of x where $g(x)$ has a point of inflection.

In order to find where $g(x)$ has a point of inflection, where $g''(x) = 0$. You know that $g'(x) = f(x)$, so $g''(x) = f'(x)$. If you look at the graph of $f(x)$, you can see that $f'(x) = 0$ at $x = -2$. Furthermore, $f'(x) > 0$ to the left of $x = -2$ and $f'(x) < 0$ to the right of $x = -2$. Therefore, $g(x)$ has a point of inflection at $x = -2$.

(d) At what value of x is $g(x)$ a minimum?

You can see from the graph that $g'(x) = f(x) = 0$ at $x = 0$ and at $x = 2$. At $x = 0$, $g'(x)$ is positive to the left and negative to the right, so at $x = 2$, $g(x)$ has a maximum. At $x = 2$, $g'(x)$ is negative to the left and positive to the right, so at $x = 2$, $g(x)$ has a minimum.

6. Bacteria is growing in a Petri dish at a rate that satisfies the differential equation $\dfrac{dA}{dt} = \dfrac{3}{10}(A - 40)$, where A is the number of bacteria and t is measured in days. There are initially 1,200 bacteria in the dish.

(a) Use the line tangent to the graph of A at $t = 0$ to estimate the number of bacteria the end of $t = 5$ days.

First, find $\dfrac{dA}{dt}$ at $t = 0$. You get $\left.\dfrac{dA}{dt}\right|_{t=0} = \dfrac{3}{10}(A(0) - 40) = \dfrac{3}{10}(1200 - 40) = 348$. The tangent line is thus $A(t) = 1,200 + 348t$. Therefore, at $t = 5$, $A(t) = 1,200 + 348(5) = 2,940$.

(b) Find $\dfrac{d^2 A}{dt^2}$ in terms of A.

$\dfrac{d^2 A}{dt^2} = \dfrac{3}{10}\dfrac{dA}{dt} = \dfrac{3}{10}\left[\dfrac{3}{10}(A - 40)\right] = \dfrac{9}{100}(A - 40)$, where $A \geq 1,200$.

(c) Find the particular solution to the above differential equation with the initial condition $A(0) = 1,200$.

First, you separate the variables: $\dfrac{dA}{A - 40} = \dfrac{3}{10}dt$. Now integrate both sides: $\displaystyle\int\dfrac{dA}{A - 40} = \dfrac{3}{10}\int dt$. You get $\ln|A - 40| = \dfrac{3}{10}t + C$. If you exponentiate both sides, you get $A - 40 = Ce^{\frac{3}{10}t}$. (Because A is always greater than 1,200, you can ignore the absolute value signs.)

Now, solve for C: $1,200 - 40 = Ce^0$, so $C = 1,160$ and $A = 1160e^{\frac{3}{10}t} + 40$.

About the Author

David S. Kahn studied applied mathematics and physics at the University of Wisconsin and has taught courses in calculus, precalculus, algebra, trigonometry, and geometry at the college and high school levels. He has worked as an educational consultant for many years and tutored more students in mathematics than he can count! He has worked for The Princeton Review since 1989, and, in addition to AP calculus, he has taught math and verbal courses for the SAT, SAT II, LSAT, GMAT, and the GRE, trained other teachers, and written several other math books.